W9-CPO-108

Bayesian Statistical Modelling

WILEY SERIES IN PROBABILITY AND STATISTICS

Established by WALTER A. SHEWHART and SAMUEL S. WILKS

A complete list of the titles in this series appears at the end of this volume

Bayesian Statistical Modelling

PETER CONGDON
Queen Mary, University of London, UK

JOHN WILEY & SONS, LTD
Chichester • New York • Weinheim • Brisbane • Singapore • Toronto

Email (for orders and customer service enquiries): cs-books@wiley.co.uk
Visit our Home Page www.wileyeurope.com or www.wiley.com

Reprinted January 2002, June 2003

Other Wiley Editorial Offices

John Wiley & Sons Inc., 111 River Street, Hoboken, NJ 07030, USA

Jossey-Bass, 989 Market Street, San Francisco, CA 94103-1741, USA

Wiley-VCH Verlag GmbH, Boschstr. 12, D-69469 Weinheim, Germany

John Wiley & Sons Australia Ltd, 33 Park Road, Milton, Queensland 4064, Australia

John Wiley & Sons (Asia) Pte Ltd, 2 Clementi Loop #02-01, Jin Xing Distripark, Singapore 129809

John Wiley & Sons (Canada) Ltd, 22 Worcester Road, Etobicoke, Ontario M9W 1L1

Wiley also publishes its books in a variety of electronic formats. Some content that appears in print may not be available in electronic books.

Library of Congress Cataloging-in-Publication Data

Congdon, P.
Introduction to bayesian statistical modelling / P. Congdon.
p. cm.—(Wiley series in probability and statics)
Includes bibliographical references and index.
ISBN 0-471-81311-7 (alk. paper)
1. Bayesian statistical decision theory. I. Title. II. Series.

QA279.5 .C65 2001
519.5′42—dc21

00-043927

British Library Cataloguing in Publication Data
A catalogue record for this book is available from the British Library

ISBN 0-471-49600–6

Typeset in 10/12pt Times by Kolam Information Services Pvt. Ltd, Pondicherry, India.
Printed and bound in Great Britain by Biddles Ltd, Guildford and King's Lynn
This book is printed on acid-free paper responsibly manufactured from sustainable forestry, for which at least two trees are planted for each one used for paper production.

Contents

Preface

For many students and researchers the main motive for learning statistical methods is to apply them to the analysis and modelling of data. There has been a considerable growth in software packages oriented to answering these needs, but relatively few offering an accessible Bayesian approach to data analysis. The present book aims to contribute to the development of accessible software methods for applying Bayesian methodology. The WINBUGS program is possibly the leading contender for this role at the moment (http://www.mrc-bsu.cam.ac.uk/bugs/). I first encountered BUGS in its early DOS versions in 1993 and have been a regular user (even addict) ever since.

The program has wide functionality in terms of possible sampling distributions, and in applicability to standard Gibbs models as well as more complex models requiring Metropolis–Hastings sampling. It develops on some of my own favourite software packages for generalised linear models (such as GLIM) in being able among other things to incorporate random errors, parameter constraints, and opportunities for robust data analysis. It has wide potential in areas as diverse as time series, spatial models, and structural equation methods. It does require some facility with programming, and has some quirks which become familiar to regular users. Its current limitations are mainly (in my view) around modelling parameters of multivariate dispersion matrices.

This said, it will undoubtedly evolve into a central feature of statistical computing. My aim has been to present a wide range of worked examples, drawing on Bayesian literature pre-and post-MCMC as well as on a range of substantive settings, with the aim of facilitating an initial approach to problems. Worked examples in WINBUGS are available via ftp at ftp://www.wiley.co.uk/pub/books/congdon. Modification by other users would then be possible and indeed is one advantage of a flexible programming approach to data modelling. I hope that it contributes to making modelling easier and accessible to those taking courses in applied statistics and Bayesian techniques and those using these methods for practical research.

Peter Congdon
Research Professor
Department of Geography
Queen Mary, University of London

Preface

Acknowledgements

My thanks go firstly to the editorial team at Wiley for their assistance and encouragement with this book. I am also grateful for the advice offered by members of the BUGS development project at the MRC Unit in Cambridge and at Imperial College. Most of all thanks to my wife Ann for her support in the writing of this book.

CHAPTER 1

Introduction: The Bayesian Method, its Benefits and Implementation

1.1 ADVANTAGES OF THE BAYES APPROACH

The practice of applied statistics undergoes continual change as a result both of methodological developments and changes in the computing environment in which statistical analysis is carried out. In the 1950s and 1960s the development of mainframe computers facilitated widespread advances in statistical estimation. These involved iterative techniques for optimising goodness of fit and obtaining maximum likelihood estimates in complex data modelling. The continuing development in computer technology has more recently facilitated computer intensive methods such as sampling-based Bayesian inference and bootstrap methods (Efron and Tibshirani, 1993).

Bayesian inference has a number of advantages. For example, the Bayes method provides confidence intervals on parameters and P-values on hypotheses which are more in line with commonsense interpretations. It provides a way of formalising the process of learning from data to update beliefs in accord with recent notions of knowledge synthesis. However in the past, statistical analysis based on Bayes theorem was often daunting because of the numerical integrations needed. Recently developed computer intensive sampling methods of estimation have revolutionised the application of Bayesian methods in fields as diverse as biostatistics, econometrics and genetic mapping (Gilks *et al.*, 1993; Albert and Chib, 1996; Heath, 1997).

Fully Bayesian methods now increasingly offer a comprehensive and robust approach to model estimation, for example in multilevel models with nested random effects, where alternative sampling densities (say) or priors on the form of the variance parameters may be adopted. They provide a way of improving estimation in sparse data sets by borrowing strength (e.g. in small area mortality studies or in stratified sampling) (Stroud, 1994). They are not usually dependent on the assumption of asymptotic normality underlying classical estimation methods such as maximum likelihood. Sampling based methods of Bayesian estimation provide a full distributional profile of a parameter (e.g. median, percentiles, mean) so that any clear non-normality is apparent. These methods permit a range of hypotheses about the parameters to be simply assessed using the sample information.

The Bayesian approach has a natural affinity with recent scientific philosophy stressing the advantages of cumulative evidence and evidence based practice (e.g. in medicine); it accords with the summarisation and cumulation of evidence via techniques such as meta-analysis. Bayesian methods may also improve on the properties of classical estimators in terms of the precision of estimates. Specifying the prior amounts to introducing extra information or data based on accumulated knowledge, and the posterior estimate in being based on the combined sources of information (prior and likelihood) therefore has greater precision. Indeed a prior can often be expressed in terms of an equivalent 'sample size'.

Bayesian analysis offers an alternative to classical tests of hypotheses and confidence interval estimation based on normality assumptions. In the classical perspective a parameter θ is regarded as an unknown constant and the currently available data are used to estimate its true value. Probabilities in the form of P-values are framed in the data space: the P-value is the probability under H_0 of data at least as extreme as that actually observed. Many users of such tests more naturally interpret P-values as relating to the hypothesis space, i.e. to questions such as the likely range for a parameter given the data or the probability of H_0 given the data. The Bayesian framework is more naturally suited to such probability interpretations.

The classical inference model poses certain problems of interpretation both in significance testing and interval estimation. For example, if the test is whether the sex ratio at birth is 0.5 (equal numbers of male and female births) and a sample of 10 births contains 9 females, then the binomial distribution tells us that the probability of this is 10/1024. A two-sided classical significance test would however be based not on this probability but on the probabilities of all 10 births being female (1/1024), and on the probabilities of 9 or 10 births being male; the overall significance level is $22/1024 = 0.02$. So probabilities of various events that did not occur are used against the null hypothesis.

The classical theory of confidence intervals for parameter estimates is that in the long run with data from many samples a 95% interval (say) calculated from each sample will contain the true parameter approximately 95% of the time, and for the remaining samples it will exclude the true parameter value. The particular confidence interval from any one sample may or may not contain the true parameter value. This goes counter to the intuitive and often held belief that the standard confidence interval from a particular sample contains the true parameter value with approximately 95% certainty.

1.2 STATEMENTS ABOUT UNCERTAINTY AND BAYESIAN UPDATING

The representation of uncertainty about parameters or hypotheses as probabilities is central to Bayesian inference. Under this framework we can calculate the probability that a parameter lies in a given interval (i.e. the everyday interpretation of a confidence interval), or the probability of a hypothesis about a parameter or set of parameters. The learning process involved in Bayesian inference is one of modifying one's initial probability statements about the parameters prior to observing the data to updated or posterior knowledge incorporating both prior knowledge and the data at hand. Thus prior subject matter knowledge about a parameter (e.g. the incidence of extreme political views or the relative risk of thrombosis associated with taking the contraceptive pill) are an important aspect of the inference process.

Let $P(H)$ denote our prior beliefs about the truth of a hypothesis H_0, for example that the excess relative risk of thrombosis for women taking the pill exceeds 2, with H_1 being that the relative risk was under 2. Suppose our actual data at hand (perhaps from a controlled trial or retrospective study) show a relative risk of 3.6. Then the probability or likelihood of the data x, given our prior beliefs is the conditional probability $P(x|H)$, with H denoting H_0 for simplicity. Bayes theorem expresses the updated probability statement about H as

$$P(H|x) = P(x|H)P(H)/P(x) \tag{1.1}$$

where $P(x)$ is the probability of the data averaged over all possible hypotheses (sometimes called the marginal likelihood). Here $P(x)$ would be the total likelihood of the data over H_0 and H_1, namely $P(x|H_0)P(H_0) + P(x|H_1)P(H_1)$.

Formula (1.1) follows from the rule for the joint probability of x and H, namely

$$P(x, H) = P(x|H)P(H) = P(H|x)P(x),$$

and the probability $P(H|x)$ denotes our updated or posterior probability beliefs about H_0 given the data. In a sense it pools the prior beliefs with the evidence at hand. More generally H can represent parameters or models as well as hypotheses. We view the model or parameters H as random, and since the divisor $P(x)$ in the expression (1.1) is independent of H, we can re-express the result as

$$P(H|x) \propto P(x|H)P(H)$$

where $\propto$ means equal to except for a constant of proportionality.

1.2.1 Prior and likelihood

In words the full result is that

$$\text{posterior distribution} = \text{likelihood} \times \text{prior distribution} / \sum(\text{likelihood} \times \text{prior}),$$

where the denominator (the likelihood accumulated over all possible prior values) is a fixed normalising factor which ensures that the posterior probabilities sum to 1. So

$$\text{posterior distribution} \propto (\text{likelihood}) \times (\text{prior distribution}).$$

This expression simply states the common-sense principle that updated knowledge combines prior knowledge with the data at hand.

The relative influence of the prior and the data at hand on the updated beliefs depends on how much weight we give to the prior (how 'informative' we make it) and the strength of the data. For example a large data sample would tend to have a predominant influence on our updated beliefs unless our prior was extremely specific. If our sample x was small and combined with a prior which was informative then the prior distribution would have a relatively greater influence on the updated belief; this might be the case for example if a small clinical trial or observational study was combined with a prior based on meta-analysis of previous findings.

A more specific and often used form of expression (1.1), especially in problems of estimation, replaces reference to a hypothesis by referring to a parameter θ or set of p parameters $\theta_1, \theta_2, \ldots, \theta_p$. Initial knowledge about the parameters is summarised by the density $p(\theta)$, the likelihood is $l(x \mid \theta)$ and the updated knowledge is contained in the posterior density $p(\theta \mid x)$ which is proportional to the product of likelihood and prior. The marginal likelihood here would be a sum over discrete values of θ of the prior times the likelihood; or if θ was a continuous variable, then the marginal likelihood would be an integral over all values of θ of the product $l(x \mid \theta)p(\theta)$. Whether θ is discrete or continuous, a relatively high value of $p(\theta)$ for a particular value of θ shows that we attach a higher weight to the chance of this value than to others, but the posterior inference modifies the prior information or belief by the information about the parameters contained in the data. The posterior estimate is therefore a compromise between the prior and the likelihood.

1.3 USING EXISTING KNOWLEDGE TO SPECIFY PRIORS

Suppose we apply the general principle of updating a prior by the data at hand by considering the prevalence of a certain disease, that is the proportion of the population exhibiting positive symptoms. Suppose a clinician is interested in π the proportion of children aged 5 to 9 in a particular population with asthma symptoms; she has some prior opinions about the likely size of π based on previous studies and knowledge of the host population which can be summarised (in a discrete way) by the prior probabilities in Table 1.1. Now a random sample of patients in the target population shows 2 from 15 with definitive symptoms.

Then the likelihoods of obtaining 2 from 15 with symptoms according to the different values of π are given by $\binom{15}{2}\pi^2(1-\pi)^{13}$. The posterior probabilities are obtained by dividing the product of the prior and likelihood by the normalising factor of 0.274. They give highest support to a value of $\pi = 0.14$. It can be seen that this inference rests only on the prior combined with the likelihood of the actual data observed, namely 2 from 15 cases. Another feature is that to calculate the posterior weights attaching to different values of π we need in fact have only used that part of the likelihood in which π is a variable and

Table 1.1 Deriving the posterior distribution of a proportion π using a discrete prior.

Prior for π proportion with asthma	Prior weight	Likelihood of π given sample	Prior times likelihood	Posterior probability for π
0.10	0.10	0.267	0.027	0.098
0.12	0.15	0.287	0.043	0.157
0.14	0.25	0.290	0.072	0.265
0.16	0.25	0.279	0.070	0.255
0.18	0.15	0.258	0.039	0.141
0.20	0.10	0.231	0.023	0.084
Total	1		0.274	1

dropped the remaining part of the likelihood; instead of using the full binomial likelihood we could have used simply the term $\pi^2(1-\pi)^{13}$ since the remaining factor $\binom{15}{2}$ cancels out in the numerator and denominator of equation (1.1).

How to choose the prior density or information is an important issue in Bayesian inference, together with the sensitivity or robustness of the inferences to the choice of prior (Berger, 1994). In Table 1.1 a mildly informative prior, favouring certain values of π above others has been used. In this example, a noninformative prior, favouring no values above any other, would assign an equal prior probability of 1/6 to each of the possible prior values of π. A noninformative prior might be used in the genuine absence of prior information, or if there is disagreement about the likely values of hypotheses or parameters. It may also be used in comparison with more informative priors as one aspect of a sensitivity analysis regarding posterior inferences according to the prior.

A more strongly informative prior in this example would have placed a higher prior weight on one or more of the values in Table 1.1 (e.g. a probability of 0.4 on the values 0.14 and 0.16 and 0.05 on the four other values). It is frequently the case that at least some prior information is available on a parameter or hypothesis though converting it into a probabilistic form remains an issue. Sometimes a formal stage of eliciting priors from subject-matter specialists is entered into (Osherson *et al.*, 1995).

If a previous study or set of studies or a range of elicited opinions is available, showing the likely prevalence of asthma in the population then these may be used directly to set up an informative prior for the current study. However, there may be limits to the applicability of previous studies to the current target population (e.g. because of differences in the socio-economic background or features of the local environment). So the information from previous studies, while still usable, may be downweighted in some way; for example, the precision of an estimated relative risk or prevalence rate from a previous study may be halved. Suppose we were using a previous study or meta-analysis to set up our prior; then down-weighting is equivalent to increasing the prior variance by a factor of $a > 1$ or reducing precision by a factor $1/a$. If we have several variables and know their variance–covariance matrix from a previous study then we can downweight this in the same way (Birkes and Dodge, 1993).

In practice, there are also mathematical reasons to prefer some sorts of prior to others (the question of conjugacy considered in Chapter 2). For example a beta density for the binomial success probability is conjugate with the binomial likelihood in the sense that the posterior has the same (beta) density form as the prior. One advantage of the new sampling-based estimation methods is that the researcher is no longer restricted to conjugate priors whereas in the past this choice was often made for reasons of analytic tractability. However, there remain considerable problems in choosing appropriate neutral or noninformative priors on certain parameters, with variance hyperparameters in random effects models a leading example (Daniels, 1999).

1.4 THE SAMPLING PERSPECTIVE: ESTIMATING THE DENSITIES OF PARAMETERS

The modern approach to Bayesian estimation has become closely linked to sampling-based estimation methods. Traditional classical estimation methods such as the Newton–Raphson method and least squares are oriented to find a single optimum estimate, such as the maximum likelihood estimate. This is broadly equivalent to finding the mode of the likelihood or the posterior density (the prior times the likelihood). The distribution of the parameter (e.g. its 95 % confidence intervals) around its mode is derived from assumptions of large sample normality.

The sampling perspective used in Bayesian estimation instead focuses on estimating the entire density or distribution of a parameter. This density estimation is based on a long run (or several parallel long runs) of samples from the posterior density. The samples are of the parameters themselves, or of functions of parameters (e.g. the difference between two binomial proportions, or the difference between a parameter and a given constant).

The sampling approach to Bayesian estimation is a considerable advance on previous approximation methods for multiple integration, which may have required an estimate of the normalising constant and became impractical with large numbers of parameters. This is not to belie the problems associated with long-run sampling as an estimation method. For example, the recent review by Cowles and Carlin (1996) highlights problems which may occur in assessing convergence, while there remain problems in setting neutral priors on certain types of parameters (e.g. variance hyperparameters in models with nested random effects).

As noted by Smith and Gelfand (1992), there is a fundamental duality between a sample from a density and the density from which the sample is generated. Provided with a reasonably large sample from a density we can approximate the mathematical form of that density via curve estimation (kernel density) methods. Whereas a data sample, the traditional object of statistical analysis, is of definite size n, there is no limit to the number of samples T of θ

which may be taken from a posterior density $p(\theta \mid y)$. The larger is T from a single sampling run or the larger is $T = T_1 + T_2 + \ldots + T_k$ based on k parallel runs of samples from the density (e.g. with different start-up values for θ), the more accurately can we recreate the posterior density and the more certainty there is about our estimate of θ.

Sampling is continued until the stationary distribution is equivalent to the posterior $p(\theta \mid y)$. If there are n observations and p parameters, then the required number of iterations T from a single sampling chain to reach stationarity is typically considerably larger than the size of the sample of observations. It will tend to increase with both p and n, and also the complexity of the model (e.g. on the number of levels in a hierarchical model, or on whether a nonlinear rather than a simple linear regression is chosen).

A variety of Markov chain Monte Carlo (MCMC) methods have been proposed to sample from posterior densities. They are essentially ways of extending the range of single-parameter sampling methods to multivariate situations, where each parameter or subset of parameters in the overall posterior density may themselves have different densities. Thus there are well-established routines for computer generation of random numbers from particular densities (Ahrens and Dieter, 1974). The usual random number between 0 and 1 is just a special case, drawing from a uniform density with lower limit 0 and upper limit 1, i.e. $\mathrm{U}(0, 1)$. There are also routines for sampling from nonstandard densities which often occur in practice (e.g. if the prior for a parameter is nonconjugate with the likelihood of the parameter).

One form of MCMC method, called Gibbs sampling, samples in turn from each parameter θ_i in the posterior density, while regarding all other parameters as fixed – so that as constants they in effect 'fall out' of the density to sample from (Casella, 1992). This sampling method is the main basis of the BUGS program used extensively in the present book. This program is intended for models which can be depicted as directed acyclic graphs (DAGs). These graphs distinguish constants (e.g. number n in the observed sample or number p of available covariates, or functions of such fixed quantities such as $n - p$), from stochastic parameters. The latter 'stochastic nodes' are variables which have a distribution within the model, either as parameters of densities (e.g. mean and variance of a normal where both are unknown) or functions of these (such as the coefficient of variation in a normal).

1.5 PREDICTIONS FROM SAMPLING

In classical statistics the prediction of out-of-sample data z (for example, data at future time points or under different conditions and covariates) often involves calculating moments or probabilities from the assumed likelihood for y evaluated at the selected point estimate θ_m, namely $p(y \mid \theta_m)$. In the Bayesian method, the information about θ is contained not in a single point estimate but in the posterior density $p(\theta \mid y)$ and so prediction is correspondingly based on averaging $p(y \mid \theta)$ over this posterior density.

If the sampling approach is used then the information about θ is contained in a long run of sampled values from this posterior density. So the prediction of out-of-sample data z given the observed data y is, for θ discrete, the sum

$$p(z \mid y) = \sum_{\theta} p(z \mid \theta) p(\theta \mid y)$$

and is an integral over the product $p(z \mid \theta)p(\theta \mid y)$ when θ is continuous. In the sampling approach, with iterations $t = T, \ldots, T + S$ after convergence, this would typically involve a sample of z_t from the same likelihood form used for $p(y \mid \theta)$ given the sampled value θ_t.

There are circumstances (e.g. in regression analysis or time series) where such out-of-sample predictions are the major interest; such predictions may be in circumstances where the explanatory variates take different values to those actually observed. In clinical trials comparing the efficacy of an established as against a new therapy, the interest may be in the predictive probability that a new patient will benefit from the new therapy (Berry, 1993). In a two-stage sample situation where m clusters are sampled at random from a larger collection of M clusters, and then respondents are sampled at random within the m clusters, prediction of population-wide quantities or parameters can be applied to allow for the uncertainty attached to the unknown data in the $M - m$ nonsampled clusters (Stroud, 1994).

1.6 THE PRESENT BOOK

The following chapters review several major areas of statistical application and modelling with a view to implementing the above components of the Bayesian perspective and to developing source code in BUGS which may be extended to similar problems by students and researchers. Any treatment of such issues is necessarily selective, emphasising particular methodologies rather than others, and particular areas of application. The goal is to contribute to developing the wide range of possibilities opened up by the BUGS software. This software is S based and offers the basis for sophisticated programming and data manipulation but with a distinctive Bayesian functionality. However, the programming flexibility offered by BUGS may be more favourable to some tastes than others – BUGS is not menu driven and pre-packaged, and does make greater demands on the researcher's own initiative.

Issues around parameter convergence or assessing robustness to alternative priors may to some viewpoints be downplayed. In general single long runs were relied on with convergence assessed informally by examining trace plots and assessing stability of parameter estimates in successive batches of iterations (Geyer, 1992). Since WINBUGS 1.2 it is possible to run multiple chains, though generating suitably overdispersed starting values remains in the hands of the researcher. Issues of convergence, optimal sampling methods and robustness are extensively discussed in the statistical literature and in the several web sites devoted to Bayesian sampling methods. By contrast some fairly routine modelling techniques have sometimes received relatively scant attention.

It is for this reason that a sacrifice on some fronts has been hopefully compensated for by introducing a WINBUGS-based analysis of a wide range of applications, including survival models, time series and dynamic linear models, structural equation models, and missing data models. Any comments on the programs, data interpretation, coding mistakes, and so on are welcome at my e-mail address p.congdon@qmw.ac.uk. The reader is also referred to the web site at the Medical Research Council Biostatistics Unit at Cambridge University, where a highly illuminating set of examples are incorporated in the downloadable software, and links exist to other collections of WINBUGS software.

CHAPTER 2

Standard Distributions: Updating, Inference and Prediction

2.1 INTRODUCTION

The general principle of Bayesian updating is that we combine prior knowledge about the density of the parameters θ with the sample data or experimental evidence y, to produce updated knowledge about the parameters. This principle can be applied to the estimation of parameters $\theta_j\ j = 1, \ldots, p$ from the major statistical distributions. More formally, we use iterative methods to take repeated samples of θ from the posterior density, $p(\theta|y)$ which combines prior assumptions $\pi(\theta)$ about the density of θ with sampling distributions applicable to different types of observational data $y, P(y|\theta) \equiv L(\theta|y)$. These distributions provide appropriate models for a wide range of data, in continuous or categorical form.

Bayesian methods of estimation via sampling are not limited to estimating the parameters of these densities. They can also be applied to estimating functions of parameters (e.g. differences in normal means or binomial probabilities between groups of observations) and testing hypotheses about them – for example, finding the probability that a parameter based on a particular data set conforms to a population-wide or reference parameter value.

Similarly sampling-based estimation provides a flexible approach to deriving the distributions of complex summary statistics which are partly functions of the data but depend also on the model parameters; examples considered below are the dependency ratio in a small urban population or the Gini coefficient of inequality of an income distribution.

Another facet of the Bayes method is that of prediction. Integrating

$$p(y_{\text{new}}, \theta|y) = p(y_{\text{new}}|\theta, y)p(\theta|y)$$

over the parameters gives a prediction or forecast in new circumstances, which is in principle verifiable since further values of y can be observed – whereas estimates of θ cannot be verified in this way (Aitchison and Dunsmore, 1975). Therefore prediction is a central feature of model assessment.

This chapter commences with an outline of the fundamental densities of statistical analysis, especially in terms of Bayesian inference and hypothesis tests regarding their parameters. This includes the univariate normal and t densities, their multivariate equivalents; the binomial and multinomial and the conjugate beta and Dirichlet densities; and the Poisson and its gamma conjugate. A final section reviews power and sample size issues for parameters from these standard densities.

2.2 CONTINUOUS DATA MODELS: UNIVARIATE NORMAL

The normal distribution is central to statistical inference and modelling, whether from frequentist or Bayesian perspectives. It is characterised by two parameters, the mean as a measure of location, and the variance (or scale parameter), measuring the extent of scatter around that central location. The central limit theorem of classical statistics and its Bayesian analogue (Berger, 1985, p.224) help justify the normal density as an approximation for the posterior distribution of many summary statistics, even those deriving from non-normal data, as sample size increases.[1,2]

Modifications of the normal distribution are available for example as heavy-tailed alternatives or discrete mixtures of normal densities with differing means and variances. These may be applicable as robust alternatives in the event of asymmetric or multi-modal data, data with nonconstant variance, or data subject to distortions by outlying observations. The same is true for prior densities used to describe the distribution of non-normal parameters. These modifications lead into the broad area of robust Bayesian methods whether in terms of non-normality in the data themselves or in the shape of the density of parameters. For the moment though, we assume the normal to be a reasonable approximation to a sample of continuous measures. The simplest case then to consider is a normal density to describe a single-outcome variable (the univariate normal).

2.2.1 Estimating a normal mean when the variance is known

Suppose first that our data consists of a single observation y from a univariate density we know is approximately normal with unknown mean μ but known variance σ^2. In updating information on the normal, it is often useful to work with the precision as well as with the variance: the precision of a normal distribution $\tau = 1/\sigma^2$ is the inverse of its variance. So the higher the precision, the more highly concentrated are observations expected to be around the mean.

Suppose the uncertainty about the parameter μ before the data are taken can also be represented in a normal form, say as $\mu \sim \mathrm{N}(\mu_0, \sigma_0^2)$, with known variance σ_0^2 and precision $\tau_0 = 1/\sigma_0^2$ The symbol $\sim$ means μ *is distributed as* a Normal. So we are given the prior distribution $p(\mu)$ proportional to

$$\exp[-0.5\tau_0(\mu - \mu_0)^2] \tag{2.1}$$

on omitting terms from the normal density not depending on μ. Similarly the likelihood $p(y|\mu) \equiv L(\mu|y)$ of the single observation y is proportional to

$$\exp[-0.5\tau(y-\mu)^2]. \tag{2.2}$$

Again constant terms not depending on μ are omitted. The posterior density of μ, the prior modified by the data y, expresses our uncertainty about μ once the data are collected. From (2.1) and (2.2) it has the form

$$p(\mu|y) \propto \exp[-0.5\{\tau_0(\mu-\mu_0)^2 + \tau(y-\mu)^2\}]. \tag{2.3}$$

Since a normal prior for the unknown mean combined with a normal sampling density leads to a normal posterior for the mean, the normal prior is known as conjugate.

On rearrangement[3] of the exponent in (2.3) we can show that the posterior for μ is a normal density with variance $\sigma_1^2 = \sigma^2/(n_0+1)$ and mean $\mu_1 = (n_0\mu_0 + y)/(n_0+1)$, where the ratio of precisions is denoted $n_0 = \tau_0/\tau$. The mean of the posterior density is thus a weighted average of y and μ_0 with weights 1 and n_0 respectively. So the ratio of precisions τ_0/τ can be seen as a measure of the 'prior sample size'.

Equivalently the variance of the posterior density of μ may be written $1/[\tau_0+\tau]$ so that its precision is $\tau 1 = \tau_0 + \tau$. The posterior mean can also be written $\mu_1 = (\mu_0\tau_0 + y\tau)/(\tau_0+\tau)$, namely as a precision weighted average of the data point and the prior mean. If we write

$$w = \frac{\tau_0}{\tau_0+\tau}$$

where w measures the ratio of prior to total precision, then μ_1 is equivalently a weighted average

$$\mu_1 = w\mu_0 + (1-w)y.$$

2.2.2 Predictions and predictive distributions

Suppose we wanted to predict the value of a future observation z and its variability on the basis both of the prior for the mean and the observed data point. This is a situation of minimal information on which to base a prediction.

In the case of a normal distribution with known precision τ, and conditional on a sampled value of μ, the value of a future observation z would be merely based on a draw from the normal density $p(z|\mu) = \mathrm{N}(\mu, \tau^{-1})$. However, the value of μ is being updated by the data y and so the prediction should reflect this. Hence the density of z conditional on the observed y, denoted

$$p(z|y)$$

is based on integrating the product $p(z|\mu, y)p(\mu|y)$ over all values of μ:

$$p(z|y) = \int p(z|\mu, y)p(\mu|y)d\mu$$
$$= \int p(z|\mu)p(\mu|y)d\mu.$$

In practice for sampling-based estimation, for a large number T of samples from the posterior density $p(\mu|y)$ we would form the empirical density of $z|y$ over the set of sampled μ. At each sample $t(= 1, \ldots, T)$ of $\mu^{(t)}$ we would draw a sample $z^{(t)}$ from a normal density with that mean and with the known variance.

We can then show that z will also be normal with mean μ_1 and variance $\sigma^2 + \sigma_1^2$. The predictive distribution of a future or out-of-sample observation z therefore has two sources of variation: that due to sampling from a normal density for given μ, and that due to the posterior uncertainty in μ itself.

2.2.3 Samples of $n > 1$ observations

Consider now a sample of $n > 1$ observations $\underset{\sim}{y} = (y_1, y_2, \ldots, y_n)$ from a univariate density and in particular the summary statistic $\bar{y}$ denoting the average of the n observations. The prior for the mean is as above, namely $\mu \sim \mathrm{N}(\mu_0, \sigma_0^2)$, with precision $\tau_0 = 1/\sigma_0^2$. We are sampling the y from a normal density with unknown mean μ and known variance σ^2. So the likelihood $p(y_1, y_2, \ldots, y_n|\mu)$ is proportional to

$$\prod_{i=1}^{n} [\exp\{-0.5\tau(y_i - \mu\}^2]$$

which, from the viewpoint of estimating μ, reduces to

$$\exp[-0.5n\tau(\bar{y} - \mu)^2]$$

since viewed as a function of μ the other terms in the likelihood are constants. Thus all the information about μ in the sample is contained in one statistic, the mean $\bar{y}$. The mean is said to be a sufficient statistic for μ in this case, in the sense that the posterior density for μ depends on the data only though $\bar{y}$.

Proceeding as for the single data point above, the posterior or updated density for μ is normal with mean

$$\mu_1 = (n_0\mu_0 + n\bar{y})/(n_0 + n)$$

and variance $\sigma^2/(n_0 + n)$. The updated mean is a weighted average of the prior mean and observed mean, with weights being the 'prior sample size' $n_0 = \tau_0/\tau$ and the number of observations in the sample. Equivalently, the posterior mean is a weighted average (with weights w and $1 - w$) of prior and observed means with weights being the prior precision τ_0 and the precision of the observed mean, namely $n\tau$:

$$\mu_1 = (\tau_0\mu_0 + n\tau\bar{y})/(\tau_0 + n\tau) \quad = w\mu_0 + (1 - w)\bar{y}$$

where the posterior precision is $\tau_0 + n\tau$.

2.2.4 Hypothesis tests

Often our concern in analysing new sample observations $\underset{\sim}{y}$ will be to assess a hypothesis about the model parameters, or about the appropriateness of a sampling distribution for the data. The choice between which of two or more hypotheses to accept involves specifying prior beliefs about their relative frequency, and a comparison (after seeing the data) of their posterior probabilities, from which we can derive the posterior odds on and against each of the hypotheses. Thus if $P(H_0), P(H_1), \ldots, P(H_M)$ are the prior probabilities regarding the relative frequency of the alternative hypotheses then their respective posterior probabilities are, via Bayes theorem,

$$P(H_i \mid \underset{\sim}{y}) \propto p(\underset{\sim}{y} \mid H_i) P(H_i), \quad i = 0, \ldots, M.$$

Suppose one were comparing two interval hypotheses about a continuous parameter θ, the first, denoted H_0, specifying that θ lies in the interval (a_0, b_0), the other, H_1, specifying θ lies in another interval (a_1, b_1) which does not overlap the first interval. Further suppose these alternatives encompass all possible values of θ. For example, if θ was a scalar, the first interval might be all negative values on the real line, the second all positive values. The prior odds on H_0 are $P(H_0)/P(H_1)$. If the alternatives cover all possible values of θ, then $P(H_0) + P(H_1) = 1$, and the prior odds on H_0 are equivalently $P(H_0)/[1 - P(H_0)]$. To obtain the posterior odds, consider

$$P(H_i \mid \underset{\sim}{y}) \propto p(\underset{\sim}{y} \mid H_i) P(H_i), \quad i = 0, 1$$

and since the proportionality constant is the same

$$P(H_0 \mid \underset{\sim}{y})/P(H_1 \mid \underset{\sim}{y}) = [P(\underset{\sim}{y} \mid H_0)/P(\underset{\sim}{y} \mid H_1)][P(H_0)/P(H_1)]. \tag{2.4}$$

So the posterior odds are equivalent to the likelihood ratio times the prior odds. The likelihoods $P(H_i \mid \underset{\sim}{y})$ are marginal in that they are obtained by integrating out parameters not associated with specifying the hypothesis.

The ratio $P(\underset{\sim}{y} \mid H_0)/P(\underset{\sim}{y} \mid H_1)$ is also called a Bayes factor, denoted B_{01}, on the relative chances of H_0 vs H_1, and converts prior odds on H_0 to posterior odds on H_0. So we can write

$$B_{01} = [P(H_0 \mid \underset{\sim}{y})/P(H_1 \mid \underset{\sim}{y})]/[P(H_0)/P(H_1)]. \tag{2.5}$$

If the alternative hypotheses cover all possible values of θ, then $P(H_0 \mid \underset{\sim}{y}) + P(H_1 \mid \underset{\sim}{y})$ also equals 1. So from (2.5), we can obtain the posterior probability of H_0 from its prior probability and the Bayes factor:

$$P(H_0 \mid \underset{\sim}{y}) = 1/[1 + (1/B_{01})\{P(H_1)/P(H_0)\}]. \tag{2.6}$$

Exactly the same procedure follows if we are comparing two models M_1 and M_2, with prior probabilities $P(M_1)$ and $P(M_2)$; for example, one regression model including a particular covariate and another excluding it. If, as is often the case, prior knowledge does not favour one hypothesis or model against another (i.e. $P(M_1) = P(M_2) = 0.5$), then the Bayes factor is equivalent to the posterior odds of the model. Indicative values of the Bayes factor have been provided by several authors as measures of increasing strength of evidence for or against hypotheses. In particular, a Bayes factor of 1 or near 1 favours neither hypothesis.

Another major strand of research is into approximations to the Bayes factor which are independent of the priors on θ in M_1 and M_2, assuming these to be non informative. For example, the Bayesian information criterion (Schwarz, 1978) is based on the usual likelihood ratio $\lambda = L(y\,|\,M_1)/L(y\,|\,M_2)$ for two models with p_1 and p_2 parameters respectively, but adding a penalty in terms of the sample size and difference in parameters:

$$\text{BIC} = -2\log\lambda - (p_1 - p_2)\log n.$$

Then $\exp(-0.5\text{BIC})$ is an approximation to the Bayes factor for large samples.

2.2.5 Interval and point hypotheses

We revert to the hypothesis test framework, especially hypotheses that can be evaluated in terms of posterior probabilities that individual parameters lie in certain regions. Alternative interval hypotheses on a parameter within a single model, such as H_0: $\theta < 0$ vs H_1: $\theta \geq 0$, can be assessed in a repeated sampling framework by cumulating the occasions when the sampled parameter values support one or other hypothesis. A long sampling run will provide evidence on $P(H_0\,|\,y)$ from which the Bayes factor may be derived.

This approach is not applicable if H_0 is a simple or point hypothesis, as often in classical testing, such as H_0: $\theta = \theta_0$ and with alternative H_1: $\theta \neq \theta_0$. In this case $P(y\,|\,H_0) = P(y\,|\,\theta_0)$, that is the posterior probability of H_0 reduces to the marginal density of y given a particular parameter value. The marginal likelihood of H_1 is the integral of the likelihood times prior, namely

$$P(y\,|\,\theta)\pi(\theta)$$

over the entire parameter space of θ, since the single point θ_0 does not affect the value of this integral. Thus $P(y\,|\,H_1) = \int P(y\,|\,\theta)\pi(\theta)d\theta \neq P(y)$, and

$$B_{01} = P(y\,|\,\theta_0)/P(y).$$

For example, consider the standard classical hypothesis test, namely whether a normal mean equals a certain value m_0. Let $(y_1, \ldots, y_n)$ be a random sample of size n from a distribution $N(\mu, \sigma^2)$, where σ^2 is known. Then the hypothesis is H_0: $\mu = m_0$, and H_1 denotes its complement. Assume the prior density of μ (or equivalently of μ given H_1) is also normal, namely $\mu \sim \text{N}(M, \phi^2)$. Then it can be shown (Migon and Gamerman, 1999, Chapter 6) that

$$B_{01} = [(\sigma^2 + n\phi^2)/\sigma^2]^{0.5}\exp[-nD/2] \tag{2.7}$$

where $D = (\bar{y} - m_0)^2/\sigma^2 - (\bar{y} - M)^2/(\sigma^2 + n\phi^2)$. Typically we take $M = m_0$, so the prior is centred on the hypothesised mean, and the term D in (2.7) reduces to

$$D = (n\tau^2/\sigma^2)(\bar{y} - m_0)^2/(\sigma^2 + n\phi^2).$$

If we write

$$z = |\bar{y} - m_0|/(\sigma n^{-0.5}),$$

analogous to a classical test statistic, then the Bayes factor in this case is, following Lee (1997, Section 4.5),

$$B_{01} = [(\sigma^2 + n\phi^2)/\sigma^2]^{0.5}\exp[-0.5nz^2\phi^2/(n\phi^2 + \sigma^2)].$$

If also we take $\phi^2 = \sigma^2$ then the Bayes factor is

$$B_{01} = (n+1)^{0.5}\exp[-nz^2/(2n+2)].$$

2.2.6 Posterior predictive checks

In certain circumstances the marginal likelihoods and hence Bayes factor may be sensitive to priors adopted on the parameters relating to the models or hypotheses being compared (Gelman *et al.*, 1995, Chapter 6). Predictive checks have been proposed as an alternative methodology, and involve sampling new data which are consistent with the model assumed to generate the actually observed data. Suppose we provisionally assume our data $y_1, \ldots, y_n$ to be generated by a Normal density with parameters $\{\theta\}$, namely a mean μ and variance V. Using MCMC techniques, we could take samples $\theta(t) = \{\mu(t), V(t)\}$, $t = 1, \ldots T$, from the posterior densities of these parameters, after having observed y. For each sampled parameter set $\theta(t)$, we could sample n replicate data points y_{rep}

$$y_{\text{rep}}.i^{(t)} \sim \text{N}(\mu(t), V(t)), \qquad t = 1, \ldots, T; \quad i = 1, \ldots, n.$$

We may then compare our observed data (whose true form is uncertain) with the replicate data which we have constructed according to the postulated model form. Specifically we construct one of a set of standard test statistics, one based on the observed sample $T(y, \theta)$ and one based on the replicate data $T(y_{\text{rep}}, \theta)$. Over a set of T iterations we would compare the calculated values of the test statistics, and derive a posterior predictive check p value somewhat analogous to classical p values:

$$\text{PPC} = \Pr(T(y_{\text{rep}}, \theta) > T(y, \theta)\,|\,y).$$

Systematic differences (e.g. in percents of extreme values or in means) between replicate and actual data indicate possible limitations in the model assumed for the observed data. Specifically values of PPC around 0.5 indicate a model consistent with the actual data, whereas extreme values (close to 0 or 1) suggest inconsistencies between the proposed model and actual data. However, it is not true that values of the PPC criterion around 0.5 show a model is the 'correct' one for the data.

Example 2.1 Systolic blood pressure Suppose we take a random sample of 20 systolic blood pressure readings y_i from a subpopulation of adult men; this might be a particular diagnostic or demographic group. We know, perhaps from national surveys, that the standard deviation of y is 13. We are interested in estimating μ, the mean blood pressure in our group and predicting its likely level in a typical 'new' patient in the same group. In the case of human physical measures there might be information on which to base an informative prior regarding the mean of the group. Suppose however, in the absence of such prior data, we select a relatively vague prior. For example, we might assume a normal density on the mean μ,

$$\mu \sim \text{N}(M, \phi^2)$$

such as

$$\mu \sim \text{N}(0, 10000) \tag{2.8}$$

centred at zero and with a very large variance. In fact, we know that blood pressure is a positive quantity so we could use a prior restricted to the positive part of the real line, such as, for example, $\text{N}(0, 10000)\text{I}(0,)$, where the indicator function $\text{I}(a, b)$ restricts sampling to values above a and below b.

The posterior inferences we may be interested in include credible interval estimates of the mean, probability statements such as the chance that blood pressure in the patient group exceeds a certain level, and comparison of the posterior evidence in support of alternative hypotheses on the mean. We may also be interested in predicting the credible interval for the blood pressure of a new patient in the diagnostic group.

These questions may be answered directly from normal probabilities, but a sampling perspective (see *Program 2.1 Systolic Blood Pressure*) is equally possible. Note that for this program, which does not estimate any variance parameters, an *inits* file is not needed – the initial values can simply be generated. Using output from the second half of a run of 100 000 iterations gives sample average and median of μ both at 128, with a posterior standard deviation around the mean of 2.9, and a 95% interval for the mean pressure in the patient group of (122.2, 133.7). The likely range in which the pressure for a new individual in the group is located is between 101.6 and 153.2; the mean of this predictive density is also 128 but its higher standard deviation reflects the uncertainty in μ and is 13.3.

Suppose we know the typical blood pressure for all adult males is 125, and we wish to test whether the particular diagnostic group has above or below average pressure. That is we wish to compare the hypotheses H_0 $(\mu > 125)$ and H_1 $(\mu < 125)$. Using the step function in BUGS we find the proportion of iterations where the sampled μ exceeds 125, this being 0.847. The Bayes factor to assess the fit of these two competing hypotheses, namely $p(y|H_0)$ and $p(y|H_1)$ is the posterior odds of the two hypotheses divided by their prior odds:

$$B_{01} = \{p(H_0|y)/p(H_1|y)/p(H_0)/p(H_1)\}.$$

Given prior (2.8) about the location of μ, the prior probability of H_0 is

$$1 - \Phi(125/100) = 0.1056,$$

where Φ is the cumulative standard normal. So the Bayes factor reduces to a comparison of the ratio 0.847/0.153 to 0.106/0.894, giving $B_{01} = 46.9$.

Example 2.2 The W particle Lee (1997) cites an example from electroweak theory of a hypothesised new particle, the W particle. This has a prior mean mass m, given by 82.4 GeV, with a hypothesised standard deviation of 1.1 GeV. Experiments confirmed the existence of the particle and provided an observed value of 82.1 GeV with SD = 1.7. Suppose we wish to test whether the mass is under 83 GeV on the basis of the prior and data (note that H_1 is then the hypothesis that the mass exceeds 83 GeV). The prior probability of H_0 is given by $\Phi((83 - 82.4)/1.1) = 0.71$, so the prior odds on H_0 are approximately $0.71/0.29 = 2.49$. The posterior probability is assessed by counting the number of times H_0 applies when the prior is updated by the data (see the step function calculation in *Program 2.2 W particle*). A long run of 100 000 iterations provides an empirical posterior probability $P(H_0 \mid y)$ of 0.7751, and odds on H_0 of $0.7751/0.2249 = 3.45$ so the Bayes factor is estimated as $B_{01} = 3.45/2.49 = 1.39$. There is no clear evidence for or against H_0.

Example 2.3 Mean weights Albert (1996, Section 6.4) considers the problem of a person's expectations about their current weight $y \sim \mathrm{N}(\mu, \sigma^2)$ (in lbs), given last year's weight was 170 lbs. A sample of ten more recent weight measures gives an average of 176 lbs. Suppose we assume both the observational (population) standard deviation σ and the prior standard deviation ϕ are known (specifically $\sigma = 3$ and $\phi = 8$). Thus the person's prior beliefs, of the form $\mu \sim \mathrm{N}(M, \phi^2)$, are that his weight is unchanged from last year, namely $m_0 = M = 170$ lbs, giving a 95 % interval of approximately 154–186 lbs.

We evaluate the Bayes factor in favour of the null hypothesis (that the current mean weight is 170 lbs) using *Program 2.3 Mean Weights* which incorporates the formula (2.7) above. Then the Bayes factor averages 0.155 with standard deviation of 0.013 and the posterior probability of H_0 is 0.135 with a standard deviation of 0.01 (see formula (2.6)). The Bayes factor $B_{10} = 1/B_{01}$ against H_0 averages 6.45 with a standard deviation of 0.56. Even with a relatively high precision on the prior mean there is little support in the data for H_0. With a larger prior variance ϕ^2, the Bayes factor against H_0 is magnified.

2.3 INFERENCE ON NORMAL PARAMETERS, MEAN AND VARIANCE UNKNOWN

It is usually the case that available information about the value of both the mean and variance of the Normal is limited, or subject to a degree of uncertainty. Suppose we focus then on the estimation of both mean and precision (inverse variance) from continuous data with a form believed approximately

Normal. The simplest option is to assume that the mean and precision can be estimated independently of each other. Consequently in specifying the priors on these parameters, we assume independent priors $p_1(\mu)$ and $p_2(\sigma^{-2})$.

Suitable prior distributions for precisions $\tau = \sigma^{-2}$ may be provided by any density confined to positive values; examples are the uniform over the positive part of the real line, the gamma and less frequently, densities such as the Pareto. A gamma density is specified in terms of two parameters, and if used as a prior for τ, its mean is the expected value of the precision. Specifically if $G(a, b)$ denotes a gamma prior density for τ, then the chance of different values of τ is proportional to

$$\tau^{a-1}\exp(-b\tau) \tag{2.9}$$

where a is known as the shape parameter (or index) and b as the scale parameter. Both a and b are restricted to be positive. The expected value of the precision τ is then a/b and its variance is a/b^2. A $G(a, b)$ prior on the precision is equivalent to an Inverse Gamma prior $IG(a, b)$ on the variance, where the Inverse Gamma density for a variate s is proportional to $s^{-(a+1)}e^{-b/s}$.

A typical non informative but proper prior for the precision is a gamma with small but positive values of a and b. For example if $a = b = 0.0001$, such that

$$p(\tau) \sim G(0.0001, 0.0001)$$

then, by substitution in (2.6), the prior of τ will be approximately[4]

$$p(\tau) \propto 1/\tau. \tag{2.11}$$

The prior on the precision in (2.11) is known as Jeffrey's prior and is a form of 'reference prior' intended to correspond to ignorance about the scale parameter. Thus suppose we want inferences to be invariant to the scale of y, i.e. to stretching or shrinking of the scale. This implies an equivalence

$$p(\tau)d\tau = p(\beta\tau)d(\beta\tau)$$

where β is a positive constant. Since $d(\beta\tau) = \beta d\tau$ this equivalence requires (2.11) to hold.

The case where a and b actually equal zero results in an improper prior density (one which doesn't integrate to unity), whereas when a and b are both just positive, the prior is 'just proper'. Special cases of the gamma density are the exponential and chi-square density and a version of the latter is frequently used in the case where the mean and precision are estimated interdependently (see below). The usual chi-square density with m degrees of freedom is equivalent to a gamma density with $a = m/2$ and $b = 1/2$.

A prior for the variance σ^2 may be typified in the form $IG(\nu_0/2, \nu_0 S_0/2)$, where S_0 is a prior guess at the variance and ν_0 is the strength of this belief (as an integer 'degrees of freedom'). Equivalently the precision $\tau = \sigma^{-2}$ is specified to have a gamma prior $G(\nu_0/2, \nu_0 S_0/2) \equiv G(\nu_0/2, \nu_0/2\tau_0)$, where $\tau_0 = 1/S_0$ is the prior guess at precision. This is the same as a chi-square for $n_0 S_0 \tau$ with ν_0 degrees of freedom. So a $G(5, 1)$ prior on τ is equivalent to a prior d.f. of $\nu_0 = 10$ on the prior beliefs regarding precision, and gives an expected precision of 5.

After observing a sample of n with mean $\bar{y}$, the posterior density of the mean, $p(\mu|\tau, y)$ has mean $(\tau_0\mu_0 + n\tau\bar{y})/(\tau_0 + n\tau)$ and precision $\tau_0 + n\tau$. The posterior density of τ given y can also be expressed in closed form (Gelman *et al.*, 1995) and so it is possible to find parameter estimates and other quantities of interest by directly sampling from these posterior densities.

Priors for precisions do not have to be of a gamma form, though a conjugate analysis for normally distributed data requires it. Alternatively, we could have a normal prior on $\log(\tau)$, which could include both negative and positive values. This has the advantage that while the precision τ itself is likely to show positive skew, $\log(\tau)$ is often approximately normal. A bivariate or multivariate normal prior on more than one log precision term allows one to study interdependence between variances.

Example 2.4 Survival times from carcinoma Aitchison and Dunsmore (1975) present data on survival times y in weeks after a combination of radiotherapy and surgery applied to a particular carcinoma. One question of interest is the length of survival expected for a new patient with this carcinoma and assigned to this type of treatment. Suppose we take a $G(0.0001, 0.0001)$ prior for the precision τ, and a similarly non informative prior for the mean, namely a normal density $N(0, 10000)$ located at zero and with low precision (high variance). The posterior estimates of μ and τ may be derived together with the predicted probability that a new patient's survival time will exceed a certain threshold. Because of the skewed form of the data we apply a log transformation, and assume that $\log(y)$ is normal (i.e. y is assumed to be log-normal).

Applying the program *Program 2.4 Carcinoma Survival* (and using the second half of a run of 100 000 iterations) gives posterior estimates for the mean, precision and variance of the log survival times as shown in Table 2.1.

Also shown is the 95% credible interval for a new patient's survival time, again in logged units. Suppose we wanted to find the probability that a new patient has a survival time exceeding 150 (namely that z exceeds 5.01, the logarithm of 150). To estimate this probability we compare the sampled value of z with the threshold of 5.01 and accumulate the number of iterations where the condition $z > 5.01$ holds as a proportion of the total iterations. The answer is 0.16, the same as obtained by Aitchison and Dunsmore using analytic methods.

Table 2.1 Posterior summary: Carcinoma survival parameters.

Parameter	Average	SD	2.5% percentile	97.5% percentile	Median
μ	3.86	0.26	3.36	4.38	3.86
τ	0.83	0.27	0.39	1.42	0.80
σ^2	1.35	0.50	0.70	2.54	1.25
z (new survival time)	3.85	1.19	1.5	6.2	3.85

It may be of interest to examine mean survival by patient characteristics such as age. In Program 2.4 we can introduce this analysis as an extra model for 'new' data also created by logging the original survival times. We accordingly dichotomise the ages of the 20 patients into those below and above 50 (groups 1 and 2), and estimate two means μ_1 and μ_2. The variance is assumed the same across both age groups.

The resulting means (Table 2.2) are clearly different, and the probabilities of surviving 150 weeks are 0.236 and 0.029 in the two groups. Allowing differentiation between group means has also reduced the scale parameter.

2.3.1 Normal Distribution: Mean and variance unknown and interdependent

In this case we assume that the prior for the mean is conditional on a previously sampled value of the precision, so the joint prior specifications involve the densities $p(\tau)$ and $p(\mu|\tau)$, with

$$p(\mu, \tau) = p(\mu|\tau)p(\tau).$$

A frequently used prior for the precision in these circumstances is the gamma $G(a, b)$ with $a = \nu_0/2$ and $b = \nu_0/(2\tau_0)$. This density has an average τ_0 expressing prior beliefs about the precision in the data. Equivalently $b = \nu_0\sigma_0^2/2$ where σ_0 is our prior belief about the variance. The degree of strength of the prior beliefs is contained in the parameter ν_0.

Given a sampled value τ, that is $1/\sigma^2$, from its prior $G(\nu_0/2, \nu_0/(2\tau_0))$, the prior for μ is of the form $N(\mu_0, \sigma^2/m_0)$. The 'prior sample size' m_0 expresses the strength of prior belief about the chosen location μ_0 for the mean of the data. This prior is analogous to the usual density for a normal mean based on the observed variance, and means the prior variance of the mean is tied to σ^2, the scale of the observations.

The posterior density of σ^{-2} after observing a sample of size n is

$$\sigma^{-2} \sim G(\nu/2, \nu\sigma_n^2/2)$$

where $\nu = \nu_0 + n$,

$$\nu\sigma_n^2 = \nu_0\sigma_0^2 + \sum(y_j - \bar{y})^2 + \rho(\bar{y} - \mu_0)^2$$

and where $\rho = m_0 n/(m_0 + n)$. The conditional posterior density of μ given the sampled σ^2 is then

$$N(\kappa, \sigma^2/(m_0 + n))$$

Table 2.2 Posterior summary: Carcinoma survival parameters (age groups).

Parameter	Mean	SD	2.5%	Median	97.5%
μ_1	4.309	0.271	3.772	4.309	4.843
μ_2	3.025	0.370	2.297	3.024	3.761
σ^2	0.963	0.362	0.490	0.889	1.874

where $\kappa = (n\tau\bar{y} + m_0\tau\mu_0)/(n\tau + m_0\tau)$ is the precision weighted average of the prior and data means, μ_0 and $\bar{y}$ respectively.

An example of a simple, though nonconjugate, option for allowing interdependence would be a bivariate normal on $\log\tau$ and μ. This enables one to actually monitor covariance between precision and mean parameters.

Example 2.5 Cancer survival times under interdependent informative priors Suppose in the carcinoma survival example, we were willing to specify a prior belief that the expected mean length of survival was 30 days (i.e. $\mu_0 = 3.4$ in log terms) with a prior sample size $m_0 = 10$. Our prior belief about the variance of the log survival time is 2, also with a prior d.f. ν_0 of 10. These are relatively informative priors regarding the normal parameters, and result in a lower posterior mean and higher posterior variance (Table 2.3, based on 10 000 iterations). The predictive density for a new survival time has a correspondingly lower mean and larger variance. The probability of a survival time over 150 days becomes 0.151.

2.3.2 Comparison of variances

Consider the situation where we have two samples $\{x_1, \ldots, x_m\}$ and $\{y_1, \ldots, y_n\}$ assumed to be distributed as follows

$$y_i \sim \mathrm{N}(\mu, \sigma_y^2) \text{ and } x_i \sim \mathrm{N}(\lambda, \sigma_x^2)$$

with λ, μ, σ_y^2, and σ_x^2 all unknown. We might wish to ascertain whether the two samples can be considered to come from a distribution with the same variance, as well as the usual question whether the means are the same. The quantity of interest is then the ratio of variances

$$\kappa = \mathrm{var}(x)/\mathrm{var}(y) = \sigma_x^2/\sigma_y^2$$

for which the empirical estimate is

$$K = s_x^2/s_y^2 = (n-1)\sum(x-\bar{x})^2/[(m-1)\sum(y-\bar{y})^2].$$

Define $\nu_y = n-1$ and $\nu_x = m-1$. Then assuming a prior

$$\mathrm{P}(\lambda, \mu, \sigma_y^2, \sigma_x^2) \propto 1/(\sigma_y^2\sigma_x^2)$$

it can be shown that the ratio $\xi = \kappa/K$ of the true ratio κ to its estimate K has posterior distribution

Table 2.3 Posterior summary from carcinoma survival with interdependent priors.

Parameter	Average	SD	2.5 % percentile	97.5 % percentile	Median
μ	3.71	0.23	3.27	4.15	3.70
τ	0.68	0.17	0.38	1.05	0.66
σ^2	1.58	0.44	0.95	2.62	1.51
z	3.70	1.27	1.22	6.23	3.69

$$P(\xi\,|\,x, y) \propto \xi^{0.5\nu_y^{-1}}[\nu_y \xi + \nu_x]^{-0.5\nu_x - 0.5\nu_y}.$$

This is an example of an F density (with ν_y, ν_x degrees of freedom). The F density occurs when variables W_1 and W_2 are independent χ^2 with n_1 and n_2 degrees of freedom respectively. Then the ratio of W_1/n_1 to W_2/n_2 has an F density with n_1, n_2 degrees of freedom:

$$\phi = W_1/n_1/[W_2/n_2] \sim F_{n_1 n_2}$$

and

$$1/\phi \sim F_{n_1 n_2}.$$

Example 2.6 Nitrogen masses Jeffreys (1961) gives data on masses in grams of nitrogen samples from the air, $\{x_i, i = 1, \ldots, 12\}$ and from a container using chemical methods $\{y_i, i = 1, \ldots, 8\}$. Then if $K = s_x^2/s_y^2$, we can be 95% certain that the true variance ratio κ lies in the range $K/F_{11,7,0.975}$ to $K/F_{11,7,0.025}$. We can use the program Density Points (see Program 2.6) to find that the 0.025 and 0.975 points of $F_{11,7}$ are respectively 0.265 and 4.68. Alternatively standard tables provide the values 0.26 and 4.52. Then we can use the actual data to estimate $\hat{\sigma}_x^{-2}$ and $\hat{\sigma}_y^{-2}$ and compare the percentiles of the resulting sampled values of κ with the expected values. It assists in estimating these precision (or variance) parameters if the original data are scaled by a factor of 100. The observed values of s_x^2 and s_y^2 are then 0.000 172 and 0.1493. So $K = 0.0015$ with expected 2.5% and 97.5% points $0.0015/4.52 = 0.000255$ and $0.0015/0.26 = 0.0044$. The sampled values of κ are close to expectation with a mean of 0.00134 with 95% credible interval $\{2.4\text{E} - 4, 0.0041\}$.

2.3.3 Comparison of several means

Suppose we are comparing a set of crop treatments, results of educational programs, etc., and wish to assess whether any treatments or programs produce better or worse results than average. The null hypothesis is then that the means of treatments or programs are the same.

Suppose the data is for a univariate outcome x with G group means $\bar{x}_i$ based on $n_1, n_2, \ldots, n_G$ observations respectively, with a total of observations $N = n_1 + n_2 + \ldots + n_G$. Assume the data are drawn from independent Normal densities

$$x_{ik} \sim N(\lambda_i, \sigma^2), \quad i = 1, \ldots G, k = 1, \ldots, n_i$$

and that we adopt correspondingly independent priors on the normal means λ_i and the homogenous variance. The null hypothesis is often that $\lambda_1 = \ldots = \lambda_G$ or equivalently that

$$\delta_1 = \ldots = \delta_G = 0$$

where $\delta_i = \lambda_i - \lambda$ and where $\lambda = \Sigma n_i \lambda_i / \Sigma n_i$ Suppose the 'reference prior'

$$\pi(\lambda_1, \ldots, \lambda_G, \sigma^2) \propto 1/\sigma^2$$

is taken. Then defining

$$S_B = \sum_i n_i(\hat{\lambda}_i - \hat{\lambda})^2,$$

and

$$S_W = \sum\sum(x_{ik} - \hat{\lambda}_i)^2,$$

the quantity

$$R_\delta = S_B(N - G)/[S_w(G - 1)]$$

has an F_{ν_1,ν_2} density with $\nu_1 = G - 1, \nu_2 = N - G$ (Lee, 1997). In a sampling estimation context, we can compare the sampled R_δ with a criterion $F_{\nu_1,\nu_2\alpha}$, (with say $\alpha = 0.01$ or $\alpha = 0.05$) as a test of the equality of the G group means.

Example 2.7 Scab index Cochran and Cox (1957) present comparison data on a 'scab index' for potato crops subject to six different treatment regimes (different dressing levels and different seasons of application) involving four plots each and a control group of eight plots. The scab index is in fact a percentage of potatoes affected by scab but is here treated as a Normal outcome. There are $N = 32$ total observations, and $G = 7$ so $\nu_1 = 6$, and $\nu_2 = 25$. Simple examination of the six treatment means against the control mean suggests treatment effects, with one treatment having only 6 % scab compared with 22 % in the control group.

The average F ratio is estimated – from a run of 20 000 with *Program 2.7 Scab Treatment* – to be 4.56 in comparison with a classical maximum likelihood ANOVA estimate of $F_{ML} = 3.6$. However, the values of F are positively skewed, so that the median F is lower at 4.32, and uncertainty about the value of F arising from small sample sizes means the probability that F exceeds the 1 % value of the $F(6, 25)$ density, namely 3.63, is only 65 %. The classical significance test would state a significance level approaching 1 %, i.e. the observed ML estimate of $F_{ML} = 3.6$ is virtually at the 1 % point of the density.

2.4 THE t DENSITY AS A HEAVY-TAILED ALTERNATIVE TO THE NORMAL DISTRIBUTION

The t density arises in the case of small samples n_j from a Normal distribution with mean μ, namely samples containing 50 or fewer cases. The means $\bar{y}_j$ of such samples have a distribution with a higher variance than applies for large n_j. The usual estimate of $\text{var}(\bar{y}_j)$ is V_j/n_j, where V_j is the estimate of the sample variance $\sum_i(y_{ij} - \bar{y}_j)^2/n_j$. This will understate the variability of the means because of variations from sample to sample in the value of V_j. In particular, standardised deviates from the mean $(\bar{y}_j - \mu)/(V_j/n_j)^{0.5}$ no longer follow the standard normal.

The t density is a heavier tailed or 'overdispersed' alternative to the Normal, bell-shaped like the Normal but with a higher chance of extreme values in the tails. In this sense it is a robust alternative to the Normal in the event of suspected outliers in the data, especially if sample sizes are small. It may also be used as a prior density to describe sets of parameters (e.g. exchangeable random effects) with potential extreme values among them.

The density has the form

$$p(y \mid \mu, \tau, \kappa) \propto (1 + \tau(y - \mu)^2/k)^{-(k+1)/2}$$

where μ and τ are the mean and precision, and the degrees of freedom parameter k determines the extent of overdispersion. Smaller values of k allow for more marked departures from normality in the tails, so a density expected to have outliers might be described by a t density with degrees of freedom under 10. Values of k over 100 lead to a density indistinguishable from the Normal.

The univariate t density with v d.f. is obtainable as a scale (variance) mixture, with variances V_i differing between individuals and obtained as $V_i = \sigma^2/z_i$, where σ^2 is the overall variance and the z_i are drawn from a gamma density with average 1 and parameters both equal to $v/2$. Thus a t density for y_i is generated by

$$z_i \sim \mathrm{G}(0.5v, 0.5v)$$
$$y_i \sim \mathrm{N}(\mu, V_i).$$

One application of the t density occurs when small samples are drawn from two Normal populations, and we are interested in inferences about the size of the difference in observed means $D = \bar{y}_1 - \bar{y}_2$, based on samples of size n_1 and n_2 respectively. Then the t density describes the distribution of the difference provided the two Normal populations are assumed to have the same variance. The classical significance test in these circumstances would form a pooled estimate of this common variance $V = \hat{\sigma}^2$ with $n_1 + n_2 - 1$ degrees of freedom.

If the assumption of equal variances cannot be made, then Welch (1938) derived a distribution for

$$(\bar{y}_1 - \bar{y}_2)/(V_1/n_1 + V_2/n_2)^{0.5} \tag{2.12}$$

which still follows the t density but with adjusted degrees of freedom. The estimated degrees of freedom depends on the data variances, namely

$$v_e = (q_1 + q_2)^2/[q_1^2/(n_1 - 1) + q_2^2/(n_2 - 1)] \tag{2.13}$$

where $q_1 = V_1/n_1, q_2 = V_2/n_2$. Welch's test has more power if the variances σ_1^2 and σ_2^2 in the two populations are substantially different.

Example 2.8 Infant weight gain Suppose we consider data on weight gain among infants from birth to day 4 after birth according to whether or not they were continuously exposed to the recorded sound of the mother's heartbeat (Salk, 1973). There are $n_1 = 20$ and $n_2 = 36$ observations in the two samples (exposed and non-exposed respectively), and the measure y is weight gain in grams. The analysis assumes two different data models: either (a) they are drawn from two Normal populations with different precisions or (b) they are drawn from t densities with 5 degrees of freedom with different precisions. The test statistic (2.12) for difference in means and Welch's formula for its degrees of freedom are calculated. The means and precisions in the two populations are given independent non informative priors.

The data set *Original Data* in Program 2.8 contains the original study observations, whereas in the amended data set *Outlier Data* the second observation in the no-heartbeat group is replaced by 1000. This might be due to a recording error, for example.

Program 2.8 Infant Weight Gain was applied for 10 000 iterations under the two assumptions about the sampling distribution and with the two data sets. Assuming two normal populations and the original study data gives a 95% credible interval for the difference in means of (30, 110), the same as obtained by Wilcox (1996) using the Welch procedure. The Welch test statistic itself is estimated to have a mean value of 3.42 which can be compared with a tabulated t density for v degrees of freedom. Here v is estimated from (2.13) to have a mean of 49.7 and median 51.3. Assuming a different data model, namely a t density with a preset 5 degrees of freedom, makes relatively little difference to the interval for the difference in means, namely (26.4, 98.5).

With the amended data including an outlier in the no-heartbeat group, the normal data model gives a posterior interval for the difference in population means which is no longer entirely positive, namely (−26, 117). So with this data model the outlier could cause a conclusion that the groups were not different. By contrast, the robust t density alternative with 5 d.f. gives an interval (27, 100).

The choice of a t density with a small d.f. parameter for the outlier data set is validated by introducing a prior for the degrees of freedom. The d.f. parameters $\{v_1, v_2\}$ are given uniform priors,

$$v_j \sim \mathrm{U}(2, 200), \quad j = 1, 2.$$

Using the *Outlier Data*, a burn-in of 5000 iterations, and a follow-up of 5000, gives an estimate of $v_1 = 48$, $v_2 = 2.9$.

2.5 CATEGORICAL DATA MODELS: BINOMIAL, POISSON AND MULTINOMIAL

With categorical rather than continuous data, the major baseline distributions are the binomial, multinomial and Poisson. As for the Normal density, robust extensions of the binomial and Poisson densities can be adopted to account for departures from their assumptions. Categorical data occur when data is only available as or recorded in discrete categories such as male/female, improvement or non-improvement after medical intervention, ethnic group, socio-economic group, and so on. Often originally continuous data may be converted to a discrete categorisation, e.g. age recorded in single years of age is grouped into ten-year intervals, which are then treated as discrete categories.

The device of converting a continuous predictor to a set of categories is often used to explore nonlinearities in regression (Woodward, 1999). It may also be used as a form of transformation if the originally continuous data shows marked skewness. Conversely a frequent question occurring in the analysis of categorical data is the plausibility or otherwise of an underlying continuous scale, which will affect the formulation of models and hypotheses on parameters. A variable with ordered categories (e.g. marked improvement after treatment, some improvement, and no improvement) can often be considered to have a continuous scale underlying it. In Bayesian applications the specification of priors for the parameters of the main discrete models also has major significance, and the prior may be modified to reflect such features as an underlying continuous scale.

2.5.1 Binomial outcomes

With binomial data there is a single parameter of interest, the probability of a certain outcome π. The data mechanism distinguishes two possible outcomes, conventionally being described as 'success' and 'failure', and with probabilities π and $1 - \pi$ respectively. Which of the two outcomes is called the success is arbitrary. The two outcomes of a binomial trial are mutually exclusive and exhaustive, in the sense that they cover the full range of possible outcomes. (Often in practice binomial variables are caused by a two-fold division of the scale of a continuous variable or by collapsing over a categorical variable with $m > 2$ categories.) The individual trials or cases producing one or other of the outcomes are called Bernoulli variables, and the binomial applies to the aggregation of such outcomes over several trials or cases.

Thus the binomial describes the distribution of x successes out of n trials. The binomial density is proportional to the product of probability π over the x successes, and of $1 - \pi$ over the $n - x$ failures. Thus

$$p(x|\pi) \propto \pi^x(1-\pi)^{n-x} \tag{2.14}$$

or

$$x \sim \text{Bin}(n, \pi).$$

The parameter of interest is the probability π, with the x successes and $n - x$ failures being the data. One way to represent our prior beliefs about the size of π is via a discrete prior as in Chapter 1, assigning probabilities to a small number of possible alternative values. But π can have an infinity of values between 0 and 1, and so its prior may also be represented by a continuous density (which is not to say that a discrete prior on π would not be a valid approach). For reasons of conjugacy, a convenient prior density for the binomial probability is the beta density with parameters a and b (both positive), denoted $\mathrm{B}(a, b)$ or $\mathrm{Beta}(a, b)$, such that

$$p(\pi) \propto \pi^{a-1}(1-\pi)^{b-1}. \tag{2.15}$$

The posterior density of π is then also a beta with parameters $a+x$ and $b+n-x$, specifically:

$$p(\pi \mid x, n) \propto \pi^{a+x-1}(1-\pi)^{b+n-x-1}. \tag{2.16}$$

So the parameters of the beta prior density in a sense amount to a previous sample with a successes and b failures. From the form of the beta density, the option $a=b=1$ is equivalent to assuming π uniform between 0 and 1.

In a beta density with parameters α and β the mean is the ratio $\pi = \alpha/(\alpha+\beta)$ of total 'successes' to total 'events'. The variance is

$$\mathrm{var}(\pi) = \pi(1-\pi)/(\alpha+\beta+1). \tag{2.17}$$

This can be seen to be analogous to the classical result that the variance of the binomial probability is that probability times its complement divided by the sample size. In the above example, $\alpha+\beta+1 = a+b+n+1$ and so $\alpha+\beta+1$ is sometimes called the extended sample size (Adcock, 1987), since it combines sample and prior sample sizes. For α and β both larger than 10 the beta density is approximated by a normal curve, so that the probability of a z score comparing a stipulated probability π_0 with π can be found using Standard Normal Tables. For relatively small α and β the approximation is better for π closer to 0.5.

Example 2.9 Danish suicide rates These properties can be illustrated by a Danish study of the incidence of completed suicide after a suicide attempt (Nordentoft *et al.*, 1993); 103 out of 974 patients admitted to hospital after self-poisoning committed suicide in a 10-year follow-up period. There are a range of studies on this overlap in attempted and actual suicide behaviour. Suppose, however, we adopt a relatively non informative prior, possibly because we doubt how far the study data from other countries or years are exchangeable with the Danish experience. One possible vague or non informative prior using the beta density for π would contain small positive values of a and b, such as $a=1, b=1$.

The posterior estimate of the suicide rate among parasuicides is then $104/976 = 0.1066$ with standard deviation from (2.17) of

$$[(0.1066 \times 0.8934)/977]^{0.5} = 0.0099.$$

Using the normal approximation the chance of the rate exceeding 12% (i.e. one in eight parasuicides go on to actual suicide) is the probability of a z score of $(0.12 - 0.107)/0.0099 = 1.313$. Such a score has a chance of occurring of just under 0.1 using tables of the Standard Normal. Alternatively we can simulate this probability: see *Program 2.9 Suicide Rate* where the number of occasions the sampled π value from the posterior B(104, 872) exceeds 0.12 is counted. This uses the step() function in WINBUGS and after 100 000 iterations gives a probability of 0.0902.

Suppose that the follow-up omitted 100 patients (through withdrawals and other nonresponse) and that this missingness is random in the sense that the completed suicide rate is not related to this loss of follow-up. Then we can predict the number of completed suicides in this group. We find the mean predicted number of suicides is 10.65 and the 95 % interval ranges from 5 to 17.

Example 2.10 Dependency on state benefits Birkin *et al.* (1994) consider the estimation of dependency rates as a proportion of total households, in the sense of total dependency on state benefits (either income supplements or unemployment benefit). Their procedure uses small area UK Census totals of households of different types and multiplies these totals by national dependency proportions, assumed known. They consider a small area in Leeds. Among 8573 total households, there are $N_1 = 1408$ households in this area (Armley ward in Leeds) with one pensioner and the national dependency rate is $r_1 = 0.606$, so dependent households are estimated as 853; for 788 two-pensioner households, and dependency rate 0.42, the estimate is 331. Households with no wage earners are calculated as households with one economically active person (2673) times the national unemployment rate 0.091, namely 243, and households with two economically active persons (3444) times the squared unemployment rate (0.0083), namely 29. One may question whether chances of unemployment within a two-person household are uncorrelated in this way, but here we focus on estimating the extent of sampling variability in the overall estimate of dependency. The Birkin *et al.* estimate is $(853 + 331 + 272)/8573 = 0.170$.

We apply the normal approximation, and for one-pensioner households take the known mean as $\mu_1 = N_1 r_1$, and known variance as $V_1 = N_1 r_1 (1 - r_1)$. Then dependent one-pensioner households are sampled as $d_1 \sim \text{N}(\mu_1, V_1)$. Using a similar procedure for all household types gives an estimated percent ratio of 17.0 %, with 95 % interval from 16.3 % to 17.6 % (see *Program 2.10 State Benefits,* with the estimate based on the second half of a run of 50 000 iterations).

2.5.2 Using prior knowledge to estimate the beta prior

If we had approximate ideas about the mean value of π and its spread about this mean we could derive suitable values of a and b for incorporating in the beta prior $\text{B}(a, b)$. Thus in Chapter 1 we might expect the mean prevalence of childhood asthma to be $\pi = 0.15$, and that the most credible range for π was between 0.1 and 0.2. So 0.05 (the difference between 0.1 and 0.15, and between 0.2 and 0.15) is approximately equivalent to two standard deviations. So our estimate of s.d.(π) is 0.025 and of $V = \text{var}(\pi)$ is 0.000 625. In the beta density with parameters a and b the mean is the ratio $\mu = a/(a + b)$ and the variance is $V = \mu(1 - \mu)/(a + b + 1)$. Turning these formulae around, we find that

$$a = \mu[\mu(1 - \mu)/V - 1] \text{ and } b = a[1 - \mu]/\mu.$$

So our prior for the childhood asthma example might be $a = 30.5, b = 172.5$. This is a relatively informative prior, and since our successes in the Chapter 1 example were 2 from a sample of $n = 15$, we can see that in this case the prior overwhelms or dominates the data. However, if we had carried out a large prevalence survey, with say, $n = 3000$ and $x = 480$, then the data would still dominate the prior. The posterior estimate of π would be $(30.5 + 480)/(172.5 + 3000) = 0.161$, and close to that based on the standard maximum likelihood, namely $480/3000 = 0.16$.

Another type of prior we can use for binomial proportions is to use a logit scale to convert the probability, restricted to $(0, 1)$ to the full real line, from $-\infty$ to $+\infty$. Thus $c = \text{logit}(\pi) = \log[\pi/(1 - \pi)]$ takes π to the real line, and we could set a prior on c, reflecting the same prior knowledge. Thus $\text{logit}(0.15) = -1.7$ and two standard deviations (to the logits of 0.1 and 0.2) are approximately 0.4. So our prior for c might be $N(-1.7, 0.04)$.

Example 2.11 Presidential actions Wilcox (1996) presents data from a 1991 Gallup opinion poll about the morality of President Bush's not helping Iraqi rebel groups after the formal end of the Gulf War. Of the 751 adults responding, 150 thought the President's actions were not moral. Suppose we assume a non informative $B(0.01, 0.01)$ prior on the probability π that a randomly sampled adult would respond 'immoral' (a reasonable choice since this is a 'one-off' event though there might be evidence from previous polls on the proportion of the population generally likely to consider a President's actions immoral). Running *Program 2.11 Presidential Actions* for 10 000 iterations we find a 95% posterior credible interval on π of $(0.172, 0.229)$. This is close to what would be obtained using the Normal approximation via classical methods. Using the alternative non informative $U(0, 1)$ prior (equivalent to a $B(1, 1)$) leads to a virtually identical interval of $(0.173, 0.230)$.

However, suppose our sample size was only $n = 8$, with $x = 0$ adults considering the Presidential action immoral. Here the 95% credible interval using a $B(0.01, 0.01)$ prior on π is $(0, 0.007)$ whereas using the uniform prior $U(0, 1)$ gives one of $(0.003, 0.33)$. Work by Blyth (1986) into the case where $x = 0$ suggests that the $(1 - \alpha)$% classical confidence interval should have upper limit $1 - \alpha^{1/n}$ rather than near 0 as would result from using the usual approximation. For $n = 8$, and $\alpha = 0.05$ this gives the upper limit of the 95% classical confidence interval as 0.31. The posterior mean for π is 0.10 with a 95% interval $(0.003, 0.336)$. So choice of prior is, as elsewhere, more important for small sample sizes and for skewed response patterns. Under the uniform prior, the credible interval for the prediction of π in a *new* survey of size 8 ranges from 0 up to 0.5.

Example 2.12 Assessing a standard life table Benjamin and Pollard (1980) consider the chi-square test of the applicability of a standard (e.g. national) life table to a smaller sample of observed deaths. Let D_x denotes the observed deaths in a year among the observed male population aged x, and E_x the exposure at age x. If P_x is the mid-year population aged x then the exposure (the start-year population) is often approximated as $P_x + 0.5D_x$. In the standard population, the death rates are known to be $q_{x,\,\mathrm{stand}}$ we wish to test whether the rates q_x in our observed population can be considered equivalent to $q_{x,\,\mathrm{stand}}$. Benjamin and Pollard suggest a chi-square test based on the normal approximation to the binomial, so that

$$\chi^2 = \sum_x (D_x - E_x q_{x.\mathrm{stand}})^2/(E_x q_{x.\mathrm{stand}} p_{x.\mathrm{stand}}) \tag{2.18}$$

where $p_{x.\mathrm{stand}} = 1 - q_{x.\mathrm{stand}}$ is the survival rate in the standard population. This gives 34.5 on 36 degrees of freedom and so the standard life table (English Life Table 10, males) is judged to be appropriate to the observed population. They also consider the chance of outlying standardised deviations at each age x, namely

$$(D_x - E_x q_{x.\mathrm{stand}})/(E_x q_{x.\mathrm{stand}} p_{x.\mathrm{stand}})^{0.5}. \tag{2.19}$$

For the 36 ages there are 7 such deviates under -1 and 5 over $+1$, and these totals are close to what would be expected given a normal approximation, namely 36×0.16 in each tail with the normal deviate above 1 in absolute terms. (The probability that a standard normal deviate is below -1 or above $+1$ is 0.1587.)

To model sampling variability in the observed D_x suppose we take

$$D_x \sim \mathrm{Bin}(E_x, q_x) \tag{2.20}$$

and set $\mathrm{B}(0.001, 0.001)$ priors on each of the q_x. Suppose we also try to estimate the unknown deaths $D_{x.\mathrm{stand}}$ in the sample population which would result from a density $\mathrm{Bin}(E_x, q_{x.\mathrm{stand}})$; that is we simulate what deaths would be in our particular population if the standard death rates did apply to it. Under each scenario we calculate the chi-square statistic (2.18) above; see *Program 2.12 Standard Life Table,* taken to 10 000 iterations, with 2000 iteration burn-in.

Under the standard death model, where we simulate deaths $D_{x.\mathrm{stand}}$ as if $q_{x.\mathrm{stand}}$ applied to the sample data, the chi-square statistic has an average of around 36 with 95% credible interval (21.4, 53.9) (see Table 2.4). By contrast the chi-square statistic comparing the deaths $E_x q_x$ from (2.20) with the standard has an average of 68 and a 98% chance of exceeding 36. The number of large deviates in equation (2.19), namely those absolutely exceeding one, averages 18 or half the 36 ages. Hence there is less support using a sampling-based approach for the applicability of the standard life table to the population at hand.

Example 2.13 Cancer and X-rays An epidemiological application using the binomial involves incidence of breast cancer cases according to exposure to multiple X-ray fluoroscopies, as analysed by Boice and Monson (1977) and Rothman (1986). See Table 2.5.

Table 2.4 Tests of life table standard.

Parameter	Posterior mean	Posterior SD	2.5 percentile	97.5 percentile	Median (50th percentile)
Number of large deviates	17.9	2.9	12	23	18
Chi-square comparing D_x with $q_{x.\text{stand}}$	68	14.6	42.8	99.8	67
Chi-square comparing simulated $D_{x.\text{stand}}$ with $q_{x.\text{stand}}$	35.8	8.4	21.4	53.9	35

Table 2.5 Breast cancer incidence among women with TB, according to person-years at risk and whether exposed to multiple X-ray fluoroscopies.

	Exposed	Non-exposed	Total
Breast cancer	41	15	56
Person-years	28010	19017	47027

The null hypothesis is that exposure is unrelated to the cancer outcome, and under this hypothesis, the probability that a cancer case will have been exposed is the ratio of exposed person-years to the total years exposed. The maximum likelihood estimate of this probability is then $p_E = N_E/N = 28010/47027 = 0.596$ with $q_E = 1 - p_E$. Under a standard testing approach, the actual total of $A = 41$ exposed cases may be compared with the expected total based on applying p_E to the total $N_C = 56$ cancer cases. Using the normal approximation, this gives a standardised deviate

$$\chi = (A - N_C p_E)/(N_C p_E q_E)^{0.5} = (A/N_C - p_E)/(p_E q_E/N_C)^{0.5} \tag{2.21}$$

for comparison with a standard normal. Alternatively squaring the deviate leads to a chi-square with an expectation of one under the hypothesis. The one-tail P value is then only 0.0174 in favour of the hypothesis of incidence unrelated to exposure.

An alternative approach may attempt to reflect the sampling variability associated with the number of cancer cases. Suppose we assume that the same binomial parameter π underlies the generation of time exposed N_E from the total exposure time N, and of exposed cases A from the number of cancer cases N_C.

$$N_E \sim \text{Bin}(\pi, N)$$

$$A \sim \text{Bin}(\pi, N_C).$$

Then as we proceed through the MCMC samples to describe the posterior density of π we generate 'replicate data' compatible with this parameter (i.e. compatible with the null hypothesis) and the total cancer cases available:

$$A_{\text{rep}} \sim \text{Bin}(\pi, 56).$$

We then assess the probability of the null hypothesis by comparing both A and A_{rep} with the estimated mean number of exposed cases under the model, namely 56π. We may compare the distributions of two sets of deviates (2.21), one where A_{rep} is compared with the expected πN_{C}, and another where A is compared with πN_{C}.

A second possible test involves estimating the unknown binomial probability p in the density $A \sim \text{Bin}(p, 56)$. We may then assess the probability of the null hypothesis by comparing (aggregating over samples) the estimate of p with the maximum likelihood estimate $p_{\text{E}} = 0.596$. This involves a run of 10 000 iterations with *Program 2.13 Cancer and X-ray Exposure* using a $\text{B}(1, 1)$ prior on p. Comparison of the estimated p and p_{E} leads to a probability of 0.0154 in favour of the hypothesis; this is approximately the same as the standard test. This test is based on accumulating samples where p is less than p_{E}, and is analogous to a one-tail test.

However, comparing the deviates (2.21) under the hypothesis that exposure is unrelated to the cancer gives a probability of 0.0225 that χ_{rep} exceeds χ_{obs} so showing slightly less evidence against the hypothesis. This is an example of a posterior predictive check, as discussed by Gelman *et al.* (1995, Chapter 6) in that the test averages over the posterior distribution of the expected number of exposed cancer cases πN_{C} among the number of cancer cases N_{C}.

2.6 POISSON DISTRIBUTION FOR EVENT COUNTS

The typical application of the binomial distribution is to take a sample of a given size and count the number of sample members characterised by a certain attribute or not. There are circumstances, however, when the number of times an event occurs can be counted without there being any notion of counting when the event did not occur. Examples are the number of goals in a football match, the number of vehicles passing a checkpoint, the number of lightning flashes in a thunderstorm, and so on. There are also many instances when there is a converse event (e.g. not being a new case of a disease) but if the event is rare then there may be a choice between a binomial or Poisson model: the greater the rarity of the event, the more appropriate the Poisson becomes.

Often the number of events can be seen against an exposure of a certain extent (e.g. a population, a geographic area or time span). For example, other things equal, we would expect the number of new cases of a rare infection to be greater for a larger population, and expect more lightning flashes in a long thunderstorm than a short one. We would have observed x events for a mean λ which is the product of an underlying rate μ and an exposure E, such that $\lambda = \mu E$. Usually E is assumed known (i.e. a fixed constant).

As an example, we can consider the number of deaths from a rare disease in a large population; suppose there are $x = 10$ deaths annually from a disease in a population of $n = 50,000$. We could possibly use the binomial with $x = 10$ and $n - x = 49990$ and estimate the underlying rate π; however, we may alternatively assume Poisson sampling with the number of deaths

$$x \sim \text{Poi}(\lambda)$$

where $\lambda = n\mu$, with μ to be estimated.

The successive terms of the Poisson density, giving the probabilities of $0, 1, 2, 3, \ldots$ events are given by the terms in the expansion

$$\mathrm{e}^{-\lambda}(1 + \lambda + \lambda^2/2! + \lambda^3/3! + \ldots).$$

Hence the likelihood of x events can be seen to be proportional to

$$\mathrm{e}^{-\lambda}\lambda^x. \tag{2.22}$$

If λ is the result of a known exposure level, with $\lambda = \mu E$, then this is in turn proportional to

$$\mathrm{e}^{-\mu E}\mu^x \tag{2.23}$$

since E^x is a constant.

If in (2.22) we observe events $x_1, x_2, x_3, \ldots, x_k$ for a sample of size k then the likelihood over the sample is proportional to

$$\mathrm{e}^{-k\lambda}\lambda^T \tag{2.24}$$

where $T = \sum_i x_i = k\bar{x}$, and $\bar{x}$ is the average of the sampled counts. If in (2.23) we observe events $x_1, x_2, x_3, \ldots, x_k$ corresponding to fixed exposures $E_1, E_2, \ldots, E_k$ and we denote $\varepsilon = \sum_i E_i$ as the total exposure in the sample, then the likelihood is proportional to

$$\mathrm{e}^{-\varepsilon\mu}\mu^T. \tag{2.25}$$

In either case the likelihood kernel can be seen to be of a gamma form and so a gamma prior for the unknown rate, λ in (2.22) or μ in (2.25) leads to a conjugate analysis. We are not restricted to conjugate priors, however, and might for example take $\log(\mu) = \beta$ where β is assigned a normal prior.

If we adopt a gamma $\mathrm{G}(a, b)$ for λ in (2.24), such that $p(\lambda) \propto \lambda^{a-1}\mathrm{e}^{-b\lambda}$ then the posterior density for λ will be of a gamma form $\mathrm{G}(a + T, b + k)$. If we assume a $\mathrm{G}(a, b)$ prior for μ in (2.25), then the posterior density for μ will be of the form $\mathrm{G}(a + T, b + \varepsilon)$.

Example 2.14 Elevated mortality in treatment group Suppose we know the population-wide death rate for a certain disease over a year is 10%. In a trial sample of 30 patients typical of the disease group and exposed to a new treatment we observe $x = 7$ deaths in a year. We may ask what evidence there is that the death rate is elevated in the treatment group (Moroney, 1965). To answer this question, we wish to estimate the death rate μ in a population of size $P = 30$, and compare the estimate with the known and fixed population rate m of 0.1 (which would produce an expected 3 deaths in a year). Suppose we adopt a G(0.1, 1) prior for μ roughly in line with the population death rate but downweighting it, and assume $x \sim \text{Poi}(P\mu)$.

Then using *Program 2.14* (with 10 000 iterations after 500 burn-in) we estimate the death rate in the treatment group to be 0.227 with a 95% credible interval (0.095, 0.42) and a chance of 0.969 that the rate in this group exceeds the fixed criterion rate of 0.10. Another possible test might be to investigate the implications of a population rate $\mu_{\text{pop}} = 0.1$ in the population of 30 patients. So we generate $x_{\text{pop}} \sim \text{Poi}(P\mu_{\text{pop}})$ and then apply a test comparing Poisson means as in (Thode, 1997),

$$z = (x - x_{\text{pop}})/(x + x_{\text{pop}})^{0.5}.$$

The average value of z is 1.32 with a 97.5% point of 2.65. This test statistic is approximately a standard normal for Poisson means as low as 2.5, and points less convincingly to there being an elevated death rate in the treatment group.

Another option would be to model x as due to a relative risk ρ multiplying the expected number of deaths $E = mP = 3$ where $m = 0.1$. Then $x \sim \text{Poi}(3\rho)$ and we would wish to assess whether ρ exceeded 1.

Example 2.15 Area mortality comparisons Another common application of the Poisson is comparing mortality between areas, hospitals, etc. after standardising for age and perhaps other factors affecting risk. Thus suppose x_{ij} denotes a vector of observed deaths in area i over J ages, and P_{ij} denote populations for age groups j in the index population in area i. If death rates in a standard (comparison) population are m_j, the expected deaths E_i in the index population are just $\sum_j m_j P_{ij}$. If actual deaths are equal to expected deaths (or nearly so) then the mortality experience in the 'index' area or hospital appears comparable to that in the standard population.

A frequently used model assumes total area deaths x_i are Poisson with mean $E_i\rho_i$ where $\rho_i = 1$ if the standard and index death rates are the same. If ρ_i is multiplied by 100 we obtain a standard mortality ratio, equalling 100 or nearly so when mortality is similar in standard and index populations, but well in excess of 100 if the local population has adverse mortality levels.

Silcocks (1994) presents data on male myeloid leukaemia deaths (1989) in Derby, denoted x_1, and in the remainder of the Trent region of England, namely x_2, of which Derby is a part. Here m_j are based on deaths in the entire Trent region (though another option is the Trent region excluding the index Derby area), and $E = 22.38$. In *Program 2.15 Trent Leukaemia Mortality* (taken to 10 000 iterations) we assume the Derby death total of 30 is Poisson with mean $E\rho$. This gives a mean estimate of the SMR ρ of 1.346 (the median is slightly lower at 1.33), with 95% credible interval of (0.907, 1.873).

While the death rates in the standard population are usually assumed fixed, there are circumstances where they may be more appropriately considered subject to sampling variation. If this alternative is adopted here, age-specific deaths x_{ij} are considered outcomes from a Poisson distribution with means $\lambda_{ij} = \theta_{ij} P_{ij}$. Here $i = 1$ denotes the index district and $i = 2$ the remainder of the standard region in which it lies, and θ_{ij} are the underlying death rates by age and area. The expected deaths in the index area are then calculated by

$$E_1 = \sum_j p_j^* (\lambda_{1j} + \lambda_{2j})$$

where p_j^* is the share of the total standard population in age group j located in the index area,

$$p_j^* = P_{1j}/(P_{1j} + P_{2j}).$$

If the index population is a relatively large share of the standard population, then there will be covariance between $x_1 = \sum_j x_{1j}$ and E_1 and the credible interval of the SMR will be narrower than if the expected deaths are treated as fixed. In Program 2.15 this approach is denoted 'standardisation method 2'. Taking 10 000 iterations, the credible interval for the Derby leukaemia SMR is then narrower, namely from 0.96 to 1.77 with a mean of 1.346. The expected deaths average 22.3 with a 95% interval ranging from 18.4 to 26.7.

2.7 POLYTOMOUS OUTCOMES VIA THE MULTINOMIAL DENSITY

The binomial distribution with two possible categories of outcome can be extended to a multinomial density, with more than two discrete levels of the outcome. These may be naturally nominal categories (such as political party choice, disease diagnosis, or religious affiliation), but may also result from categorisation of originally continuous outcomes. Combining continuous observations into categories may be useful in lessening the impact of outliers (Berry, 1996) or in the handling of large numbers of observations. For example national population data on age structure are commonly presented for grouped ages, either just grouping into single years of age, or for five-year or ten-year age groups. Similarly national data on incomes are frequently grouped. Converting a continuous explanatory variable to a categorical variable may also be a relatively simple way of examining nonlinear relationships to an outcome (e.g. via log-linear analysis or general linear modelling where the explanatory variable becomes a categorical 'factor').

Let $x_1, x_2, \ldots, x_k$ denote counts from $k > 2$ categories of the outcome. Then the multinomial likelihood specifies

$$p(x_1, x_2, \ldots, x_k | \theta_1, \theta_2, \ldots, \theta_k) \propto \prod_j \theta_j^{x_j}$$

where the θ_j, the probabilities of belonging to one and one only of the k classes, sum to 1. Just as the binomial is conditioned on the sample size n, with x as the number of 'successes', then the multinomial is conditioned on the sum of the x_j, denoted X. The multinomial can equivalently be represented as the product of k independent Poisson variables with $x_1 \sim \text{Poi}(\mu_1), x_2 \sim \text{Poi}(\mu_2), \ldots, x_k \sim \text{Poi}(\mu_k)$, also subject to the condition[5] that $\sum_j x_j = X$. Then the multinomial probabilities are obtained as

$$\theta_j = \mu_j / \sum \mu_j.$$

The conjugate prior density for the multinomial is the multivariate extension of the beta density, namely the Dirichlet density. This is a density for $\theta_1, \ldots, \theta_k$ specified in terms of positive parameters $\alpha_1, \ldots, \alpha_k$, namely

$$p(\theta_1, \theta_2, \ldots, \theta_k \mid \alpha_1, \alpha_2, \ldots, \alpha_k) \propto \prod_j \theta_j^{\alpha_j - 1}.$$

Suppose we assign initial values $c_1, c_2, \ldots, c_k$ to the $\alpha_1, \ldots, \alpha_k$ then the posterior density of the $\theta_1, \ldots, \theta_k$ is just a Dirichlet again with parameters $c_1 + x_1, c_2 + x_2, \ldots, c_k + x_k$. So the total of the assigned values $\sum_j c_j = C$ is equivalent to a 'prior sample size'.

From the properties of the Dirichlet, the posterior means of the multinomial probabilities, for $j = 1, \ldots, k$ are then

$$(x_i + c_j)/(X + C).$$

Equivalently these are a weighted mean of prior and data estimated proportions, namely

$$\{X/(X + C)\} x_j / X + \{C/(X + C)\} \eta_j$$

where $\eta_j = c_j / C$.

Often the c_j are assumed equal to each other, i.e. $c_j = C/k$ for all j. The choice is then how to select an appropriate total C. Bishop *et al.* (1975, Chapter 12) discusses estimating C in this case, but using the observed data. This amounts to an empirical Bayes approach, since the prior is estimated from the data. Adcock (1987) presents an alternative method, based on the assumption that before the data are observed, there are two separate and independent vector 'estimates' e_1 and e_2 of the unknown $\theta_1, \theta_2, \ldots, \theta_k$.

Suppose $k = 3$, and we are considering the outcome of a US presidential election, with a Democrat, Republican and one Other Party candidate. We might assume, say on the basis of pre-election polls, that the Democrat share of the vote is either 0.40 or 0.43 respectively, and the Republican 0.47 or 0.45 respectively; so that the other candidate will receive 0.13 and 0.12 in each case. Then the averages of e_{1i} and e_{2i} are respectively 0.415, 0.46 and 0.125. If these averages are taken as central prior estimates η_i of each multinomial probability, then the sum of the squares of the differences $0.43 - 0.4 = 0.03, 0.47 - 0.45 = 0.02$ and $0.13 - 0.12 = 0.01$, namely $0.0014 = 0.0009 + 0.0004 + 0.0001$ has expectation

$$2\left(1 - \sum \eta_i^2\right)/(C + 1).$$

So we estimate C in this case as $\{2(1-0.39945)/0.0014\}-1 \approx 857$, and our prior on the multinomial parameters would be $c=\eta C=(356, 394, 107)$.

Example 2.16 Cancers in women There is concern about the increased incidence of breast cancer in women, though in the UK the actual numbers of deaths from this disease are more or less static. Taking both trends together implies a lowering of 'case fatality', the death rate among the new cases of the disease. In 1995, breast cancer in the UK accounted for 18 % of female cancer deaths. Suppose we were interested in the credible interval of this rate, in a situation where we have $k=11$ cancer types (breast, lung, colon, ovary, pancreas, stomach, oesophagus, rectum, lymphoma other than Hodgkins, bladder and other). We assume a noninformative Dirichlet, with $c_i=1$, and $C=11$, so the 1995 female cancers data, for which $X=76680$, will tend to 'swamp' the prior. Other choices of prior might have been to take the female cancer pattern for the UK in 1994, possibly downweighted in some way (e.g. the prior sample size equal to a tenth of the 1994 total); or to base the prior on the female cancer pattern of other developed countries.

Apart from the true multinomial procedure, the other option is to estimate the multinomial parameters by taking each of the 11 outcomes as independent Poissons. Under this option we may assume noninformative gamma priors G(0.001, 0.001) on each Poisson mean, and then calculate (for each sample from the posterior forms of these 11 distributions), the estimates $\theta_j=\mu_j/\sum \mu_j$. *Program 2.16 Female Cancers* is taken to 10 000 iterations, with different *inits* files for the true multinomial and Poisson-multinomial methods. Under the multinomial option the 95 % credible interval for θ_1 (proportion of female cancer deaths due to breast cancer) is 0.1808 to 0.1863, with both mean and median at 0.1835. The Poisson analysis differs only in the central location estimate with 0.1836 for mean and median.

Suppose for a particular health agency we were interested to explore the resource and health needs associated with a 10 % rise in overall female cancer incidence. The data (for a health agency in east London) is for 1991, when there were $X=847$ new female cancers registered and we are interested in the distribution over cancer types of a larger forecast total of 932. Of the 847 new cases in 1991, 247 were breast cancer (i.e. about 29 % of the total and higher than the national proportion in 1991 of 25 %). The predictive density adopts the Poisson multinomial method, and with England and Wales incidence used as the basis an informative prior. In each health district the numbers of cancer cases from year to year can be quite small and subject to sampling fluctuation; a 'national prior' is used to smooth the local data. The 12 cancer types are slightly modified as compared with the preceding analysis (see the listing in Program 2.16).

There were 104 718 new cases of female breast cancer in England and Wales in 1991, and their proportionate distribution over the 12 types is used to set priors for the local analysis. If $R_1, \ldots, R_{12}$ denotes the national percents (per 100), then the prior for each local Poisson mean is $G(S_i, \rho)$ where

$S_i = \rho R_i X/100$. Increasing ρ means a greater influence of the national data (a more informative prior) in smoothing the local data. If $\rho = 1$ then the 95 % credible interval for breast cancer cases in a larger total of 932 cases is from 221 to 288, or from 24 % to 31 % of the total. If $\rho = 0.1$ then the interval is from 230 to 308 (i.e. from 25 % to 33 % of new cases).

Example 2.17 The Gini coefficient of inequality: UK income before tax The multinomial is useful for inequality indicators based on grouped data from an underlying continuous variable or ranking, such as income or health (Wagstaff and Vandoorslaer, 1994). Bartholomew (1996) presents aggregated data on UK incomes before tax in 1991/92, with $k = 16$ groups (see Table 2.6).

We assume multinomial sampling for the observed frequencies x_i from each income stratum ($k = 16$ groups) with the total being $X = 25891$, in thousands (see *Program 2.17 Gini Inequality*). The multinomial probabilities p_i are assigned a noninformative Dirichlet prior, $c_i = 1$ for $i = 1, \ldots, 16$; though more informative options are possible. To calculate the Gini coefficient we estimate the numbers in each stratum

$$\hat{f}_i = X\hat{\theta}_i$$

using the sampled θ_i at each iteration, and form cumulative totals F_i of these estimated frequencies, i.e. $\hat{F}_2 = \hat{f}_1 + \hat{f}_2, \hat{F}_3 = \hat{f}_1 + \hat{f}_2 + \hat{f}_3$, etc. We then multiply the terms

$$\hat{F}_i(\hat{F}_k - \hat{F}_i)/\hat{F}_k^2$$

Table 2.6 Distribution of personal incomes, 1991/92, UK.

Group	Mid income	Cumulative frequency	Frequency	Relative frequency
1	3398	258	258	0.010
2	3750	879	621	0.024
3	4250	1692	813	0.031
4	4750	2530	838	0.032
5	5250	3475	945	0.036
6	5750	4251	776	0.030
7	6500	5901	1650	0.064
8	7500	7611	1710	0.066
9	9000	10941	3330	0.129
10	11000	13931	2990	0.115
11	13500	17501	3570	0.138
12	17500	21421	3920	0.151
13	25000	24341	2920	0.113
14	40000	25461	1120	0.043
15	75000	25787	326	0.013
16	150000	25891	104	0.004
Total			25891	1.000

by the average income increment between adjacent categories, $y_{i+1} - y_i$. (y_i is the midpoint income for stratum i.) The sum of these terms over i gives the numerator of the Gini coefficient. The denominator is the estimate of average income, using estimates $\hat{Y}_i$ of total income accrued in each stratum, obtained by multiplying average stratum income y_i by the $\hat{f}_i$.

Our estimate of the Gini coefficient in 1991/92 (from the second half of 10 000 iterations) is 0.375 with a 95% credible interval from 0.370 to 0.380. By comparison, Bartholomew quotes a UK Gini index in 1959/60 of 0.352, so there is an apparent growth in inequality. Bartholomew (1996) gives a central estimate for 1991/92 of 0.376. Average income is estimated at £14 720 with a credible interval from £14 550 to £14 890. Both parameters have symmetric posterior densities (closely approximating normality).

The above calculations do not incorporate uncertainty regarding the average stratum incomes. This will be slight for $i = 2, \ldots, 15$ but may be important for the lowest and highest bands. Bartholomew allows for this by taking alternative values of £200 000 and £300 000 as the average in the highest band, leading to revised Gini coefficients in his work of 0.383 and 0.397. Here we adopt a uniform prior over the interval £150 000 to £300 000, giving a Gini coefficient averaging 0.388, and a wider credible interval, from 0.374 to 0.403. Average income is then estimated at £15 030.

Another question of interest is how far income distributions for income y approximate the log-normal density, with parameterisation $E(\log y) = \mu$ and $V(\log y) = \sigma^2$. Expressed in the original income scale, $E(y) = \exp(\mu + \sigma^2/2)$ and $V(y) = \exp(2\mu + 2\sigma^2)$. Hence the coefficient of variation $[V(y)]^{0.5}/E(x)$ in the original scale is $\exp(\sigma^2 - \sigma^2/2)$. The Gini coefficient in the original scale is $2\Phi(\sigma/\sqrt{2}) - 1$ where Φ is the standard normal distribution function. With the above uniform prior on the upper stratum average we estimate the Gini coefficient consistent with a log-normal to be less than that based on the actual data, namely 0.356 with $E(y)$ estimated as 14 400.

2.8 MULTIVARIATE CONTINUOUS DATA: MVN AND MULTIVARIATE *t* DENSITIES

The multinomial density illustrates how many of the properties of univariate distributions extend to their multivariate equivalents. The most commonly used multivariate distribution for *continuous* outcomes is the multivariate normal describing the association between a vector x of q continuous variates. For a single observation of such a vector the likelihood is

$$f(x) \propto |\Sigma|^{-0.5} \exp[-0.5(x - \mu)P(x - \mu)]$$

where μ is the vector of means, Σ is the covariance matrix of order $q \times q$, symmetric and positive definite, and the precision matrix is $P = \Sigma^{-1}$. This distribution is often denoted $\mathrm{N}_q(\mu, \Sigma)$. The positive definiteness condition is the equivalent of the variance in the univariate normal distribution being necessarily positive. If the variates $x_1, \ldots, x_q$ are standardised then Σ reduces to the correlation matrix between the variates; and so if Σ is just diagonal then the variates $x_1, x_2, \ldots, x_q$ are uncorrelated.

An important special case is the bivariate normal, when two variables show a jointly normal interdependence. In this case the covariance matrix can be expressed as

$$\begin{pmatrix} \sigma_1^2 & \rho\sigma_1\sigma_2 \\ \rho\sigma_1\sigma_2 & \sigma_2^2 \end{pmatrix}$$

with ρ as the correlation between the two variables.

2.8.1 Multivariate normal with unknown mean and variance

In the multivariate case, the conjugate prior for the covariance matrix Σ is the multivariate generalisation of the inverse gamma. Similarly, the multivariate analogue of the gamma is known as the Wishart density and is conjugate for a precision matrix $P = \Sigma^{-1}$ which is symmetric and positive definite. The Wishart is specified in terms of two parameters. One is a degrees of freedom parameter v which must be equal to or exceed q if the prior is to be proper. The other parameter is a scale matrix B of order $q \times q$, symmetric and positive definite. The Wishart density has the form

$$f(P) \propto |B|^{v/2}|P|^{(v-q-1)/2}\exp^{-0.5\text{tr}(B'P)}$$

where tr( ) denotes the trace of the matrix product (i.e. the sum of its diagonal elements). The exponent to which the determinant of P is taken makes it clear why v must be at least equal to the order of B. Then

$$E(P) = (B/v)^{-1}.$$

v is typically equal (actually or nearly) to q, and B/v amounts to a prior estimate of the dispersion matrix based on v observations, and B to an estimate of the sum of squares matrix.

For data assumed to be of multivariate normal form, a conjugate prior distribution for (μ, Σ) jointly can be parameterised in terms of

(a0) a Wishart density prior for Σ^{-1} with scale matrix V_0 and with v_0 degrees of freedom, where larger values of v_0 represent stronger beliefs;

(b0) for Σ sampled from (a0), a mean generated from $\mu \sim N_q(m_0, \Sigma/\kappa_0)$ where κ_0 (analogous to the number of prior measurements) is a measure of the prior strength of knowledge about the mean.

For vague prior knowledge, v_0 and κ_0 might be small integers.

Suppose $\bar{y}$ and S are respectively an observed vector of means and a sum of squares and cross-products matrix. Let $w_0 = \kappa_0/(\kappa_0 + n)$ denote the ratio of prior 'sample size' to total sample size, and $v = v_0 + n$ denote the total degrees of freedom for the dispersion matrix. Then the posterior density has vector mean $m = w_0\mu_0 + (1 - w_0)\bar{y}$ and sum of squares matrix

$$W = B + S + nw_0(\bar{y} - \mu_0)'(\bar{y} - \mu_0).$$

To draw samples from the joint posterior density of (μ, Σ) given observed data $y_1, \ldots, y_n$ (or $\bar{y}$ and S as sufficient statistics) we first draw P from a Wishart with parameters W and $v = n_0 + n$, then draw μ from a multivariate normal with mean m and precision vP. To draw a new data value y_{new} (i.e. a prediction) we draw from a multivariate normal using the sampled values of μ and P.

In BUGS a multivariate normal with mean `mu.Y` and precision matrix `Prec.Y` for n observations on q variables Y_{ij} (say $q = 2$) could be specified as

```
for (i in 1 : n){Y[i, 1 : 2] ~ dmnorm(mu.Y[], Prec.Y[, ]);}
```

with appropriate priors on both `mu.Y` and `Prec.Y`, which are respectively of order q and qxq. For example a vague (just proper) prior mean and precision for `mu.Y[1:2]` could be:

```
          mu.Y [1:2] ~ dmnorm (mean.mu[], prec.mu[,])
for (k in 1:q) {mean.mu [k] <- 0.0; prec.mu [k, k]<- 0.001;
for (l in (k + 1) :q)
          {prec.mu [l, k] <- 0.0; prec.mu [k,l]<- prec.mu [l, k];}}
```

The Wishart prior for `Prec.Y[,]` could be specified:

```
Prec.Y [1:q, 1:q ] ~ dwish (B[,], nu);
nu <-q;
```

with the elements of B set to values which allow for a wide choice of dispersion matrices (cf. Equation (2.10)). For example:

```
for (k in 1:nu) B {[k, k]<- 1;
for (l in (k + 1) :nu) {B [k, l] <- 0; B [l, k] <- B [k, l];}}
```

We could try alternative values in the elements of `B[,]` to assess sensitivity of posterior inferences to the prior. Note that these may alternatively be included in the data file. The estimate of the covariance matrix would be obtained from the code

```
for (i in 1:q) {for (j in 1:q) {sigma [,] <- inverse (Prec.Y [,], i, j);}}
```

Partitioning multivariate priors

Just as knowledge of the mean and variance completely specifies a univariate normal distribution, so knowledge of the means and variances of each of q variables, and of the covariances between them, is sufficient to specify a multivariate normal density,

$$y \sim \mathrm{N}_q(\mu, \Sigma).$$

Further the marginal distribution of any subset of the y_j, $j = 1, \ldots, q$, has an MVN distribution with covariance defined by the appropriate submatrix of Σ. In particular suppose y is partitioned into two sets of variables $y_{(1)} = \{y_1, \ldots, y_r\}$ and $y_{(2)} = \{y_{r+1}, \ldots, y_q\}$. Then we may partition μ and Σ as follows

$$\begin{pmatrix} y_{(1)} \\ y_{(2)} \end{pmatrix} \sim N_q\left(\begin{pmatrix} \mu_{(1)} \\ \mu_{(2)} \end{pmatrix}, \begin{pmatrix} \Sigma_{11} & \Sigma_{12} \\ \Sigma_{21} & \Sigma_{22} \end{pmatrix}\right)$$

where Σ_{11} is $r \times r$, Σ_{12} is $r \times (q-r)$, Σ_{21} is $(q-r) \times r$ and Σ_{22} is $(q-r) \times (q-r)$. Σ_{12} is the matrix of covariances between the variables in the two subsets of y. Then the conditional distribution of $y_{(1)}$, when $y_{(2)}$ has a known value A, is MVN with mean

$$\mu_1 + \Sigma_{12}\Sigma_{22}^{-1}(A - \mu_2)$$

and $r \times r$ covariance

$$\Sigma_{11} - \Sigma_{12}\Sigma_{22}^{-1}\Sigma_{21}.$$

However if nothing were known about $y_{(2)}$ then $y_{(1)}$ has a marginal MVN distribution

$$y_{(1)} \sim N_r(\mu_1, \Sigma_{11}).$$

In terms of Bayesian prior specifications, this property means that a new family of prior distributions can be derived by considering the transformation of Σ to the parameters of the conditional distribution $y_{(1)} \mid y_{(2)}$, namely

$$B_1 = \Sigma_{12}\Sigma_{22}^{-1}$$

and

$$B_2 = \Sigma_{11} - \Sigma_{12}\Sigma_{22}^{-1}\Sigma_{21}$$

together with the parameter Σ_{22} of the marginal normal of $y_{(2)}$. Specifically (see Brown *et al*., 1994) Σ can be written

$$\Sigma = \begin{bmatrix} B_2 + B_1\Sigma_{22}B_1 & B_1\Sigma_{22} \\ \Sigma_{22}B_1 & \Sigma_{22} \end{bmatrix}.$$

The prior on the elements of (μ, Σ) may then be expressed in a series of conditional multivariate models, or even as a sequence of conditional univariate models. Thus for $q = 3$ and y_{ji}, $j = 1, q$ not necessarily centred, we may write

$$\begin{aligned} y_{1i} &\sim N(\mu_{1i}, V_1) \\ y_{2i} &\sim N(\mu_{2i}, V_2) \\ y_{3i} &\sim N(\mu_{3i}, V_3) \end{aligned} \tag{2.26}$$

where $\mu_{1i} = \alpha_1$, $\mu_{2i} = \alpha_2 + \beta_2(y_{1i} - \mu_{1i})$ and $\mu_{3i} = \alpha_3 + \beta_{31}(y_{1i} - \mu_{1i}) + \beta_{32}(y_{2i} - \mu_{2i})$.

The multivariate *t* density

A robust alternative to the multivariate normal density for multivariate data $x = (x_1, \ldots, x_q)$ is provided by the multivariate t density, with mean vector $\mu = (\mu_1, \ldots, \mu_q)$, precision $P = V^{-1}$ and degrees of freedom v. Thus, in a straightforward extension of the univariate t density,

$$f(x|\mu, P, v) \propto [1 + 1/v(x-\mu)P(x-\mu)]^{-0.5(p+v)}$$

with covariance for x given by $vV/(v-2)$. A similar partitioning as above for the multivariate normal may be applied to its Student t counterpart. Thus suppose $x = (x_1, \ldots, x_r, x_{r+1}, x_q)$ is partitioned into sub-vectors $x^{(1)}$ and $x^{(2)}$ of dimension r and $p = q - r$, and P and V are correspondingly partitioned:

$$V = \begin{bmatrix} V_{11} & V_{12} \\ V_{21} & V_{22} \end{bmatrix}, \quad P = \begin{bmatrix} P_{11} & P_{12} \\ P_{21} & P_{22} \end{bmatrix}.$$

Then the marginal distribution of $x^{(1)}$ is also multivariate t with mean $\mu^{(1)} = (\mu_1, \ldots, \mu_r)$ and covariance

$$vV_{11}/(v-2) = v(P_{11} - P_{12}P_{22}^{-1}P_{21})^{-1}/(v-2).$$

The conditional distribution of $x^{(1)}$ given $x^{(2)}$ is also multivariate t with mean

$$\mu^{(1)} + P_{11}^{-1}P_{12}(x^{(2)} - \mu^{(2)})$$

and degrees of freedom $v + p$.

The multivariate t density has wide applicability in the linear regression model with k dependent variables. For example consider a univariate linear $(k = 1)$ model, with q covariates, $x_1, \ldots, x_q$

$$y = x\beta + e,$$

with errors $e \sim \mathrm{N}(0, \pi^{-1})$. For priors

$$g1(\pi) = \mathrm{G}(0.5v_0, 0.5v_0V_0)$$

$$g2(\beta|\pi) = \mathrm{N}(\mu.\beta, (\pi A)^{-1})$$

the posterior density of the regression parameter β is multivariate t as above but with x replaced by (the estimate of) β, μ replaced by (the estimate of) $\mu.\beta$, v replaced by $N + v_0$, $p = k = 1$, and P replaced by

$$V_0^{-1}(A + X'X).$$

There are different ways to sample a multivariate t, $T_q(\mu, \Sigma, v)$, a relatively simple one involving a sample of z from a gamma, $\mathrm{G}(0.5v, 0.5v)$ and then a sample from a multivariate normal $y \sim \mathrm{N}_q(\mu, V/z)$. This mode of generation reflects the scale mixture form of the multivariate t density, parallel to that of the univariate t. In BUGS the multivariate normal is parameterised by the precision matrix P, so we program a multivariate t density to *generate* a sample of n cases, for `Sigma[,]`, nu.2 and `mu[]` known, as follows:

```
for (i in 1:n) {z[i] ~ dgamma(nu.2, nu.2)
                y[i, 1:q] ~ dmnorm(mu, P.sc[,]}}
for (i in 1:q) {for (j in 1:q) {P[i,j] <- inverse(Sigma[,], i,j)
                                P.sc[i,j] <- z[i]* P[i,j]}}
```

If we have observed multivariate data and wish to assume multivariate Student t sampling, then in BUGS the dmt() form is available:

```
for (i in 1:n) {y[i,1:q] ~ dmt(mu[], P[,],nu)}
```

where nu is assumed known.

Example 2.18 Bivariate normal data with partial missingness Tanner (1996) presents 12 data points from a bivariate normal density $\{y_1, y_2\}$ with known mean $\mu_1 = \mu_2 = 0$, but unknown dispersion matrix Σ. The data contains four fully observed pairs $\{y_{i1}, y_{i2}\}$, with the remaining observations being partially missing: values on one or other of y_1 and y_2 are not available. Two of the fully observed pairs are consistent with a population-wide correlation ρ of -1, the other two with a correlation of 1. As noted by Tanner (1996, p.96) the posterior density of ρ is bimodal, with modes close to the 'observed' values of $+1$ and -1. The true posterior of the correlation is obtainable analytically under the vague prior

$$\pi(\Sigma) \propto |\Sigma|^{-(p+1)/2}$$

where $p = 2$ for a bivariate normal. This posterior is proportional to $(1 - \rho^2)^{4.5}(1.25 - \rho^2)^8$ and is also bimodal. Here we wish to retain the information provided by the 8 data points subject to missingness even if they do not add directly to knowledge about the covariance σ_{12}. They will however add to knowledge of the variances σ_1^2 and σ_2^2 and so contribute to estimating ρ.

To estimate the dispersion matrix and values for the missing data, we may use partitioned sampling. Thus for cases with y_{i1} observed but y_{i2} missing we sample y_{i2} from $f(y_{i2} | y_{i1})$ which is a univariate normal with mean

$$\mu_2 + \rho\sigma_2(y_{il} - \mu_1)/\sigma_1$$

and variance

$$\sigma_2^2 - \sigma_{12}^2/\sigma_1^2 = \sigma_2^2 - \rho^2\sigma_2^2.$$

The term $\beta_{2.1} = \rho\sigma_2/\sigma_1$ is familiar as the regression coefficient in a linear model relating y_2 and y_1. If we assume $\mu_1 = \mu_2 = 0$ is already known in this example, then the mean of $f(y_{i2} | y_{il})$ reduces to $\rho\sigma_2 y_{il}/\sigma_1$. An analogous density is defined for cases where y_{i2} is observed but y_{i1} missing. In BUGS this would require direct sampling from the full conditional densities of the variances and covariance.

Alternatively, as in (2.26), we may define a marginal model for y_1, then a conditional regression for y_2 given y_1. The correlation is then estimated from the observed and imputed data through its part in defining the regression coefficient $\beta_{2.1}$. The parameter cycles through positive and negative values, with long-run average zero, but two distinct modes. The posterior density for ρ shown in Figure 2.1 is based on a sample of 100 000.

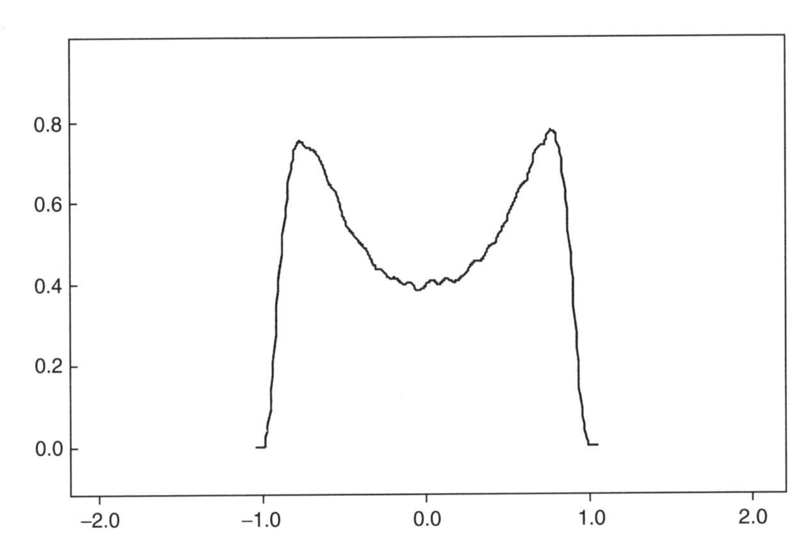

Figure 2.1 Posterior for bivariate correleation.

Example 2.19 Income and dependency in cities Birkin *et al.* (1994) present data for 14 selected UK cities on households wholly or partly dependent on state benefit (y_1 and y_2 respectively), on income per head (y_3) and on the income/capital ratio (y_4). We illustrate an application of the multivariate normal model to study the interdependence of these outcomes. To illustrate predictions to new data we omit variables y_2 and y_4 for the last city, Glasgow, though in fact they are known. The predictions are then based on the missing-at-random principle: that missingness on y_2 or y_4 is not related to the underlying missing value of the variable (for example, the principle would be violated if the dependency ratio was more likely to be missing for high values of dependency). There are certain characteristics of the data that might suggest initial transformation of the variables: y_1 and y_2 are proportions, and y_3 and y_4 are necessarily positive, so logit and log transforms might be applied, at least in a sensitivity analysis.

We here analyse the untransformed data, and adopt a multivariate normal prior with means $\mu = (\mu_1, \ldots, \mu_4)$ and 4×4 dispersion matrix Σ. From the latter we derive correlations among the outcomes. So the model (see *Program 2.19 Incomes and Dependency Ratio*) is

$$y_{[i,\,1:4]} \sim \mathrm{N}_4(\mu_{[1:4]}, \Sigma_{[1:4,\,1:4]})$$

where Σ^{-1} is assigned a Wishart prior with 4 degrees of freedom and identity scale matrix, while the μ_j are assigned vague priors with mean zero and large variances. Missing values y_{i2} and y_{i4} will have long-run averages the same as the corresponding means μ_2 and μ_4. The second half of a run of 20 000 iterations yields predictions of y_2 and y_4 for Glasgow of 36.5 and 118.5 with credible intervals which do in fact include the true values of 38.4 and 111.0. The correlations between y_1 and y_2 and between y_3 and y_4 are both around 0.8.

The predictions for Glasgow do not however use the correlation structure and may be improved on.

Applying a multivariate t density with 5 d.f. leads to slightly lower average correlations but heavier tailed densities for them; for example the correlation between y_3 and y_4 has mean 0.785 under the multivariate t option but 2.5th and 97.5th percentiles $\{0.43, 0.95\}$. Under the MVN option this interval is $\{0.52, 0.94\}$.

If we adopt the conditioning procedure involving multivariate regression, as described earlier, then variables known for all cases (y_1, y_3) are separated from the variables missing on one or more cases (y_2, y_4). The redefined variables $z[1:4]$ thus map to the old variables as $(z_1, z_2, z_3, z_4) \equiv (y_2, y_4, y_1, y_3)$. In terms of predictions for Glasgow, this procedure works best for predicting complete dependence on state benefit: now estimated with a posterior mean of 37.3, and 95% credible interval 36.9 to 37.8. Two regression coefficients, representing the relations between (a) complete and partial dependence and (b) income and income/capital ratio, are clearly positive in their effect. The correlation structure is reproduced, though the estimated correlations are slightly lower than in the previous (simple imputation) method (Table 2.7).

Adopting more robust multivariate t densities for both sets of variables (with 5 degrees of freedom) changes the predictions hardly at all, though further reduces the correlations.

Example 2.20 Multiple tests Westfall *et al.* (1997) consider the formulation of appropriate priors in the multiple testing situation, where there are k simultaneous probability tests p_i being carried out on the null hypotheses

Table 2.7

	Mean	SD	2.5%	97.5%
Regression coefficients for redefined variables (z)				
B13	2.03	0.56	0.91	3.08
B14	0.09	1.13	−2.18	2.29
B23	0.01	0.04	−0.07	0.07
B24	0.25	0.07	0.12	0.40
Correlations between redefined variables (z)				
1,2	−0.62	0.17	−0.87	−0.22
1,3	0.75	0.12	0.45	0.92
1,4	−0.44	0.21	−0.78	0.03
2,3	−0.46	0.20	−0.78	−0.01
2,4	0.74	0.12	0.45	0.91
3,4	−0.63	0.16	−0.86	−0.24
Predictions for Glasgow				
Dependency	37.3	0.2	36.9	37.8
Capital-income	116.8	0.4	115.9	117.6

$H_{0i}(i = 1, \ldots, k)$. Then the usual frequentist rejection rule, namely $p_i \leq 0.05$ for each test $(i = 1, \ldots, k)$ may lead to a collective type I error rate greater than 5% if many or all the null hypotheses are true. The simplest Bonferroni adjustment uses an adjusted collective error rate

$$p_i^{\mathrm{C}} = kp_i,$$

which provides a conservative upper bound to the adjustment required.

In a Bayesian version of this adjustment consider observed (and centred) data $z = (z_1, \ldots, z_k)$ distributed according to $\mathrm{N}_k(\Theta, I)$ where the vector of effects $\theta_i, i = 1, \ldots, k$ has mean 0, variance σ^2 and correlation ρ. (Thus the covariance matrix of θ_i has diagonal σ^2 and off-diagonal terms $\rho\sigma^2$.) The actual prior on the effects does not refer directly to this postulated dispersion pattern, but indirectly in terms of the implied prior probability that the hypotheses will all be true. For a single test the neutral prior on H_0 being true (if there is only one alternative hypothesis) is $\pi_{0i} = \Pr(H_{0i} \text{ true}) = 0.5$. The multivariate dispersion pattern relating to the joint probability

$$\Pi_0 = \text{Prob (all } H_{0i} \text{ true)}$$

then implies

$$(0.5)^k \leq \text{Prob(all } H_{0i} \text{ true)} \leq 0.5$$

with the extremes of the interval applying for complete independence $(\rho = 0)$ and perfect correlation between tests $(\rho = 1)$. Negative correlation is inconsistent with the Bonferroni adjustment procedure which is that there is some increase in the joint probability as compared with independence. Thus if the hypotheses are regarded as independent with each $\pi_{0i} = 0.5$, then the prior probability that they are all true is $\Pi_0 = (0.5)^k$, which may be smaller than the experimenter intends as the prior collective probability of the hypotheses. On the other hand setting $\tilde{\pi}_{0i} = (0.5)^{1/k}$ so that $\Pi_0 = 0.5$ may overstate the collective probability. These priors do, however, set bounds on the more likely situations (somewhere between these extremes) and so define limits on the posterior probabilities of H_{0i} and the Bayes factors B_{01i} for each test.

The data Westfall *et al.* consider relate to a study of extrasensory perception. Five remote viewers were asked to provide descriptions of videos or photos exhibited in another room. The accuracy of the description by viewer i on experiment j was summarised in five-point ratings R_{ij}, with a rating of 1 for the most accurate match and 5 for the least. The null hypothesis H_0 of no ESP corresponds to random matching of remote descriptions to actual exhibits with $E(R_{ij}) = 3, \mathrm{Var}(R_{ij}) = 2$. There are thus $k = 5$ null hypotheses to evaluate simultaneously. The number of experiments n_i for each of the subjects ranged from 40 to 70, and the results are summarised in overall scores

$$\bar{R}_i = \sum_{j=1}^{n_i} R_{ij}/n_i.$$

To these correspond z scores,

$$z_i = (\bar{R}_i - 3)(n_i/2)^{0.5}.$$

The effect sizes are

$$\theta_i = (\gamma_i - 3)/2^{0.5}$$

where $\gamma_i = E(R_{ij}) = n_i E(\bar{R}_i)$, so that θ_i is estimated by $z_i/n_i^{0.5}$. The alternative hypothesis H_{1i} is that underlying individual viewer effects θ_i are significantly less than zero, corresponding to better than expected match of remote description to exhibit, and the null hypothesis H_{0i} is that $\theta_i = 0$.

We compute posterior probabilities p_0 of the hypothesis H_0 being true on the basis of the prior probability π_0 and the Bayes factor:

$$P(H_0 \mid y) = 1/[1 + (1 - \pi_0)/(\pi_0 B_{01})]$$

(see *Program 2.20 ESP Bayes factor*). The priors on the mean effects θ_i follow those used by Westfall *et al.*, who distinguish two subgroups within which exchangeability applies (viewers 1 and 2 classed as highly skilled and the remainder classed as skilled). The respective $N(v, \tau^2)$ priors for these groups are taken here as $N(-0.5, 0.1^2)$ and $N(-0.3, 0.15^2)$, and correspond approximately to those presented in Figure 1 of Westfall *et al.* The Bayes factors are derived from the formula (2.7) above

$$B_{01} = [(\tau^2 + \sigma^2)/\sigma^2]^{0.5} \exp[-D/2]$$

where $D = (\hat{\theta}_i - m_0)^2/\sigma^2 - (\hat{\theta}_i - v)^2/(\sigma^2 + \tau^2)$, and $m_0 = 0, \sigma^2 = 1/n_i$, $\hat{\theta}_i = z_i/(n_i)^{0.5}$.

The Bayes factors are lowest for the first two viewers, namely 0.023 and 0.019. However, the posterior probability of H_0 depends on the prior adopted for each H_{0i} and whether it is adjusted for the collection of all H_{0i}. For example, taking $\pi_{0i} = \Pr(H_{0i} \text{ true}) = 0.9$ for each hypothesis separately gives a joint prior probability of 0.59 if all five hypotheses are regarded independently. If however, we adjust the individual π_{0i} to reflect prior beliefs about $\Pr(\text{all } H_{0i} \text{ true})$, then the most extreme option is to take the adjusted priors $\tilde{\pi}_{0i}$ to equal $(\pi_{0i})^{0.2}$, which in this case is 0.979. Under this scenario, as detailed in Table 2.8, all the mean posterior probabilities of H_0 exceeed 0.10.

Example 2.21 Bivariate screening In medical and quality control applications, we may have two correlated measures X and Y, with X less expensive to apply. The means of Y and X are μ_Y and μ_X respectively. Suppose the Y measure must exceed a threshold $Y.\tau$, for example for a screened patient to be deemed at risk or not at risk, or for a product to be deemed defective or of acceptable quality. Using the properties of the bivariate normal we can specify a limit on X, say $X.\tau$ such that, with a probability δ, we will know that Y exceeds its own threshold if X in turn exceeds the limit $X.\tau$. For a bivariate normal with dispersion matrix

$$\Sigma = \begin{bmatrix} \sigma_y^2 & \rho\sigma_y\sigma_x \\ \rho\sigma_y\sigma_x & \sigma_x^2 \end{bmatrix}$$

Table 2.8

	Mean	SD	2.5%	Median	97.5%
Bayes factors on H_0					
Viewer 1	0.023	0.126	0.000	0.002	0.169
Viewer 2	0.020	0.149	0.000	0.001	0.127
Viewer 3	0.822	0.882	0.036	0.529	3.315
Viewer 4	1.528	1.643	0.040	0.957	6.130
Viewer 5	0.364	0.635	0.003	0.137	2.095
Posterior probs of H_0					
Viewer 1	0.215	0.255	0.001	0.099	0.887
Viewer 2	0.171	0.238	0.000	0.054	0.855
Viewer 3	0.926	0.099	0.624	0.961	0.994
Viewer 4	0.945	0.096	0.650	0.978	0.997
Viewer 5	0.765	0.246	0.106	0.865	0.990

the predictive density of a new Y value, Y_{n+1}, given a new X value, X_{n+1}, is a univariate normal with mean $\mu_Y + \rho\sigma_Y(X_{n+1} - \mu_X)/\sigma_X$ and variance $\sigma_Y^2(1-\rho^2)$. This is analogous to a least squares regression estimate.

Hence the predictive density of Y_{n+1} given only X_{n+1} can be found by sampling Y_{n+1} from this density at each sampled value of the five bivariate normal parameters: $\mu_Y, \mu_X, \rho, \sigma_Y$ and σ_X. Suppose, following Wong *et al.* (1985), we apply dissolution testing to measure the active ingredient in a pharmaceutical product at times 1 and 2 (denoted X and Y). We observe both Y and X for a sample only, and require that at time 2 the cumulative release of Y exceeds $Y.\tau = 1500$ with a probability $\delta = 0.99$. If X and Y are highly correlated, we can avoid taking a full set of repeated measurements at time 2 by using the sample data on the association of X and Y.

Suppose a sample of 10 measures at times 1 and 2 gives $\bar{X} = 1256$, $\bar{Y} = 1969, s_X = 133, s_Y = 177$, and $r = 0.975$. To allow for the uncertainty in these sample estimates we assume a prior scale Λ_0 based on ν_0 degrees of freedom, giving a prior distribution on the precision

$$\Sigma^{-1} \sim \mathrm{W}[\Lambda_0, \nu_0],$$

We also assume a bivariate normal prior on μ_X and μ_Y according to

$$\mathrm{N}_2(\mu_0, \Sigma/\kappa_0).$$

Here μ_0 is a vector of assumed prior means μ_{Y0} and μ_{x0}, and κ_0 is a measure of the quantity of prior information on these means. The relevant posterior densities, combining these priors with the observed sample values, are as given by Gelman *et al.* (1995, Chapter 3) and are incorporated in *Program 2.21 Dissolution Testing*. From these densities we can sample $\mu_X, \mu_Y, \sigma_X, \sigma_Y$ and ρ. For comparability with Wong *et al.*, Λ_0 is taken as a null matrix, and $\nu_0 = \kappa_0 = 0$. We set default prior mean values $\nu_{Y0} = 1950$, $\mu_{X0} = 1250$, though these values are not used unless $\kappa_0 > 0$.

We can set up a range of new values X_{n+1} and assess the breakpoint $X_{n+1,\tau}$ at which Y_{n+1} exceeds $Y.\tau$ with 99% certainty. Acceptance of new batches at time 2 can then be based on the amount released at time 1. We use the step() function in BUGS to test whether `Y.new` ($\equiv Y_{n+1}$) exceeds 1500, given `X.new` and the estimates of the bivariate normal parameters. These tests are accumulated in the vector `MeetSpec[]` in Program 2.21. An initial run with values of `X.new` at intervals of 25 between 900 and 1100 narrows the likely range to between 975 and 1025. A second run then takes values of `X.new` at intervals of 5 between 975 and 1025. This yields a range of values from 96.7% to 99.8% of samples of `Y.new` exceeding the threshold of 1500, with the threshold of 99% occurring between `X.new = 995` and `X.new = 1000`.

2.8.2 Hypotheses on multivariate means

Suppose we wish to test whether an observed set of means from sample of size n

$$\bar{x} = (\bar{x}_1, \bar{x}_2, \ldots, \bar{x}_p)$$

drawn from a multivariate normal density $\mathrm{N}_p(\mu, \Sigma)$ are equal to a hypothetical mean, $\mu_0 = (\mu_{01}, \ldots, \mu_{0p})$. We can adapt the classical Hotellings test to this situation, while taking into account uncertainty in the estimates of the parameters of the multivariate density and hence uncertainty about the value of the test statistic. A multivariate test whether $\hat{\mu} = \mu_0$ rests on finding a transformation $z = \alpha' x$ such that the value of α is that which is most unfavourable to the hypothesis.

This results in taking $\alpha \propto \hat{\Sigma}^{-1}(\hat{\mu} - \mu_0)$, and in the test statistic

$$\varphi = (n-p)T/[pn-p] \tag{2.27}$$

where

$$T = n(\hat{\mu} - \mu_0)\hat{\Sigma}^{-1}(\hat{\mu} - \mu_0)$$

is the Hotellings statistic (Krzanowski, 1988, p.241). The statistic φ follows an $\mathrm{F}(p, n-p)$ density if the hypothesis is true. An equivalent test is to compare T/n with a χ^2 density with p degrees of freedom, and so assess

$$\Pr[\chi^2_p > (\hat{\mu} - \mu_0)\hat{\Sigma}^{-1}(\hat{\mu} - \mu_0)]$$

leading to a posterior probability of μ values with lower posterior density than μ_0 (Box and Tiao, 1973).

Suppose we are comparing two samples $\underset{\sim}{x} = (x_1, \ldots, x_m)$ and $\underset{\sim}{y} = (y_1, \ldots, y_n)$ of p-variate data, and we assume possibly different means but a common dispersion matrix

$$\underset{\sim}{x} \sim \mathrm{N}_p(\mu_1, \Sigma) \text{ and } \underset{\sim}{y} \sim \mathrm{N}_p(\mu_2, \Sigma).$$

The comparable statistic to (2.27) for testing the hypothesis $\mu_1 = \mu_2$ is

$$\varphi = (n+m-p-1)\psi/[pn+pm-2p] \tag{2.28}$$

where $T = nm(\hat{\mu}_1 - \hat{\mu}_2)'\hat{\Sigma}_{-1}(\hat{\mu}_1 - \hat{\mu}_2)/(n+m)$. The φ statistic here follows an $F(p, n+m-p-1)$ density when the two means are equal.

Example 2.22 Growth measures Chatfield and Collins (CC, 1980) present data on height, chest circumference and mid-upper-arm circumference for $m = 6$ two-year-old boys from a high-altitude Asian region, and assume

$$x_{i,1:3} \sim N_p(\mu_x, \Sigma), \quad i = 1, \ldots, 6.$$

The observed means in centimetres are $(82, 60.2, 14.5)$ and the hypothesized value is the lowland average $(90, 58, 16)$. With the small sample size, estimates about μ_x and Σ are subject to uncertainty. We adopt vague priors on Σ and μ, and compare the sampled value of φ with a value drawn from an $F_{3,3}$ density. Posterior estimates of the MVN parameters from the second half of a run of 20 000 iterations (using *Program 2.22 Growth Measures*) show the uncertainty especially with regard to Σ_{ij}. The classical test statistic φ has value 84.1 and is highly significant ($P = 0.002$) in the sense of being incompatible with the hypothesis. This is confirmed by a sampling-based test of the hypothesis, though less decisively so: the proportion of 50 000 iterations at which a point from the $F_{3,3}$ density exceeds the sampled value of φ is 0.0047.

CC present comparable data $y_{i,1:3}$ on a set of $n = 9$ girls, and assume

$$y_{i,1:3} \sim N_p(\mu_y, \Sigma), \quad i = 1, \ldots, 9.$$

The classical test of hypothesis $\mu_x = \mu_y$ involves comparing the observed value of the φ statistic with the 5% point of an $F(3, 11)$ density, namely 3.59. This gives a probability of 0.27, suggesting that the two sexes have effectively the same mean vector. In a sampling-based test we compare the sampled value of φ with a point drawn from the $F(3, 11)$ density. The posterior estimates of Σ, now based on 15 cases, are more precise, but the test statistic

$$\varphi = (n+m-p-1)T/[pn+pm-2p]$$

shows positive skew and has a low precision; its average value of 2.65 exceeds the median of 2.35, and the posterior density has a standard deviation of 1.7. There is an approximately 21% chance that a point from the $F(3, 11)$ density exceeds the sampled value of φ, so confirming the classical result.

2.9 POWER AND SAMPLE SIZE CALCULATIONS

2.9.1 Sample size specification to achieve a desired accuracy

Very often we wish to estimate a parameter (e.g. a prevalence rate for a disease) with a certain degree of accuracy. For example, a 10% coefficient of variation may be considered the maximum allowable for some baseline parameter (and narrower coefficients of variation are applied in clinical contexts). The level of accuracy required is often a major input for calculating the desired sample size and hence a major consideration in the costs of a survey. There may be benefits in establishing a survey design using existing prior knowledge about the parameters, since in the case of social and behavioural variables there may be extensive prior knowledge. If we use such information, we may require a smaller sample size than otherwise.

Suppose following Adcock (1988) we can specify an accuracy range $\bar{y} - d$ to $\bar{y} + d$ within which we want a survey average to be located as an estimate of the population mean μ. We need this accuracy to apply with a probability $1 - \alpha$, where α might be 0.05 or 0.1. That is we need a sample size n large enough so that the probability is at least $1 - \alpha$ that the absolute difference $|\bar{y} - \mu|$ is less than d. Thus we require

$$P(|\bar{y} - \mu| \leq d) \geq 1 - \alpha.$$

If y is normal with distribution $N(\mu, \sigma^2)$, then d is given approximately by $Z_{\alpha/2}\sigma/\sqrt{n}$, where $Z_{\alpha/2}$ is the upper $\alpha/2$ point of the standard normal. Alternatively we may specify accuracy as a proportion of the variance. Thus d should be less than $r\sigma^2$ with probability $1 - \alpha$, where r is a small value such as $r = 0.1$ or $r = 0.05$.

One option is to assume σ^2 is known and use no prior knowledge about the mean. In this case (method A) n must exceed $Z^2_{\alpha/2}\sigma^2/d^2$ or $Z^2_{\alpha/2}/r^2$ under these alternative ways of specifying accuracy. Suppose, instead, we both include prior knowledge about the mean, and allow for its uncertainty, by assuming a prior density for μ. We attach a measure m to this prior belief, namely an equivalent sample size so that $\mu \sim N(\mu_0, S_0/m)$, where μ_0 is a prior guess for the mean of y and S_0 is a guess for the variance. Usually m will be a small integer (1, 5 or 10 say) but may be substantially more if, say, we had national survey data and were carrying out a local survey. The posterior distribution of μ given the data value $\bar{y}$ then has mean $(n\bar{y} + m\mu_0)/(n + m)$ and variance $\sigma^2/(n + m)$. So method A is effectively just modified to take account of the prior sample size pertaining to the prior mean (method B). In this method we specify n such that $(n + m)$ exceeds $Z^2_{\alpha/2}\sigma^2/d^2$.

Suppose we allow fully for uncertainty by introducing prior densities for both μ and the precision $\tau = \sigma^{-2}$ (method C). A commonly used approach for specifying the prior for the precision is to assume $\tau_1 = \nu S_0 \tau$ follows a chi-square density with ν degrees of freedom. S_0 is a prior estimate of the variance (usually very approximate), and ν is a measure of the strength of these prior beliefs (typical values would be small integers such as 1 or 5). So given a draw of τ_1 from the chi-square density, τ is given by $\tau = \tau_1/\nu S_0$. With a known draw of the variance $\sigma^2 = 1/\tau$, the prior for μ is then $\mu \sim N(\mu_0, \sigma^2/m)$.

When the strength of prior knowledge about the variance is considerable (say ν in excess of 20) then the required sample size is similar to method B, but with smaller ν the sample size needed for a specified precision is increased.

Example 2.23 Sample size to achieve given accuracy We continue the systolic blood pressure example (Example 2.1), and so take $\mu_0 = 125$, $S_0 = 169$. Suppose the prior sample size (strength of belief) on the prior mean is set at $m = 20$, and that for cost reasons the sample size is set at $n = 40$. Suppose we express accuracy as a proportion of the variance, i.e. we want $|\bar{y} - \mu|$ to be less than $r\sigma^2$ with 0.95 probability, and with $r = 0.025$. Suppose method C above is adopted (see *Program 2.23 Sampling Size for Accuracy*) with $v = 1$. Then the condition

$$|\bar{y} - \mu| < r\sigma^2$$

occurs with probability 0.941. If, however, $v = 5$, the condition holds with probability 0.953; and if $v = 10$, with probability 0.956.

2.9.2 Sample size specifications for testing hypotheses

An alternative basis for sample size calculations occurs in evaluating a scientific model or in group comparisons to test a certain hypothesis. Often such hypotheses are framed in terms of differences between group means. Suppose then that we wish to test a hypothesis of a null difference between group means, explicitly or implicitly against an alternative hypothesis concerning the likely size of the effect. We are then faced with matching our conclusion from the test against the true situation, presumed usually to be unknown.

The size of a test, or its type I error rate α, refers to the chance of rejecting the null hypothesis H_0 when this hypothesis is in fact true. We may set stringent levels on α to ensure against a type I error. For example if $\alpha = 0.01$, we only reject the hypothesis when the data are widely at variance with it. However, the more stringent the level adopted for α, the lower is the chance of rejecting the null hypothesis and providing evidence in support of the alternative hypothesis even when that alternative is true (i.e. H_1 holds as a description of reality). Thus for small α, the test may not reject H_0 even though H_1 is true.

The probability that a test will lead to a rejection of a false null hypothesis at a given significance level is called its power. For a particular test, the power will be determined by the sample size, the actual difference or effect present, and the significance level chosen. While a high significance level may seem equivalent to rigour in fact it reduces the power to detect a real effect. The complement of the power, $\beta = 1 -$ power, is also an error, since it represents the chance of failing to reject a false null hypothesis (this is a type II error). For example, if a very low significance level of $\alpha = 0.001$ leads to a power of only 0.10, then the ratio of the two risks β/α is 900 to 1. So the researcher is saying that rejecting a null hypothesis when it is true is 900 times more serious than mistakenly accepting it. A power of 80 or 90 % and type I error rate of 5 % give risk ratios β/α of 4:1 and 2:1 respectively.

Incorporating power into a study design requires the researcher to consider what the alternative hypothesis might be, e.g. what direction the effect is, or what effect size is expected. The power with a given size of sample, or the required sample size to obtain a certain power, will be related to the direction of the effect or effect size postulated under the alternative hypothesis (H_1).

Classical test statistics

Consider the case of comparing two means $\bar{x}_1$ and $\bar{x}_2$ between distinct populations 1 and 2 (e.g. separate geographical populations, or treatment and control groups). The null hypothesis H_0 is then often that the underlying means are the same, and the alternative H_1 is that the underlying means are different, specifically

$$H_1: \quad \mu_1 = \mu_2 + \delta, \quad \delta \neq 0. \tag{2.29}$$

Suppose $\bar{x}_1$ and $\bar{x}_2$ are based on samples of size n_1 and n_2. Assuming the observations on x_1 and x_2 are approximately normal, with the same known variance σ^2, then an appropriate test statistic, assuming H_0 holds, is

$$T = (\bar{x}_1 - \bar{x}_2)/\eta$$

where

$$\eta^2 = \sigma^2(1/n_1 + 1/n_2)$$

is the variance of the difference in means. If the sample size n_1 is g times the size of the sample n_2, then $\eta^2 = \sigma^2(g+1)/gn_2$.

The size of a test, denoted α, is the probability that H_0 is rejected when it is true. In this case

$$\alpha = \Pr(\text{reject} H_0 \,|\, H_0 \text{true}) = \Pr(T > z_\alpha \,|\, H_0 \text{true})$$

where the cutoff point z_α is the appropriate point of the cumulative distribution for a standard normal. For example, if $\alpha = 0.05$ then $P(T < z_\alpha) = 1 - \alpha = 0.95$. This analysis assumes T has a standard normal density when H_0 is true. With some test statistics, including Bayes criteria, it may be necessary to establish a cutoff point by simulation (Weiss, 1997).

The probability of type II error is the probability of 'accepting' H_0 when in fact H_1 is true (more precisely we might say 'failing to reject' H_0). So an error occurs if an analysis falsely concludes there is no difference between the underlying means when in fact there is. The power is the converse of the chance of a type II error, namely the probability of rejecting H_0 when H_1 is true (i.e. the power is the chance of coming to the correct conclusion). In the case of the comparison of two means

$$\beta = \Pr(\text{accept } H_0 \,|\, H_1 \text{ true}) = 1 - \text{power} = \Pr(T < z_\alpha \,|\, H_1 \text{ true}).$$

So the power is

$$1 - \beta = \Pr(\text{reject } H_0 \,|\, H_1 \text{ true}) = \Pr(T > z_\alpha \,|\, H_1 \text{ true}).$$

If we write

$$R = T - \delta/\eta = (\bar{x}_1 - \bar{x}_2 - \delta)/\eta$$

then

$$1 - \beta = \Pr(R > z_\alpha - \delta/\eta \,|\, H_1 \text{true})\cdot$$

We can therefore establish power by sampling $\bar{x}_1$ and $\bar{x}_2$ from the respective populations, defined by a real world in which H_1 holds. If H_1 is true these distributions will be $x_1 \sim \mathrm{N}(\mu_2 + \delta, \sigma^2)$ and $x_2 \sim \mathrm{N}(\mu_2, \sigma^2)$. The sample means will then be drawn from normal densities with precisions $n_1\tau$ and $n_2\tau$, where $\tau = 1/\sigma^2$. So for preset values of n_1 and n_2 we can calculate the power empirically as the proportion of samples in which $R > z_\alpha - \delta/\eta$.

If H_1 describes reality, then R rather than T is approximately standard normal (Woodward, 1999). Therefore a common sample size rule is based on the approximate equality

$$z_\alpha - \delta/\eta \approx z_{1-\beta} = -z_\beta.$$

Then

$$\delta/\eta = z_\alpha + z_\beta,$$

and using the formula for η we obtain an indicated sample size

$$N = n_1 + n_2 = (g+1)^2(z_\alpha + z_\beta)^2\sigma^2/(g\delta^2).$$

The Bayes factor as a test statistic

A Bayesian approach to hypothesis tests involving group comparisons may be based on an alternative test statistic, such as the Bayes factor: namely the ratio of posterior to prior odds in favour of H_0 rather than H_1, given respective likelihoods $f(|H_j), j = 0, 1$. For example, following Weiss (1997), assume n paired subjects with observations $[x_{1i}, x_{2i}]$ on each pair, and an outcome variable contrasting x_1 and x_2, for example $d = x_1 - x_2$. With H_0 being that the mean of d is zero, the Bayes factor for alternative hypotheses on d is

$$B_{01} = f(d|H_0)/f(d|H_1),$$

where f is the assumed density for data $d = (d_1, \ldots d_n)$, and it is also assumed that H_0 and H_1 have equal prior probability. It is often more convenient to work with the test statistic $b_{01} = \log(B_{01})$. So values of B_{01} below one (i.e. b_{01} negative) give more support to H_1 than H_0, and so point to rejection of H_0.

To establish whether or not H_0 is supported by the data, and the sample size to achieve a given power, one option is termed the relative approach by Weiss (1997). This involves presetting a cutoff point such as $k = 0$, and rejecting H_0 if $b_{01} < k$. Then we establish the sample size needed to guarantee a certain power (e.g. 90 %) under this rejection rule. So for trial values of n and k we establish the power

$$\Pr(\text{reject } H_0 | H_1 \text{ true}) = \Pr(b_{01} < k | H_1 \text{ true}).$$

This is done by empirically sampling from the two populations under H_1. The other option is analogous to a classical testing approach, and adopts as the type I error

$$\alpha = \Pr(\text{reject } H_0 | H_0 \text{ true}) = \Pr(b_{01} < b_{01,\alpha} | H_0 \text{ true}).$$

This option would involve establishing the distribution of b_{01} under H_0 via simulation. Then the power is given by

$$\Pr(\text{reject } H_0 \,|\, H_1 \text{ true}) = \Pr(b_{01} < b_{01,\alpha} \,|\, H_1 \text{ true}).$$

Consider a difference between means $\bar{d} = \bar{x}_1 - \bar{x}_2$ for outcomes on paired subjects x_{1i}, x_{2i}. Paired outcomes are often incorporated in clinical evaluations and other experimental situations. Alternatively we might compare pairs of observations randomly drawn from different populations. Then if $x_1 \sim \mathrm{N}(\mu_1, \sigma^2/2)$ and $x_2 \sim \mathrm{N}(\mu_2, \sigma^2/2)$, $x_1 - x_2$ is $\mathrm{N}(\mu_1 - \mu_2, \sigma^2)$.

Under H_0, the prior for $\delta = \mu_1 - \mu_2$ is $\mathrm{N}(0, \rho_0)$ while under H_1, δ is $\mathrm{N}(\theta, \rho_1)$ with $\theta \neq 0$. If H_0 is taken as a sharp null hypothesis, then $\rho_0 = 0$ effectively. The prior variance ρ_1 of the mean under H_1 can be assumed known or remain a free parameter. The data here are random, so our posterior density is in fact a predictive density. So having 'observed' data d via randomly generating it, the predictive posterior densities of $d = (d_1, \ldots d_n) = (x_{11} - x_{21}, x_{12} - x_{22}, \cdots, x_{1n} - x_{2n})$ under the two hypotheses are

$$f(d \,|\, H_1) = \mathrm{N}(\bar{d} + \theta, \sigma^2/n + \rho_1)$$
$$f(d \,|\, H_0) = \mathrm{N}(\bar{d}, \sigma^2/n + \rho_0) = \mathrm{N}(\bar{d}, \sigma^2/n).$$

The log of B_{01} in this case can be shown (Weiss, 1997) to equal

$$b_{01} = A - B(\bar{d} + C)^2$$

where $A = 0.5\log(1 + n\rho_1/\sigma^2) + 0.5\theta^2/\rho_1, B = 0.5\rho_1/(\sigma^4/n^2 + \sigma^2\rho_1/n)$ and $C = \mu_1\sigma^2/n\rho_1$.

Suppose we wished to assess which was the highest of the means μ_1 and μ_2 of two different Normal populations, under the situation where the variances σ_1^2 and σ_2^2 in the two populations may differ. Then we know that the sample means $\bar{x}_1$ and $\bar{x}_2$ (from samples of size n_1 and n_2 respectively) are sufficient statistics. So to test the alternative hypotheses H_a: $\mu_1 > \mu_2$ and H_b: $\mu_2 > \mu_1$ we can sample from the likelihoods for the means:

$$\bar{x}_j \,|\, \mu_j, \sigma_j^2 \sim \mathrm{N}(\mu_j, \sigma_j^2/n_j) \text{ for } j = 1, 2.$$

We then obtain the posterior density of $\mu_1 - \mu_2$ by which we can test the two hypotheses. The opportunity to incorporate prior information on the location of μ_1 and μ_2 may result in different inferences from those obtained under a classical significance test that $\mu_1 = \mu_2$. If additionally, we allow for uncertainty in the variances, then we can assume inverse variances distributed according to $\mathrm{G}(\nu_0/2, \nu_0 S_0/2)$ where S_0 is a prior guess at the variance.

Example 2.24 Systolic blood pressure samples by city Smith *et al.* (1990) compare systolic blood pressure (SBP) in mm Hg in males aged 40–44 between two Scottish cities. These are generally affluent Edinburgh (population 2) and the more deprived N. Glasgow (population 1), so we might expect $\mu_1 > \mu_2$. Accumulated clinical knowledge (Smith *et al.*, 1989) suggests the mean SBP in population 2 is about 130 and the standard deviation σ in both populations to be about 15.6. Suppose a researcher wishes to be 90% certain of detecting that average SBP in N. Glasgow is 3 mm Hg higher than in Edinburgh (i.e. a 95% power with $\beta = 0.05$). So with $\delta = 3$ under H_1 as in (2.29), μ_1 equals 133. We sample n_1 and n_2 random cases under the assumption H_1 holds, and monitor the power

$$1 - \beta = \Pr(R > z_\alpha - \delta/\eta \,|\, H_1 \text{ true}).$$

We try sample sizes in Edinburgh of $n_2 = 300$, 400 and 500 and with $g = 2$, $n_1 = 600, 800, 1000$ and so $N = 900, 1200, 1500$. The powers for these values of the total sample size are 0.86, 0.93 and 0.97 respectively. So the required sample size in Edinburgh is between 400 and 500 (see *Program 2.24* S*ystolic by City*).

Example 2.25 Sample size via Bayes factors Weiss (1997) gives an example of sample size selection using the Bayes factor approach with $\delta = \mu_1 - \mu_2 = 0$ under H_0, while under H_1, δ is $\text{N}(\theta, \rho_1)$. The trial values are $\rho_1 = 1, \theta = 2$, and $\sigma = 5$ (the standard deviation of the difference between x_1 and x_2). Setting $n = 20$ in *Program 2.25 Bayes Factor Sample Size Selection (Model 1)* produces a density for $P(b_{01} \,|\, H_0)$. This has an average value 0.96 and an $\alpha = 0.05$ quantile of -1.03.

We can then use the latter as the cutoff value in the power calculation

$$\Pr(b_{01} < b_{01,\alpha} \,|\, H_1 \text{true}).$$

Alternatively we can take a default such as $k = 0$, with a correspondingly different type I error value. With $n = 20$, this break has a type I error value of approximately 0.15. This type I rate can be obtained by ranking the b_{01} values obtained from running *Program 2.25 Model 1* and finding the proportion below zero.

Using these two possible cutoff points *Program 2.25 Model 2* gives empirical powers of 0.54 and 0.72 respectively for $n = 20$ (from the second half of a run of 25 000 iterations). Weiss obtained comparable values of 0.543 and 0.73, and showed that for $n = 66$, the power based on the criterion

$$\Pr(b_{01} < b_{01,\alpha} \,|\, H1 \text{ true})$$

is 80%.

This approach may be applied to the systolic blood pressure example, with equal sample sizes in each city $n_1 = n_2 = 650$. We assume that under H_1, the difference in city means

$$\delta \sim \text{N}(\theta, \rho_1)$$

where $\rho_1 = 1, \theta = 3$. These assumptions lead to a mean of b_{01} under H_0 of 2.7 and an $\alpha = 0.05$ quantile of -0.24. Note that we take $\sigma^2 = 2(15.6)^2$ since the analysis using the log Bayes factor relates to differences between observations in two populations. Then the power of the b_{01} test, when mean systolic blood pressure x_1 in N. Glasgow is in fact 3 mm HG higher, is 0.916 (see *Program 2.25 BF Sample Size Selection, Models 1 and 2*).

Example 2.26 Comparing mortality rates Suppose our hypothesis is that two small areas s and t have the same mortality rates for coronary heart disease (CHD), that both have a population of 700, and observed CHD annual deaths rates per 100 000 population of $\gamma_s = 1500$ and $\gamma_t = 3000$. These rates correspond to multiples of one and two times the regional average (these multiples being denoted $\eta_s = 1$ and $\eta_t = 2$ respectively). We have a one-sided alternative to the null hypothesis of equality, namely that γ_t exceeds γ_s, and so wish to assess the power to detect that the death rate in area t is higher than that in area s (Thode, 1997).

Assume the populations exposed are p_s and p_t (in units of hundreds of thousands) then the observed deaths d_s and d_t will be Poisson with means $\lambda_s = p_s\gamma_s$ and $\lambda_t = p_t\gamma_t$. Possible test statistics comparing γ_t and γ_s are, following Thode (1997),

$$T_1 = (p_s d_t - p_t d_s)/(p_s p_t [d_s + d_t]^{0.5})$$
$$T_2 = (d_t/p_t - d_s/p_s)/[d_t/p_t^2 + d_s/p_s^2]^{0.5}$$
$$T^3 = (d_t - d_s)/(\lambda_s + \lambda_t)^{0.5}.$$

With a one-sided alternative we assess the probability that these statistics exceed $z_\alpha = 1.64$. The three tests give probabilities of 0.600, 0.600 and 0.57 respectively (see *Program 2.26 SMR Comparison*). In finding them one can either use the step function in BUGS to monitor the proportion of iterations where the test statistic T_1, T_2 or T_3 exceeds $z_\alpha = 1.64$; or the usual $1 - \Phi()$ where the argument of the cumulative normal is the difference between the actual T_1, T_2 or T_3 and the threshold 1.64. In the case of the latter the median of the values of $1 - \Phi()$ is close to the proportion using the step function. If we reduce the type I significance level α to 0.025, then our power will be diminished, with T_1 and T_2 yielding a power of 0.48.

If the populations of the two areas are doubled to 1400, the powers under a one-tailed alternative with $\alpha = 0.05$ rise to 0.846, 0.846 and 0.826 under T_1, T_2 and T_3 respectively. Suppose in fact there is a threefold mortality difference in two populations of 700 (i.e. set $\eta_t = 3$ in *SMR Comparison*, with η_s still at 1) such that $\gamma_s = 1500$ and $\gamma_t = 4500$. Since the effect size is increased, the power to detect the higher mortality in area t rises to around 0.96. If, by contrast, we assume a smaller effect size with $\eta_t = 2$ and $\eta_s = 1.5$ then the power under T_1 and T_2 falls to 0.13.

Example 2.27 Failure rate of aircraft components Shiue and Bain (1982) give an example comparing the frequency of failure for a new airplane component t with that of an existing component s (see *Program 2.27 Airplane Components*). The frequency of failures is to be assessed from h flying hours per plane among two fleets, one of $N_t = 10$ planes with the new component, the other with $N_s = 20$ planes and using the old component. Hence the exposures e_t and e_s are given by hN_t and hN_s. The respective failure rates are $\gamma_t = 0.02$ (2 per 100 flying hours) and $\gamma_s = \rho\gamma_t$ where ρ is expected to exceed 1. So the power relates to the required exposures to assess an effect size of this magnitude.

In statistics T_1 and T_2 of the previous example, the exposures e_s and e_t replace p_s and p_t. If, following Thode (1997) we assume $h = 85.7$ hours per plane then statistics T_1 and T_2 (as in example 2.26) give median powers under the cumulative normal method of 0.84 and 0.90 respectively, and 0.85 and 0.87 under the step function method. If, following Shiue and Bain (1982), we assume $h = 97.5$ and take all parameters as fixed (known already with certainty) the median powers under the cumulative normal method are 0.876 and 0.933 and under the step function method are 0.915 and 0.931. The second test implies a smaller sample size to attain a given power, though the differential is less marked using the step function approach.

The powers obtained are reduced if we specify a G(20, 10) prior on ρ and assume γ_s known, so that $\gamma_t = \gamma_s/\rho$. Power is further reduced if we allow uncertainty in both rates, e.g. using $\gamma_s \sim$ G(4, 100) and $\gamma_t \sim$ G(2, 100); to 0.66 under T_1 and 0.67 under T_2 using the step function method.

NOTES

1. For continuous data y about which prior information provides both mean and variance but nothing else, the principle of maximum entropy also leads to the assignment of a normal density (Sivia, 1996, Chapter 5).
2. The Bayesian version of the central limit theorem for a parameter vector θ is expressed in the multivariate normal approximation (see Section 2.8)

$$(\theta - \hat{\theta}) \sim N_p(0, V)$$

where V is the p $\times$p dispersion matrix for $\theta - \hat{\theta}$, with $\hat{\theta}$ regarded as a fixed quantity. This is based on a Taylor series of the log-likelihood $l(\theta|y) = \log L(\theta|y)$ about the maximum likelihood estimate $\hat{\theta}$,

$$l(\theta|y) = l(\hat{\theta}|y) + (\theta - \hat{\theta})^T S(\hat{\theta}|y) - 0.5(\theta - \hat{\theta})^T I(\hat{\theta}|y)(\theta - \hat{\theta}) + r(\theta|y)$$

where $S(\hat{\theta}|y)$ is the score function at $\hat{\theta}$ defined by

$$S(\hat{\theta}|y) = \delta l(\hat{\theta}|y)/\delta\theta,$$

and $I(\hat{\theta}|y)$ is the observed information defined by

$$I(\hat{\theta}|y) = \delta^2 l(\hat{\theta}|y)/\delta\theta'\delta\theta.$$

The value of $l(\hat{\theta}|y)$ is fixed and $S(\hat{\theta}|y) = 0$ by definition. So providing the remainder term $r(\theta|y)$ is negligible and the prior for θ is flat in the region of the ML estimate, the posterior distribution of θ has a density

$$p(\theta|y) \propto \exp[-0.5(\theta - \hat{\theta})^T I(\hat{\theta}|y)(\theta - \hat{\theta})].$$

This has the form of a of multivariate normal density of dimension p and with a $p \times p$ covariance matrix $V = I^{-1}(\hat{\theta}|y)$.

3. The exponent of this expression may be re-arranged as a sum of a quadratic function of μ, and terms not involving μ, namely as -0.5 times

$$\mu^2[\tau_0 + \tau] - 2\mu[\mu_0\tau_0 + y\tau] + \text{terms not involving } \mu.$$

Denoting the ratio of precisions as $n_0 = \tau_0/\tau$, the function of μ may in turn be expressed as

$$[n_0\tau + \tau]\{\mu^2 - 2\mu(n_0\mu_0\tau + y\tau)/(n_0\tau + \tau)\}$$

$$= [n_0\tau + \tau]\{\mu^2 - 2\mu(n_0\mu_0 + y)/(n_0 + 1)\}.$$

The latter term is equivalent to

$$[1/\{\sigma^2/(n_0 + 1)\}][\mu - \{(n_0\mu_0 + y)/(n_0 + 1)\}]^2 + \text{terms not involving } \mu$$

and this provides the terms in the exponent of a normal density for μ.

4. A prior with a and b both actually equal to zero, and with that on μ flat, results in an improper density

$$p(\mu, \sigma^2) \propto 1/\sigma^2.$$

This is often used as a standard reference (i.e. minimally informative) joint prior for the variance and mean. It is in fact equivalent to a density uniform over $(\mu, \log\sigma)$ (Gelman *et al.*, 1998). This particular form of joint prior for μ and σ^2 results in simplifications in posterior inferences:

(a) The marginal posterior density $p(\sigma^2|y)$ is an inverse gamma with $a = (n-1)/2$ and $b = (n-1)s^2/2$ where n is the number of data points and $s^2 = \sum_i (y_i - \bar{y})^2/(n-1)$ is the sample variance. It follows that $(n-1)s^2/\sigma^2 = (n-1)\tau s^2$ is a chi-square with $n-1$ degrees of freedom.

(b) The posterior density of μ given y is a t density with $n-1$ degrees of freedom, mean zero and scale 1. Similar simplifications hold when μ is replaced by the regression parameter β in the general linear model $y = x\beta + e$, where $e_i \sim N(0, \sigma^2)$ and the joint prior for (β, σ^2) is proportional to $1/\sigma^2$ (Tanner, 1996, Chapter 2).

5. X is the sum of the k Poisson variates and therefore is Poisson with mean $\sum \mu_j$. So the distribution of $x_1, \ldots, x_k$ conditional on X is

$$\begin{aligned} & p(x_1, \ldots, x_k)/p(X) \\ & = \exp(-\textstyle\sum \mu_j)\prod(\mu_j^{xj}/x_j!)/\{\exp(-\sum \mu_j)(\sum \mu_j)^X/X!) \\ & = X!\textstyle\prod(\theta_j^{nj}/n_j!) \end{aligned}$$

where $\theta_j = \mu_j/\sum \mu_j$.

CHAPTER 3

Models for Association and Classification

3.1 INTRODUCTION: ASSOCIATION AND CLASSIFICATION

Frequently, problems in social, health and natural sciences involve the estimation of associations between X and Y on the basis of a postulated causal link between the two. For example, in epidemiology the goal is to identify risk factors which are causal in the sense that exposure to them leads to an increased risk of disease, though susceptibility may vary between individuals. In the case where all individuals with a disease Y (e.g. heart disease) have been subject to a particular exposure X1 (e.g. smoking), but not all exposed individuals are diseased, we can say the risk factor is necessary but not sufficient for the disease. If the disease may appear for other reasons X2 (e.g. hereditary factors) then the factor X1 is neither sufficient nor necessary.

The complications of inferring causation from association have long been known to philosophers: for example J. S. Mill noted that the regularity of Y following X could be the consequence of a third variable or event Z preceding both X and Y. So the regularity of Y following X might only be due to confounding. Commonly adopted notions of causality among statisticians and other scientists in fact involve complex implicit assumptions. For example, the manipulability definition is essentially that X is a cause of Y if and only if Y can be changed by manipulating X and X alone. To establish that X is the cause of Y would then require an ideal experiment in which all other relevant variables were held fixed while the postulated causal variable alone is manipulated. The resulting change in Y if any would then have to be observed.

One aspect of such an ideal experiment would be to compare how the same individual subject exposed to X would have responded had they not been subject to X. So the idea of producing Y by doing X has a counterfactual aspect attached to it (Greenland, 1998). A related principle is that of reversibility: if a causal factor X is removed or diminished, then the effect Y should be reduced or eliminated. This has relevance in policy applications or health interventions where one might proceed on the assumption that if a factor X were reduced or diminished by a certain amount then an outcome Y would correspondingly be reduced.

The idea of controlled experiments such as randomised trials is that they attempt, by randomisation, to create treatment and control groups that are homogenous except in differing in their allocation to levels of the causal variable. Potential confounders and competing influences on the outcome are hopefully removed by random assignment of subjects to control and treatment groups. In medicine, clinical trials are 'the gold standard' scientific method used to compare the effect of two or more treatments on a health outcome – for example, to assess whether a new surgical technique or drug is more effective than conventional therapies. If sources of bias are controlled in a scientific manner then the differences between treatment success rates can be more clearly regarded as a causal consequence of the exposure or treatment. Bradford-Hill (1965) and others, such as Evans (1976) and Fletcher *et al.* (1982), have set out guidelines for inferring causality from association. These are partly based on the design adopted, partly on criteria such as temporal precedence (cause precedes effect) and the scientific plausibility of the association being causal. Compared with a trial, weaker levels of evidence for causality attach to case-control and many other types of observational studies.

However, not all potentially causal associations can be studied by a controlled experiment. The occurrence of disease, for example, may in part be related to remote causes such as environmental, behavioural and genetic factors, which occur early in the chain of causation of the disease. So there may be no clear pattern of exposure followed by incubation period. Necessary reliance on less than ideal designs in such circumstances then raises potential issues about causation mentioned above, such as the possibility of confounding or of multiple causation. In epidemiology the concern is often to establish the enhanced risk of a certain outcome associated with cumulative or historic exposure to a particular risk factor or predictor. An illustration is provided by studies into the relative risk of endometrial cancer associated with high-oestrogen contraceptive pill use (Rothman, 1986) A recent review has cast doubt on the original information based on observational studies suggesting that thrombo-embolism occurred twice as often among third generation pill users. Spitzer (1998) suggests a number of sources of bias which unduly elevated the observed estimate of risk.

Usually the assessment of a causal relationship between the outcome Y and the principal input variable X (allocated therapy or exposure) will be confounded by other possible influences Z on the outcome. A whole range of regression methodologies are designed to investigate these issues. However, observational and experimental design may also be used to the same end. This chapter deals with research methods based on the major distributions of Chapter 2 to assess hypotheses and estimate effects of individual risk factors (if necessary adjusted for confounders) without formally involving regression analysis.

There are a range of techniques used in the analysis of observational studies and clinical trials, with a simple example being measures of association in 2×2 tables. These are commonly based on randomisation in clinical trials, and the use of controls in observational studies, both intended to partial out the effect of confounding variables Z. We might then, for example, be interested in the odds ratio of a certain exposure as a putative risk factor, given the presence of an adverse outcome. Sometimes the values of the third variable are used as the basis for stratification of the population, and one aim then might be to assess the strength of association within strata and over all strata combined.

Classification using a diagnostic variable or screening test is an essentially similar problem: hypothesised cause X and effect Y are replaced by test X and condition Y. The screening tests or characteristics are used to classify individuals and we are trying to assess the accuracy of the predicted classification against the true state of the individual. Our test or classification should be such that the odds of in fact being affected are considerably higher among those assessed as high risk. Measuring their effectiveness or predictive strength (both among those actually affected and those in fact unaffected) is the main focus in comparative analysis of screening procedures. The assessment of risk (e.g. of disease, criminality, divorce) may be based on a single indicator such a rating scale. Often though we may have multidimensional measures to be combined in making a classification, or we want to compare the effectiveness of more than one screening tool.

3.2 ANALYSING ASSOCIATION IN OBSERVATIONAL STUDIES: FOLLOW-UP AND CASE CONTROL STUDIES

While the natural sciences may be often able to study causality using experimental methods, in the social and health sciences it is often the case that there are no unambiguous means to establish causality. A crucial distinction in trying to assess causation from observational studies may be drawn with regard to the time sequence with which the outcome and the influences on it are observed. In a prospective study we observe or choose a sample according to certain explanatory variables, and then follow up the selected subjects to see what response or outcome occurs. To increase validity, subjects may be matched at the start of the study on a range of potential influences on the outcome. This reduces the scope for confounding of the exposure with other influences on the outcome. For example, parental smoking and household irritants may be confounded as influences on childhood asthma prevalence; if the interest were focused on the effect of exposure to household irritants, then the exposed (high irritant level) group would be matched to the non-exposed group (low irritant level) in terms of having the same proportion of parents smoking. Such prospective studies are often termed follow-up, panel and cohort designs.

By contrast, in a retrospective study we observe or choose a sample according to the outcome of interest (e.g. whether a patient has an elevated blood chloresterol measure) and then investigate what values of potential explanatory variables had previously existed in the patient's history. In a sociological context, we might consider current income or social group in relation to social background; or we might examine educational achievement at age 16 in relation to earlier educational and social variables. In medical settings we may formally compare the case group (where the outcome, usually an adverse health event, has occurred) with the control group, where the outcome is absent. While some form of matching may be used in case-control studies, it may also introduce confounding if the matching is on factors (e.g. demographic and social) which are linked to the main exposure variable as well as to the outcome (Rothman, 1986).

Evidence in observational studies is often reduced to the form of a 2×2 table, cross-classifying two binary variables. In epidemiological studies, for example, we commonly want to assess the enhanced disease risk associated with a particular factor. The disease outcome may be unambiguous and naturally binary (e.g. death vs survival) but is often binary only by using a threshold point, for example survival more than or less than 30 months after the study's commencement.

In a follow-up study we are comparing the relative risk of disease between an 'exposed' group A and a non-exposed group B. These form two populations of size n_A and n_B, with possibly differing rates of incidence of the outcome, p_A and p_B. The number of outcomes in the two populations over the follow-up period are then described by two binomial distributions:

$$r_A \sim \text{Bin}(p_A, n_A)$$
$$r_B \sim \text{Bin}(p_B, n_B).$$

One possible null hypothesis is that $p_A = p_B = p$, i.e. that the risk factor is not associated with an enhanced incidence rate. Rothman (1986, p. 159) illustrates the ambiguity which can attach to common measures of association in this situation, for given totals n_A and n_B of exposed and non-exposed cases. However, ambiguity in the central estimate may be resolved to some extent by considering credible intervals for these association measures and Bayesian significance tests comparing these measures between samples.

Example 3.1 Comparing follow-up samples Rothman's example is a case of two samples, in both of which there are $n_A = 20$ exposed subjects and $n_B = 100$ non-exposed. In the first data set there are ten exposed cases and 45 unexposed cases. In the second there are 16 and 76 respectively. For the first data set, the central maximum likelihood estimates are $\hat{p}_A = 0.5$, and $\hat{p}_B = 0.45$; so the corresponding estimates of the risk difference, risk ratio and odds ratio in this data set are respectively

$$\hat{p}_A - \hat{p}_B = 0.5 - 0.45 = 0.05$$
$$\hat{p}_A/\hat{p}_B = 0.5/0.45 = 1.11$$
$$\hat{p}_A(1 - \hat{p}_B)/[\hat{p}_B(1 - \hat{p}_A)] = 1.22.$$

For the second data set, the risk difference and risk ratio are lower than for the first data set (namely 0.04 and 1.05) but the odds ratio is higher, namely 1.26, giving an apparent inconsistency.

If we set noninformative beta priors on the probabilities p_A and p_B, our sampling model first of all shows the wide uncertainty attached to these binomial probabilities, especially in view of the small number of exposed cases. This uncertainty in turn impacts on measures of association. *Program 3.1 Follow Up Study Example* combines the two data sets in one likelihood. After 5000 iterations, the 95% credible interval for p_A in the first data set is from 0.29 to 0.71 (with a mean of 0.5) and in the second from 0.6 to 0.94 (with a mean of 0.8).

We can then monitor the values of the above association measures themselves and assess their posterior distributions. In the first data set, the risk difference $\hat{p}_A - \hat{p}_B$ (RD) appears approximately normal with an average of 0.05 and a 95% credible interval from −0.18 to 0.28. However, the odds and risk ratios (RR and OR) tend to be positively skewed (i.e. non-normal): thus the 95% credible interval for the odds ratio in data set 1 is from 0.46 to 3.28 with an average of 1.39 but a median of 1.24. The latter figure is slightly above the maximum likelihood estimate: the median is a preferable estimate of location for a skewed distribution (the mean is distorted by a few extreme values).

It is preferable to monitor log(RR) and log(OR) and then exponentiate the mean of the distribution, since these log transforms better approximate normality (Ashby *et al.*, 1993). The credible interval for the odds ratio in data set 1 on this basis is still $(0.46, 3.28)$ but the mean is $\exp(0.203) = 1.23$.

In neither data set and for none of the association measures is there clear evidence of association between exposure and caseness, as would be shown by a credible interval of entirely positive or negative values for the risk difference, for log(OR), or for log(RR), or a credible interval entirely below or above 1, in the case of the OR and RR. We can also monitor the proportion of iterations for which log(OR), log (RR), or RD exceed zero (positive values indicate a positive association between exposure and incidence). Not surprisingly we get the same results from each of these quantities, since when one is positive so are the others; for example, the proportion of times which log(OR) exceeds zero is 0.67 in data set 1 and 0.68 in data set 2. A 'significant' positive association would require a proportion exceeding 0.90 or 0.95.

Since we have combined the two data sets in our likelihood we can compare the association measures between data sets 1 and 2. This relates to the question of ambiguity across the data sets in these measures. Thus $RD_1 - RD_2$ averages 0.0096 (roughly equivalent to $0.05 - 0.04 = 0.01$) but has a wide credible interval, from −0.28 to 0.32. The difference in log odds ratios $\log(RR_1) - \log(RR_2)$ averages 0.039 so RR_1 is on average 4% higher than RR_2, but the 95% interval for this difference is −0.56 to 0.58, so the ratio of RR_1 to RR_2 could be anything between 57:100 to 179:100.

Example 3.2 Congenital heart defects: Case control study In a case control study the two binomial populations are the numbers n_A in the case series and n_B in the control series. The number of cases with a positive exposure among the cases is then binomial with

$$r_A \sim \text{Bin}(p_A, n_A),$$

and the number of exposed controls is binomial with

$$r_B \sim \text{Bin}(p_B, n_B).$$

Rothman *et al.* (1979) quote data on a study where the cases were 390 mothers whose children were born with congenital heart defects, of whom four mothers had used chlordiazopoxide in early pregnancy; of the $n_B = 1254$ mothers in the control series (with normal children at birth), four also had a history of using this drug. With a non informative beta prior on p_A and p_B, the log odds relating 'caseness' and exposure is monitored over 10 000 iterations (see Program 3.2).

The log odds averages 1.162, giving an average for the OR of 3.22. The 95% credible interval is from 0.86 to 11.63 so there appears to be a slight doubt about a significant effect (which would require the 90 or 95% interval for the OR to be entirely above 1). This is in part due to the small numbers of cases, and despite this small number the posterior distribution of the OR consists predominantly of positive values. The proportion of iterations (25 000 to 50 000) for which the log(OR) exceeds zero is 96%, or 4% in terms of the two-tail 'significance level' that incidence is higher among exposed women. By contrast, Rothman (1986) finds the probability of four or more exposed cases as 0.096 (total one-tail P value) and 0.059 (one-tail mid-P value). He assumes a hypergeometric model appropriate to a contingency table with fixed margins (in preference to the two-binomial model).

Example 3.3 Leukaemia following Hodgkins disease Ashby *et al.* (1993) consider a case control study where the outcome (defining a case) is leukaemia following Hodgkins disease. The exposure suspected of being causal is chemotherapy as sole or partial treatment as against no exposure to chemotherapy. As mentioned above the log of the odds ratio (of chemotherapy given leukaemia) is often a 'better behaved' variable than the odds ratio itself. There are $n_A = 149$ cases of whom $r_A = 138$ had chemotherapy, and $n_B = 411$ controls of whom 251 were exposed to chemotherapy. The empirical value of the log of the odds ratio is $g = \log\{138 \times [411 - 251]/(251 \times [149 - 138])\}$, with precision $1/\text{var}(g)$. The observations $r_j, n_j - r_j$ in the cross-classification of exposure and caseness are such that the normal approximation will be adequate. We therefore assume this empirical value of the log odds is a draw from a normal density with unknown mean γ but known precision τ, equal to the inverse of the estimated variance. This is a case of a single observation from a normal distribution as considered in Chapter 2. The estimated variance is obtained via the formula

$$\begin{aligned}\hat{\text{var}}(g) &= 1/r_A + 1/[n_A - r_A] + 1/r_B + 1/[n_B - r_B] \\ &= 1/138 + 1/11 + 1/251 + 1/160. \end{aligned} \tag{3.1}$$

With a non informative prior on γ, namely $\gamma \sim \text{N}(0, 100)$, we obtain a posterior mean for the odds ratio $\text{OR} = \exp(\gamma)$ of 8, from the second half of a run of 100 000 iterations. Exponentiating the 95% credible interval estimates for γ gives an interval on the OR of 4.2 to 15.1.

A Bayesian analysis enables one to use informative prior information when it is available. Ashby *et al.* use results from a cohort study by Kaldor *et al.* (1987) which reported a value for $g = \log(\text{OR})$ of 2.361 with variance 1/106. This forms an informative normal prior with mean 2.361 and precision 106. In this case the prior tends to dominate the data, and the posterior mean for $\text{OR} = \exp(\gamma)$ is $\exp(2.34) = 10.4$ with a 95 % interval from 8.65 to 12.42.

Example 3.4 Follow-up study normal approximation With sufficient cell totals in a follow-up study the normal approximation may also be applied, though with a different formula for the variance estimate. Thus suppose r_1 out of n_1 exposed subjects develop a disease in the follow-up period, as against r_2 of n_2 non-exposed subjects, and that $s_1 = n_1 - r_1$, $s_2 = r_2 - n_2$. Then following Katz *et al.* (1978) the estimated variance of the log relative risk is

$$\hat{\text{var}}(\ln\text{RR}) = s_1/(r_1 \times n_1) + s_2/(r_2 \times n_2). \tag{3.2}$$

Consider data from the Framingham study on the development of myocardial infarction (MI) among men in a follow-up period of 16 years. Of 135 men with serum cholesterol over 250 mg %, 10 developed MI while of 470 men with measured cholesterol under 250 mg %, only 21 developed MI.

Using the full binomial likelihood we find the RR has median value of 1.69 and mean 1.80, with 95 % credible interval from 0.81 to 3.34. This is close to the range {0.79, 3.48} cited by Kahn and Sempos (1989) using a quadratic approximation. It is of interest here to compare the odds ratio and relative risk for this relatively rare outcome: the OR has a 95 % interval {0.80, 3.66} with median 1.75.

The normal approximation here takes the single datum $x = \log(10/135 \div 21/470)$ as normal with unknown mean but variance given by (3.2). The estimated RR then has 95 % credible limits 0.81 and 3.45, with mean 1.78 and median 1.66.

3.2.1 Simulating Controls through Historic Exposure

The facility for a Bayesian approach to incorporate existing knowledge extends to situations where only data on cases may be available. Usually the goal of a case-control study is to accumulate a set of cases and investigate whether their exposure to a suspected causal factor is unusual. The control group is used to derive the posterior distribution of exposure. Zelen and Parker (1986) argue that in some cases there may be extensive information about exposure levels in the population (e.g. on average levels of health behaviours or use of common drugs). This knowledge can be used to set an informative prior for the exposure in the control group. In this situation collecting data from a control group may be unnecessary, in that the posterior distribution of exposure will typically be very similar to the prior distribution.

Let Y denote the dichotomous variable which has value 1 for cases and zero for controls, and X denote whether or not the individual is exposed to the causal agent. Then the case control study considers the distribution

$$f(X|Y) = \mathrm{e}^{\alpha X + \beta YX}/(1 + \mathrm{e}^{\alpha + \beta Y}) \tag{3.3}$$

where X and Y are independent only if $\beta = 0$. Suppose the observed data are as given in Table 3.1.
Then the likelihood (3.3) is

$$L(s, r, c, m|\alpha, \beta) = \mathrm{e}^{(\alpha+\beta)s+\alpha r}/\{(1 + \mathrm{e}^{\alpha})^m(1 + \mathrm{e}^{\alpha+\beta})^c.$$

The conjugate prior $\pi(a, b)$ has the same form and relates to equivalent 'prior data' s', r', c' and m'. Then the posterior is

$$f(\alpha, \beta|s, r, c, m) \propto \mathrm{e}^{(\alpha+\beta)S+\alpha R}/\{(1 + \mathrm{e}^{\alpha})^M(1 + \mathrm{e}^{\alpha+\beta})^C\}$$

where $S = s + s', R = r + r', C = c + c'$ and $M = m + m'$ and $T = C + M$.

Zelen and Parker propose a method to derive 'prior data' for controls, namely r' exposed individuals among m' individuals without the disease or condition. These simulated control data constitute the entire control data in the analysis (i.e. so $m = r = 0$ and $M = m'$, $R = r'$) and are based solely on knowledge about population exposure. For example suppose $A = 30\%$ of a nation's female population are smokers, then

$$r'/m' = 0.3. \tag{3.4}$$

Suppose the probability that this proportion exceeds $H = 35\%$ is put at 0.05. Then using the normal approximation

$$\log[A/(1 - A)] + 1.64\sigma = \log[H/(1 - H)]. \tag{3.5}$$

σ is also derivable from the formula

$$\sigma^2 = [1/r' + 1/(m' - r')]. \tag{3.6}$$

So using the value of σ derived from (3.5) we can solve (3.4) and (3.6) to provide prior values for r' and m'.

Example 3.5 Adenocarcinoma in young women Herbst *et al.* (1971) report on cases of adenocarcinoma of the vagina in eight young US women, seven of whom had been exposed *in utero* to a drug intended to prevent pregnancy complications. Use of this drug (diethylstilbestrol, DES) in a pregnancy was indicated for threatened abortion (e.g. when repeated bleeding in pregnancy had been observed). Historical data indicated a maximum exposure rate of 10 % of the population: circa 10 % of women were subject to such complications so this provides a (maximal) possible exposure rate to DES. A prior mean

Table 3.1

	Exposed	Non-exposed	Total
Cases	s	$c - s$	c
Controls	r	$m - r$	m
Total	e	n	t

exposure of 10% with an upper limit of 20% is likely to overstate exposure to DES, so we expect only a 0.05 chance that exposure exceeds 20%. Using the normal approximation with $A = 10\%$ and $H = 20\%$ gives $r' = 4.6$ and $m' = 45.7$.

Using the actual case data and simulated control data gives an estimated log-odds ratio (beta in Program 3.5) of 4.6 with 95% interval from 2.58 to 4.45. This compares closely with the normal approximation, namely

$$\beta_{\text{NA}} = \log(SR) - \log\{(N - S)(M - R)\} = 4.14$$

with a standard deviation

$$(1/S + 1/(T - S) + 1/R + 1/(T - R))^{0.5} = 1.17.$$

Zelen and Parker use the normal approximation to assess the strength of evidence in favour of $\beta = 0$. The posterior probability ratio is based on comparing the probability of $\beta = 0$ against the probability that an observed value of β_{NA} would occur if the actual value of β were zero. A hypothesized value of $\beta = 0$ corresponds to a normal deviate of $Z = 4.14/1.17 = 3.54$ and an ordinate 0.000 82. Therefore the value of the posterior probability ratio is $0.399/0.00082 = 486$. The numerator is simply $1/(2\pi)^{0.5}$, the ordinate corresponding to β actually equalling 0. This provides overwhelming evidence in favour of an association between the outcome and exposure to DES *in utero*.

3.2.2 Accounting for stratification in study design

Stratification is included in a study design or analysis to counter confounding and effect modification (Rothman, 1986). Confounding variables are associated (not necessarily causally) both with the outcome and the risk factor. For example, in Table 3.2, age of woman is associated causally with an increased risk of heart attack (myocardial infarction), because risk of this medical event usually increases with age. But for these data, age is also associated with the pattern of oral contraceptive usage, in that 38 out of 124 younger women use this method (31%), but only 12% of older women; insofar as controls represent the common 'source population' which also gave rise to the cases, the discrepancy in usage is 22% among younger controls (17 from 76)

Table 3.2 Case control study: Oral contraceptives use and myocardial infarction, by maternal age group.

	Age under 40		Age 40–44		All ages	
	User	Nonuser	User	Nonuser	User	Nonuser
Cases of infarction	21	26	18	88	39	114
Controls	17	59	7	95	24	154
Odds ratio	2.8		2.8		2.2	

vs 7% among older controls. Confounding will be apparent in differences in the odds ratios OR_k specific to levels k of the confounder and the overall odds ratio formed by aggegating the data over k. The maximum likelihood estimate of the latter is $(39 \times 154)/(114 \times 24) = 2.2$, but the estimates for the age-specific tables are both 2.8.

Another example is provided by Bishop *et al.* (1975, p. 41), in which clinic is a confounder in the relationship between amount of prenatal care and mortality. The overall mortality rates differ between the two clinics (i.e. clinic is associated with the outcome), and clinic is also associated with prenatal care, in that clinic A has a higher proportion of 'more' prenatal care than clinic B. The odds ratio relating mortality and prenatal care in the collapsed table is $(20 \times 316)/(6 \times 373) = 2.8$ but in the separate tables is only 1.2 and 1.0 (Table 3.3).

Example 3.6 Smoking, tenure and CHD We need a method to combine the separate OR_k or RR_k which avoids the pitfalls of collapsing the data itself over the confounder. The goal is a summary measure of the association between exposure and outcome which is not distorted by the confounder. We may assume the log(OR) for each level of the confounder is taken from an underlying normal (as in Example 3.3), and use the inverse of the empirical variance formula (3.1) as the precision. In line with the procedure suggested by Mantel and Haenszel (1959), we then assume the underlying normal has the same mean γ_O over strata – this provides an estimated pooled mean for the $\log(OR_k)$ which can then be exponentiated to provide an estimate of the pooled OR. A summary estimate of the relative risk in follow-up studies can be made on the same lines (Tarone, 1981), using the normal approximation approach of Example 3.4 to derive a common mean γ_R for the $\log(RR_k)$.

We consider follow-up data for males from the Scottish Heart Health Study, presented by Woodward (1999, p. 169). This considers smoking status (Z) as a confounder in the relationship between the risk factor, household tenure (X) and the outcome, coronary heart disease (Y). Household renting as against ownership is considered as the exposure. Of 1770 non-smoking owner occupiers, 48 had CHD (2.7%) as compared with 33 out of 956 non-smoking renters (3.5%). Among 707 smoker owner-occupiers, 29 or 4% had CHD while among

Table 3.3 Mortality by clinic and care received.

Place where care received	Amount of prenatal care	Infant survival		Total	Mortality Rate (%)
		Died	Survived		
Clinic A	Less	3	176	179	1.7
	More	4	293	297	1.4
Clinic B	Less	17	197	214	7.9
	More	2	23	25	8.0
Both clinics	Less	20	373	393	5.1
	More	6	316	322	1.9

Table 3.4

	Mean	SD	2.5%	Median	97.5%
RR	1.32	0.21	0.96	1.30	1.78
γ_R	0.26	0.16	−0.04	0.26	0.58
OR	1.34	0.22	0.96	1.32	1.82
γ_O	0.28	0.16	−0.04	0.28	0.60

the 950 smokers who also rented their housing, 52 or 5.5% had CHD. For these data we can calculate both odds ratios and relative risks within the subtables formed by smoking status. Using the Mantel–Haenszel assumption for the odds ratio and relative risk leads to the estimates given in Table 3.4 for γ_O and γ_R and their exponentials.

3.3.3 Attributable Risk: Effects of eliminating postulated causes

Attributable risk is used to measure the health consequences of an association between a risk factor and a disease, and so assess the potential consequences of health prevention programs which aim to reduce or eliminate the risk factor from the population (Northridge, 1995). Similar approaches may be envisaged in social interventions. Attributable risk combines in a single measure the following concepts (a) the relative risk RR of a disease D following exposure to the risk factor, and (b) P, the proportion of the total population exposed to the risk factor.

There are alternative definitions of attributable risk, but they generally reduce to *either* attributable risk in the exposed subpopulation (ARE) *or* attributable risk in the total population (ARP). The first definition (the attributable risk fraction in the exposed subpopulation) measures the proportion of new disease cases which is attributable to exposure to the risk factor:

$$\text{ARE} = \frac{\text{Risk (or rate) for exposed} - \text{Risk (or rate) for unexposed}}{\text{Risk (or rate) for exposed}}.$$

Let I_1 be the incidence rate among those exposed and I_0 be the incidence rate among those not exposed, so that $\text{RR} = I_1/I_0$. Then this form of attributable risk can be written

$$\text{ARE} = (I_1 - I_0)/I_1.$$

If we divide both numerator and denominator by I_0 we obtain the proportion of the incidence rate, among those with the risk factor, which is due to the risk factor. This is simply

$$\text{ARE} = (\text{RR} - 1)/\text{RR}$$

as long as $\text{RR} > 1$. If $\text{RR} < 1$, the ARE will lie between 0 and 1, and will increase as the strength of the association between the risk factor and the disease increases.

The incidence rate due to the risk factor among those exposed to it is then

$$I_1(\mathrm{RR} - 1)/\mathrm{RR}.$$

So if the population is of size N, then NP persons will be exposed, and NPI_1 new cases will occur among these exposed persons. Then the number of new cases of the disease due to (associated with or caused by) the risk factor is

$$(NPI_1)(\mathrm{ARE}) = NPI_1\mathrm{RR}/(1 - \mathrm{RR}).$$

The attributable risk in the population, ARP, is defined as new cases related to the risk factor divided by all new cases

$$\mathrm{ARP} = \frac{NPI_1\mathrm{RR}/(1 - \mathrm{RR})}{NPI_1 + N(1 - P)I_0}$$

where $N(1 - P)I_0$ is the total of new cases among those not exposed. Dividing by NI_0 gives

$$\mathrm{ARP} = \frac{P(\mathrm{RR} - 1)}{1 + P(\mathrm{RR} - 1)}.$$

This definition of attributable risk increases both with the relative risk and with the population exposure rate, P. It has a zero value when the relative risk is 1 or there is no population exposure ($P = 0$).

As an example of the first formula (the ARE) may be cited the Doll and Hill study of lung cancer and smoking among British doctors. The incidence of lung cancer among smokers was 133 deaths for 102 600 person-years of risk (Doll and Hill, 1964), giving an annual mean rate per 1000 persons of 1.30. The incidence rate for lung cancer among nonsmokers was 0.07 per 1000 person-years. Therefore the attributable risk is $(I_1 - I_0)/I_1 = (1.30 - 0.07)/1.30 = 0.95$. Thus 95% of lung cancer deaths in the exposed group are attributable to smoking. This estimate will be subject to uncertainty because it is based on a relatively small number of deaths, especially among the nonsmokers.

Estimates using the second definition will additionally be influenced by the threshold for deciding exposure. For example, the ARP for lung cancer associated with smoking more than 60 cigarettes a day will be smaller than the ARP associated with smoking more than 5 cigarettes per day.

The ARP can be derived from both case-control and cohort studies, as well as from cross-sectional studies. For case-control studies P has to be estimated from the proportion exposed among the controls, and the estimate of the ARP rests on using the 'rare disease' assumption under which odds ratios approximate relative risks, giving

$$\mathrm{ARP} = P(\mathrm{OR} - 1)/[1 + P(\mathrm{OR} - 1)]. \tag{3.7}$$

Alternatively, the ARP can be written in terms of the probability $\Pr(E|D)$ of exposure given disease, namely

$$\mathrm{ARP} = \Pr(E|D)(\mathrm{OR} - 1)/\mathrm{OR}. \tag{3.8}$$

This probability is estimable from the proportion of cases who were exposed, while the RR can again be approximated by the odds ratio using the rare disease assumption.

Example 3.7 Lung cancer mortality among US veterans A prospective study reported by Kahn (1966) compared death rates occurring in 701 768 person-years observed for a set of current cigarette smokers among US veterans and 1 015 999 person-years for nonsmokers. Observed deaths in the two groups – over the follow-up period – are 1116 and 426 respectively. Denoting the two death totals as O_j ($j = 1$ for smokers, $j = 2$ for nonsmokers) and person-years exposed as Y_j, we adopt a Poisson model for the death counts (see *Program 3.7 Attributable Risk Lung Cancer*):

$$O_j \sim \text{Poi}(\gamma_j)$$
$$\gamma_j = \mu_j Y_j.$$

Non informative priors are used for the unknown mean rates μ_j. This yields a posterior mean for the ARE of 0.736 with 95 % credible interval from 0.705 to 0.764. This estimate may be contrasted with the earlier population study of Doll and Hill (1964). The probability of exposure can be modelled as binomial and using the person-year totals we obtain

$$Y_1 \sim \text{Bin}(p, Y)$$

where $Y = Y_1 + Y_2$. A vague Beta$(1, 1)$ prior is used for p, though in fact exposure levels (i.e. population smoking rates) are reported from several public surveys. The ARP is then estimated as 0.533 with 95 % credible interval from 0.49 to 0.57. Thus just over a half of the lung cancer cases in the US veteran population are associated with cigarette smoking.

Example 3.8 Birth defects and paternal smoking As a case-control example, consider data reported by Zhang *et al.* (1992) on birth defects and paternal smoking in Shanghai in 1986–87. For 1012 infants with defects, 639 had fathers who smoked, whereas among 1012 controls (selected from problem-free births) there were 593 paternal smokers.

To estimate attributable risk in the population we need to establish (a) that birth defects are rare and (b) that control selection is representative of the population of all births. Since all deliveries in Shanghai were hospital based, we can be relatively sure about (b), while the evidence during the observation period was that around 2 % of births had defects (Woodward, 1999). We can estimate the ARP by either of the above formulae, with a resulting estimate that 10.8 % of birth defects are attributable to smoking by fathers, with a 95 % credible interval from 0.7 % to 20.2 %.

3.4 CONTROLLED EXPERIMENTS AND TRIALS

A controlled trial or experimental design typically involves the comparison of one or more groups of subjects. These groups are differentiated in terms of intervention or risk they are exposed to (e.g. new vs standard medical treatment, or trial drug vs placebo). The goal of the trial is to assess particular probabilities such as that a new treatment is more effective than an existing therapy, or more effective by a certain margin. Features of the Bayesian method have a particular relevance in such trials, including the use of existing or elicited information from specialists and previous trials, and the role of updating from prior to posterior in framing predictions which might (for example) indicate early cessation of the trial.

The researcher exercises control over confounding factors in a trial by assigning the subjects to the groups in a randomised fashion. The randomisation is intended to control for the range of extraneous influences on the outcome other than the input variable (e.g. treatment type) of interest. Ideally it should involve double masking or blinding in the sense that neither researcher nor subject knows which treatment is being applied to which subject. It may still be necessary to check the characteristics of the groups to establish comparability; for example, in medical applications this comparison would be in terms of major prognostic variables, and chance imbalances would be corrected by a secondary regression analysis (Grant, 1989). Moreover the study must be large enough to have sufficient power to establish the effect or hypothesis of interest. A major problem with experiments which have too few subjects is that they ascribe real differences between treatments to chance (a type II error, when the null hypothesis is 'accepted' but is in fact not true) (Pocock *et al.*, 1987).

Other experimental options are possible. For example, both the new and control treatment may be applied to the same subject but at different times, with the order of application decided randomly. This is known as a cross-over design with the subject forming his or her own control; this lessens the effect of variability between subjects and may require a smaller set of subjects to confirm the hypothesis.

The Bayesian approach to controlled experiments, particularly randomised clinical trials, has been discussed in several studies. For example, Bayesian inference may provide a more flexible basis for interim analysis, namely applying an early stopping rule in a trial (Fayers *et al.*, 1997). In many trials length of survival or improved prognosis may be a central outcome measure, and if early results provides strong evidence for or against the intervention being tested then it would be unethical to continue recruiting patients to the trial. If the new treatment is clearly beneficial, then it is unethical to withhold it from the control group, while if it has unexpected harmful side-effects then it would be unethical to continue administering it to the treatment group. The frequentist approach to testing whether a trial should be prematurely terminated is based on a number of looks at the data as it is being accumulated, with testing for significance at a specified level (Pocock, 1982). The Bayesian approach to interim analysis facilitates continuous monitoring of the data as it accumulates and posterior inferences are modified, with the decision to continue in principle being sequential; after each new observation, a decision is made whether to continue or not (Berry, 1993).

Bayesian predictions of likely success or survival rates in as yet unrecruited patients also provide an approach to stopping. Suppose two treatments, new and standard, are being compared, and the sizes of the experimental groups are the same, with patients being randomised in pairs. A 'success' occurs if the new-treatment patients survive longer than those on the standard treatment. Then for the new treatment to be judged successful its success rate will need to exceed 0.5. Data monitoring to assess stopping or continuation would continually examine the success rates of the two treatments. Thus if s successes have been observed in n patient pairs, and indicate a success rate exceeding 0.5, then the relevant prediction relates to the number of successes in m new patient pairs and the overall probability of success in the total of $T = n + m$ pairs. The minimum acceptable number of observations $T = n + m$ may have been decided beforehand in terms either of achieving an agreed power or clinical credibility, or in terms of cost acceptability.

Since the randomised trial approximates the best scientific method, cumulation of evidence from relevant previous trails should be summarised in the prior for the current trial. This leads on to questions of combining information from studies, for example via meta-analysis (Chapter 5). Only if there is no relevant information available should a flat prior be chosen by default. The circumstances of previous trials may differ in terms of design features, patient mix and so on, and may need to be downweighted or partially discounted in terms of their applicability to the current trial. The Bayesian approach may then be advantageous in facilitating a form of sensitivity analysis based specifying a range of priors. Spiegelhalter *et al.* (1994) and Fayers *et al.* (1997) discuss how to specify

(a) reference or neutral priors;
(b) clinical priors based on elicitation;
(c) priors sceptical about the benefits of the new treatment, and so less favourable towards it than the reference prior;
(d) enthusiastic priors, erring towards belief in the efficacy of the new treatment and more favourable towards it than the reference prior.

Example 3.9 Survival in acute leukaemia Berry (1993) discusses an example where 21 pairs of patients were entered into a trial of a new drug for patients with acute leukaemia; the drug was called 6–mercaptopurine (abbreviated as 6–MP). In $s = 18$ out of the $n = 21$ pairs the patient stayed in remission longer under 6–MP than under a placebo. So the outcome is a dichotomy, although actual times of remission may also be considered as an outcome. Hence the probability of success with 6–MP, denoted p, is distributed

$$18 \sim \text{Bin}(p, 21).$$

We adopt first a standard reference prior for p, namely $p \sim \text{Beta}(1, 1)$. The posterior probability that 6–MP is better than the placebo is the probability that p exceeds 0.5, which is 0.999. This probability is monitored by the comparison `test.6MP` in Program 3.9 and is equivalent to the probability that 6–MP is better than a placebo for a randomly selected future patient. The predictive probability that a future patient will stay in remission longer under 6–MP is then sampled as `s . new` $\sim$ `Bernoulli(p)`, and the mean of `s.new` is estimated as 0.8256.

An alternative sceptical prior is framed in terms of a 'point' hypothesis that $p = 0.5$. We assume *a priori* that $p = 0.5$ with probability $\pi = 0.7$. The remaining probability $1 - \pi$ is spread uniformly over $(0, 1)$. Then the posterior probability that $p = 0.5$ is

$$W = \pi\text{Prob}(s, n-s, p = 0.5)/\{\pi\text{Pr}(s, n-s, p = 0.5) + (1-\pi)\int_0^1 u^s(1-u)^{n-s}du\}$$

$$= 0.7(0.5)^n/\{0.7(0.5)^n + 0.3\int_0^1 u^s(1-u)^{n-s}du\}.$$

We use Simpson's rule to evaluate the integral and find $W = 0.0315$. Under this sceptical prior, the predictive probability of success is 0.5. So the unconditional predictive probability of success (`sceptic.6MP` in Program 3.9), averaging over the two priors, is

$$0.0315(0.5) + 0.9685(\widehat{\texttt{s.new}})$$

A run of 100 000 iterations (discarding the 1st 5000) gives an average unconditional predictive probability of success of 0.8153.

Example 3.10 Bladder cancer therapy Suppose p_a and p_b denote probabilities of effective therapy under new and standard therapies respectively. Then the null comparison typically relates to $\delta = p_a - p_b$. Spiegelhalter and Freedman (1986) propose an alternative to the usual null health gain δ as the basis for a power calculation to establish whether a new treatment is more effective than an established therapy. Thus power is normally understood to be the probability of rejecting the hypothesis $H_0{:}\delta_A = p_a - p_b = 0$ at a specified significance level. They suggest instead that the 'minimum clinically important difference' be used as the base for the null hypothesis. This difference is the smallest clinically worthwhile improvement δ_w needed to recommend a new treatment and could be established on the basis of prior elicitation from relevant specialists. The distribution $\pi_1(\delta_w)$ reflects the demands that they have about the new therapy in order to adopt it.

A new treatment or drug may show a slight improvement but have disadvantages (in wider efficiency terms and not just health effectiveness), for example in cost, toxicity, or patient compliance, which in practice reduce the true gain. There might be a range of equivalence (in terms of net gain of adopting the new treatment), only which when exceeded is the new treatment adopted. Alternatively a single-point summary δ_{w0} from $\pi_1(\delta_w)$ may be used, and the power comparison is then against the null hypothesis $H_0{:}\delta_A = \delta_{w0}$. The specialists will also have a range of opinions about what gain a new therapy will actually show; these form the prior $\pi_2(\delta)$ on δ.

Spiegelhalter and Freedman (1986) compare an established surgical intervention for bladder cancer with a new therapy combining surgery with instillation of a drug called Thiotepa. The health gain δ criterion is the increase in two-year cancer recurrence-free rate, namely the difference in two-year recurrence rates after each treatment has been given to n patients. If the recurrence-free rates under the established and new therapies are $p.b$ and $p.a = p.b + \delta$ respectively then the number of patients still cancer free after two years will be distributed as $n.b \sim \text{Bin}(p.b, n)$ and $n.a \sim \text{Bin}(p.a, n)$ respectively. The test statistic D is the observed difference $n.a/n - n.b/n$ divided by its standard deviation. For large n the difference is approximately normal with variance $\text{Var}(D) = \sigma^2/n$ where

$$\sigma^2 = p.b(1 - p.b)/n + p.a(1 - p.a)/n.$$

The recurrence-free rate $p.b$ under the surgery-alone treatment for bladder cancer is known to be 50%. Plausible values $\pi_2(\delta)$ for expected gain δ were assessed from a sample of 18 consultants, and average $\delta = 0.08$; see Figure 1 in Spiegelhalter and Freedman (1986). The minimum expected health gain δ by these specialists was -0.05 and the maximum 0.3, and the prior may be summarised in terms of a grid of eight discrete values (see Program 3.10).

For a 5% significance level (type I error) we compare the test statistic $(D - \delta_A)/\text{SE}(D)$ with 1.96. Spiegelhalter and Freedman (1986) suggest a single-point prior (with probability of 1) on the value of $\delta = 0.10$, as a known value for the 'true difference'. This is close to the average of the specialist opinions. For the usual $\delta_{\text{A}} = 0$ null hypothesis and for a sample size of $n = 150$ the power against this null hypothesis and in favour of the new treatment is 0.42. A sample of around 380 is needed to give an 80% power. However, the average health gain demanded by the specialists to make them adopt the new therapy was $\delta_{w0} = 0.05$ (i.e a 5-point gain in the percent of patients cancer free after two years). With a sample of 150 the power against the null hypothesis $H_0{:}\delta_{\text{A}} = \delta_{w0}$ and in favour of the true value is only 0.13, and about 1550 patients are needed for 80% power.

Example 3.11 Survival times and interim analysis George *et al.* (1994) and Berry (1995) report on a series of interim analyses on a phase III clinical trial for lung cancer treatment. The design involved random assignment of 240 patients as they became eligible, either to radiotherapy alone (RT), or to a combination of radiotherapy and chemotherapy (RT + CT). The trial was in fact stopped at the 5th interim analysis, when information was available on 105 patients (51 on RT and 54 on RT + CT), though 155 had been entered by this time. The basis of trial curtailment involved differential mortality rates λ_{1t} for RT and λ_{2t} for RT + CT at each of the five interim analyses and subsequently in 1992, when a follow-up on all 155 patients was made (so providing $t = 1, \ldots, 6$ time points). Under exponential survival the inverse of these rates provides an estimated average survival time under the two treatments.

In fact the results suggest an initial survival time advantage for the combined therapy which was, however, diminishing by the later interim analyses. George *et al.* used a $\text{G}(2, 20)$ prior on λ_{1t} and then rather than additionally taking λ_{2t} as a parameter, adopt a prior for the difference in rates $\delta_t = \log(\lambda_{2t}/\lambda_{1t})$.

Specifically they assume a standard normal. However, we would expect the precision of δ_t to be related to the sample sizes (number of patients, n_{1t} and n_{2t}) in the two arms of the trial; so here we take a prior

$$\delta_t \sim \mathrm{N}(0, 1/\tau_t)$$
$$\tau_t = \gamma(n_{1t} + n_{2t})$$

where γ is assigned a flat gamma prior. With this approach we find the average survival times S_{1t} and S_{2t} under the two treatments at each of the time points. We also find the posterior probabilities that the δ_t are less than three possible thresholds, namely 0, –0.25 and –0.5 (Table 3.5). The latter criterion (δ_t under –0.5) places more stringent demands on evidence of the relative benefit of the combined therapy. We find that by 1992 the probability is under 0.5 that $P(\delta < -0.5)$. On the other hand, there is a 0.996 probability of some survival advantage, i.e.$\delta < 0$ for RT + CT.

3.5 CLASSIFICATION, SCREENING AND DIAGNOSTIC ACCURACY

Often it is necessary to determine whether a characteristic or condition exists in a subject on the basis of a screening procedure, or set of indicators. As noted

Table 3.5

Analysis		Probability			
		$d < 0$	$d < -0.25$	$d < -0.5$	
1		0.95	0.89	0.77	
2		0.99	0.96	0.88	
3		0.98	0.89	0.68	
4		0.98	0.88	0.60	
5		0.99	0.89	0.57	
6		1.00	0.90	0.44	
Survival times by treatment and analysis					
RT		Mean	St devn	2.5%	97.5%
	1	19.8	7.1	9.8	37.4
	2	23.0	6.9	13.2	40.1
	3	15.9	3.5	10.5	23.9
	4	16.5	3.4	11.2	24.2
	5	14.5	2.6	10.3	20.2
	6	16.4	2.0	13.0	20.6
RT + CT					
	1	53.7	32.7	20.7	139.2
	2	71.8	35.4	31.4	161.1
	3	30.4	8.1	18.7	49.6
	4	29.2	6.8	18.6	45.2
	5	25.2	5.0	17.2	36.9
	6	26.5	3.3	20.9	33.9
γ		0.023	0.022	0.003	0.077

above, a screening test is broadly analogous to using a predictor (the test) with outcome being the condition. We wish to classify subjects into one or more categories of the characteristic D using the information provided by the indicators or test. We also wish to establish a decision rule with the present set of subjects so that future subjects can be classified correctly.

In the simplest situation, the characteristic is binary (e.g. does a person have a disease or not), with outcomes D and $\bar{\mathrm{D}}$ and the test result is also binary. The results of the test are denoted P and N, positive and negative, with the positive result indicating the characteristic is present.

The sensitivity of a test is the probability the test will give a positive result given that the condition (e.g. disease) is present. Let $\Pr(\mathrm{D}) = \pi$ denote the probability that an individual drawn at random from the population has the characteristic. For example, in epidemiology this would be known as the prevalence of the disease in the population. Then $\Pr(\mathrm{P}|\mathrm{D}) = \eta$ is the sensitivity of the test, and $\Pr(\mathrm{P}|\mathrm{D})\Pr(\mathrm{D}) = \eta\pi$ is the joint probability of having the condition and being identified as such. We may also be interested in the probability that the test correctly identifies that an individual is disease free. So given $\bar{\mathrm{D}}$ is true, $\Pr(\mathrm{N}|\bar{\mathrm{D}}) = \theta$ is the probability the test will say they are disease free. This is called the specificity of the test.

We are also interested in how far the predictions of positive or negative tests actually match reality. So given that a test says the individual is diseased, the probability is required that the individual is actually diseased. This probability, $\Pr(\mathrm{D}\,|\,\mathrm{P}) = \Psi$ is known as the predictive value of a positive test. The probability $\Pr(\bar{\mathrm{D}}\,|\,\mathrm{N}) = \Lambda$ is the predictive value of a negative test, and is the chance that an individual classed as free of the disease actually is.

Thus Gastwirth *et al.* (1991) consider the prevalence of AIDS in the population from the viewpoint of screening out donated blood which is infected (i.e. contains antibodies to the HIV virus). In this case $1 - \Lambda$ is the probability that an individual classed as HIV free, and hence believed to be giving uninfected blood, is in fact donating infected blood.

Geisser (1993) considers the situation where two tests are available for a binary outcome, and the possible decision rules which may be adopted to combine the information they provide. Suppose we are given information about the costs consequent on the four possible outcomes of a test as against actual outcome (e.g. in terms of costs of incorrect treatments following mistaken diagnosis). We can then evaluate which is the best decision rule. The situation may be represented in terms of a loss function associated with the outcomes as shown in Table 3.6.

A generalisation of the above notation regarding detection of a condition by a single test may also be used. Thus η_{11} is the probability $\Pr(\mathrm{P}_1, \mathrm{P}_2\,|\,\mathrm{D})$, when the condition D is present, that both tests detect it. η_{10} and η_{01} are the probabilities that the first test alone and the second test alone detect it when it is present, while η_{00} is the chance $\Pr(\mathrm{N}_1, \mathrm{N}_2\,|\,\mathrm{D})$ that neither test detects the condition when it is present. Hence $\eta_{11} + \eta_{10} + \eta_{01} + \eta_{00} = 1$. When the condition is absent, then θ_{00} denotes the probability $\Pr(\mathrm{N}_1, \mathrm{N}_2\,|\,\bar{\mathrm{D}})$ that both tests find it to be absent. Analogous notation follows where either one or both

Table 3.6 Loss function.

	Condition	
Test	Present (D)	Absent ($\bar{D}$)
Positive (P)	L_{11}	L_{10}
Negative (N)	L_{01}	L_{00}

tests register the condition as present when it is not (i.e give a false positive), with θ_{11} denoting the probability that both tests yield a false positive. Thus $\theta_{00} + \theta_{01} + \theta_{10} + \theta_{11} = 1$. The situation may be summarised as in Table 3.7.

Suppose we adopt four possible decision rules, R_1–R_4, with regard to deciding whether D is present on the basis of the two test results. Thus, under rules 1–4, D is assumed present if

1. test 1 is positive, regardless of test 2 (P_1)
2. test 2 is positive, regardless of test 1 (P_2)
3. both tests 1 and 2 are positive ($P_1 \cap P_2$)
4. either test 1 or 2 is positive ($P_1 \cup P_2$).

The respective sensitivities and specificities under these rules, which we denote S_i and T_i ($i = 1, \ldots, 4$), are then as shown in Table 3.8.

Let A_i denote the administrative costs of each test administered separately ($i = 1, 2$) and A_{12} the cost of administering both. The total losses incurred given rule R_i are then

$$\pi S_i(L_{11} - L_{01}) + (1 - \pi)T_i(L_{00} - L_{10}) + \pi L_{01} + (1 - \pi)L_{10}$$

so total costs consist of these losses plus administrative costs.

Table 3.7 Conditional probabilities for outcomes of two tests.

	Condition present (D) 2nd test		Condition absent ($\bar{D}$) 2nd test	
1st test	Positive (P_2)	Negative (N_2)	Positive (P_2)	Negative (N_2)
Positive, P_1	η_{11}	η_{10}	θ_{11}	θ_{10}
Negative, N_1	η_{01}	η_{00}	θ_{01}	θ_{00}

Table 3.8

Rule	S_i	T_i
1	$\Pr(P \mid D) = \eta_{11} + \eta_{10}$	$\Pr(N \mid \bar{D}) = \theta_{00} + \theta_{01}$
2	$\Pr(P \mid D) = \eta_{11} + \eta_{01}$	$\Pr(N \mid \bar{D}) = \theta_{00} + \theta_{10}$
3	$\Pr(P \mid D) = \eta_{11}$	$\Pr(N \mid \bar{D}) = \theta_{00} + \theta_{01} + \theta_{10}$
4	$\Pr(P \mid D) = \eta_{11} + \eta_{10} + \eta_{01}$	$\Pr(N \mid \bar{D}) = \theta_{00}$

Example 3.12 Two tests for detecting AIDS antibodies in blood Two commercial preparations were applied to testing serum specimens to detect antibodies to the AIDS virus by Burkhardt *et al.* (BME, 1987) in a Canadian study. To evaluate costs we first require information about the prevalence π, namely the contemporary proportion of contaminated samples in Canadian blood donations available for transfusion. A survey by Nusbacher *et al.* (1986) found 14 out of 94 496 blood samples positive by the Western blot test.

BME then cite accuracy data from two serum tests, ELISA-A and ELISA-D, as shown in Table 3.9.

The condition being tested for here, and its converse, are respectively D = blood contaminated, and $\bar{\text{D}}$ = blood safe. The largest loss was assigned to the outcome (D present, test N) in Table 3.9, where the condition is present but the test misses it, resulting in an individual having a transfusion of contaminated blood. The medical losses consequent on this outcome are set at $L_{01} = \$100\,000$. The costs of a false positive (finding a sample contaminated when it is pure) are set at $L_{10} = \$25$. The other outcomes are assigned cost zero. Administrative costs are set at $A_1 = A_2 = 1$ and $A_{12} = 2$.

The two sets of probabilities $(\eta_{11}, \eta_{10}, \eta_{01}, \eta_{00})$ and $(\theta_{00}, \theta_{01}, \theta_{10}, \theta_{11})$ are assigned Dirichlet priors with total 'prior sample' size of 5 as follows

$$(\eta_{11}, \eta_{10}, \eta_{01}, \eta_{00}) \sim \text{D}(3.9, 0.5, 0.5, 0.1)$$
$$(\theta_{00}, \theta_{01}, \theta_{10}, \theta_{11}) \sim \text{D}(3.9, 0.5, 0.5, 0.1).$$

These priors reflect a belief that simultaneous correct screening results are more likely than incorrect results (see *Program 3.12 Elisa Tests*). The resulting estimates, from the second half of a run of 10 000 iterations relate to the detection rates with two tests combined and costs of the four rules. They are as shown in Table 3.10.

The costs are conditional on the prevalence data and the relatively small samples involved in the ELISA tests results, and so exhibit wide variability.

Example 3.13 Testing for strongyloides infection with no gold standard test Joseph *et al.* (JGC, 1995) consider the problem of using the results of one or more diagnostic tests to make inferences about test detection rates (sensitivity, specificity) and prevalence in a situation where there is no gold standard diagnosis. They present results of stool and serologic tests of strongyloides infection on 162 refugees from Cambodia arriving in Canada between July 1982

Table 3.9 Test results.

	Condition present (D) 2nd test		Condition absent ($\bar{\text{D}}$) 2nd test	
1st test	Positive	Negative	Positive	Negative
Positive, P	92	0	8	9
Negative, N	1	0	23	370

Table 3.10 Costs, positive predictive value (PPV), sensitivity and specificity by rule.

	Mean	SD	2.5%	97.5%
Cost(R1)	4.66	2.17	2.14	10.25
Cost(R2)	3.88	1.37	2.53	7.74
Cost(R3)	5.89	2.54	2.88	12.32
Cost(R4)	4.65	0.64	3.85	6.24
PVP[1]	0.0039	0.0014	0.0018	0.0072
PVP[2]	0.0021	0.0007	0.0011	0.0037
PVP[3]	0.0089	0.0041	0.0035	0.0191
PVP[4]	0.0016	0.0005	0.0008	0.0027
Sens[1]	0.984	0.013	0.952	0.999
Sens[2]	0.994	0.008	0.970	1.000
Sens[3]	0.979	0.015	0.943	0.997
Sens[4]	0.999	0.003	0.989	1.000
π	0.00016	0.00004	0.00009	0.00025
Spec[1]	0.958	0.010	0.937	0.975
Spec[2]	0.924	0.013	0.896	0.947
Spec[3]	0.980	0.007	0.965	0.991
Spec[4]	0.901	0.015	0.870	0.928

Table 3.11 Observed data and elicited priors for two tests of strongyloides infection.

Data			Stool test		
			+	−	Total
	Serology	+	38	87	125
		−	2	35	37
		Total	40	122	162
Elicited Priors		Serology		Stool	
		2.5%	97.5%	2.5%	97.5%
Sensitivity (%)		65	95	5	45
Specificity (%)		35	100	90	100
Beta parameters					
Sensitivity (a, b)		(22,5.5)		(4.4,13.3)	
Specificity (c, d)		(4.1,1.8)		(71.2,3.8)	

and February 1983 (Table 3.11). The sample prevalence using the stool test is around 25% (40 out of 162) while from serology alone it is 77%. It is desired to estimate the sensitivity (η), specificity (θ) and population prevalence (π) from the results of each test separately, or from both test results combined.

In this situation, drawing useful inferences may require substantive prior information on these parameters to be introduced. Such prior information may be based on accumulated experience in applying the tests and on factors such as the prevalence of the disease being tested for. Thus if the prevalence in the population is high then there is relatively little information contained in the data regarding specificity, since there will be small numbers of negative results on which to make estimates. There is in fact substantial accumulated knowledge about these two well-established parasitological tests, in terms of their estimation of prevalence and their accuracy. Stool examination is known to understate population prevalence, and have lower sensitivity than serology but to yield high specificity (over 90%). Serology results in overestimates of prevalence but has accordingly higher sensitivity.

Joseph *et al.* developed a clinical prior based on elicitation in terms of equally tailed 95% probability intervals and converted these to corresponding 'prior sample' sizes in beta densities. Suppose these priors are denoted

$$\eta \sim \text{Beta}\ (s, t)$$
$$\theta \sim \text{Beta}\ (c, d).$$

A vague prior was used for the unknown prevalence $\pi \sim$ Beta $(1, 1)$ of the disease in the refugee population.

The prior information is combined with the observed data, consisting of observed positive and negative test counts on one test, A and B, or of counts $\{A_1, A_2;\ B_1, B_2\}$ on two tests. For a single test, let T_1 and T_2 be the unobserved numbers of true positives and false negatives among the observed totals, A and B. These latent totals are additional parameters to estimate. So $B - T_2$ is then the number of correctly identified patients with a negative diagnosis.

The total probability of being identified by a single test as positive is

$$\text{Prob(D)Prob(P}\,|\,\text{D)} + \text{Prob}(\bar{\text{D}})\text{Prob(P}\,|\,\bar{\text{D}}) = \pi\eta + (1 - \pi)(1 - \theta).$$

So T_1 is distributed as a binomial from a total of A total positives and with probability

$$\pi\eta/\{\pi\eta + (1 - \pi)(1 - \theta)\}.$$

Similarly the total probability of being identified by the test as negative is

$$\text{Prob(D)Prob(N}\,|\,\text{D)} + \text{Prob}(\bar{\text{D}})\text{Prob(N}\,|\,\bar{\text{D}}) = \pi(1 - \eta) + (1 - \pi)\theta.$$

Hence T_2 is distributed as a binomial from a total of B total negatives with probability

$$\pi(1 - \eta)/\{\pi(1 - \eta) + (1 - \pi)\theta\}.$$

Given sampled values T_1 and T_2 at a given iteration the prevalence then has an updated density

$$\pi \sim \text{Beta}(T_1 + T_2 + 1, A + B - T_1 - T_2 + 1),$$

the sensitivity has an updated density

$$\eta \sim \text{Beta}(T_1 + s, T_2 + t),$$

and the specificity an updated density

$$\theta \sim \text{Beta}(B - T_2 + c, A - T_2 + d).$$

Table 3.12

	Mean	SD	2.5%	Median	97.5%
θ	0.610	0.203	0.233	0.625	0.943
η	0.834	0.051	0.737	0.833	0.930
T_1	107.4	24.1	26.0	117.0	125.0
T_2	20.7	9.2	3.0	21.0	37.0
π	0.787	0.188	0.197	0.834	0.990

An extension of this procedure can be applied jointly to the observations provided by both tests.

Here we consider the parameters implied by a single test and by the prior data elicited on it. The estimation is set in motion by initial guesses at T_1 and T_2, namely arbitrary numbers between 0 and A, and 0 and B respectively. Here we use the serology test results, with $A = 125$, $B = 37$, to obtain the posterior estimates (from the second half of a run of 20 000 iterations) given in Table 3.12.

Closely comparable results are obtained by Joseph *et al.* (1995).

3.5.1 Multivariate Discrimination

We also frequently face the problem of tests or decision rules based on combined indicators. In the multivariate situation, dealing with several continuous variables which are indicators of an underlying condition, the potential number of joint distributions increases combinatorially with the number q of such variables. However, in many data sets the marginal and conditional distributions are sufficiently near normal that the multivariate normal is a suitable model for classification and prediction. In some instances prior transformation may be used to improve the normal approximation – for example, in the Downs syndrome screening methodology described below (see, e.g. Wald and Kennard, 1998).

Suppose we wish to discriminate between two populations on the basis of q continuous markers. Our goal is to derive a decision rule from the observed cases where we know the diagnosis (this is the 'estimation' or 'training' sample). We then wish to use this decision rule to predict the classification in a new population (the 'validation' sample or samples). We consider in particular the medical diagnosis of high-risk individuals from individuals not at risk, though generalisation to more sensitive categorisations such as high/medium/low risk is possible.

The two populations are likely to differ in their average levels on the risk factors, and also possibly on the way they interrelate. Suppose the high-risk population has mean μ_h and dispersion matrix Σ_h, while the lower risk population has mean μ_l and dispersion matrix Σ_l. Let P_h and P_l denote the inverse dispersion, or precision, matrices. Then for subject i with measures

$$X_i = (x_{i1}, x_{i2}, \cdots, x_{iq})$$

the likelihood ratio relating to the subject's probability of being in one or other group may be used to discriminate between high and low risk. This is the difference in log-likelihoods $\log(f_{\text{h}}) - \log(f_{\text{l}})$, or the log of the likelihood ratio (LLR), where $f(\mu, \Sigma)$ is often taken as the multivariate normal density. The assumption of equal dispersion matrices between the two populations is often made. If, however, the dispersion matrices are unequal, the LLR is a quadratic function

$$0.5\{\log(\det\Sigma_{\text{l}}/\det\Sigma_{\text{h}}) - (X_i - \mu_{\text{h}})' P_{\text{h}}(X_i - \mu_{\text{h}}) + (X_i - \mu_{\text{l}})' P_{\text{l}}\ (X_i - \mu_{\text{l}})\}.$$

A cutoff value of this ratio or its exponential is chosen with a view to maximising screening accuracy in as yet unscreened subjects. If the dispersion and precision matrices are equal, i.e. $\Sigma_{\text{h}} = \Sigma_{\text{l}} = \Sigma$ and $P_{\text{h}} = P_{\text{l}} = P$, then the LLR function is linear. The function $(\mu_{\text{l}} - \mu_{\text{h}})' P(\mu_{\text{l}} - \mu_{\text{h}})$ is then known as the Mahalanobis distance between the two means, and the coefficients of the Fisher discriminant procedure are derived as

$$\alpha_{[q \times 1]} = P(\mu_{\text{h}} - \mu_{\text{l}})\cdot$$

Allocation of new cases with known markers $z_{i1}, \ldots, z_{iq}$, but unknown conditions, is then based on scores

$$\alpha_1 z_{i1} + \alpha_2 z_{i2} + \ldots + \alpha_q z_{iq}.$$

The criterion to discriminate between high or low risk scores may be adjusted to allow for unequal prior probabilities (e.g. if the condition were a rare one then the prior probability π_l would exceed π_h) and for costs of misallocation. Thus let the cost of a false negative be $C(\text{l}|\text{h})$ and the cost of a false positive be $C(\text{h}|\text{l})$. Then the subject would be allocated to the high-risk population if

$$f_{\text{h}}(x)/f_{\text{l}}(x) > (\pi_{\text{l}}/\pi_{\text{h}})\{C(\text{h}|\text{l})/C(\text{l}|\text{h})\}.$$

Example 3.14 Discriminators for sudden infant death syndrome (SIDS) Everitt (1994) presents data on 16 SIDS cases and 49 controls in terms of birthweight and a variable based on electrocardiogram and respiratory readings (called FACTOR68). These are denoted `B[1:65]` and `F[1:65]` in Program 3.14 with `G[1:65]` denoting cases ($G_i = 2$) and controls ($G_i = 1$). Following the Fisher linear discriminant model, we assume multivariate normality with different means on $x_1 = B$ and $x_2 = F$ in the two groups (μ_1 and μ_2), but the same covariance matrix P^{-1} in cases and controls. Following Everitt, define $v = \mu_2 - \mu_1$. Then the coefficients of the discriminant function, with $q = 2$ criteria, are obtained as

$$\alpha[q \times 1] = P[q \times q]v[q \times 1].$$

The classification or allocation rule is then derived from the group scores

$$Z_1 = \alpha_1 \mu_{11} + \alpha_2 \mu_{12}$$
$$Z_2 = \alpha_1 \mu_{21} + \alpha_2 \mu_{22}$$

with the cutoff value being $Z.\kappa = (Z_1 + Z_2)/2$, and individual scores being calculated from

$$z_i = \alpha_1 x_{i1} + \alpha_2 x_{i2}.$$

Assume for the moment equal prior probabilities π_1 and π_2 on the two groups. Then if $Z_1 > Z_2$ (so that infants at high risk of SIDS have low scores) the optimal allocation rule is to assign an infant to the high-risk group if his or her score z_i is under the $Z.\kappa$. The data in the present example are such that SIDS infants have lower birthweight but higher FACTOR68 scores, and the discriminant function then assigns low scores to SIDS infants if α_1 is positive and α_2 negative. Note that there is an implicit multiple regression here, and in fact we can code a dependent y_i in such a way as to derive the linear discriminant scores via regression.

We find apparently significant effects α_1 and α_2, with group scores of 1.81 and −0.83 for normal and SIDS infants respectively. The allocation rule could be adjusted to take account of unequal prior probabilities, and specifically by the ratio of π_1 to π_2 – though unless the priors are functions of B or F, this just shifts the cutoff value by a fixed amount. In the present case we might shift the cutoff by a factor $Z.\pi = \log(49/16) = 1.12$. The impact of a series of cutoff rules on the sensitivity and specificity of a test can be shown by an operator–receiver curve.

In Program 3.14 we simply compare an equal priors allocation rule and the unequal priors rule just described. The sensitivity and specificity under equal priors average 0.77 and 0.81. Raising the cutoff point by 1.12 means the risk of false positives is increased, and so the specificity falls to just over 50 %, while the sensitivity is increased to 90 %.

Example 3.15 Screening for Downs syndrome One approach to discriminating high-risk populations (typically in medical screening or diagnosis) is to select several screening measures and assess their interdependence via a distributional model such as the multivariate normal. Royston and Thompson (RT, 1992) give an application involving screening for Downs syndrome on the basis of three biochemical measures. Royston and Thompson give the means and covariance matrix for the markers in an observed sample of affected and unaffected populations. From these parameter estimates, 100 high-risk and 100 low-risk maternities were generated (see the Data Generation Program within Program 3.15 which is run only once).

With the randomly generated sample data, the Analysis Program calculates the LLRs and their exponentials. A high-risk threshold for discriminating high risk in the 'training' sample, and for future screening of new maternities for Downs syndrome, may be set at the upper quartile or 80th percentile of the posterior LLRs (RT, 1992). To reduce the calculation load it may be advisable to derive an estimate of the break point first and then assess detection accuracy in a subsequent run. Thus we find the 80th percentile to average 1.64, with a 95 % interval from 0.95 to 2.45.

Taking the mean break as 1.64, the sensitivity then averages 0.46 with a 95% interval from 0.33 to 0.57. The positive predictive value (probability that a child identified as high risk actually has Downs) averages 0.87 (with interval 0.82 to 0.94), and the specificity (proportion of non-affected identified as such) is 0.93 (interval from 0.89 to 0.98). Taking a lower percentile of the LLR (e.g. the 75th) increases sensitivity but reduces specificity.

Example 3.16 Kangaroo skull measurements Geisser (1993) gives an example drawing on Andrews and Herzberg (1985) who studied 18 skull measures on three species of male and female kangaroos. Geisser selects $q = 2$ skull measures, basilar length and zygomatic width, which are used to discriminate between males and females of one species (*M. giganteus*). A training sample of 10 males and 10 females is chosen randomly from a larger sample of 49 kangaroos, 25 of which are male. So the prediction involves using the training sample of 20 to predict gender in the remaining 29 kangaroos.

We apply a multivariate normal model with unequal dispersion matrices to distinguish between two populations. We firstly examine the distribution of the log likelihood ratios in the training sample (see *Program 3.16 Kangaroo Skulls*). The second half of a run of 20 000 iterations yields the following mean estimates for the training sample members given in Table 3.13.

We select breakpoints at LLR $= 1$ and LLR $= 1.5$ and apply them to predicting sex in the remaining 25 kangaroos. The overall accuracy rates are

Table 3.13

Sex	X_1	X_2	LLR
M	1439	824	1.78
M	1413	823	0.69
M	1490	897	1.59
M	1612	921	1.65
M	1388	805	1.05
M	1840	984	10.82
M	1294	780	−0.09
M	1740	977	3.09
M	1768	968	6.14
M	1604	880	6.45
F	1464	848	0.87
F	1262	760	−0.07
F	1112	702	0.09
F	1414	853	0.69
F	1427	823	1.26
F	1423	839	0.30
F	1462	873	0.71
F	1440	832	1.08
F	1570	894	1.82
F	1558	908	0.95

given in Table 3.14, and show the usual tradeoff between sensitivity and specificity as the risk threshold is varied.

For an LLR threshold of 1.5 we obtain the probabilities that the kangaroos are male and their actual sexes (the latter are not used in the predicted classification) as given in Table 3.15.

Table 3.14

	Mean	SD	2.5%	Median	97.5%
$LLR = 1$					
Sensitivity training sample	0.68	0.21	0.30	0.70	1.00
Specificity training sample	0.57	0.31	0.00	0.60	1.00
Sensitivity new sample	0.63	0.22	0.20	0.60	1.00
Specificity new sample	0.69	0.22	0.29	0.71	1.00
$LLR = 1.5$					
Sensitivity training sample	0.59	0.21	0.20	0.60	1.00
Specificity training sample	0.72	0.27	0.10	0.80	1.00
Sensitivity new sample	0.53	0.20	0.13	0.53	0.93
Specificity new sample	0.80	0.19	0.36	0.86	1.00

Table 3.15

Sex	X_1	X_2	Prob(Gender =M)
M	1312	782	0.08
M	1378	778	0.77
M	1315	801	0.14
M	1090	673	0.35
M	1377	812	0.04
M	1296	759	0.39
M	1470	856	0.08
M	1575	905	0.40
M	1717	960	0.84
M	1587	910	0.46
M	1630	902	0.95
M	1552	852	0.94
M	1595	904	0.72
M	1846	1013	0.91
M	1702	947	0.90
F	1373	828	0.13
F	1415	859	0.41
F	1374	833	0.23
F	1382	803	0.31
F	1575	903	0.44
F	1559	920	0.44
F	1546	914	0.43
F	1512	885	0.17
F	1400	878	0.72
F	1491	875	0.13
F	1530	910	0.46
F	1607	911	0.70
F	1589	911	0.46
F	1548	907	0.33

CHAPTER 4

Normal Linear Regression, General Linear Models and Log-Linear Models

4.1 THE CONTEXT FOR BAYESIAN REGRESSION METHODS

The Bayesian approach to univariate linear regression has long been of interest in areas such as econometrics (Zellner, 1971). Bayesian methods have more recently played a major role in developments in general linear models (e.g. with discrete or survival time outcomes) and in models with complex nonlinear structures, such as in pharmacokinetics (Gelman *et al.*, 1996). Advantages of a Bayesian specification in regression modelling include the ease with which parameter restrictions or other prior knowledge about regression parameters are incorporated, and the ready extension to robust regression methods (for example, ridge regression).

In a regression model we usually specify a probability distribution for the data $y_1, \ldots, y_n$ such as a member of the exponential family (normal, Poisson, etc.). In the Bayesian approach we additionally need to specify the prior distributions of the regression parameters and the error variances (and possibly other unknowns such as selection indices or degrees of freedom), and use the data to update our prior specification. For example, consider a simple linear regression with a univariate Normal outcome y and a single predictor x

$$y_i = a + bx_i + e_i$$

with $e_i \sim \mathrm{N}(0, \sigma^2)$. Our priors then will usually need to specify the form of density assumed for $\theta = (a, b, \sigma^2)$. In BUGS the equivalent centred form

$$y_i \sim \mathrm{N}(\mu_i, \sigma^2)$$
$$\mu_i = a + bx_i$$

is used instead for a Normal outcome variable. For a Student t density the equivalent centred form, including a degrees of freedom parameter v, is

$$y_i \sim \mathrm{t}(\mu_i, \sigma^2, v)$$
$$\mu_i = a + bx_i.$$

The application of regression methods involves a range of issues, including selecting a model appropriate to the form of outcome, according to whether it is a continuous quantity, a count, or binary outcome (for when an event occurs or not). Sometimes an outcome may be alternatively modelled by more than one sampling distribution, for example by adopting one of several different transformations of the outcome. Thus a proportion based on large sample sizes may be modelled as normal as well as via a form (logit, probit, etc.) designed for proportions. In BUGS, for a count of successes T and number of trials N both large, the coding `T ~ dbin (p,N)` with p assigned a beta prior (say) is equivalent to specifying p via a coding such as

```
T ~ dnorm(m,TAU)
TAU <- 1/V ; V <- (T/N)*(1-T/N)*N
m ~ dnorm(0,0.00000001)
p <- m/N
```

A second major issue is specifying the nature of the explanatory or descriptive relationship between the outcome and predictors, ideally in terms of a substantively valid causal model, and then actually identifying it with the data at hand. This leads to further questions about the parameterisation assumed, about model selection in the face of dependencies between predictors, and how to incorporate existing knowledge into priors on the regression parameters. This might be informative prior knowledge, for example, in specifying the sign of a regression effect or its range. This is often the case with economic analysis, for example with coefficients representing marginal propensities to consume or invest.

A major question in regression and other statistical models is that of identifiability; namely, are the data sufficient to precisely identify a complex multifactorial model involving several influences on the response? Lack of fit or poor identification for a model may be apparent in slow convergence or low parameter precisions. This occurs despite the model being appropriately specified in subject matter terms, and may reflect overparameterisation. This in turn may occur because the available data are based on a small sample and are insufficient to identify the theoretical model in its entirety. There may also be mis-specification, due for example, to the omission of an important independent variable, or not allowing for endogeneity between predictors and outcome.

Convergence in MCMC regression applications may also depend on the form of the parameters and variables; for example, a positively skewed regression parameter β is usually better monitored in the transformed form $\log(\beta)$, and may be more approximately normal when transformed. However, when there are real departures from asymptotic normality in the distribution of regression parameters, the Bayesian sampling approach will better represent the actual or exact posterior density of the parameter. Thus Zellner and Rossi (1984) show how the asymptotic normality assumption may be violated in small sample estimation of logit regression, so that the maximum likelihood standard errors will be incorrect.

Correlations between regression parameters may be reduced (and MCMC convergence improved) by simple expedients such as taking centred forms of the independent variables, i.e. $z = x - \bar{x}$ where $\bar{x}$ is the sample average of the original independent variable. In the normal linear model above we would reparameterise as follows:

$$y = A + b(x - \bar{x}) + e$$

where $A = a + b\bar{x}$. Such transformations (which reduce correlation between parameters) are sometimes known as orthogonalising transformations (Naylor and Smith, 1988).

Another familiar problem is that of dependencies between multiple predictors ($p > 1$) and consequent difficulty of selecting a parsimonious model based on a subset of regressors. In particular, multicollinearity refers to the situation in multiple regression when because of a strong relationship between the explanatory variables, it is difficult to disentangle their separate effects on the outcome variable. The ability to interpret the coefficients is diminished, their precision is reduced, and estimates of unknown parameters may not appear to be significantly different from zero simply because the sample is inadequate to identify their effect. In a Bayesian estimation setting, multicollinearity will delay convergence to the stationary distribution. In this situation it may be advantageous to adopt multivariate priors (e.g. multivariate normal) to allow for intercorrelations between covariates, and hence between the associated regression parameters.

Exact multicollinearity exists when the X matrix of dimension nxp has rank less than p, which occurs if there are exact linear relations between the explanatory variables; in this case the matrix $X'X$ has determinant zero and cannot be inverted. In practice what is often observed is that the matrix $(X'X)$ is close to singularity and slight changes in the X matrix, for example by omitting one or two observations or by omitting an explanatory variable, can produce large changes in the regression coefficients.

The Bayesian approach allows incorporation of knowledge from previous studies or derived from theoretical considerations, such as to improve identifiability and reduce the impact of collinearity. For example, the latter might include restrictions on the signs (direction of effect) of covariates in econometric models. Such prior knowledge, if incorporated in 'informative' priors, has the effect of increasing sample size and enhancing the precision of estimated regression parameters. Specialised forms of prior have been proposed for certain regression problems. For example, in a logit regression

$$y_i \sim \text{Bin}(\pi_i, n_i)$$
$$\text{logit}(\pi_i) = \beta_0 + \beta_1 x_i$$

it may be easier (in terms of eliciting priors) to specify prior expectations in terms of expected success probabilities $\tilde{\pi}_1, \tilde{\pi}_2$ specified for two different values of x. This is an alternative to specifying a prior directly for $\{\beta_0, \beta_1\}$. It amounts to specifying prior data points and so is denoted the 'data augmentation prior' or (DAP) by Christensen (1997).

Regression results may also be affected by influential observations and outliers, which are aberrant in terms of the majority associations between outcome and predictors. Special techniques such as mixture regressions (see Chapter 6) may then be used. Alternatively robust regression methods using heavier tailed densities than the normal may be used to identify such cases and reduce their influence on parameters.

4.2 THE UNIVARIATE LINEAR REGRESSION MODEL

The linear regression model describes the relation between an outcome y_i and one or more predictors variables $x_i = (x_{i1}, x_{i2}, \ldots, x_{ip})$ with $i = 1, \ldots, n$ observations. This model has wide applicability for situations where the predictors are either (a) levels, or functions of levels, of continuous variables such as height, income, etc.; or (b) binary indicators taking the value 0 or 1 according to the presence of an attribute; or (c) categorical factors, indicating which of one of several categories case i belongs to (e.g. of a treatment variable), and so generalising binary indicators. In this way applications such as analysis of variance and covariance amount to forms of regression model.

In the Normal linear regression model the conditional distribution $p(y_i \mid \beta, x_i)$ is Normal, with expectation given by a linear combination of the explanatory variables, including intercept $x_{io} = 1$, namely

$$E(y_i \mid \beta, x_i) = \beta_0 + \beta_1 x_{i1} + \ldots + \beta_p x_{ip} = \beta x_i$$

and a conditional variance σ^2 fixed over all i. The centred normal version of the model is then

$$y_i \sim \mathrm{N}(\beta x_i, \sigma^2).$$

The first major interest focuses on updating prior knowledge about parameters with the data at hand, and then drawing samples from the posterior density of the regression parameters β, in order to assess whether the effects $\beta_1, \ldots, \beta_p$ of all or particular predictors on y_i can be judged significant in substantive terms. The second main goal of the model is prediction, for example of future outcomes *y.new* based on future values of x, either known or hypothetical, denoted *x.new*.

4.2.1 Posterior for β, precision known

Suppose first that the variance σ^2 is known, with $\tau = 1/\sigma^2$ denoting the precision. Also let Y denote a $n \times 1$ vector, and X a $n \times (p+1)$ matrix of covariates $x_{ij} (i = 1, \ldots, n; j = 1, j = 0, \ldots, p)$ for a sample of size n. Then the Normal linear model likelihood function is proportional to

$$\exp[-0.5\tau(Y - \beta X)'(Y - \beta X)]. \tag{4.1}$$

Suppose we set

$$b = (X'X)^{-1}X'Y$$

as in the classic least squares regression estimate. Then writing $Y - \beta X = Y - bX + bX - \beta X = Y - bX + X(b - \beta)$, the likelihood is equivalently proportional to

$$\exp[-0.5\tau\{(Y - bX)'(Y - bX) + (\beta - b)X'X(\beta - b)\}] \tag{4.2}$$

since the cross product term $(Y - bX)'(\beta - b)$ is zero from the definition of b. Regarded as a function of the variable β, the last expression is proportional to a multivariate normal density function for β with mean b and covariance $(X'X\tau)^{-1}$. This form for the posterior density of β holds if we assume the flat prior $p(\beta\,|\,\sigma) \propto$ const, as shown by Box and Tiao (1992).

If, however, a prior for β is selected which is also normal with mean b_0 and covariance matrix B_0 and precision $T_0 = B_0^{-1}$ then the product of prior and likelihood will be normal after regrouping terms in the exponent. This product has an exponent equal to -2 times

$$\begin{aligned}
&\tau(\beta - b)X'X(\beta - b) + (\beta - b_0)B_0(\beta - b_0)\\
&\quad = \beta(X'X\tau)\beta - 2\beta(X'X\tau)b + b(X'X\tau)b + \beta T_0\beta - 2\beta T_0 b_0 + b_0 T_0 b_0\\
&\quad = \beta(X'X\tau + T_0)\beta - 2\beta(X'X\tau b + T_0 b_0) + b(X'X\tau)b + b_0 T_0 b_0\\
&\quad = (\beta - \mu_\beta)(X'X\tau + T_0)(\beta - \mu_\beta) + \text{terms not involving } \beta
\end{aligned}$$

where

$$\mu_\beta = (X'X\tau + T_0)^{-1}(X'X\tau b + T_0 b_0) \tag{4.3}$$

is a precision weighted average of b and b_0. So now the posterior density of β is normal with mean μ_β and precision $(X'X\tau + T_0)$.

It can be seen from the form of (4.3) that multicollinearity may potentially be reduced, either by incorporating prior information from previous studies, or by incorporating general theoretical and subject-matter considerations. The matrix $X'X\tau + T_0$ may then be less subject to singularity than $X'X$ if sufficient information is included in the prior, especially by adding precision on the diagonal.

4.2.2 Posterior for $\boldsymbol{\beta}$ and $\boldsymbol{\tau}$, precision unknown

Suppose, following the usual situation in practice, that the precision τ is unknown and so to be considered as a variable parameter as well. Then from (4.2), the likelihood $l(\beta, \tau\,|\,y)$ is proportional to

$$\begin{aligned}
&(s^2\tau)^{n/2}\exp[-(Y - bX)(Y - bX)\tau/2]\exp[-(\beta - b)X'X(\beta - b)\tau/2]\\
&= (s^2\tau)^{n/2}\exp[-(n - p)s^2\tau/2] \quad \exp[-(\beta - b)X'X(\beta - b)\tau/2]
\end{aligned}$$

where

$$s^2 = (Y - Xb)(Y - Xb)/(n - p)$$

is the standard estimate of the residual variance.
A possible reference noninformative prior in this case is one for which

$$p(\beta, \sigma^2) \propto 1/\sigma$$

equivalent to $p(\beta, \tau) \propto 1/\tau^{0.5}$, so that $\log(\sigma)$ and β are locally uniform. Sometimes the prior

$$p(\beta, \sigma^2) \propto 1/\sigma^2$$

is used instead so that the log of σ^2 is essentially uniform on $(-\infty, \infty)$. The corresponding joint posterior distribution $p(\beta, \tau \mid y)$ is then proportional to

$$\tau^{(n+1)/2}\exp[-(n-p)s^2\tau/2]\exp[-(\beta - b)X'X(\beta - b)\tau/2]. \qquad (4.4)$$

The second term in (4.4) above can be seen to be a multivariate normal for the conditional posterior $p(\beta \mid b, \tau)$, with mean b and precision $(X'X)\tau$. The first term is a marginal posterior for τ which is a scaled chi-square $vs^2\chi^2_v$ with $v = n - p$ degrees of freedom. So the joint posterior can be factored:

$$p(\beta, \tau \mid Y) = p(\tau \mid s^2)p(\beta \mid b, \tau).$$

Integrating out τ, it can be shown that the marginal posterior density of β is a multivariate t with mean b, precision $(X'X)\tau$, and $v = n - p$ degrees of freedom. If preferred, this approach may be implemented in BUGS by sampling directly from this multivariate t form, without specifying a prior on β or τ. More usually though we would specify a noninformative prior

$$\beta \sim \mathrm{N}_p(b_0, B_0)$$

with precision $T_0 = B_0^{-1}$. The matrix B_0 is often taken as diagonal, so that we may specify the prior on the regression coefficients in terms of separate univariate normals, remembering that in BUGS a normal prior is specified in terms of precisions. For the prior on τ we typically take just proper priors, such as $\tau \sim \mathrm{Gamma}(a, a)$ where a is a small positive constant. The goals of our analysis would then include testing hypotheses about β, making predictions for new circumstances, and assessing the sensitivity of these inferences to alternative prior specifications on both β and $\tau = 1/\sigma^2$.

Example 4.1 Weighing method Box and Tiao (BT, 1992) present data on a micro method for weighing light objects, which has an error assumed to be approximately Normal with expectation zero. Data are available for 18 weighings of two objects A and B. Two weighings are for specimen A only, the next nine for specimen B only and the remainder are for both specimens. The covariates are defined as $x_1 = 1$ if object A was weighed (i.e. for cases 1–2 and 12–18), and $x_2 = 1$ if specimen B was weighed (for cases 3–18). The analysis adopts the reference prior

$$p(\beta, \sigma^2_y) \propto 1/\sigma^2_y$$

and is conducted by directly sampling from the joint posterior. This is a multivariate t density, with mean being the OLS estimate, with $n-2=16$ degrees of freedom and precision matrix $(X'X)/s^2$.

Taking the second half of a run of 10 000, this gives 95 % credible intervals for β_1 and β_2 of (88.1,110) and (115.7,132.5). BT present a full contour plot of the bivariate density. Results obtained with a separate prior and likelihood specification (see Program 4.1) are virtually identical, again from the second half of a run of 10 000 iterations.

Example 4.2 York rainfall Lee (1997) illustrates linear regression with data on rainfall in millimetres, in successive months (November, December) over $n=10$ years, 1971–1980. That is, y is December rainfall and x is November rainfall. Perhaps contrary to expectation, the association tends to be negative: a wet November is typically followed by a dry December. Here we specify prior for both the regression coefficient and the precision. The data model is simply

$$y_i|x_i \sim \mathrm{N}(\mu, 1/\tau_y)$$

with $\mu_i = \alpha + \beta(x_i - \bar{x})$ and $\tau_y = 1/\sigma_y^2$ denoting the precision.

Lee adopts the same reference prior as in Example 4.1. Under this prior the posterior density of α and β is bivariate t (with $n-2$ degrees of freedom) around the conventional least squares estimates. He also examines a prediction *y.new* for December given a new November observation *x.new* of 46.1 mm.

Here we adopt a vague but proper prior on all parameters. In a small data set like this prior assumptions are likely to influence posterior inferences. We initially adopt vague N(0, 10000) and N(0, 1000) priors for α and β respectively. Note that it is good practice to assign a prior for the intercept which will be sufficiently noninformative to enable a match to the scale of the data. If the outcomes are centred or standardised this become less necessary. The intercept here is around 40 mm. We also assume a just proper gamma prior, G(0.001, 0.001), prior for τ_y.

The second half of a run of 50 000 iterations gives a 50 % credible interval (i.e. from lower to upper quartile) for σ_y^2 of (150, 302). This compares with an interval of (139, 277) obtained by Lee. The prediction of December rainfall for the new November observation is 42.5 with a standard deviation of 16.7; Lee has a smaller standard deviation of the prediction, namely 14.6. By contrast, the 50 % interval for the slope is (–0.246, – 0.077), the same as obtained by Lee.

A closer match to the analysis of Lee (1997) is obtained by taking $\tau_y \sim \mathrm{G}(1, 1)$, when the 50 % credible interval for σ_y^2 becomes (123, 230) and the standard deviation of the prediction is 14.6. Under this prior, however, the 50 % credible interval for the slope changes to (–0.236, – 0.086).

Example 4.3 Path analysis of job satisfaction Path analysis is an extension of multiple regression in relation to explicitly stated causal models, though it cannot conclusively establish causality. The aim is to derive estimates of the causal connections between sets of variables, most simply on the assumptions of causal connections in one direction and assuming no measurement errors. More complex models might introduce reciprocal causation or latent (unobserved) variables. The path diagram is used to depict the causal connections.

For example, Bryman and Cramer (BC, 1994) provide job survey data on age, income and various scales attached to working and occupational environment (job satisfaction, job autonomy, job routine, etc.) for 68 workers. They propose a path model with $y_1 (= \text{age})$ influencing three other variables and being causally prior to them; these are $y_2 =$ autonomy, $y_3 =$income, and $y_4 =$satisfaction. Autonomy is assumed to influence income and satisfaction, but not vice versa. To the degree that felt autonomy is a proxy for job seniority and responsibility, then the path from autonomy to income seems appropriate. Finally all three variables age, income and autonomy are assumed to impact on satisfaction.

We firstly standardise all variables and estimate the appropriate models, for cases $i = 1, \ldots, 68$, omitting intercepts

$$\begin{aligned}
y_{2i} &\sim \text{N}(\mu_{2i}, v_2) \\
y_{3i} &\sim \text{N}(\mu_{3i}, v_3) \\
y_{4i} &\sim \text{N}(\mu_{4i}, v_4) \\
\mu_{2i} &= b_1 y_{1i} \\
\mu_{3i} &= b_2 y_{1i} + b_3 y_{2i} \\
\mu_{4i} &= b_4 y_{1i} + b_5 y_{2i} + b_6 y_{3i}.
\end{aligned}$$

The resulting b_j estimate direct effects, but we might also wish to consider total effects of, say, age on satisfaction. The path from age to income to satisfaction would be calculated as $b_2 b_6$, the path from age to autonomy to satisfaction as $b_1 b_5$, and the path from age to autonomy to income to satisfaction would be represented by $b_1 b_3 b_6$. The total effect of age on satisfaction is then $t_1 = b_4 + b_2 b_6 + b_1 b_5 + b_1 b_3 b_6$. The total effect of autonomy on satisfaction includes its direct effect b_5, plus an indirect effect $b_3 b_6$ through income. The total effect of income is simply b_6. Comparing the estimates of the three total effects (see *Program 4.3 Job Satisfaction Path Analysis*) shows that autonomy has the highest effect on satisfaction, averaging 0.66 compared with 0.46 for income and 0.36 for age (Table 4.1).

Table 4.1 Total effects (t_1 to t_3 of age, autonomy and income on satisfaction.

	Mean	SD	2.5%	Median	97.5%
t_1	0.36	0.11	0.14	0.37	0.59
t_2	0.66	0.09	0.49	0.66	0.84
t_3	0.46	0.09	0.28	0.47	0.65

This analysis follows BC in case deletion for two observations with income missing; potentially more robust procedures are considered in Chapter 6.

4.3 MULTICOLLINEARITY AND ROBUST REGRESSION METHODS

In observational studies, the data generated by uncontrolled mechanisms may be subject to limitations not present in controlled experiments where the design ensures that sufficient data are available to estimate required parameters with precision. The most common problem present in observational data is caused by interrelationships among the independent variables that hinder precise identification of their separate effects. In such circumstances, regression parameters will tend to exhibit large sampling variances, perhaps leading to incorrect inferences regarding their significance, and there will be high correlations between parameters.

A number of specialised methods have been proposed to identify multicollinearity. For example, one common technique involves deriving variance inflation factors: if the original X variables are standardised to new variables Z with mean 0 and standard deviation 1, then variance inflation will be apparent in the inverse of the correlation matrix $Z'Z$. Possible solutions to multicollinearity include

(1) the introduction of extra information, for example via prior restrictions on the parameters based on subject matter knowledge;
(2) the multivariate reduction of the set of covariates (e.g. by principal components analysis) to a much smaller set of uncorrelated predictors;
(3) ridge regression, in which the parameters are a function of a 'shrinkage parameter' $k > 0$, $\beta(k) = (X'X + kI)^{-1}X'y$.

The latter procedure will necessarily induce bias but also yield a more precise estimate of the regression parameter.

In fact the ridge regression estimator is closely related to a version of the standard posterior Bayes regression estimate, but with an exchangeable prior distribution on the elements of the regression vector. Thus we assume the elements of β are drawn from a common normal density

$$\beta_i \sim \mathrm{N}(0, \phi^2), i = 1, \ldots, p. \tag{4.5}$$

In such situations a preliminary standardisation of the variables $x_1, \ldots, x_p$ may be needed to make this prior assumption more plausible, with $z = (x - \bar{x})/s$ where $\bar{x}$ and s are sample mean and standard deviation. This is because in ridge regression each explanatory variable is treated the same in that the same constant k is added down the diagonal of the cross-product matrix of the predictors. If the observational variance is σ^2, the mean of the posterior distribution of β is then

$$\beta = (Z'Z + kI)^{-1}Z'y$$

where $k = \phi^2/\sigma^2$. The most commonly used ridge estimator corresponds to a prior on β with mean zero, but suppose we additionally had prior information on the location of the regression coefficient. Then we may specify the prior

$$\beta \sim \mathrm{N}(\gamma, \phi^2 V)$$

instead of (4.5), so that the posterior mean of β becomes

$$\beta = (Z'Z + kV^{-1})^{-1}(Z'y + kV^{-1}\gamma).$$

We can set a prior on ϕ^2 so that it is updated by the data, or on the ratio of ϕ^2 to σ^2, or assess sensitivity to prespecified fixed values. Estimates for k have also been suggested based on the least squares regression coefficients b_s of y on the standardised predictors z (Birkes and Dodge, 1993). These might be used to form the basis for a prior on k.

Issues of robustness in Bayesian modelling extend beyond collinearity to include questions of the identifiability and parameterisation, and how far these are related to the information included in the priors on the model parameters. Thus a regression model with a flat likelihood over one or more parameters can be made more identifiable by adding more information in the priors – ridge regression being a particular example of this. In this way robustness of regression modelling extends to broader questions of robustness to prior specification, and to the specification of a realistic range of alternative models to describe the data, and on to the central question of choosing the best model or best set of regressors to include.

We are here concerned with a particular aspect of robustness, namely regression estimates which are satisfactory under a variety of data generating mechanisms. The true distribution of the regression errors is never known with certainty, and much research has focused on estimation methods which are robust to non-normal errors. For example, when the regression errors have heavier tails than the normal, then the method of least squares can be improved on. In a regression situation, robustness may also refer to estimation of regression parameters in the face of outlying or influential observations, or skewed errors. In the extreme, the errors may have infinite variance, as would be the case if they were generated by a t- density with degrees of freedom 1 or less. A robust estimator will place lower weight on outliers than procedures based on normality.

This is a feature inherent in the use of the Student t density, as can be seen when it is expressed in a scale mixture form. Thus we can write the Student t density sampling form

$$x_i \sim \mathrm{t}(\mu_i, V, \nu),\ i = 1, \ldots, n \tag{4.6}$$

as a mixture of normal densities with variable scale. These densities are governed by parameters λ_i, that multiply the overall scale V, and are drawn from an inverse gamma with parameters $\nu/2$ and $\nu/2$. Thus the t density of (4.6) can equivalently be written

$$\begin{aligned} x_i &\sim \mathrm{N}(\mu_i, \lambda_i V) \\ \lambda_i &\sim \mathrm{IG}(0.5v, 0.5v). \end{aligned} \tag{4.7}$$

In particular, West (1984) discusses interpretation of the weight parameters

$$\kappa_i = 1/\lambda_i$$

as outlier indicators. Lower values of κ_i (especially those under 1) indicate either outliers or bimodality. The bimodal interpretation would only be feasible if a large proportion of weights (e.g. over 20% of all weights) were small. In BUGS the Normal is parameterised via the precision, so we could write

```
for (i in 1:n) { x[i] ~ dnorm(mu,tau[i])
          tau[i] <- Prec*kappa[i]
          kappa[i] ~ dgamma(nu.2,nu.2)}
          Prec ~ dgamma( , )
          nu.2 <- nu/2
```

If additionally v was a free parameter then we could assign it a gamma prior such as G(1, 0.1), having a prior mean of 10 and variance of 100; for identifiability this may require that the variance is assumed known.

While the analysis of heavy tailed densities has received major attention, Fernandez and Steel (1998) note the relative absence of methods addressed to tackling both skewness and fat tails together. They adopt a method involving differential scaling of a baseline variance according to whether the regression error term $\varepsilon_i = y_i - \mu_i = y_i - x_i\beta$ is negative or positive. Thus suppose the outcome y_i is initially taken as Normal with mean μ_i and denote the overall variance conditional on x_i as σ^2 with precision $\tau = \sigma^{-2}$. More generally we can take ε with any unimodal density, symmetric around 0.

Then for error terms exceeding zero we scale the precision τ by a positive factor $1/\gamma^2$, with $\gamma = 1$ corresponding to a symmetric density, and values of γ exceeding (less than) 1 corresponding to positive (negative) skewness. For negative error terms we scale by a factor γ^2. So when there is positive skewness in the errors ε, values of $\gamma > 1$ are selected since they reduce the precision (i.e. increase variance) for positive ε and increase it for negative ε. This model for skewness can be combined with a Student t density for y_i allowing the modelling of both skewness and fat tails.

Example 4.4 US consumption and income: Analysing collinear predictors Judge *et al.* (1988) present data originally analysed by Klein and Goldberger (KG, 1955) on the relation of total US domestic consumption (y) to wage income (x_1), nonwage-nonfarm income (x_2) and farm income (x_3). The time series spans 1921–41 and 1945–50. Least squares estimates of the regression coefficients show an incremental effect β_1 of wage income on consumption of 1.06, implying that a one dollar rise in income generates more than one dollar extra spending, whereas we would expect the marginal propensity to consume to be between 0 and 1. The effects of the other two variables appear nonsignificant though subject matter knowledge would suggest otherwise.

One approach to obtaining more precise estimates is to introduce restrictions on the parameters. Thus KG assumed that the wage effect on consumption (β_1) exceeds the other effects, and that $\beta_2 > \beta_3$. (In fact they assumed $\beta_2 = 0.75\beta_1$ and $\beta_3 = 0.625\beta_1$.)

Introducing only the order constraint $\beta_1 > \beta_2 > \beta_3$ (see *Program 4.4 US Consumption and Income*) does not improve the estimation. In fact, the coefficient β_1 becomes more in excess of one. However, introducing also the knowledge that income–consumption effects are positive and lie between 0 and 1 leads to posterior estimates on all the coefficients (Table 4.2) in accordance with economic theory expectations (using the second half of a single run of 20 000 iterations).

Introducing an exchangeable prior on the $\beta_j, j = 1, \ldots, 3$, also leads to substantively more sensible estimates, but with low precision for β_2 and β_3 (Table 4.3).

Example 4.5 Robust regression of data with t1 errors Judge *et al.* (1988) present 40 sample observations generated via the linear model

$$y_i = \beta_1 + \beta_2 x_{1i} + \beta_3 x_{2i} + e_i$$

where x_{1i}, x_{2i} are known, and $\beta_1 = 0, \beta_2 = \beta_3 = 1$. The regression errors e_i are drawings from a t density with $\nu = 1$ degree of freedom and scale parameter $\tau^{-1} = \sigma^2 = 4$ (i.e. a Cauchy density), namely

$$f(e_i) = c[\nu\tau^{-1} + e_i^2]^{-(\nu+1)/2}.$$

The available data are y, x_1 and x_2 as generated by Judge *et al.* (1988). The original errors and parameters can be reproduced in BUGS using a direct linear regression with Student t errors and $\nu = 2$ degrees of freedom, though the actual Cauchy errors have neither mean nor variance. To achieve the latter a

Table 4.2 US consumption and income.

Parameter	Mean	SD	2.5%	Median	97.5%
β_1	0.95	0.04	0.86	0.96	1.00
β_2	0.71	0.18	0.29	0.74	0.95
β_3	0.39	0.22	0.03	0.39	0.80

Table 4.3 US consumption and income.

Parameter	Mean	SD	2.5%	Median	97.5%
β_1	0.97	0.18	0.60	0.97	1.32
β_2	0.60	0.69	−0.77	0.60	1.99
β_3	0.57	1.12	−1.60	0.55	2.83
$\kappa = \phi^2/\sigma^2$	0.3	0.37	0.02	0.19	1.25

mixture formulation as in (4.6) and (4.7) is needed – see Program 4.5. Similar results are achieved by the two methods. With vague priors on the precision, the 'direct' method with 2 d.f. gives predicted errors, original errors, and regression coefficients as in Table 4.4.

The estimated regression coefficients compare with those obtained by trimmed least squares and by L-estimation (i.e. estimators that are linear combinations of the order statistics $y[1], y[2], \ldots, y[N]$).

Example 4.6 Student data location West (1984) discusses an example from Lindley (1979) concerning the estimation of the mean of a Student t density with $\sigma^2 = 1$, and $n = 7$ observations

$$\{-1, -0.3, -0.1, 0.4, 0.9, 1.6, 3\}.$$

We take $v = 2$ and the precision set to 1, and frame the example in terms of a 'regression' with a single covariate $x = 1$ for all cases. Posterior estimates of the weight parameters $\kappa_i = 1/\lambda_i$ (from the second half of a run of 10 000 iterations) show the last observation having a very low weight, suggesting it is a probable outlier. West provides formulae to derive the estimated mode, and it can be seen that the posterior mode and mean are very close (both about 0.38) once Student t sampling is adopted (see *Program 4.6 Student Data Location with Outlier*). The simple arithmetic average of the data is 0.64 and is distorted by the outlier.

Example 4.7 ANBS share price changes Fernandez and Steel (FS, 1998) consider changes in the daily share price of the Abbey National Building Society (ANBS). A series of 50 prices p_i is observed and their percent relative changes $y_i = 100(p_{i+1} - p_i)/p_i$ are the outcome variable here. The analysis of FS is in terms of the simple relative changes $(p_{i+1} - p_i)/p_i$ but taking the outcome values on this scale may cause numerical difficulties.

FS cite the danger that a regression model with intercept only and neglecting positive skewness will shift the location $\mu_i = \beta_0$ to the right. In the ANBS example they quote a mean for β_0 under a simple Student t, with unknown d.f., of –0.0012 (with SD = 0.0018). Note that this value is in terms of the scale they use, and would be –0.12 in the percent change scale.

Their mean value for the baseline scale σ^2 is 0.0103 (with SD = 0.0018), and the degrees of freedom v has a modal value of 4. However, when the Student t is combined with a model for skewness they find a more negative value for the location, namely $\beta_0 = -0.0064$ (0.0028), and smaller variance $\sigma^2 = 0.0091$ (0.0018). In their implementation they use the scale mixture approach to the Student t with weights that are drawn from a $G(0.5v, 0.5v)$ density. Using this method they find the degrees of freedom parameter is unchanged when the skew component is added to the Student t model. They conclude that the model for skewness is effectively independent from that for heavy tails.

Table 4.4 Actual and predicted errors and regression coefficients, t1 data.

Parameter	Mean	St devn	2.5%	Median	97.5%	Judge *et al.*
β_1	1.69	5.71	−11.26	2.43	11.88	
β_2	1.18	0.16	0.87	1.18	1.49	
β_3	0.89	0.15	0.62	0.88	1.2	
e_1	−0.31	1.1	−2.71	−0.25	1.67	0.18
e_2	0.04	1.28	−2.39	0.04	2.6	−0.6
e_3	−74.96	0.74	−76.4	−74.97	−73.450	−74.5
e_4	7.64	0.83	5.95	7.67	9.23	8.7
e_5	52.4	1.43	49.66	52.36	55.28	54.2
e_6	2.17	1.19	−0.03	2.09	4.76	3.2
e_7	1.29	1.5	−2.06	1.41	3.91	1.7
e_8	2.54	1.72	−0.73	2.48	6.06	4.6
e_9	−11.6	1.16	−14.18	−11.53	−9.53	−10.6
e_{10}	−2.49	1.34	−4.94	−2.6	0.41	−1.4
e_{11}	−0.99	1.26	−3.75	−0.91	1.26	0.26
e_{12}	0.78	1.39	−1.72	0.69	3.78	0.64
e_{13}	5.91	0.74	4.38	5.93	7.36	6.8
e_{14}	3.84	0.81	2.3	3.82	5.52	4.06
e_{15}	−0.49	1.47	−3.25	−0.47	2.43	−1.37
e_{16}	−0.68	2.48	−5.4	−0.8	4.31	2.1
e_{17}	3.09	1.25	0.78	3.01	5.78	4.2
e_{18}	0.86	1.43	−1.74	0.78	3.8	0.21
e_{19}	−0.9	1.19	−3.06	−0.97	1.79	−0.55
e_{20}	32.78	1.51	29.42	32.89	35.41	33.36
e_{21}	−5.88	1.25	−8.18	−5.91	−3.33	−6.38
e_{22}	−2.48	1.68	−5.91	−2.4	0.56	−0.18
e_{23}	0.62	1.35	−2.21	0.73	3.1	0.6
e_{24}	−1.91	0.98	−3.77	−1.9	0.05	−2.1
e_{25}	11.65	1.47	8.39	11.76	14.23	12.12
e_{26}	−0.91	0.91	−2.8	−0.88	0.79	0.33
e_{27}	−4.38	1.53	−7.81	−4.28	−1.72	−3.69
e_{28}	1.24	0.94	−0.68	1.28	3.04	1.41
e_{29}	−4.68	1.2	−6.86	−4.76	−2.04	−4.08
e_{30}	12.28	1.43	9.41	12.35	14.91	14.31
e_{31}	168.3	0.75	166.8	168.2	169.7	169
e_{32}	−3.52	1.16	−5.77	−3.56	−1.14	−2.1
e_{33}	1.11	1.13	−1.35	1.18	3.14	2.3
e_{34}	−2.29	1.37	−5.34	−2.21	0.11	−1.66
e_{35}	−6.57	0.78	−8.1	−6.56	−5.05	−5.64
e_{36}	0.49	0.78	−1.15	0.52	1.99	1.44
e_{37}	0.28	0.71	−1.1	0.27	1.68	0.85
e_{38}	0.37	0.78	−1.2	0.38	1.89	0.64
e_{39}	1.87	1.15	−0.21	1.79	4.43	2.24
e_{40}	−1.29	0.99	−3.46	−1.24	0.54	−0.15
τ	0.09	0.04	0.03	0.08	0.19	

Here we adopt direct sampling from a Student t and with ν as a free parameter. In the percent change scale and without allowing for skewness, we obtain similar results to those of FS, with $\beta_0 = -0.14$. However, in the model combining skew and heavy tails we obtain a value of γ greater than 1, and estimated at 1.40, corresponding to positive skewness (from the second half of a run of 50 000 iterations using *Program 4.7 ANBS Shares*). This skew parameter was assigned a gamma prior G(0.5, 0.5), following FS. FS themselves obtain a modal value of $\gamma \cong 1.55$ for this parameter. The location is shifted left with a median β_0 of -0.27, a less marked change than obtained by FS, though the difference may be related to their use of a scale mixture version of the Student t. The baseline scale parameter σ^2 and degrees of freedom parameter are both virtually unchanged.

4.4 NONPARAMETRIC REGRESSION VIA SPLINES

Questions of robustness also occur in the face of nonlinear relationships, for example if $x_1, \ldots, x_{p-1}$ have conventional linear effects but the effect of x_p is known to be nonlinear. We may also apply regression methods to smoothing an originally 'ragged' relationship between y and a single predictor x. Splines are used for interpolating functions, in smoothing ragged curves, and in modelling nonlinear components of regression models.

An example of the first use is presented by McNeil *et al.* (1977) in demographic applications where a smooth curve of an age-specific schedule (e.g. of fertility or mortality) is needed but the only data available are rates by aggregated five- or even ten-year age band. The second type of application often involves recovering a true function or 'signal' from observed data subject to large random errors or 'noise'. Smith and Kohn (1996) present examples of artificially generated data of this type, and illustrate the impact of adding different levels of noise on recovering the true function. Examples of this type also occur in demographic graduation, for example of mortality data (Benjamin and Pollard, 1980). The regression application typically involves sifting out the effect of one variable with a highly nonlinear or noisy effect on the outcome, in order that the effect of other covariates can be identified.

The most frequently used spline functions are cubic splines, and they are most often applied to smoothing y in terms of a single covariate x. In these circumstances, cubic splines are piecewise cubic polynomials which interpolate a function $f(x)$ at selected points $\widetilde{x}$ within the range of the variable x. These points are called knots. In a univariate regression setting with x as the independent variable the model would take the form

$$y_i = f(x_i) + u_i,\ i = 1, \ldots, n$$

where the errors u_i are typically taken as independent normal with variance σ^2, and $f(x)$ is a smooth function. We assume the points x_i are arranged in ascending order of size. The m knots $\widetilde{x}_k$ are then placed within the range of x, such that $\min(x_i) < \tilde{x}_1 < \tilde{x}_2 < \ldots < \tilde{x}_m < \max(x_i)$. If we approximate $f(x)$ by a cubic regression spline, then the regression could have the form

$$\begin{aligned} y_i &\sim \mathrm{N}(\mu_i, \sigma^2) \\ \mu_i &= P(\beta, x) + S(\gamma, x) \\ &= \beta_0 + \beta_1 x + \beta_2 x^2 + \beta_3 x^3 + \sum_{k=1}^{m} \gamma_k \delta(x_i - \tilde{x}_k)(x_i - \tilde{x}_k)^3. \end{aligned} \quad (4.8)$$

The term $\delta(x_i - \tilde{x}_k)$ is an indicator function, equal to 1 if x_i exceeds the kth knot $\tilde{x}_k$, and zero otherwise. In BUGS this entails using the step() function. So one additive cubic term only is introduced in the model for μ_i when $x_i > \tilde{x}_1$ but is less than the remaining knots. If x_i exceeds both $\tilde{x}_1$ and $\tilde{x}_2$ then two cubic terms are introduced into the model for μ_i. Usually the polynomial $P(\beta, x)$ has the same degree as the spline $S(\gamma, x)$ in the model for μ_i above – but other forms are possible. Since spline regression has uncertainty with regard to how many knots to include and where to locate them it is useful to monitor the relative importance of different knot points in the overall smooth. If there are too few knots or they are poorly located, details of the true curve will be missed, while a fitted spline based on too many knots will have high local variance. Hence the usual set of regression variable selection methods come into play, and may be used in an iterative procedure to improve location of the knots.

Example 4.8 Aggregated fertility data An example of the interpolating role of splines is provided by data on fertility in Italy in 1955 (McNeil *et al.*, 1977). The data consist of cumulated fertility C_x at ages $x = 20, 25, 30, \ldots, 50$. If f_x is the fertility rate at single year of age x (namely births at age x divided by numbers of women at age x), then cumulated fertility at age 20 is defined as the total over five single years of age, $f_{15} + f_{16} + f_{17} + f_{18} + f_{19}$. From the cumulated data we can form initial estimates of our input data; thus we could estimate

$$\begin{aligned} &f_{20}, f_{21}, \ldots, f_{24} \text{ as } (C_{25} - C_{20})/5, \\ &f_{25}, f_{26}, \ldots, f_{29} \text{ as } (C_{30} - C_{25})/5, \text{etc.} \end{aligned}$$

Then our input data consists of 36 initial estimates of the schedule f_x for ages of women 15 to 50. Since only the original cumulated rates are supplied, not the births and women at risk, we assume the f_x are normal, though a binomial data model would be more appropriate. We use a polynomial $P(\beta, x)$ of degree 5 in ages x and a cubic spline $S(\gamma, x)$, with knots at ages 20, 25, 30, 35, 40 and 45. We obtain a smoothed schedule as shown in Table 4.5, with peak estimated fertility at around ages 27–28. The problem of slight negative rates at the lowest age may be avoided by restricted sampling, or by a higher degree in the spline.

Note that in the example in Table 4.6, simply taking $P(\beta, x) = \beta_0 + \beta_1 x$ leads to an acceptable interpolation, with well identified parameters β and γ.

An example of a variable selection approach to spline regression in the present problem involves specifying dummy indicators g_j which are 1 if the

Table 4.5 Predicted fertility rates by single year of age (*P* quintic).

Mean fertility at ages 15 to 50 ($x = 1, \ldots, 36$)	Mean	SD	2.5%	Median	97.5%
$\mu[1]$	−0.012	0.015	−0.040	−0.012	0.018
$\mu[2]$	0.007	0.014	−0.017	0.007	0.034
$\mu[3]$	0.025	0.012	0.004	0.025	0.050
$\mu[4]$	0.042	0.011	0.022	0.042	0.065
$\mu[5]$	0.058	0.010	0.040	0.058	0.079
$\mu[6]$	0.073	0.009	0.056	0.073	0.092
$\mu[7]$	0.087	0.009	0.071	0.087	0.104
$\mu[8]$	0.100	0.008	0.084	0.099	0.115
$\mu[9]$	0.110	0.008	0.095	0.110	0.125
$\mu[10]$	0.119	0.008	0.104	0.119	0.134
$\mu[11]$	0.126	0.008	0.111	0.126	0.141
$\mu[12]$	0.131	0.008	0.115	0.131	0.146
$\mu[13]$	0.133	0.008	0.117	0.133	0.149
$\mu[14]$	0.133	0.009	0.116	0.133	0.150
$\mu[15]$	0.131	0.009	0.112	0.131	0.148
$\mu[16]$	0.127	0.010	0.107	0.127	0.145
$\mu[17]$	0.120	0.010	0.100	0.121	0.139
$\mu[18]$	0.112	0.010	0.091	0.112	0.130
$\mu[19]$	0.103	0.010	0.083	0.103	0.121
$\mu[20]$	0.093	0.010	0.074	0.093	0.111
$\mu[21]$	0.084	0.009	0.065	0.084	0.102
$\mu[22]$	0.075	0.009	0.057	0.075	0.093
$\mu[23]$	0.067	0.010	0.048	0.067	0.086
$\mu[24]$	0.059	0.010	0.039	0.059	0.079
$\mu[25]$	0.051	0.011	0.031	0.052	0.072
$\mu[26]$	0.043	0.011	0.022	0.043	0.065
$\mu[27]$	0.034	0.012	0.013	0.034	0.059
$\mu[28]$	0.025	0.013	0.002	0.025	0.052
$\mu[29]$	0.017	0.014	−0.008	0.016	0.047
$\mu[30]$	0.010	0.014	−0.016	0.010	0.040
$\mu[31]$	0.006	0.013	−0.019	0.005	0.034
$\mu[32]$	0.004	0.014	−0.024	0.004	0.031
$\mu[33]$	0.003	0.017	−0.031	0.004	0.036
$\mu[34]$	0.003	0.019	−0.033	0.004	0.038
$\mu[35]$	0.003	0.017	−0.032	0.003	0.035
$\mu[36]$	0.000	0.028	−0.057	0.001	0.055

variable associated with the *j*th knot is selected for the regression; i.e. $\gamma_j \delta(x_i - \tilde{x}_j)(x_i - \tilde{x}_j)^3$ is included in the regression if $g_j = 1$ and not included if $g_j = 0$. We assume *a priori* that all terms are needed and adopt a prior which favours inclusion; thus

$$g_j \sim \text{Bernoulli}(\pi_j)$$
$$\pi_j \sim \text{Beta}(19, 1)$$

Table 4.6 Predicted fertility rates by single year of age (P linear).

	Mean	SD	2.5%	Median	97.5%
$\mu[1]$	0.004	0.015	−0.028	0.004	0.032
$\mu[2]$	0.016	0.014	−0.011	0.017	0.042
$\mu[3]$	0.029	0.012	0.004	0.029	0.052
$\mu[4]$	0.041	0.011	0.020	0.042	0.062
$\mu[5]$	0.054	0.010	0.035	0.054	0.073
$\mu[6]$	0.066	0.009	0.049	0.067	0.084
$\mu[7]$	0.079	0.009	0.062	0.079	0.096
$\mu[8]$	0.091	0.009	0.074	0.091	0.108
$\mu[9]$	0.103	0.009	0.085	0.103	0.121
$\mu[10]$	0.114	0.010	0.095	0.114	0.133
$\mu[11]$	0.124	0.010	0.105	0.123	0.145
$\mu[12]$	0.132	0.011	0.113	0.132	0.155
$\mu[13]$	0.139	0.011	0.119	0.138	0.163
$\mu[14]$	0.143	0.012	0.122	0.142	0.168
$\mu[15]$	0.143	0.012	0.122	0.142	0.168
$\mu[16]$	0.138	0.011	0.117	0.138	0.162
$\mu[17]$	0.129	0.011	0.107	0.129	0.150
$\mu[18]$	0.116	0.011	0.094	0.116	0.137
$\mu[19]$	0.101	0.011	0.079	0.101	0.124
$\mu[20]$	0.087	0.012	0.063	0.087	0.111
$\mu[21]$	0.075	0.013	0.048	0.075	0.099
$\mu[22]$	0.065	0.013	0.038	0.066	0.088
$\mu[23]$	0.059	0.012	0.032	0.060	0.082
$\mu[24]$	0.054	0.013	0.027	0.054	0.078
$\mu[25]$	0.049	0.013	0.023	0.049	0.075
$\mu[26]$	0.044	0.013	0.019	0.044	0.069
$\mu[27]$	0.037	0.012	0.013	0.036	0.061
$\mu[28]$	0.029	0.013	0.003	0.029	0.055
$\mu[29]$	0.021	0.015	−0.009	0.021	0.050
$\mu[30]$	0.013	0.016	−0.019	0.014	0.045
$\mu[31]$	0.008	0.015	−0.023	0.008	0.037
$\mu[32]$	0.004	0.015	−0.025	0.004	0.031
$\mu[33]$	0.002	0.018	−0.032	0.002	0.037
$\mu[34]$	0.001	0.020	−0.037	0.001	0.042
$\mu[35]$	0.002	0.018	−0.034	0.002	0.038
$\mu[36]$	0.002	0.030	−0.057	0.002	0.062

and then consider the posterior probabilities that γ_j is included against these prior odds. The resulting analysis (see Program 4.8) shows a largely similar schedule to that above except that peak fertility is slightly later, around age 30. The spline coefficients with highest posterior probabilities of selection are those for the knots at ages 25 and 30; this suggests that a revised model might locate more knots within this age range (e.g at ages 27.5 and 32.5).

Example 4.9 Normal mixture Smith and Kohn (1996) investigate a Bayesian approach to the semi-parametric estimation of additive regression models. They consider how to select the best knot points from an initial large number. Suppose we are modelling y_i in relation to a single predictor x_i and that the data are ranked in ascending order of x. Knots may initially be placed at every 3rd or 4th observation in the x series.

Here we follow one of their analyses of simulated data, generated by functions with known nonlinearity. Specifically, 100 points are generated from the mixture

$$f(x) = \phi(x, 0.15, 0.05)/4 + \phi(x, 0.6, 0.2)/4$$

where $\phi(x, \mu, \kappa)$ is the normal density with mean μ and standard deviation κ. A normal random error with mean 0 and variance 1 was added to the point generated from the normal mixture to give a noisy version of the true function $f(x)$. The true (mixture) curve when plotted peaks at $f(0.175) \cong 2$, rises sharply from $f(0) = 0$ at $x = 0$ and tails off rapidly, being flat at $f() \cong 0.3$ after $x = 0.25$. The generated pairs $(x_i, f(x_i))$, including noise, are sorted in ascending order of x_i and knots are placed at every fourth observation.

Selector indices γ_k that allow the spline coefficients to be included or excluded are defined for each knot. They are drawn from a Benouilli prior as follows

$$g_k \sim \text{Bernoulli}(0.5).$$

The full model (see *Program 4.9 Semiparametric Normal Mixture*) assumes $P(\beta, x)$ quartic, so that

$$y_i \sim \text{N}(\mu_i, \sigma^2)$$

$$\mu_i = \beta_0 + \beta_1 x_i + \beta_2 x_i^2 + \beta_3 x_i^3 + \beta_4 x_i^4 + \sum_{k=1}^{m} g_k \gamma_k \delta(x_i - \tilde{x}_k)(x_i - \tilde{x}_k)^3.$$

Knots are taken at every 4th observation giving $m = 25$ knots. Estimates of μ_i are based on the second half of a run of 10 000 iterations. They show that the procedure recovers the essentials of the true curve, but has a lower peak, and detects modes at higher values of x (Figure 4.1). This in fact mirrors certain features of the randomly generated sample data.

The gamma coefficients in Table 4.7 are selected more than 50 % of the time, with average inclusion rates as shown.

4.5 GENERAL LINEAR MODELS: LOGIT AND POISSON REGRESSION

The general linear model (GLM) approach extends the Normal linear model for metric outcomes to discrete outcomes. It is used when the assumption of normal sampling variation no longer applies or when a linear relation between

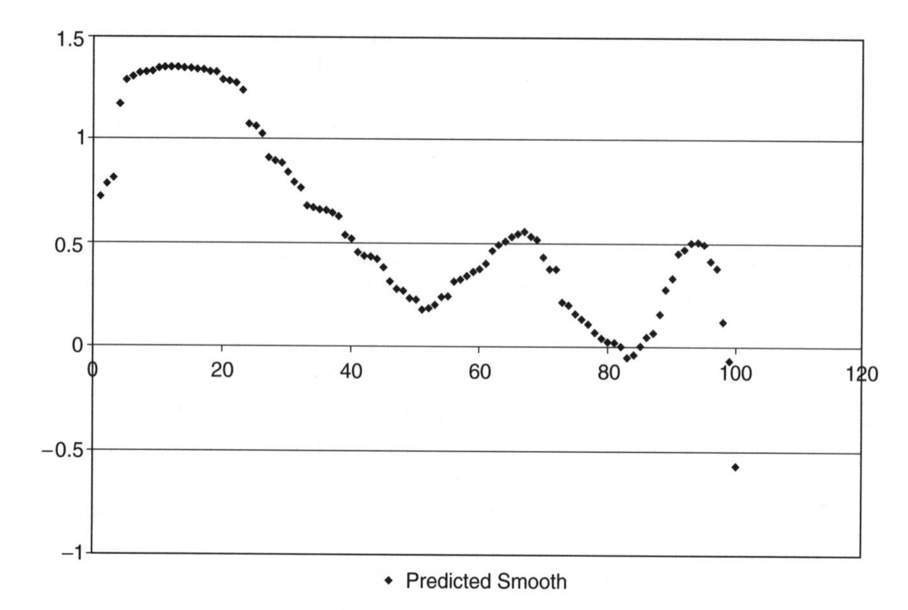

Figure 4.1 Smooth against $100x$.

Table 4.7

g_{11}	0.97	g_{23}	0.51
g_{17}	0.73	g_{25}	0.50
g_{22}	0.62	g_{24}	0.50
g_{14}	0.52		

outcome and predictors is not appropriate – though the standard linear model for metric outcomes is a subclass of GLMs. Assume that continuous or discrete outcome data $y_1, \ldots, y_n$ follow a distribution drawn from the exponential family

$$f(y|\theta, \phi) = \exp\{[y\theta - b(\theta)]/a(\phi) + c(y, \phi)\}$$

where $a(), b()$ and $c()$ are monotonic. The mean $E(y)$ and variance $V(y)$ are given by $\mu = b'(\theta)$ and $b''(\mu)$ respectively, where V is expressed in terms of μ rather than θ. The parameter θ is denoted the canonical parameter, and ϕ is a fixed non-negative scale parameter. Thus the Normal density with mean θ and variance ϕ can be written

$$\begin{aligned} N(y|\theta, \phi) &= (2\pi\phi)^{-0.5} \quad \exp[-0.5(y - \theta)^2/\phi] \\ &= \exp(-0.5\log(2\pi\phi)\exp[-0.5(y^2 + \theta^2 - 2\theta y)/\phi] \\ &= \exp\{[y\theta - 0.5\theta^2]/\phi - 0.5(y^2/\phi + \ln 2\pi\phi)\} \end{aligned}$$

so that $b(\theta) = 0.5\theta^2, a(\phi) = \phi$ and $c(y, \phi) = -0.5(y^2/\phi + \log(2\pi\phi)$. Then the mean is $b'(\theta) = \theta$ and the variance function is $b''(\theta) = 1$. For the Poisson density, a comparable procedure gives $b(\theta) = \mathrm{e}^{\theta}$, and $c(y, \phi) = -\log(y!)$, so that $b'(\theta) = \mu = \mathrm{e}^{\theta}$ and $\mathrm{var}(y) = \mu$ also.

This re-expression of a range of densities in terms of exponential family is useful for modelling the mean μ_i for case i as a function of p covariates $x_i = (x_{i1}, x_{i2}, \ldots, x_{ip})$. Thus we can express the outcome as

$$\mu_i = g^{-1}(\eta_i)$$

where η_i is the linear predictor, expressed for general linear models in terms of the covariates as follows

$$\eta_i = \sum_{k=1}^{p} \beta_k x_{ik}.$$

$g^{-1}()$ is the inverse of a link function g which relates the linear predictor to the means μ_i of the outcome. More generally we can envisage nonlinear predictors, or random effect terms in addition to the effects of covariates in the predictor term η_i. To specify a general linear model, it is thus necessary to define the nature of the predictor term, the sampling distribution (Normal, Poisson, etc.) and the link function.

Bernoulli and binomial outcomes

A major class of general linear model is for outcomes measured on a binary scale. Suppose y_i denotes a binary outcome for case i, $i = 1, \ldots, n$, coded 1 for one of the two possible outcomes (the 'success') and zero otherwise. Also let π_i denote the probability that y_i equals 1, written

$$\pi_i = \Pr(y_i = 1).$$

This type of data model, with a single case at risk of a binary outcome, is known as a Bernoulli density with parameter π_i.

For a sample of n such outcome variables (e.g. with binary outcomes observed over n cases), the likelihood is the product

$$\prod_{i}^{n} \pi_i^{y_i}(1 - \pi_i)^{1-y_i}.$$

If covariates are available for individuals we may model at this level. If a set of individuals are subject to the same predictors then we would work with aggregated Bernoulli data: for r successes in n cases we have binomial sampling $r \sim \mathrm{Bin}(\pi, n)$.

It remains to specify the link adopted for the unknown probability. In a so-called quantal response model for binary or binomial outcomes, it is assumed that $\pi_i = F(x_i\beta)$ where $F(.)$ is a distribution function and so lies between 0 and 1. The vector x_i contains any known explanatory variates. We choose a distribution function as the inverse link in order to model probabilities because then our prediction of the probability will lie between 0 and 1. This type of model has been widely used in econometrics (Zellner and Rossi,

1984) and in bioassay (Finney, 1973). The inverse of F is the link function relating the probability of success to the regression term. Thus

$$F^{-1}(\pi_i) = x_i\beta.$$

There are many possible cumulative density functions F and associated link functions which could be used. The most frequently used form for F is the standard normal cumulative density where

$$f(t) = 1/(2\pi)^{0.5}\exp(-t^2/2).$$

So

$$\pi_i = F(x_i\beta) = 1/(2\pi)^{0.5}\int_{-\infty}^{x_i\beta}\exp(-t^2/2)dt = \Phi(x_i\beta) \tag{4.9}$$

where Φ denotes the cumulative probability function of a standard normal variable. The coefficients $\beta_1, \beta_2, \ldots$ in (4.9) represent the change in standard units of the normally distributed variable per unit change in x_i. The link function F^{-1}, the inverse cumulative Normal probability, is known as the probit, so that

$$\text{probit}(\pi_i) = x_i\beta.$$

For example, in the BUGS program, with a series of binary outcomes y_i on n cases, and with just one covariate apart from the constant we could write:

```
for (i in 1:n) {y[i] ~ dbern (π[i]); probit (π[i]) = beta 1 + beta 2*
x[i]}
```

or equivalently

```
for (i in 1:n) {y [i] ~ dbern(π[i]); π[i] = phi(beta1 + beta2* x[i])}
```

Also frequently used to model a binary outcome is the distribution function of the logistic density

$$F(t) = e^t/(1 + e^t)$$

so that the predicted probability is

$$\pi_i = F(x_i\beta) = 1/(1 + e^{-x_i\beta}). \tag{4.10}$$

The link function F^{-1} is here the log odds or logit

$$\text{logit}(\pi_i) = \log(\pi_i/1 - \pi_i) = x_i\beta.$$

In BUGS this just involves the coding, again for one covariate

```
for (i in 1:n) {y[i] ~ dbern(π[i]); logit(π[i]) = beta1 + beta2* x[i]}
```

The coefficients $\beta_1, \beta_2, \ldots$ in (4.10) represent the change in the log odds for a unit change in each x_i.

Also sometimes used is the link function derived from the cdf of the extreme value distribution,

$$F(x) = 1 - \exp(-\exp(x)).$$

The inverse of F is then the complementary log-log function

$$\log[-\log(1 - \pi_i)] = \beta x_i.$$

Coefficients obtained by these three links will differ partly because of different scaling between the distribution functions.[1] Also whereas the normal and logistic are symmetric around zero, this is not so for the extreme value distribution. It may be noted that it is possible to use links which do not ensure that probabilities lie between 0 and 1; for example with death rates for an infrequent disease but based on large populations we might use a log transform, which ensures the predictions are positive. We would rely on the empirical characteristics of the data to ensure that the predicted p_i are under 1.

Priors for regression coefficients

Priors for Bernoulli and binomial regression coefficients may be effectively noninformative or flat, in the absence of prior expectation about direction or size of covariate effects. In BUGS this may be approximated by taking β_j to be Normal with mean zero and large variance. Note that priors on regression parameters permitting a wide range of values may lead to numerical problems if a large change in value of the total regression term results from certain combinations of parameter and covariate values.

Alternatively a data augmentation prior may be based on subjective information about the likely success rate associated with various combinations of covariate values. Suppose there is an intercept and p covariates $x_1, \ldots, x_p$. Then we consider the likely success rate $s_i = P(y_i = 1)$ for $r = p + 1$ combinations of values of the covariates, and solve to find the implied parameter values (see the O-ring example below). If there is a single covariate, then we would select two values from the range of the covariate. For each of these values we would guess the probability of success, s_1 and s_2, and a measure of certainty on these guesses (prior sample size), C_1 and C_2. The latter is equivalent to adding $C_1 + C_2$ prior data points. Suppose we were predicting the annual risk of heart attack on the basis of a binary covariate for hypertension ($x = 1$ for systolic blood pressure over 160 mm Hg, $x = 0$ otherwise). Suppose we estimate that the risk is $s_1 = 0.1$ for $x = 1$ and $s_2 = 0.02$ for $x = 0$, and that we regard these estimates as worth one data point each, $C_1 = C_2 = 1$. This information is converted into a prior beta density for r probabilities, with respective parameters $C_i s_i$ and $C_i(1 - s_i), i = 1, \ldots, r$.

Latent variable sampling

Underlying differences in the chance of a positive binary outcome, we may posit a continuous latent variable Z_i such that $y_i = 1$ if Z_i is positive, and $y_i = 0$ if Z_i is negative. Thus suppose

[1] Thus the standard deviation of the extreme value distibution is $\pi/\sqrt{6}$ while that of the logistic distribution function is $\pi/\sqrt{3}$

$$Z_i = \beta x_i + u_i$$

where the u_i are independent and identically distributed in a way consistent with the selected distribution function F. Then a success occurs according to

$$\Pr(y_i = 1) = \Pr(Z_i > 0) = 1 - F(-\beta x_i).$$

For forms of F that are symmetric about zero, this expression restates the equality $\pi_i = F(\beta x_i)$, and so is equivalent to the above formulations.

If F is the cumulative Normal, then sampling of Z may be based on draws from a truncated normal: truncation is to the right (i.e. zero is the ceiling value) if $y_i = 0$, and to the left by zero if $y_i = 1$ (Albert and Chib, 1993). In BUGS this involves setting different sampling limits according to whether the observed outcome y_i is 1 or 0. Thus sampling of the latent variables underlying a probit link would involve normal sampling as follows, with the predicted probabilities determined by the normal means (with one covariate in the following):

```
        Z[i] ~ dnorm(mu[i], 1) I(low[y[i] + 1], high[y[i] + 1]);
        probit(p[i]) <- mu[i];
        mu[i] <- b0 + b1* x[i]}
# sampling bounds
        low[1] <- -20; low[2] <- 0; high[1] <- 0; high[2] <- 20;
```

To approximate a logit link, Z_i can be sampled from a t density with 8 degrees of freedom – since a $t(8)$ variable is approximately 0.634 times a logistic variable (Albert and Chib, 1993). This can be done by direct t sampling or by retaining the normal sampling and introducing additional latent scale mixture variables λ_i, such that $Z_i \sim \mathrm{N}(\beta x_i, \lambda_i^{-1})$, with λ_i sampled from a gamma density $G(4, 4)$. The regression coefficients b will then need to be scaled from the $t(8)$ to the logistic, with b.logistic $=$ b.t(8)/0.634.

One useful diagnostic feature resulting from this latent variable approach is that the residuals

$$Z_i - \beta x_i$$

are nominally a random sample from the distribution F (Johnson and Albert, 1999). There are certain problems with testing goodness of fit for binary outcome data with classical analysis of deviance (for Bernoulli data, the deviance reduces to a function of the maximum likelihood estimate). So the Bayesian approach offers this method (among others) for assessing fit and detecting outliers.

Example 4.10 Proportional mortality from leukaemia Dobson (1990) presents data on deaths during 1950–59 from leukaemia and other cancers among survivors from the Hiroshima atomic bomb, and aged 25–64 in 1950 (Table 4.8).

Radiation dose x for the grouped intervals (1–9 rads, 10–49 rads, etc.) is taken as the midpoint, though another option is to use the start point of each interval

Table 4.8 Radiation dose and cancer deaths.

Dose in rads	Midpoint	Leukaemia deaths	Other cancer deaths	All cancer deaths
0	0	13	378	391
1–9	5	5	200	205
10–49	30	5	151	156
50–99	75	3	47	50
100–199	150	4	31	35
Over 200	250	18	33	51

(i.e. $0, 1, 10, 50, \ldots$). The binomial outcome is an example of a *proportional mortality rate*; leukaemia deaths are not compared with survivors in a general population at risk, but with deaths from other cancers.

In Program 4.10 three different forms of link are applied: the logit, probit and complementary log-log. The best fit is obtained with the latter, with mean numbers of leukaemia deaths in the six radiation bands being predicted as 10.9, 6, 5.9, 3, 4.6, 17.3. Monitoring fit using the usual binomial deviance shows an average deviance of 2.9 (using 10 000 iterations after a burn-in of 5000). The logistic model gives an average deviance of 3.1 and the probit one of 3.6. All three show a significant positive relationship between dose and leukaemia deaths, though the data are rather 'sparse' and may be supplemented by informative prior information. It should be noted that a log transform of the independent variable, such as $z = \log(1 + x)$, results in a worse fit.

Example 4.11 O-ring failures by temperature Christensen (1997) presents an analysis of 23 binary observations of O-ring failures y_i in relation to temperature x_i in Fahrenheit. A logit link model for these data is

$$y_i \sim \text{Bernoulli}(\pi_i)$$
$$\text{logit}(\pi_i) = \beta_0 + \beta_1 x_i.$$

Program 4.11 performs a standard logistic analysis of these data and obtains predictive probabilities of O-ring failure as x varies from $30, 32, 34 \ldots$ up to 80 degrees. The DAP prior proposed by Christensen is that for a low temperature of 55 degrees the probability of failure is $\tilde{\pi}_1 \sim \text{Beta}(1, 0.577)$. This gives an approximate probability of 2/3 that the failure risk $\tilde{\pi}_1$ exceeds 0.5. For a higher temperature of 75 °F the prior probability is assumed to be distributed as $\tilde{\pi}_2 \sim \text{Beta}(0.577, 1)$. These two prior probabilities can be used to determine β_0 and β_1 via the expression

$$\text{logit}(\pi_i) = \beta_0 + \beta_1 x_i.$$

If there were three regression parameters and another covariate w_i then the DAP would involve three probabilities at different values of x and w.

Here we adopt slightly less informative DAP priors on $\tilde{\pi}_1$ and $\tilde{\pi}_2$, namely $\text{Beta}(0.1, 0.058)$, and $\text{Beta}(0.058, 0.1)$. These priors lead to more variability about the mean prior probabilities of 0.63 and 0.37 and have the advantage that the modal prior probabilities are not at the extremes 0 and 1. We centre temperatures and obtain estimates of β_0 and β_1 (Table 4.9) clearly showing the fall in risk of O-ring failure at higher temperatures.

We also make a conventional logit analysis with $\beta_0 \sim N(0, 100)$ and $\beta_1 \sim N(0, 10)$. The posterior summary of the regression parameters is then as shown in Table 4.10.

The temperature (LD50 in Program 4.11) below which the chance of an O-ring failure is at least 50% is estimated as 61.2 °F. In general the formula for these 'lethal dose' parameters in dose–response relationships is $LD_\alpha = [F^{-1}(\alpha) - \beta_0]/\beta_1$ where α is between 0 and 1.

For low temperatures, the posterior distribution of the predictive probabilities is highly skewed to the left, with median exceeding mean.

The sensitivity of inferences to one or two cases can be examined in a number of ways. Here, as an example, we consider the predictive divergence for predicting new observations resulting from deletion of one case at a time from the full set of 23 observations. Thus we estimate the model with case omitted and predict for 11 'new' values of x (at 31, 33, ..., 51 degrees Fahrenheit), for the data $y_2, \ldots y_{23}; x_2, \ldots, x_{23}$. The predictions of O-ring failure (with case k omitted) for the 11 new points are denoted $P^{[k]}_{*j}$ for $j = 1, \ldots, 11$ (Table 4.11). These are compared with the baseline predictions based on retaining all 23 cases, and denoted $P_j (j = 1, \ldots, 11)$. Omitting case 2 gives predictions $P^{[2]}_{*j}$, and so on. Then the symmetric Kullback–Leibler diagnostic for case i is

$$D_i = \sum_{j=1}^{11} K(P^{[i]}_{*j}, P_j)$$

where $K(r, s) = (r - s)[log(r - rs) - log(s - rs)]$. This procedure is illustrated with statistics D_1, D_{10} and D_{18}. Case 18 has a high temperature but an O-ring failure is observed. The divergence statistic confirms it as a potential outlier.

Table 4.9 O-ring regression parameters, DAP prior.

	Mean	SD	2.5%	97.5%
β_0	−0.45	0.59	−1.63	0.67
β_1	−0.28	0.06	−0.40	−0.17

Table 4.10 O-ring regression parameters, standard prior.

Parameter	Mean	SD	2.5%	97.5%
β_0	−1.26	0.62	−2.58	−0.07
β_1	−0.29	0.13	−0.59	−0.09

Table 4.11 Mean predictions of O-ring failure at 11 new temperature values (31 °, 33 °, ..., 51 °), retaining all cases and with one case at a time deletion.

Predictions (all cases)	Predictions omitting cases		
P_j	$P^{[1]*}$	$P^{[10]*}$	$P^{[18]*}$
0.972	0.953	0.982	0.992
0.969	0.948	0.979	0.991
0.965	0.942	0.976	0.989
0.960	0.936	0.972	0.986
0.954	0.928	0.968	0.983
0.947	0.919	0.962	0.979
0.938	0.908	0.954	0.974
0.926	0.895	0.944	0.967
0.912	0.878	0.932	0.957
0.893	0.858	0.916	0.944
0.869	0.833	0.896	0.925
Divergence for case omission (at mean predictions)	0.128	0.059	0.341

Example 4.12 SAT scores for maths students An example of the latent variable approach to binary outcomes is the data presented by Johnson and Albert (JA, 1999) on grades obtained by 30 university students of maths. The grades are dichotomised such that grades C or higher correspond to success ($y_i = 1$) and failure ($y_i = 0$) corresponds to grade D or below. The binary outcomes are then related to a Math SAT score on college entry (SAT-M). Substantively, we assume that each student is characterised by a continuous latent performance variable Z_i with a logistic or normal distribution centred on a linear function of the SAT-M score.

JA pay particular attention to outlier detection, with observation 4 a potential outlier: the student failed despite having a relatively high SAT-M score. We apply the latent variable form with both logistic and probit links. Thus the logit model is

$$\pi_i = \Pr(y_i = 1) = \Pr(Z_i > 0) = 1 - F(-\mu_i)$$
$$\mu_i = \beta_0 + \beta_1 \text{SATM}_i$$

where F is the logistic distribution function and SATM scores are centred (see *Program 4.12 SATM Scores*). This gives (from a single run of 50 000 iterations) $\beta_0 = 1.36$, and a SATM coefficient of $\beta_1 = 0.069$, with a standard error of 0.026. Examining the residuals $Z_i - \mu_i$ shows observation 4 as having an average residual of nearly -2.5(s.d. $= 0.85$), and a 0.33 probability of being the lowest residual. However, observations 18 and 19 both have probabilities of around 0.14 of being the lowest residual. The latent variable probit model gives $\beta_0 = 0.74$, and $\beta_1 = 0.038$. The probit residuals $Z_i - \mu_i$ show observation 4 as having a 0.32 probability of being the lowest residual

Example 4.13 Logit discrimination in medical screening Dunsmore and Boys (DB, 1988) consider the problem of using an already derived logit discriminant score to predict the probability of survival after radiotherapy for new patients of Hodgkins disease. Typically we have a collection of binary outcomes and tolerances $S = \{y_i, t_i\}$ for a set of already screened patients and we wish to assess the category y_i of a hypothetical future patient given a tolerance value of t_i.new for this patient.

In the Hodgkins disease example of DB, data was available for 42 patients who had this form of therapy and then either died ($y_i = 1$) or survived ($y_i = 0$) within three months of completing treatment. Four screening variables were available, patient age and three biochemical markers: erythrocyte sedimentation rate (ESR), lymphocyte count (LC) and haemoglobin level (HB). The linear discriminant function, in terms of the logarithms of these variables was

$$t = -9.71 + 0.082\log(\text{age}) - 0.72\log(\text{ESR}) + 0.08\log(\text{LC}) + 5.10\log(\text{HB}).$$

Because we are working with a logistic model, this is a positive survival function: higher values of t_i predict higher survival rates. We seek to identify a threshold of t_i (a 'specification region') above which patients are likely to survive more than three months with a certain probability. We seek a guarantee that the predictive probability of survival for a new patient $P(y_i.\text{new} = 1 \mid t_i.\text{new}, S)$ for at least 3 months exceeds 0.95. We set up a predictive logistic

$$\text{logit}P(y = 1 \mid t, \kappa) = \kappa_0 + \kappa_1 t$$

estimated over the existing sample of 42 patients. Then the prediction with a set of new tolerances, t.new, involves averaging over values of κ_0 and κ_1.

Two approaches are most easily implemented. One involves setting a range of trial values of t.new for a hypothetical set of new patients. Here we set t.new at intervals of 0.1 from -3 to 5.5 inclusive, so that there are 86 'new patients'. We then identify that value of t.new where the predicted survival probability exceeds 0.95. Survival is worse for low values of t.new.

The other approach is to specify the tolerances t.new such that $\hat{\kappa}_0 + \hat{\kappa}_1$ t.new exceeds the log of $0.95/(1 - 0.95)$. This may be solved from $W = \{\log(0.95/0.05) - \hat{\kappa}_0\}/\hat{\kappa}_1$.

The two approaches give similar results when using the 86 trial values of t.new, though the posterior density of W is skewed (with mean of 2.87 exceeding a median of 2.7). The predicted probability exceeds 0.95 when t.new is 2.9 (the 60th new patient).

4.5.1 Poisson Regression

Count data may be related to covariates x_i and exposures E_i via a Poisson regression model. Suppose y_i are Poisson counts with mean $\gamma_i = E_i\mu_i$, and that μ_i is predicted using regressors x_i. Since the Poisson mean is positive we use a log link with $\mu_i = \exp(\beta x_i)$ and $\log(\mu_i) = \beta x_i$. The full model, including possible exposure terms, is then

$$\begin{aligned} y_i &\sim \text{Poi}(\gamma_i) \\ \gamma_i &= E_i \mu_i \\ \log(\mu_i) &= \beta x_i. \end{aligned}$$

It follows (when we have exposure totals as part of the data) that we can use a simpler form

$$\begin{aligned} & y_i \sim \text{Poi}(\gamma_i) \\ & \log(\gamma_i) = \log(E_i) + \log(\mu_i) = \log(E_i) + \beta x_i. \end{aligned}$$

An example where this setup would be appropriate would be when we observed counts of deaths y_i by area, and exposures E_i were expected deaths. The latter are obtained using the usual demographic methods to correct for age structure differences and based on either an internal or external standard schedule of death rates.

Other applications of Poisson regression include change- point problems in time series. Thus suppose $y_1, \ldots, y_N$ is a time series of counts, and $y_i \sim \text{Poi}(\gamma_i)$. A change-point problem involves finding an unknown point k, such that $y_1, \ldots, y_k$ and $y_{k+1}, \ldots, y_N$ follow different models. One possibility is a change in level, such that

$$\begin{aligned} y_i &\sim \text{Poi}(\gamma_1), i = 1, \ldots, k \\ y_i &\sim \text{Poi}(\gamma_2), i = k + 1, \ldots, N \end{aligned}$$

where $\gamma_1 \neq \gamma_2$. Standard classical methods are impeded by the fact that the likelihood function cannot be differentiated with respect to k. In a Bayesian approach to this problem, Carlin *et al.* (1992) put independent gamma priors on γ_1 and γ_2 and a discrete uniform prior for k on $(1, \ldots, N)$. So $\gamma_1 \sim \text{G}(a_1, b_1), \gamma_2 \sim \text{G}(a_2, b_2)$ where a_1 and a_2 are known constants and an additional gamma prior stage is put on b_1 and b_2, namely $b_1 \sim \text{G}(g_1, h_1)$ and $b_2 \sim \text{G}(g_2, h_2)$.

A variant of this is the regression switching model, with earlier points $y_1, \ldots, y_k$ following a different regression model from the later observations. If we expect a change-point in the parameter for a single covariate x, then

$$\begin{aligned} y_i &\sim \text{Poi}(\gamma_{i1}), i = 1, \ldots, k \\ y_i &\sim \text{Poi}(\gamma_{i2}), i = k + 1, \ldots, N \end{aligned}$$

where

$$\begin{aligned} \log(\gamma_{i1}) &= \beta_0 + \beta_{11} x_i \\ \log(\gamma_{i2}) &= \beta_0 + \beta_{12} x_i \end{aligned}$$

and $\beta_{11} \neq \beta_{12}$.

In this and preceding examples a robust alternative to the Poisson (e.g. in cases of excess variability compared with the Poisson assumption) is provided by the negative binomial. We might also allow for additional variation (extra-Poisson variation) by adding an error term in the regression (see Chapter 5).

Example 4.14 Coalmining disasters Carlin *et al.* (1992) analyse counts of coalmining disasters for $N = 112$ years from 1851 to 1962. There seems to be a generally lower rate of disasters from the late 19th century. We adopt a change-point model and analyse the yearly Poisson means via a log-link:

$$\begin{aligned}
&Y_t \sim \text{Poi}(\mu_t)\\
&\log(\mu_t) = \beta_1 + \beta_2 \times \delta(t - \tau)\\
&\beta_j \sim \text{N}(0, a_j), j = 1, 2\\
&\tau \sim \text{U}(1, N).
\end{aligned}$$

The function $\delta()$ is defined as 1 if its argument is zero or positive, and zero otherwise. Using Program 4.14, the second half of a run of 10 000 iterations shows a mean change year (τ) estimate of 38.9 (i.e. the year 1890). However, a plot of the posterior density of this parameter (Figure 4.2) shows some bimodality.

This reflects uncertainty as to whether the change occurred before or after the sequence '1, 3, 2,' for $t = 38 - 40$ (years 1889 to 1891).

Example 4.15 Inpatient events We are often interested in analysing aggregate count data, and in the absence of individual level covariates it is preferable to

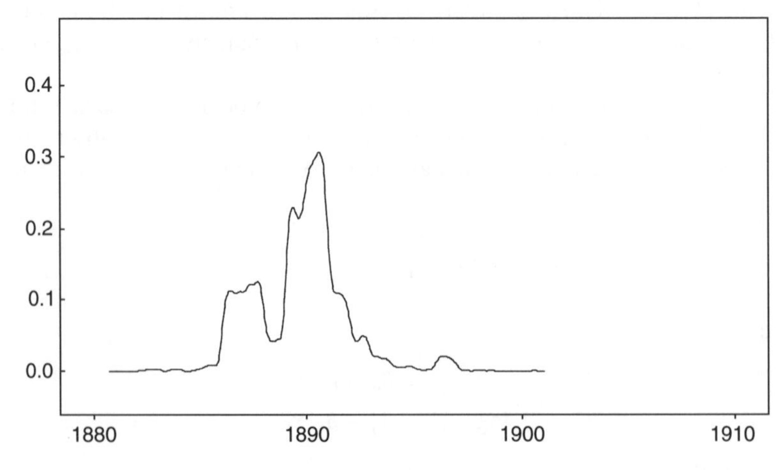

Figure 4.2 Posterior for change year.

model the totals T_i of $y_i = 0, 1, 2, \ldots$ events. An example of one possible approach to aggregate counts in BUGS is provided by data on hospitalisation events in a set of patients surviving for a certain period (Lancaster and Intrator, 1998). The observed counts T_i of patients with $y_i = 0, 1, 2, \ldots$ events are $975, 213, 94, \ldots$ etc. So we calculate $G_i = y_i T_i$ and model G_i as Poisson with mean μT_i. Using *Program 4.15 Inpatient Events* we find that for these data, the mean is 0.55 and the variance is 1.35, so there is overdispersion relative to the Poisson assumption. The Poisson fitted values overpredict 1 and 2 events and underpredict zero and large event counts (Table 4.12).

4.5.2 General Nonlinear Models

The linearity or nonlinearity of a model is determined by the way a change in the value of a predictor operates via the regression parameter to change the response. In a linear model, such as $\mu = \beta_0 + \beta_1 x$, a unit change in the coefficient β_1 leads to the same change in μ whatever the original value of the parameters β_0 and β_1. Thus if $\mu' = \beta_0 + (\beta_1 + 1)x$, then $\mu' - \mu = x$ regardless of the original value of the parameters. However, consider the model for the mean response defined by

$$\mu = \alpha + \beta e^{-\gamma x} \tag{4.11}$$

Suppose β increases in (4.11) by one unit to give

$$\mu' = \alpha + (\beta + 1)e^{-\gamma x}.$$

Then $\mu' - \mu = e^{-\gamma x}$ which depends on the value of γ. The change in mean response is not then independent of the original values of the parameters.

Certain apparently nonlinear models may be linearised by transforming; for example, if we have the multiplicative error model

$$y_i = \exp(\alpha_0 + \alpha_1 x_i) e_i$$

Table 4.12 Frequency of hospital events in patients surviving for 15 months: actual and Poisson fit.

Number of events	Actual	Poisson mean	SE
0	975	787.50	15.95
1	213	435.90	7.11
2	94	120.80	6.39
3	47	22.35	2.00
4	21	3.11	0.39
5	6	0.35	0.06
6	4	0.03	0.01
7	3	0.003	0.001
8	5	0.00018	0.00005
9	2	0.00001	0.00000

then its linearised form is

$$\log(y_i) = \alpha_0 + \alpha_1 x_i + \log(e_i).$$

The same is not possible if the original model has additive errors, as in $y_i = \exp(\alpha_0 + \alpha_1 x_i) + e_i$.

As noted by McCullagh and Nelder (1989), it may be unwise to include more than a few nonlinear parameters, especially when the covariates themselves are correlated. Estimates of nonlinear parameters may be highly correlated with each other and with linear parameters, especially when the regression term includes sums of exponentials. For example, if

$$y_i \sim \mathrm{N}(\mu_i, \sigma^2)$$

and

$$\mu_i = \alpha_0 + \alpha_1 \mathrm{e}^{\beta_1 x_{1i}} + \alpha_2 \mathrm{e}^{\beta_2 x_{2i}}$$

then coefficients α_1 and β_1, and α_2 and β_2 may tend to be highly correlated. This 'ill-conditioning' is a common difficulty in estimating models in pharmacokinetics (chemical absorption and metabolism) where decay times are defined by mixtures of exponentials (Gelman *et al.*, 1996). The parameters are difficult to estimate simultaneously and may only be estimated by (a) fixing some parameters at guessed 'indicative' values, or (b) ensuring parameters have substantive meaning in relation to the process being modelled and can be assigned informative priors.

Despite potential issues of identifiability, nonlinear generalisations of standard models have been proposed on grounds of greater realism. For binomial outcomes, Prentice (1976) suggested a generalised logit model for y successes in n trials, namely

$$y_i \sim \mathrm{Bin}(\pi_i, n_i)$$

with π_i modelled in terms of predictors via

$$\log(\pi_i) = m[x\beta - \log(1 + \mathrm{e}^{x\beta})]$$

or equivalently,

$$\pi_i = [\mathrm{e}^{x\beta}/(1 + \mathrm{e}^{x\beta})]^m. \tag{4.12}$$

In a similar vein, Breslow and Storer (1985) propose a general relative risk function in the logistic regression model. The standard logistic and proportional hazards model assume effects of various risk factors combine multiplicatively. More generally, however, suppose $\rho(x) = x\beta$ denotes a regression function, and y indicates a binary outcome or disease status ($y = 1$ for disease present, $y = 0$ for disease absent). Then take the standard logit form

$$\begin{aligned} y_i &\sim \mathrm{Bernoulli}(\pi_i) \\ \mathrm{logit}(\pi_i) &= \alpha + \rho(x) \end{aligned} \tag{4.13}$$

and let

$$R(x) = \exp\{\rho(x)\}$$

express the total relative risk associated with exposure variables x. Then a more general model for the relative risk function is

$$\log R(x) = [(1 + x\beta)^{\lambda} - 1]/\lambda, \qquad \text{for } \lambda \neq 0$$
$$\log R(x) = \log(1 + x\beta), \qquad \text{for } \lambda = 0. \qquad (4.14)$$

The parameter λ describes the shape of the relative risk function with $\lambda = 1$ corresponding to the usual multiplicative model, and $\lambda = 0$ giving an additive structure such that $R(x) = 1 + x\beta$. For identifiability, the term $x\beta$ should exceed -1 for all values of λ. Therefore the exposure factor level with the lowest risk should be selected as the baseline (i.e. with relative risk, $R(x)$, equal to 1). A Bayesian sampling perspective has the advantage of permitting substantively based priors constraining the regression effects of increased levels of risk so as to raise the occurrence rate of the outcome.

Box–Cox transformations

Similar transformations to (4.14) were first proposed for the linear regression model. For responses and predictors on a continuous scale, the Box–Cox transformation is frequently adopted when the original data y is subject to skewness such that after a transformation, such as $z = \log(y)$ or $z = y^{0.5}, z$ is more approximately normal. More generally in a regression situation with predictors x_i also available, we seek a transformation to minimise skewness and produce approximate normality in the error term. Thus for $\lambda \neq 0$, the transform is

$$z_i = (y_i^{\lambda} - 1)/\lambda$$

with the transformed variable related to covariates

$$z_i \sim \mathrm{N}(\beta x_i, \sigma^2)$$

via a standard normal regression. For $\lambda = 0$, this transformation reduces to $z_i = \log(y_i)$.

The likelihood in this model can be written

$$f(y_i \mid \lambda, \beta, \tau) = (2\pi\tau)^{0.5} y_i^{\lambda-1} \exp[-0.5\tau(z_i - \beta x_i)^2]$$

where $\tau = \sigma^{-2}$, and with appropriate modification for $\lambda = 0$. If we set

$$b_0 = 1 + \lambda\beta_0,$$
$$b_j = \lambda\beta_j \quad (j > 0),$$

and

$$\varsigma = |\lambda|\sigma,$$

then this likelihood can be re-expressed as

$$f(y \mid \lambda, b, \upsilon) = (2\pi\upsilon)^{0.5} |\lambda| y^{\lambda-1} \exp[-0.5\upsilon(y_i^{\lambda} - bx_i)^2]$$

where the precision is $\upsilon = \varsigma^{-2}$.
Either likelihood form can be implemented in BUGS as a non standard sampling density using the device of creating dummy data values $\delta_i = 1$ for all i, with likelihood probabilities

$$\delta_i \sim \text{Beta}(p_i)$$

where for example, we set $p_i = (2\pi\upsilon)^{0.5}|\lambda|y^{\lambda-1}\exp[-0.5\upsilon(y_i^{\lambda} - bx_i)^2]$.

As for any nonlinear model identifiability is an issue, with correlation likely between the exponent λ on the one hand, and the intercept and the other regression parameters on the other. It may also be necessary in the prior specification to confine λ to either positive or negative values since the regression models for $z_1 = y^{\lambda}$ (e.g. square root) and $z_2 = y^{-\lambda}$ (e.g. inverse square root) may tend to be approximate inverses of one another.

This model is nonlinear in the response only, but we might consider models in which the x_i were also potentially subject to skew minimising transformations, such that

$$W_{1i} = (x_{1i}^{\eta 1} - 1)/\eta_1$$
$$W_{2i} = (x_{2i}^{\eta 2} - 1)/\eta_2$$

and so on. Then the full regression model becomes

$$z_i \sim \text{N}(\gamma W_i, \sigma^2).$$

Example 4.16 Bermuda grass yields McCullagh and Nelder (MN, 1989) discuss modelling the growth-promoting effects of three nutrients on coastal Bermuda grass. The design was a 4×3 factorial experiment defined by four replications involving different application levels of three nutrients: nitrogen (N), phosphorus (P) and potassium (K). In each of the replications 0, 1, 2 and 3, the inputs of nitrogen in lb/acre were successively 0, 100, 200 and 400; for phosphorus they were 0, 22, 44, and 88; and for potassium they were 0, 42, 84, 168. Thus the outcome Y_i (yield of Bermuda grass in tons/acre averaged over 3 years) is defined by 64 different settings. If (N, P, K) denotes the inputs then the replication settings are $(0,0,0), (0,0,1), \ldots, (0,1,0), (0,1,1), \ldots, (3,3,0), (3,3,1), \ldots, (3,3,3)$ with inputs $(0,0,0), (0,0,42), \ldots, (0,22,0), \ldots, (0,22,42), \ldots, (400,88,0), (400,88,42), \ldots, (400,88,168)$.

We assume the outcome Y is gamma with mean $(\nu, \nu\varepsilon_i)$ where $\nu \sim G(c_1, c_2)$ with c_1 and c_2 known, and

$$1/\varepsilon_i = \beta_0 + \beta_1/(N_i + \alpha_1) + \beta_2/(P_i + \alpha_2) + \beta_3/(K_i + \alpha_3).$$

The parameters α_1, α_2 and α_3 can be regarded as background nutrient levels in the soil. Informative priors are obtained for these by taking reciprocals $1/Y_i$ of the data and plotting them against $u_{1i} = 1/(N_i + \alpha_1), u_{2i} = 1/(P_i + \alpha_2)$ and $u_{3i} = 1/(K_i + \alpha_3)$ for trial values of α_1, α_2 and α_3. This gives suggested values $\alpha_1 = 40, \alpha_2 = 22$ and $\alpha_3 = 32$, and we adopt normal priors centred at these values with variance 100. Non informative priors are used for v and the growth effects $\beta_j, j = 1, \ldots, 3$, except that we assume these to be positive. (The latter constraint is likely to improve precision of estimates but is not strictly necessary.)

Estimates and standard deviations obtained from the second half of a run of 10 000 iterations and using *Program 4.16 Bermuda Grass* are as follows

Intercept:	$\beta_0 = 0.092\ (0.0066)$		
Background levels:	$\alpha_1 = 44.7\ (3.8)$	$\alpha_2 = 24.3\ (7.3)$	$\alpha_3 = 34.4\ (8.8)$
Growth effects:	$\beta_1 = 13.2\ (1.2)$	$\beta_2 = 1.2\ (0.47)$	$\beta_3 = 1.45\ (0.48)$

These are comparable to those cited by MN, though the estimates of α_3 and β_3 (relating to potassium) are notably more precise.

Example 4.17 Flour beetle mortality Carlin and Louis (CL, 1996) analyse data $\{y_i, n_i\}$ on beetle mortality in relation to a dosage x_i of insecticide, using the generalised logit model in (4.12). Dosages are subject to a centralising transformation, defined by parameters μ and σ, namely

$$z = (x - \mu)/\sigma$$

with corresponding model

$$y_i \sim \text{Bin}(\pi_i, n_i)$$
$$\pi_i = [\text{e}^Z/(1 + \text{e}^Z)]^m.$$

After 10 000 iterations, this transformed scale model (see *Program 4.17 Flour Beetles*) gives the parameter estimates in Table 4.13.

CL noted intercorrelations among the parameters in their estimation and resultant slow mixing. The possibility of excess parameterisation here could be reduced by fixing one or both of μ and σ at their 'empirical' values defined by the observed mean and standard deviation of the doses.

Example 4.18 Case-control study of endometrial cancer Breslow and Storer (BS, 1985) illustrate the generalised relative risk approach of equations (4.13) and (4.14) with a Connecticut case-control study data for endometrial cancer in relation to replacement estrogens. The risk factors are a woman's weight, WT, with three categories based on grouped weights (under 57 kg, 57–75 kg, and over 75 kg) and estrogen use, EST, arranged as no/yes. This ordering of categories (with 1 as baseline) provides the lower risk as baseline. Let EST(2) denote the yes response to estrogen use, and WT(2) and WT(3) the two higher weight bands. As Breslow and Storer note, the log-likelihood is distinctly non-normal.

Hence the regression function is

$$\rho(x) = \beta_1 \text{EST}(2) + \beta_2 \text{WT}(2) + \beta_3 \text{WT}(3)$$

Table 4.13

Parameter	Mean	SD	2.5%	Median	97.5%
M	0.50	0.21	0.25	0.46	1.00
μ	1.80	0.01	1.77	1.80	1.83
σ	0.022	0.004	0.015	0.022	0.032

where we take each β_j normally distributed but constrained to positive values (see *Program 4.18 Estrogen Use*). A uniform prior $U(-2, 2)$ is adopted for the exponent λ. As BS also find, the best fitting model involves a negative exponent, and shows a greater risk attaching to estrogen use than the results from a multiplicative model with $\lambda = 1$. There are 2 degrees of freedom and the median x^2 shows a close fit (Table 4.14).

Example 4.19 Wood from cherry trees Aitkin *et al.* (1989) consider data on 31 black cherry trees, relating usable wood volume y (cubic feet) to covariate data $x = \{D, H\}$, where D is tree diameter in inches, and H is tree height in feet. They consider the profile likelihood method for a transformed outcome $z = y^\lambda$, involving a grid search method for the optimal λ using the function

$$P(\lambda) = -0.5n \sum_i (y_i^\lambda - \beta x_i)^2 + n\log|\lambda| + (\lambda - 1) \sum_i \log y_i.$$

They select $\lambda = 0.33$ at which point $P(\lambda) = 152.2$.
In BUGS we apply the case likelihood approach discussed above with $m_i = \beta x_i$ and

$$p_i = (2\pi \upsilon)^{0.5} |\lambda| y^{\lambda - 1} \exp[-0.5\upsilon(y_i^\lambda - m_i)^2].$$

Table 4.14

Parameters	Mean	SD	2.5%	Median	97.5%		
β_1	31.2	18.2	6.6	27.7	75.9		
β_2	1.6	1.4	0.1	1.3	5.2		
β_3	28.2	17.4	5.5	24.4	71.2		
λ	−0.53	0.17	−0.89	−0.51	−0.23		
χ^2	3.2	2.5	0.1	0.4	2.6		
Weight	Estrogen use	Total	Cases observed	Cases fitted	SD	Rel risk	SD
<57	N	195	12	12.0	3.2	1.0	
	Y	81	20	19.2	2.0	5.1	1.7
57–75	N	423	45	47.4	6.6	2.1	0.7
	Y	150	37	36.0	3.6	5.3	1.8
>75	N	182	42	42.2	4.6	5.0	1.7
	Y	32	9	8.3	0.9	5.9	2.4

Table 4.15

	Mean	SD	2.5%	Median	97.5%
Profile log-likelihood	155.5	2.22	152.6	155	160.9
λ	0.35	0.03	0.30	0.36	0.38
β_0	−0.15	0.29	−0.66	−0.14	0.43
β_1	0.17	0.03	0.12	0.17	0.21
β_2	0.014	0.003	0.008	0.014	0.021
σ^2	0.009	0.004	0.004	0.009	0.018

The regression model has the form

$$m_i = \beta_0 + \beta_1 D_i + \beta_2 H_i$$

with priors $\beta_j \sim \mathrm{N}(0, 10)$ and $\lambda \sim \mathrm{N}(0, 1)$.

The second half of a run of 20 000 using *Program 4.19 Cherry Wood* shows both diameter and hardness as relevant to usable volume, with the optimal power around 0.3–0.4 (Table 4.15).

Similar results were apparent in the work of Aitkin *et al.*

4.6 GENERAL LINEAR MODELS FOR SURVEY DATA

In classical linear regression and general linear modelling, analysis proceeds on the assumption that outcomes y and regressors x are measured on individuals forming a simple random sample from a larger population. However, data obtained from surveys often result from differential sampling and response rates in different strata of the population which lead to different probabilities of selection for each individual. These selection and response biases may invalidate simple assumptions of representativeness in making inferences (Smith, 1983).

The inference process from survey data is also affected by the perspective adopted. The goal may be to make inferences about the parameters of the population distribution which generated the sample measurements, or alternatively to estimate a quantitative characteristic of the particular population from which the sample derives (the population values are taken as fixed). These are known as 'analytic' and 'descriptive' perspectives respectively (Smith, 1983; Nordberg, 1989; Breckling *et al.*, 1994). In descriptive inference, whether for regression or other purposes, one considers only random variation from the sampling and response mechanisms. In an analytic framework for survey regression, the survey relation between response and predictors reflects a broader super population model relating the measures. Allowing for this broader relationship forms an additional element of random variation.

To reflect differential chances of inclusion in the survey, suppose inclusion is determined by auxiliary or design variables Z which are already known for the population (for example, area of residence). A small set of known design variables $Z_1, Z_2, \ldots$ is often used in designing a stratified sample. Under this framework, the sample data are assumed to be generated as follows:

(a) A population of N elements is governed by the superpopulation model $f(y)$. In its fullest form the density of y would be specified conditional both on x and Z, and on the parameters β and γ describing their relation to y, thus $f(y \mid \beta, x, \gamma, Z)$.
(b) A sample $i = 1, \ldots$,n is drawn from this population according to a particular design, specified by sample inclusion indicators S_i. For random sampling schemes inclusion depends only on Z but some sampling methods (e.g. quota sampling) may induce additional dependence of inclusion on y.
(c) A sampled case from step (b) may or may not respond (on y and possibly x also) with this recorded by response indicators R_i.

Nordberg (1989) and Dumouchel and Duncan (1983) consider application of unweighted general linear model (GLM) inference to data generated from stratified random samples. If the survey design, as expressed by inclusion probabilities

$$\delta_i = \Pr(S_i = 1), \qquad i = 1, \ldots, N$$

contain no information on y beyond that contained in the regressors x then unweighted GLM inference is valid.

If the design does contain additional information then one option is Horvitz–Thompson weighting such that the likelihood for survey cases with low δ_i is given a higher weight. This is an example of a 'design-based' approach to variable inclusion probabilities. An alternative is to incorporate in the regression additional variables W which are highly correlated with the inclusion probabilities; examples include the design variables Z themselves, so that $W = Z$. This approach can reduce design bias to such a degree that unweighted inference is still valid.

Example 4.20 Milk production in sweden Nordberg (1989) reports on a simulation study which draws from a study of structural change among small Swedish dairy farmers, and of factors influencing the decision ($y = 1$ or 0) to abandon or continue milk production between 1983 and 1984. A population of 12195 farmers was 'generated' on the basis of the study results, and subdivided by 36 categories according to the age group of farmer 0–49, 50–59 and 60+ (A with 3 levels); region (G with 3 levels); cow herd size, 4–9 vs 1–3 ($Z = 0$ vs 1), and type, T (1= major production branch, 0=otherwise). The following regression (significant coefficients only) was found to fit the decision well and used as a 'superpopulation' model to regenerate the outcome data $y_i, i = 1, \ldots, 12\,195$ from known values of A_i, G_i, Z_i and T_i:

Table 4.16

Effect of intercept	Mean	SD	2.5%	97.5%
	−0.962	0.116	−1.192	−0.738
A_2	−0.294	0.135	−0.552	−0.025
A_3	0.213	0.111	−0.003	0.438
G_2	−0.222	0.101	−0.407	−0.018
G_3	−0.562	0.115	−0.782	−0.334

$$y_i \sim \text{Bernoulli}(\pi_i)$$
$$\text{logit}(\pi_i) = -2.5 + 1.59S_i - 0.3A_{2i} + 0.8A_{3i} - 0.8A_{3i} \times Z_i + 1.01T_i$$
$$- 0.3G_{2i} \times Z_i - 0.5G_{3i} \times Z_i.$$

A sample of 3001was then taken by Nordberg from the full population with random sampling within four strata defined by combinations of (Z, T): namely $(0, 0)$, $(0, 1)$, $(1, 0)$ and $(1, 1)$. The sample sizes in the respective strata were 840, 521, 920 and 720 corresponding to inclusion probabilities of 10, 100, 60 and 42%.

The design vector for the survey is thus a function of actual explanatory variables, and the object is to show the effect of incorporating or omitting such design variables when they carry important information about the outcome y.

Here a population of 12 195 and sample of 3037 was generated by the same procedures, and two models applied to predict the binary outcome y_i (abandonment of milk production). The first (model A in Program 4.20) predicts y on the basis of age and region alone, i.e. without the stratifying variables. This can be seen (Table 4.16) to lead to biased estimates of the superpopulation regression parameters for A and G, especially in terms of the overall rate of milk production abandonment and the effect of being over 60.

By contrast a second model includes the design variables, with main effects of age, region, size and type, and with age × size, region × size, type × size, type × region and type × age interactions (age by region interactions are regarded as implausible). This reproduces the coefficients in the superpopulation mechanism generating y for all 12 195 farmers, except that the region by size parameters are not precisely estimated (Table 4.17).

4.7 LOG-LINEAR MODELS FOR CROSSED CATEGORICAL OUTCOMES

In many research areas, especially the social sciences, the information available for statistical modelling is obtained from sample surveys or official statistics, and is often nonmetric in nature, with the observed variables being categorical or presented in grouped form, even though originally metric (e.g. income bands). The interest then focuses on modelling counts accumulated in the crossed categories formed by two or more of these categorical or qualitative

Table 4.17

Effect of intercept	Mean	SD	2.5%	97.5%
	−2.543	0.284	−3.105	−2.010
A_2	−0.238	0.211	−0.646	0.186
A_3	0.710	0.214	0.301	1.132
Z	1.630	0.268	1.106	2.126
$A_3 \times Z$	−0.894	0.194	−1.271	−0.511
T	0.768	0.301	0.172	1.339
$Z \times T$	0.083	0.221	−0.336	0.534
G_2	0.099	0.238	−0.373	0.584
G_3	0.010	0.263	−0.499	0.507
$Z \times G_2$	−0.354	0.245	−0.844	0.147
$Z \times G_3$	−0.461	0.289	−1.005	0.104

variables. Such accumulated counts are variously entitled frequency cross-tabulations, contingency tables, or cross-classifications. We generally adopt a logarithmic transform of the Poisson mean to model such data, to ensure the predicted count is positive. In fact, though, the data are multinomial frequency data, and we are using the equivalence of the multinomial with the conditional distribution of Poisson variables given their sum.

Sometimes there is a clear response variable in contingency tables (or indeed more than one response) with the remaining categorical variables predicting that categorical response. In other situations there is no clear distinction between predictor and outcome, and the focus is more on elucidating the structure of the cross-tabulation, with all the variables being regarded as responses. A further important distinction is between binary outcomes on the one hand, and multinomial or polytomous outcomes on the other (with three or more possible response categories). In the case of multinomial outcomes, there may be ordering of the response categories which is necessary to model, in order to fully utilise the sampling information.

Log-linear models also occur when there is a categorical response and a mixture of categorical and continuous predictor variables. In all these situations the basic response is a count variable and its prediction typically uses a log-link model, though in some cases fitting can equally well be done without adopting a log transform.

4.7.1 The two-way contingency table

Suppose we have a two-dimensional table with I row categories and J column categories, with y_{ij} as the count of respondents having attribute i of the row variable and attribute j of the column variable, and $n = \sum\sum y_{ij}$ is the total of respondents, or total sample size. In some situations there may also be an exposure E_{ij} defined for each table cell, if say the y_{ij} were cumulations of health events by age i and sex j over different lengths of exposure time (a mortality example is considered below).

The aim is to model the structure of the IJ counts without necessarily assuming one variable is an outcome and the other a predictor. The focus in this chapter is on fixed effects models, i.e. we do not suppose the parameters describing the impact of separate row or column categories come from some form of common density. For such count data, a Poisson model is usually the starting point, so that

$$y_{ij} \sim \text{Poi}(\mu_{ij})$$

where μ_{ij} is the Poisson mean. The log-linear model to explain this mean uses the log-link for the mean μ_{ij} and in a two-way table specifies four types of influence on it: the overall level of the counts, the differential effects of rows (α_i), the effects of columns (β_j), and the effect of each cell combination (i,j), called the interaction effect (γ_{ij}). An alternative data model would be that the y_{ij} are multinomial observations with cell probabilities π_{ij}

$$y_{ij} \sim \text{Mult}(\pi_{ij}, n).$$

The results under this assumption are the same as under the Poisson.

The row (column) effects express the relative frequency of each row (column) category: for example, a larger row (column) effect α_i (β_j) will attach to a more frequently occurring category. These are called the 'main effects'. If the best fitting model contains only main effects then the variables are effectively independent of one another. The interaction effects describe the degree of association between the row and column variables, and from the point of view of exploring associations between the variables are the major focus of interest (Upton, 1991). A model including all possible interactions and main effects represents all possible features of the data (i.e. results in a perfect fit) and is called a saturated model.

For example, suppose the two variables in a cross- classification measure father's status and son's status respectively using the same status ranking system. These cross-classifications are called social mobility tables, and a positive association between the two status variables ('occupational inheritance') will show if the interaction terms along the diagonal tend to be larger than the remaining interaction terms. The main effects describe the status distribution of fathers and sons.

The saturated log-linear model for the two-way table incorporates both possible main effects and the interaction effect:

$$\log(\mu_{ij}) = c + \alpha_i + \beta_j + \gamma_{ij}. \tag{4.15}$$

The number of parameters here, assuming they are all fixed effects, is $1 + I + J + IJ$, and exceeds the number of cells IJ. To identify the parameters constraints must be imposed. There are two ways of setting up these constraints.

The corner system imposes constraints by fixing one (usually the first) row effect and one column effect to a constant term, such as $\alpha_1 = \beta_1 = 0$. Also fixed are the first row and first column of the interaction parameters so that $\gamma_{1j} = 0$ for all j, and $\gamma_{i1} = 0$ for all i. The centred system stipulates that row and column effects sum to zero, so that one is a linear combination of the others and is not a free parameter. Also interaction parameters in each separate row and in each separate column sum to zero, $\sum_i \gamma_{ij} = \sum_j \gamma_{ij} = 0$. In a 2×2 table for example, the centred constraint would mean $\alpha_1 = -\alpha_2, \beta_1 = -\beta_2$ and $\gamma_{11} + \gamma_{21} = \gamma_{12} + \gamma_{22} = \gamma_{11} + \gamma_{12} = \gamma_{21} + \gamma_{22} = 0$. From this it follows that $\gamma_{11} = \gamma_{22} = -\gamma_{12} = -\gamma_{21}$. In estimation via repeated sampling it is possible to estimate parameters subject to one form of constraint (e.g. the corner system) but calculate the equivalent parameters which would have been estimated had a centred system been used.

Subject to sampling fluctuations, the predictions μ_{ij} under a saturated model such as (4.15) will more or less exactly reproduce the actual counts y_{ij}. In fact, the Bayesian sampling framework will provide the full distribution of each μ_{ij}, and there may be slight deviations of actual counts from fitted posterior means; by contrast the maximum likelihood analysis produces an exact match. A saturated model, or a nearly saturated model with several sets of interactions included, may achieve a close fit at the expense of parameter redundancy ('overfitting') with a number of parameters being poorly identified (e.g. in terms of the ratio of posterior means to posterior standard deviations).

4.7.2 Modelling choices and strategies

There are two major modelling choices in log-linear analysis. First, to simplify the saturated model and assess the comparative goodness of fit of a range of reduced models. For two-way tables these simpler models are comparatively few, but for higher dimensional tables, the model choice can become complex. A four-dimensional $I \times J \times K \times L$ table (e.g. political affiliation by sex by age by social class) would have four sets of main effects $\{\alpha_{1i}, \alpha_{2j}, \alpha_{3k}, \alpha_{4l}\}$, six sets of two-way interactions $\{\beta_{1ij}, \beta_{2ik}, \beta_{3il}, \beta_{4jk}, \beta_{5jl}, \beta_{6kl}\}$, four sets of three-way interactions $\{\gamma_{1ijk}, \gamma_{2ijl}, \gamma_{3jkl}, \gamma_{4ikl}\}$, and a four-way interaction term δ_{ijkl}.

An examination of parameters in the saturated model which are clearly identified as against those which are poorly identified is often the first step in such a model scanning and selection procedure. More complex procedures might involve comparing summary measures of fit as one type of parameter is omitted (e.g. Aitkin, 1979; Upton, 1981). For example, in a four-way table we might first exclude the four-way interaction and then all three-way interactions.

The other main modelling question is to what extent parameters can be expressed in a way which mirrors substantive features of the data; for example, if the rows and columns are ordered socio-economic categories or are geographic areas, then the interaction term may express social or geographic distance effects. One may also want to model the underlying continuous variable if there is an ordering to one or more of the dimensions of the table; neglecting this information means the model is less precise than it would otherwise be.

In the two-way model (4.15) the most obvious simplification is to assume away interactions between the row and column variable. A complete absence of interaction means that the two variables are independent of each other. The goodness of fit of various models may then be assessed for model choices between the saturated and independence model. These options may take account of substantive features of the type of tabulation being analysed.

The interaction term may be retained but in a simplified 'intermediate' form, leading to 'quasi-independence' models – see Leonard (1975) and Laird (1979) for Bayesian treatments. For example γ_{ij} might be expressed as the product of a row and column effects or scores (note that these are distinct from the main row and column effects), such as

$$\gamma_{ij} = \delta_i \varepsilon_j$$

so that instead of $(I-1)(J-1)$ free parameters describing the interaction pattern there are only $I+J-2$.

4.7.3 Fitting log-linear models for a two-way model

Suppose initially that we completely eliminate the interaction term in (4.15). There are now $1+(I-1)+(J-1)$ parameters to estimate, assuming that we are applying a fixed effects model. The Bayesian approach then specifies priors appropriate to the fixed effects included in the log-linear model. Assuming a log-link, these effects might typically range over the entire real line, so for example normal or uniform priors might be used. The corner constraint implies fixing the first row and column effect to zero; if a noninformative normal prior (with a large variance and no direction of effect) is assumed for the remaining effects then the prior might be something like

$$\alpha_1 = 0, \beta_1 = 0,$$
$$\alpha_i \sim \mathrm{N}(0, 100), \ i = 2, \ldots, I;$$
$$\beta_j \sim \mathrm{N}(0, 100), \ j = 2, \ldots, J.$$

In the BUGS program the *inits* file with initial values of *free* parameters would, in the case of $I = J = 3$ and an independence model, have the code `list (alpha=c(NA,0,0), beta=c(NA,0,0))`. This is on the assumption that zero initial values for $\alpha_2, \alpha_3, \beta_2$ and β_3 are the most appropriate (there may well be a basis for more sensible starting values of these parameters).

Example 4.21 Social mobility In a social mobility table, the case of no interactions between social origin i (parental social group) and respondent social group j is known as the 'perfect mobility' model. Under this model

$$\log(\mu_{ij}) = c + \alpha_i + \beta_j$$

or in multiplicative form

$$\mu_{ij} = a_i b_j$$

where $a_i = \exp(\alpha_i + 0.5c)$ and $b_j = \exp(\beta_j + 0.5c)$. Program 4.21 fits this model to British Social Mobility Data from Glass (1954), as in Table 4.18.

Table 4.18 Intergenerational social mobility.

	Son's status				
Father's status	1	2	3	4	5
1	50	45	8	18	8
2	28	174	84	154	55
3	11	78	110	223	96
4	14	150	185	714	447
5	0	42	72	320	411

There is an exact correspondence between the row and column ranks which can often be ordered with regard to prestige, status, etc. (A table with row and column variables corresponding in this way is often called a square table.)

Fitting the independence model gives a likelihood ratio G^2 statistic averaging 808 with a minimum of 799 (this compares the fit of the actual model to the saturated model). Fit along the main diagonal is not good, with status retention over generations underpredicted. The transition (1, 1) from high parental to high current status is predicted as $\mu_{11} = 37.6$, and the other diagonal predicted means are 69.2, 68.0, 617.2 and 246. So a more satisfactory model might treat the main diagonal differently from the rest of the table.

This is the basis of the 'quasi-perfect mobility' (QPM) model of Goodman (1981) which has

$$\mu_{ij} = a_i b_j, \qquad \text{if } i \neq j$$

but

$$\mu_{ij} = n_{ii}, \qquad \text{if } i = j.$$

The log-linear equivalent of this model involves $L = 2(I - 1) + (J - 1) + 2$ parameters, so that $IJ - L$ degrees of freedom remain, and has the form

$$\log(\mu_{ij}) = r + s_i + t_j, \qquad i \neq j$$
$$\log(\mu_{ii}) = u + v_i.$$

Fitting this model gives an average G^2 of 258 with a minimum of 245. Note that the fitted means along the main diagonal do not exactly reproduce the observed values, due to sampling variability. The full set of fitted means under the QPM model, and their 95 % credible intervals, are given in Table 4.19.

The fit off the main diagonal is improved but discrepancies still remain, for example in the predicted pattern of downward mobility from origin status 1 to status 2, 3, 4 and 5. Longer distance downward mobility is overpredicted and short distance mobility (from status 1 to 2) is underpredicted. Here distance refers to social distance in the sense of prestige gap between origin and destination status.

Another approach (the quasi-symmetry model) to these transition tables is considered below.

Table 4.19

	Mean	SD	2.5%	97.5%
μ_{11}	51.2	7.2	37.6	65.7
μ_{12}	9.7	1.2	7.5	12.2
μ_{13}	11.2	1.3	8.7	14.0
μ_{14}	38.9	4.3	30.9	47.8
μ_{15}	20.2	2.3	16.0	25.1
μ_{21}	6.7	1.0	4.9	8.8
μ_{22}	173.7	13.3	148.4	199.9
μ_{23}	49.8	3.9	42.7	57.6
μ_{24}	173.8	10.8	154.0	195.2
μ_{25}	90.3	6.3	78.5	103.0
μ_{31}	8.7	1.3	6.2	11.4
μ_{32}	56.1	4.2	48.1	64.8
μ_{33}	109.8	10.4	90.6	130.8
μ_{34}	225.9	12.8	201.3	252.2
μ_{35}	117.4	7.6	102.8	132.9
μ_{41}	28.0	4.1	20.7	36.6
μ_{42}	180.9	11.2	159.9	203.9
μ_{43}	208.7	12.2	185.9	233.9
μ_{44}	714.2	26.4	663.2	766.4
μ_{45}	378.2	17.4	345.4	413.3
μ_{51}	10.6	1.6	7.7	13.9
μ_{52}	68.6	5.2	59.0	79.1
μ_{53}	79.2	5.8	68.3	91.0
μ_{54}	276.1	14.7	248.3	305.6
μ_{55}	410.8	20.4	371.3	451.5

Example 4.22 Inter-sib marriage: A table with structural zeros The zero cell in Table 4.18 with value zero is known as a sampling zero, arising purely from the rareness of long-distance mobility from status band 5 to band 1. By contrast some cross- tabulations contain entries which are zero by definition, i.e structural zeros. Bishop *et al.* (1975) analyse data on intermarriage between sibs (clans related by kinship) among the Purum people of India. Not only is exogamy (marriage outside the sib) the rule, but certain other inter-sib marriages are not allowed. Table 4.20 contains observations on 128 Purum marriages with the forbidden marriage links denoted NA. There is one intra-sib marriage but in fact this was between partners in different sub-sibs. The table also contains two sampling zeros (e.g. there were no marriages between Makan and Parpa sibs although they were not forbidden).

The model applied to these data is one of independence but confined to permitted ties (i.e a form of quasi-independence). A 5×5 matrix $D[,]$ is defined with elements 1 for permitted ties, and 0 for structural zeros. Then the sampling is confined to permitted cells using the equals(,) command in BUGS to differentiate between $D_{ij} = 1$ and $D_{ij} = 0$. The estimates of posterior Poisson means

Table 4.20 Purum marriage ties (observed and posterior mean estimates).

		Sib of husband				
Sib of wife		Marrim	Makan	Parpa	Thao	Kheyang
Marrim	Obs	NA	5	17	NA	6
	Est	NA	10.86	10.71	NA	6.58
Makan	Obs	5	NA	0	16	2
	Est	4.86	NA	6.54	7.58	4
Parpa	Obs	NA	2	NA	10	11
	Est	NA	8.35	NA	9.52	5.05
Thao	Obs	10	NA	NA	NA	9
	Est	10.35	NA	NA	NA	8.65
Kheyang	Obs	6	20	8	0	1
	Est	5.75	7.83	7.71	8.95	4.71

using *Program 4.22 InterSib Marriage* are similar to those of Bishop *et al.* (1975), and the deviance statistic has a minimum of 75.9, close to the $G^2 = 76.2$ value of Bishop *et al.*

4.7.4 The quasi-symmetry model

One attempt to allow for off-diagonal patterns such as those in social mobility tables of Example 4.21 is the quasi-symmetry model (QSM) of Caussinus (1965). The basis of this is that patterns of upward and downward status change tend to be parallel in the sense that short-distance moves outnumber longer distance moves; this would be expected to produce (approximate) symmetry in a square table, namely $\mu_{ij} \approx \mu_{ji}$. In fact, exact symmetry implies that row marginal totals μ_{i+} are equal to column marginal totals μ_{+i}, a pattern known as 'marginal homogeneity'.

The quasi-symmetry model therefore re-introduces interaction parameters γ_{ij} but assumes that they are equal in the off-diagonal cells, namely that $\gamma_{ij} = \gamma_{ji}$. Then the log-linear version of the model is

$$\log(\mu_{ij}) = \delta + \alpha_i + \beta_j + \gamma_{ij} \tag{4.16}$$

with $\gamma_{ij} = \gamma_{ji}$ and with the usual corner or zero sum constraints applying to α_i and β_j. The identification constraint on the interaction terms applies just to the rows of γ_{ij}, for example that $\sum_i \gamma_{ij} = 0$ under a zero sum constraint. The quasi-symmetry model can also be stated in multiplicative form as

$$\mu_{ij} = a_i b_j e_{ij}, \quad i \neq j$$

where $e_{ij} = e_{ji}$ and

$$\mu_{ii} = a_i.$$

A particular type of QSM model is the diagonal parameters model for off-diagonal cells, namely

$$\mu_{ij} = a_i b_j d_k \tag{4.17}$$

where $k = i - j$ for $i \neq j$, and k would have values 1, 2, 3, 4 and -1, -2, -3, -4 in a 5×5 table. In the social mobility context, the parameters d_k would measure social distance impacts and the expected decline in mobility as k increases in absolute size. It is usually assumed that downward and upward effects are the same, i.e. that $d_k = d_{-k}$. As in quasi-perfect mobility, the diagonal parameters are intended to exactly reproduce the cells n_{ii}.

An epidemiological application of the quasi-symmetry model is to case-control data with equal numbers of controls for each case (Lovison, 1994). In particular, suppose there are n matched pairs (one control to each case) and a polytomous exposure variable with I levels. Then the data can be represented as an $I \times I$ 'concordance' table with y_{ij} the number of pairs in which a case is exposed to exposure level i and a control is exposed to level j. Thus the row and column variables are the same but observed on two members of a pair. The expected frequencies μ_{ij} can be modelled as follows

$$\mu_{ij} = n\pi_{ij}\Psi_{ij}/(1 + \Psi_{ij}) \tag{4.18}$$

where π_{ij} is the probability that one member of a pair is exposed to risk level i and the other to level j; and where Ψ_{ij} is the (i,j)th exposure odds ratio, namely

$$\begin{aligned}\Psi_{ij} = {} & \text{Prob(exposure at level } i|\text{case)Prob(exposure at level } j|\text{control)}/ \\ & \text{Prob(exposure at level } j|\text{case)Prob(exposure at level } i|\text{control)}.\end{aligned}$$

If the Ψ_{ij} terms are constant over the matching variables they satisfy the condition

$$\Psi_{ij} = \Psi_{ib}/\Psi_{jb} \tag{4.19}$$

where b is the baseline exposure (Breslow and Day, 1980, p. 183). Hence the $I(I-1)/2$ odds ratios can be expressed as $(I-1)$ parameters $\Psi_{ib} = \alpha_i$. So there is an effect of exposure on the disease outcome and its effect depends on the level of exposure – this is to be expected if the matching variables are appropriate. The equivalent log-linear model is

$$\begin{aligned} y_{ij} &\sim \text{Poi}(\mu_{ij}) \\ \mu_{ij} &= M + \delta_{ij} + \alpha_i, \quad i \neq j \\ \mu_{ii} &= M + \gamma_i \end{aligned}$$

with $\delta_{ij} = \delta_{ji}$, and the corner constraints $\alpha_1 = 0, \gamma_1 = 0$. The hypotheses of no effect and constant effect respectively correspond to $\alpha_i = 0$ and $\alpha_i = \alpha$.

Example 4.23 Social mobility (continued) The quasi-symmetry model of (4.16) applied to the data in Table 4.18 gives an average G^2 of 28.4 and minimum of 12.6 (see Program 4.23). This is a considerable improvement over the perfect or quasi-perfect models.

The minimum G^2 fit of the 'social distance' model of (4.17) (see *Program 4.23 Social Distance Model*) is 20.7 with an average of 35.5, compared with a maximum likelihood value of 19.1 obtained by Bishop *et al.* (1975, p. 228). The d_k parameters (and their 95% credible intervals after 20 000 iterations) are respectively $d_1 = 1$, $d_2 = 0.59\ (0.53, 0.66)$, $d_3 = 0.26\ (0.21, 0.32)$ and $d_4 = 0.084\ (0.035, 0.158)$. There is the expected decline with social distance, and in fact an approximately geometric progression. Note that we may fit this model in multiplicative rather than log-linear form, so the priors on free parameters are then expressed in terms of gamma densities, namely G(0.01, 0.01). The fitted means under the diagonal parameters and quasi-symmetry models are given in Table 4.21.

Example 4.24 Matched pairs by blood group Lovison (1994) analyses data on 301 matched pairs classified by the risk variable blood group with four levels (groups O, A, B and AB). There is no 'exposure absent' category and instead group O is the reference category. The contingency table is then a concordance table (Table 4.22).

Table 4.21

	Quasi-symmetry		Diagonal parameters	
	Mean	SD	Mean	SD
μ_{11}	46.9	6.4	51.6	6.6
μ_{12}	43.3	6.2	35.7	4.8
μ_{13}	11.5	2.7	15.4	2.2
μ_{14}	18.4	3.2	21.4	3.3
μ_{15}	7.1	1.6	5.4	1.9
μ_{21}	30.3	5.4	24.2	3.5
μ_{22}	174.0	12.9	174.1	13.1
μ_{23}	78.2	7.4	85.8	7.1
μ_{24}	155.5	11.2	155.9	11.1
μ_{25}	56.9	6.2	55.3	5.8
μ_{31}	8.5	2.2	11.2	1.7
μ_{32}	83.4	7.7	92.2	7.2
μ_{33}	109.9	10.6	109.9	10.5
μ_{34}	215.2	13.4	208.2	12.6
μ_{35}	101.2	8.6	96.8	8.0
μ_{41}	12.4	2.8	13.8	2.4
μ_{42}	148.7	11.0	148.5	10.7
μ_{43}	193.0	12.4	184.6	11.8
μ_{44}	714.6	27.4	712.9	26.5
μ_{45}	441.7	20.1	449.1	20.4
μ_{51}	3.5	0.9	2.6	1.0
μ_{52}	40.0	4.7	38.7	4.3
μ_{53}	66.8	6.3	63.0	5.8
μ_{54}	324.7	17.1	329.6	17.0
μ_{55}	410.7	20.4	410.3	20.0

Table 4.22

	Control			
Case	O	A	B	AB
O	64	18	8	3
A	66	74	14	6
B	4	2	4	2
AB	12	10	12	2

Table 4.23

	Mean	SD	2.5%	Median	97.5%
Ψ_1	3.68	0.94	2.21	3.55	5.85
Ψ_2	0.58	0.26	0.21	0.54	1.21
Ψ_3	5.41	2.42	2.21	4.90	11.51

The posterior estimates of the exposure odds ratios for groups A, B and AB assuming model (4.19) are obtained from the second half of a run of 10 000 iterations (Table 4.23).

As can be seen from Table 4.23, skew in the densities leads to the mean odds ratios exceeding the medians. The posterior medians are closer to those reported by Lovison, namely $\Psi_2 = 3.50$, $\Psi_3 = 0.56$ and $\Psi_4 = 4.67$.

4.8 VARIABLE AND MODEL SELECTION IN REGRESSION AND LOG-LINEAR MODELS

In a log-linear model for a contingency table, certain terms such as global intercept and main effects are necessarily included, but the inclusion of others (such as second and higher order interactions) is open to doubt. This raises the question of selection among a range of possible models and of Bayesian perspectives on model selection. Similarly in univariate regression with response variable y and a set of p potential predictors, $x_1, \ldots, x_p$, there is the question of selecting the best possible regression model. The constant is always assumed present, though may be neglected if both dependent and independent variables are centred. There are 2^p potential regression models; for example with two predictors x_1 and x_2, and denoting x_0 as the constant, the models are (x_0), (x_0, x_1), (x_0, x_2) and (x_0, x_1, x_2).

In a Bayesian approach to selecting the best log-linear or regression model, we may set a discrete prior on the chance attached to each of the possible models that we are considering. For a univariate regression this may not necessarily be all 2^P possibilities but a smaller set, perhaps based on initial examination using the usual forward or backwards election methods. Suppose we have K possible models, then a default prior on the different models is just $\pi_1 = \pi_2 = \ldots = \pi_K = 1/K$. Initial probabilities on the models will be modified to posterior probabilities $P_j, j = 1, \ldots, K$, and we can compute Bayes factors on different models. If we just have $K = 2$ then the prior odds favouring model 1 over 2 are π_1/π_2, and the Bayes factor favouring model 1, namely B_{12}, is $(P_1/P_2)/(\pi_1/\pi_2) = (P_1\pi_2)/(\pi_1 P_2)$.

4.8.1 Selection among regression models

A two-stage methodology for variable (and hence model) selection is suggested by George and McCulloch (1993). In its full version this approach assigns a latent inclusion variable δ_i to every predictor, except usually for the intercept always being assumed present. We may have prior reasons, however, to suppose some of the predictors apart from the intercept should necessarily be included. In the situation where all p predictors may be included or excluded, each component of $(\beta_1, \ldots, \beta_p)$ is modelled as a normal mixture

$$\beta_i \,|\, \delta_i \sim \delta_i \mathrm{N}(0, \eta_i \Phi_i^2) + (1 - \delta_i)\mathrm{N}(0, \Phi_i^2). \tag{4.20}$$

If $\delta_i = 1$ then β is distributed as a normal with mean zero and variance $\eta_i \Phi_i^2$. We take η_i as a large constant premultiplying a small default variance term Φ_i^2. If inclusion of β_i is not supported by the data then the prior with this default variance, namely $\mathrm{N}(0, \Phi_i^2)$, will tend to be selected more often. For example, if $\eta_i = 1000$ and $\Phi_i^2 = 0.01$, then the prior variance of β_i if $\delta_i = 1$ is 10. The choice $\delta_i = 1$ then corresponds to retaining predictor x_i, while $\delta_i = 0$ means that β_i is distributed around zero with a small variance, $\beta_i \sim \mathrm{N}(0, 0.01)$. The latter selection therefore means in effect that the data provide little support for a nonzero β_i.

The prior is then specified in terms of $M \leq 2^p$ discrete patterns each corresponding to different combinations of predictors $\{x_1, \ldots, x_p\}$ being included in the regression. We apply this discrete prior and see which of the distinct patterns is selected most often. If the data clearly support a particular combination of covariates more often than the others then our second stage is simply to fit this selection of covariates. Brown *et al.* (1998) extend this model selection procedure to multivariate dependent variables and also assess different models by their ability to predict outside the sample.

Joint space procedures

Direct joint space procedures have also been proposed for selection between two or more competing regression models. These involve jointly searching on both model and model parameters, with possible jumps between models at each MCMC iteration. In the procedure of Carlin and Chib (1995), this involves setting both the standard 'true priors' for each model assuming it is chosen to be better fitting, as well as 'pseudo-priors' for the parameters in the other models which are not (currently) chosen. These are linking densities needed to completely define the joint model. For example, if we had parameters β for model 1 and parameters γ for model 2, we would need the standard estimation priors, $\pi_1(\beta)$ and $\pi_2(\gamma)$ that we would use to estimate models 1 and 2 in the absence of any model choice problem. But we would also need a pseudo-prior $T_1(\beta)$ to cover the situation when model 2 is chosen, and another $T_2(\gamma)$, on the model 2 parameters, when model 1 is chosen. These priors are needed even if the component is not meaningful in the other model.

Both sets of priors may be chosen from initial single model alone runs, for example via least squares in a normal regression example. For $M = 2$, the initial runs could be via least squares run just on model 1, and then just on model 2. Carlin and Chib suggest using the results, namely $\hat{b}$ and SE($\hat{b}$) from these initial fits as the basis for the pseudo-priors on the regression parameters β. That is, they are just taken without modification to provide the linking densities. The standard priors may be taken as the standard non informative ones (e.g. mean zero, low precision), or centred on the least squares estimates but with reduced precision.

Suppose a single model run OLS estimate of $\beta, \hat{b}$ had precision $\hat{\tau}(\hat{b})$. Adapting George and McCulloch (1993) we may assess sensitivity to priors by various scalings of the single model regression coefficient precisions. To obtain the pseudo-prior $T_1(\beta)$, we may scale $\hat{\tau}(\hat{b})$ by a factor f set close to unity (note that $f = 1$ in Carlin and Chib, 1995). For the standard prior $\pi_1(\beta)$ we reduce precision by applying a factor $g < 1$ to $f\hat{\tau}(\hat{b})$, i.e. the precisions on the regression parameters in $\pi_1(\beta)$ are $fg\hat{\tau}(\hat{b})$. We can vary the choices (f, g) to identify sensitivity to prior specification; a typical pair of values might be $\{1, 0.001\}$.

Example 4.25 Two-stage variable selection with simulated data We follow George and McCulloch (1993) in generating a sample of 60 Normal linear outcomes y_i as follows:

$$y_i = x_{4i} + 1.2x_{5i} + e_i$$

where x_1, x_2, x_3, x_4 and x_5 are distributed as N(0, 1), and the e_i are N(0, 6.25) (see *Program 4.25 Regression Selection, Simulated Data*). We then fit a model with all five predictors potentially included or excluded, namely:

$$y_i \sim \mathrm{N}(\mu_i, \sigma^2)$$
$$\mu_i = \beta_1 x_{1i} + \beta_2 x_{2i} + \beta_3 x_{3i} + \beta_4 x_{4i} + \beta_5 x_{5i}.$$

The 'true' model should include at least one of x_4 and x_5 though randomly generated patterns for a relatively small sample may lead to other configurations being present in the data. We adopt the discrete mixture from (4.20) but instead of the full version with $2^5 = 32$ possible choices confine our choice to $M = 12$ options:

all included (i.e. x_1, x_2, x_3, x_4, x_5);
none included;
x_4 and x_5 only;
x_4 only;
x_5 only;
(x_1, x_2, x_3) but neither x_4 nor x_5;

and then the six options formed by retaining either one or two from (x_1, x_2, x_3) in addition to (x_4, x_5).

We adopt a uniform prior probability of 1/12 for each of these options. We carry out a 10 000 iterations run, and in the second half find relative sampling frequencies on each of these options which enable calculation of Bayes factors on the possible models. The combination (x_3, x_4, x_5) is selected in 27 % of the iterations, (x_4, x_5) in 34 %, and x_4 alone in 13 %. However due to the vagaries of the random sample, the model with x_5 alone is supported only infrequently. The option (x_1, x_2, x_3) is selected only 7 times out of 5000.

Example 4.26 Joint space model choice with the Hald data The Hald data on heat evolved in a chemical reaction are often used in studies of variable selection; they are reproduced in Draper and Smith (1980) who also give results on a range of possible models for the data. There are four predictors x_1, x_2, x_3 and x_4 denoting inputs to the reaction, and Draper and Smith identify two models with just two predictors which have high explanatory power. These are, with constant included, (x_0, x_1, x_2) and (x_0, x_1, x_4), and constitute models 1 and 2 in the program *Program 4.26 Chemical Reaction* with respective parameters β and γ. This program enables initial single model only runs taking $\pi_1(\beta) = 1$, $\pi_2(\gamma) = 0$ (giving model 1 estimates in isolation) and $\pi_1(\beta) = 0, \pi_2(\gamma) = 1$. The estimated means (SDs) of the conditional precision τ_y emerging from these two runs were 0.17 (0.07) and 0.13 (0.06), so we set pseudo-priors on the precisions to be $G(5, 29)$ and $G(5, 36)$ densities.

Priors for regression parameters may draw on the single model runs (though Draper and Smith also provide regression coefficients and standard errors for a range of models). For example, pilot runs on model 1 give the following estimates, with standard errors, for the regression parameters on (x_0, x_1, x_2): 53 (2.7), 1.47 (0.12) and 0.66 (0.05). We then set the standard prior $\pi_1(\beta_0)$ on the intercept β_0 as

$$\beta_0 \sim N(53, \tau_{01})$$

where $\tau_{01} = [fg/(2.7^*2.7)]$. The pseudo prior $T_1(\beta_0)$ on β_0 is $N(53, \upsilon_{01})$ where $\upsilon_{01} = [f/(2.7^*2.7)]$. We take $f = 1$ and $g = 0.2$ initially on all these regression priors.

Taking prior model probabilities $\pi_1 = \pi_2 = 0.5$ in a model choice run of 20 000 iterations (and burn-in of 5000) results in a posterior probability on model 2 of 0.223. Changing the parameters (f, g) to (1,0.001), (0.25,0.001) and (0.25, 0.2) gave model 2 probabilities of 0.223, 0.175 and 0.175. So there appears to be slight evidence in favour of model 1 with (x_1, x_2) as predictors, but this is hardly decisive. This is broadly consistent with the least squares evidence in Draper and Smith (1980) which gives model 1 an R^2 of 97.9 % and model 2 an R^2 of 97.2 %. Note that the standard deviation of the model 2 probability can be obtained from the binomial formula as $(0.223 \times 0.777/15000)^{0.5} = 0.0029$.

4.8.2 Selection of log-linear models

A similar methodology to that of McCulloch and Rossi has been proposed by Albert (1996) for model selection in the case of log- linear models. The particular application in Albert (1996) was to contingency tables, where certain terms of the log-linear model, such as global intercept and main effects, are assumed to be included, but the inclusion of others (such as second and higher order interactions) is open to test. For example, let y_{ij} denote counts in a two-way table, with

$$y_{ij} \sim \text{Poi}(\mu_{ij}),$$
$$\log(\mu_{ij}) = u_0 + u_{1i} + u_{2j} + u_{12ij}.$$

Our baseline model includes the main effects but the interaction term is not certain to be necessary. We therefore test the model without this term against the model including it. Albert proposes the usual fixed effects priors for the included main effects, with the corner constraint $u_{11} = u_{21} = 0$, and $u_{1i} \sim \text{N}(0, T_1^{-1}), i = 2, \ldots, I$; $u_{2j} \sim \text{N}(0, T_2^{-1})$, where typically we take T_1 and T_2 known and equal to a small number T.

For the interaction term at issue, Albert proposes an exchangeable normal hierarchical prior (discussed further in Chapter 5), namely $u_{12ij} \sim \text{N}(0, P_h^{-1})$ where P_h denotes a precision parameter under model h. The null model with $u_{12ij} = 0$ corresponds to $P_0 = \infty$, since then the u_{12ij} are zero with certainty. The prior for the model with nonzero interaction terms, by contrast, has a relatively small precision P_1 on the value 0 thus allowing real (nonzero) effects to emerge. Albert proposes a mixture prior which chooses between these alternatives. Either we can take P_1 as fixed and experiment with different values (the Bayes factor may be sensitive to choice of P_1) or we can establish a prior on P_1 itself.

It may be noted that an alternative prior for the zero interaction null model is to take it to correspond to $\text{N}(0, P_0^{-1})$, where the precision P_0 is large enough to make the interactions effectively being estimated as zero. This is similar to the George and McCulloch approach in normal linear regression.

In these circumstances analytical forms for the Bayes factor are obtainable, and Albert investigates their sensitivity to varying values of T and P_1. A possible BUGS coding for this approach, including a prior on P_1, is as follows:

```
for (i in 1:N) {for (j in 1:N) { y[i,j] ~ dpois(m[i,j]);
  log(m[i,j]) <- u0 +u1 [i] +u2 [j] +u12[k,i,j] }}
# priors on main effects
  T < -0.001; u0 ~ dnorm (0,T)
  u1[1] < - 0; u2[1] <-0; for (i in 2:N) {u1[i] ~ dnorm(0,T); u2[i] ~
dnorm (0,T)}
# discrete prior on model choice
  prior[1] <- 0.5; prior[2] <- 0.5; k ~ dcat(prior[]);
# priors on interaction terms under H0 and H1
```

```
for (i in 1:N) {for (j in 1:N){ u12[1,i,j] ~ dnorm(0,P1)
                                u12[2,i,j] <- 0}}
P1 ~ dgamma (0.01,0.01)
```

Example 4.27 Trial contingency table Albert (1996) analyses the 3×3 data $\{14, 3, 5; 7, 20, 10; 4, 8, 12\}$. Applying the coding in Program 4.27 gives 0.084 as the proportion of 100 000 iterations for which the independence hypothesis is selected. Since the prior probabilities on the alternative hypotheses are equal, the Bayes factor against independence is 10.9.

Scale mixture prior

As a more general approach to robust log-linear model selection, Albert proposes a scale mixture prior for the parameters whose inclusion is in doubt. Thus for a two-way table, we would have $u_{12ij} \sim \mathrm{N}(0, P_1^{-1}ij)$ and with $P_1^{-1}ij$ taken from a $\mathrm{G}(\nu/2, \nu t/2)$ density. Taking $\nu = 1$ leads to a Cauchy with scale parameter t. For the Cauchy we can be 75% certain that the density is between $-t$ and $+t$ (i.e. these amount to prior expectations about the location of the lower and upper quartile respectively).

Albert investigates a data set also analysed by Raftery *et al.* (1993) and Raftery (1996) concerning the impact of oral contraceptive use and age on a woman's chance of myocardial infarction. The $2 \times 5 \times 2$ data consist of observations y_{ijk} on contraceptive use i (= 1 for No, = 2 for Yes), age group j (25–29, 30–34, up to 45–49), and infarction ($k = 1$ for No, $k = 2$ for Yes). The terms in doubt are the second order interactions, u_{13ik} between contraceptive use and infarction, and the third order terms, u_{123ijk}. The other second order interactions are assumed to be necessary. So there are four possible hypotheses (models 1 to 4) to assess:

1. $u_{13} = u_{123} = 0$ (i.e. no extra terms are needed in the model, in the sense that these parameters are effectively zero).
2. $u_{13} \neq 0, u_{123} = 0$.
3. $u_{13} = 0, u_{123} \neq 0$.
4. $u_{13} \neq 0, u_{123} \neq 0$.

The third hypothesis does not, of course, conform to the usual hierarchical assumptions made in testing log-linear models.

Example 4.28 Contraceptive use As a prior for the nonzero interaction alternative we adopt the scale mixture of Albert, but take $t = 1$ as the scale parameter for the second order parameters, and $t = 0.5$ for the third order case. Thus, we expect the second order terms u_{13} to be symmetric about zero and with upper and lower quartiles, 1 and -1, and also expect the third order interactions to be less widely scattered than the second order ones. For the alternative zero interaction hypothesis, we take $\mathrm{N}(0, 10^{-1})$ as representing a prior effectively equivalent to zero effects. It must be emphasised that these are only one among a range of choices and Bayes factors for discriminating between nonzero and zero interactions will be influenced by the prior selected. However, there will generally be consistency in selecting the most appropriate model.

A run of 50 000 iterations (see *Program 4.28 Contraceptive Use and MI*) shows around 6:4 support in favour of model 1 against model 2, with models 3 and 4 only occasionally chosen. Albert adopts the approach outlined above where there is infinite precision $P_0 = \infty$ on the cases where one or both of u_{13} and u_{123} are 0. He obtains more support for models 2 and 4 (and none for model 1). Nevertheless the estimated log-odds ratios for different age groups of women obtained here are very close to those cited by Albert.

A second analysis assumes the prior effectively equivalent to zero interaction effects as $N(0, 20^{-1})$. This gives a higher weight, around 7:3 on model 2. The age effects given by this and the $N(0, 10^{-1})$ option are shown in Table 4.24.

4.9 LOG-LINEAR MODELS FOR CUMULATED EVENT DATA

In epidemiological and medical applications the dependent variable is often an event count (such as hospital admissions or deaths) accumulated over sub-groups which are defined by combinations of risk factors. For each count there may be a corresponding total of exposure (e.g. person-years at risk). For example, we might have cancer deaths accumulated by age group, sex, ethnicity

Table 4.24 Estimated log-odds ratios (of MI by age group of woman).

Age band	Unequal risks (ML) (saturated model)	Equal risks (ML)	Albert prior (Bayes)
25–29	1.98 (0.88)	1.38 (0.25)	1.33 (0.57)
30–34	2.18 (0.48)	1.38 (0.25)	1.83 (0.44)
35–39	0.43 (0.57)	1.38 (0.25)	0.71 (0.48)
40–44	1.31 (0.54)	1.38 (0.25)	1.22 (0.45)
45–49	1.36 (0.62)	1.38 (0.25)	1.16 (0.48)

	Program 4.28 Analysis Prior N(0, 0.1) on zero interactions		Prior N(0, 0.05) on zero interactions	
Age band	Mean	SD	Mean	SD
25–29	1.38	0.55	1.32	0.47
30–34	1.81	0.43	1.66	0.40
35–39	0.71	0.44	0.85	0.40
40–44	1.22	0.45	1.21	0.42
45–49	1.23	0.49	1.23	0.43

and by grouped measures of one or more exposure variables. Such exposure variables might relate to dietary composition, smoking level and years of smoking, industrial exposures, etc. In disease mapping applications, the observations may be small-area counts of deaths or other adverse health events (e.g. new cancer cases), aggregated over age and other demographic categories. Here the expected values (exposures) would be obtained by demographic standardisation methods. These control for different age structures between areas and possibly for socio-economic confounding also. With large data sets derived from official mortality records or large observational studies, such an aggregate or grouped analysis of the data is the most feasible way of proceeding (or the only permissible way given confidentiality restrictions, for example on registered mortality data).

Suppose we generically indicate the subgroups defined by combinations of risk factors by an index i with a maximum of I subgroups (e.g. if the aggregation were just by age and an exposure, and if there were 10 age groups, and 10 grouped levels of the exposure variable, then $I = 100$). In effect the outcome is multinomial with I categories, but the Poisson model with log link is usually more feasible.

If the observed count in subgroup i is y_i and the exposure is E_i then the expected number of events in subgroup i is a Poisson variable with mean $E_i\mu_i$ where μ_i is predicted using a log link. Suppose x_i denotes the risk factors expressed as categorical predictors. Then the regression effect describing the separate and joint effect of these factors follows the same rules as for any multidimensional table (e.g. for 10 ages and 10 grouped exposures the saturated log-linear model has a grand mean, 18 free main effects, and 81 interaction parameters). In disease mapping, the covariates would often be continuous, describing geographic location or aggregate risk scores (e.g. area deprivation scores).

Example 4.29 Cigarette smoking and lung cancer deaths Doll (1971) discusses the results of a large follow-up study of British physicians, on the relation between lung cancer deaths and

(A) years of smoking with 9 groups (grouped as physician age minus 20, so that physicians aged 35–39 are assumed to be exposed for 15–19 years);
(B) grouped smoking with 7 levels (nonsmoker, 1–9 per day,... 35+ per day).

Suppose we use separate indices for each categorical predictor, so that $i = 1, \ldots, 9$ and $j = 1, \ldots, 7$. Then Table 4.25 contains exposures in person-years E_{ij} for each of the 63 risk groups and the number of nonzero deaths y_{ij} (in brackets).

The response is deaths $y_{ij} \sim \text{Poi}(E_{ij}\mu_{ij})$ where μ_{ij} is predicted by the row (year smoking) effect α_i, the column (smoking level) effect β_j, and also possibly by an interaction term γ_{ij}. Suppose we initially fit the independence model,

$$\log(\mu_{ij}) = c + \alpha_i + \beta_j.$$

Table 4.25 Exposure and lung cancer deaths (bracketed) by years of smoking and smoking level.

Years smoking*	Cigarettes per day						
	None	1–9	10–14	15–19	20–24	25–34	35+
15–19	10366(1)	3121	3577	4317	5683	3042	670
20–24	8162	2937	3286(1)	4214	6385(1)	4050(1)	1166
25–29	5969	2288	2546(1)	3185	5483(1)	4290(4)	1482
30–34	4496	2015	2219(2)	2560(4)	4687(6)	4268(9)	1580(4)
35–39	3512	1648(1)	1826	1893	3646(5)	3529(9)	1336(6)
40–44	2201	1310(2)	1386(1)	1334(2)	2411(12)	2424(11)	924(10)
45–49	1421	927	988(2)	849(2)	1567(9)	1409(10)	556(7)
50–54	1121	710(3)	684(4)	470(2)	857(7)	663(5)	255(4)
55–59	826(2)	606	449(3)	280(5)	416(7)	284(3)	104(1)

* Years exposed for nonsmokers

It may be noted that we can equivalently express the model as $y_{ij} \sim \text{Poi}(\lambda_{ij})$ where independence between risk factors gives

$$\log(\lambda_{ij}) = \log(E_{ij}) + c + \alpha_i + \beta_j.$$

This is the form used in *Program 4.29 Lung Cancer Deaths,* and which in epidemiology is known as the 'product model'. We adopt noninformative priors on c, α_i and β_j since when it was done it was a pioneering study, though now there is substantial evidence on this dose–response relation to make priors more informative. It may be noted that the above model amounts to assuming the age-specific 'hazard rate' with respect to years of smoking alone is a piecewise exponential, and so is consistent with the proportional hazards assumption (see Chapter 9).

After 10 000 iterations, we obtain the estimates of the age effects per 100 000 person-years of exposure, namely 100000exp$(c + \alpha_i)$, and the smoking effects, $\exp(\beta_j/\beta_1)$, as in Table 4.26.

The posterior medians are close to values obtained by Frome (1983) by 'plugging in' the maximum likelihood estimates of c, α_i and β_j. The Bayesian sampling approach enables one to monitor the exact distribution of these effects; one can see they show both low precision and skewness, with the medians lower than the means. The median relative risk of cancer attached to smoking over 35 cigarettes per day is about 39 times that associated with no smoking. The average value of the deviance for this model is 48.5 with a minimum (over 10 000 samples) of 33.2. This compares with a figure of 51.5 quoted by Frome (1983)

It is of interest here to compare the fit of the categorical risk factor model with a nonlinear regression based on using approximate quantitative values attached to each row and column factor. For the years variable these are 17.5, 22.5,... up to 57.5 while the averages for 1–9, 10–14, 15–19, ... etc. cigarettes per day are 5.2, 11.2, 15.9, 20.4, 27.4 and 40.8. If the analysis were confined to smokers then a model of the form, for cigarettes per day (d) and years smoking (t)

Table 4.26

Age	Mean	SD	Median
15–19	0.4	0.4	0.3
20–24	0.9	0.7	0.7
25–29	1.9	1.3	1.6
30–34	8.6	5.0	7.7
35–39	8.9	5.2	7.9
40–44	23.5	13.4	21.0
45–49	29.9	17.1	26.6
50–54	46.8	26.8	41.6
55–59	77.8	45.0	69.5
Cigarettes/day			
None	1.0		
1–9	4.7	4.4	3.5
10–14	11.4	9.8	8.6
15–19	14.0	12.0	10.6
20–24	25.3	20.3	19.5
25–34	31.4	25.3	24.2
35+	51.4	41.9	39.6

$$\mu(t, d) = \zeta d^{\beta} t^{\alpha}$$

would be appropriate. Doll (1971) proposed that the mortality rate among smokers is approximately proportional to d and to the fourth power of years of smoking t (so that $\beta = 1$ and $\alpha = 4$).

To include nonsmokers the daily cigarette variable must be modified. Frome fits a nonlinear model as follows

$$\mu(t, d) = (\delta + \zeta d^{\beta}) t^{\alpha}$$

and obtains a deviance 8.1 higher than under the product model (i.e. 59.6) though with 11 fewer parameters. An alternative model, with 12 fewer parameters,

$$\mu(t, d) = (\delta + d)^{\beta} t^{\alpha}$$

is fitted here, with t rescaled as (years of smoking)/42.5. So δ^{β} is the baseline incidence for a nonsmoker at age 62.5 (for whom $d = 0$ and $t = 1$ since exposure is then 42.5 years).

Fitting this model – see the program *Nonlinear Model* in Program 4.29 – illustrates the use of a grid search on a parameter which makes the posterior density non-log-concave. However, with δ fixed the model is log-concave. A wide grid can be set at the first stage on δ (e.g. 20 bins spaced 10 apart) to find where the density is concentrated, and then the analysis repeated with a narrower grid interval. Equal prior probabilities 1/20 are set on each possible value. Here a second stage 100 value grid in steps of 0.25 from 0.25 to 25 was used and $\log(\delta)$ monitored, as the distribution of δ itself is positively skewed. Note that a Metropolis search is also possible in WINBUGS with a more standard parametric prior, though grid priors may have advantages for skewed parameters.

The second half of a 15 000 iteration run provides posterior means and standard deviations as follows: $\alpha = 4.47$ (0.33), $\beta = 1.76$ (0.41), $\log(\delta) = 1.77$ (0.73). The average and minimum value of G^2 were 51.6 and 41.8 respectively. The log of the baseline incidence δ^β averages 3.37 with a standard deviation of 1.96, so is rather imprecisely estimated. Exponentiating this gives a baseline rate of 29.1 compared with 18.9 per 100 000 quoted by Frome. The parameter β has 95 % credible interval (1.09, 2.71) and so the effect of daily smoking on cancer appears nonproportional.

4.10 MARKOV CHAIN MODELS

The Markov chain model generalises the log-linear model to account for dependence through time in the states of a categorical variable. For example, we might follow choice among K political parties by a cohort of persons voting in successive elections. Hence the data are sequences of occupied states, parties chosen, etc. and

$$y_{it} \in (1, \ldots, K)$$

for persons $i = 1, \ldots, N$ and times $t = 1, \ldots, T$, and where y_{it} may fall into one of K categories. We might generalise this to trace the interaction between two or more categorical variables, for example, political choice and economic status, at successive time points.

Under the first order Markov chain assumption, the likelihood in this situation is multinomial with the choice of future state $k \in 1, \ldots, K$ depending only on current state j (and not any preceding states), as well as possibly also on characteristics of the individual i and on time itself t. For a single categorical variable

$$\Pr(Y_{i,t+1} = k \mid Y_{it} = j) \sim \text{Mult}(\underset{\sim}{\pi}_{itj})$$

where

$$\underset{\sim}{\pi}_{itj} = (\pi_{itj1}, \pi_{itj2}, \ldots, \pi_{itjK})$$

and $\sum_{k=1}^{K} \pi_{itjk} = 1$.

The homogenous and stationary first order Markov chain model additionally assumes that the transition probabilities π_{itjk} are fixed over individuals i and times t, so that

$$\underset{\sim}{\pi}_{itj} = (\pi_{j1}, \pi_{j2}, \ldots, \pi_{jK}).$$

Parameter estimation then reduces to $K(K-1)$ transition probabilities given (as multinomial sampling totals) aggregated flows over times and persons, but using detail on the origin and destination states. For example, this may involve multinomial logit regression specific to origin and destination, but with a corner constraint for each origin.

The basic computing principles are illustrated by the following coding with $N = 3$ subjects, $T = 5$ periods, and $K = 3$ states.

```
model { ≠ Likelihood
for (i in 1:N){ for (j in 2:T){ Y[i,j] ~ dcat( p[i,j, Y[i,j−1],1:K ])
≠ Time-stationary model, transition probabilities specific to cases
and times
        for (k in 1:K){ for (l in 1:K){ p[i,j,k,l] <- e[i,j,k,l]/
               sum(e[i,j,k,])
log(e[i,j,k,l]) <- beta[k,l] } } }}
≠ Priors
for (k in 1:K){ beta[k,1] < - 0
    for (l in 2:K){ beta[k,l] ~ dnorm(0,1.0E-2) }}}
Data
list( N = 3, T = 5, K = 3,
Y = structure(.Data = c(1,3,2,2,3,1,3,1,2,1,3,1,3,2,1),.
  Dim = c(3,5)))
```

Monitoring `p[1:N,1:T,1:K,1:K]` will give the same results for all persons i and times t.

Often, however, we may be interested in situations which depart from the homogeneity and stationarity assumptions in that transition behaviour is influenced by individual characteristics, either constant, x_i, or time varying, x_{it}. Then covariates specific to individuals or times may be added to predict `e[i,j,k,l]` in the above coding.

Example 4.30 Changes in health status Grigsby and Bailey (1993) report changes in health status (including deaths) among old people in the US, as aggregated over three periods, 1984–86, 1986–88 and 1988–90. These data draw from four follow-up surveys of a nationally representative sample of over-70s. Health status while alive is determined by ability to perform Activities of Daily Living, either Instrumental Activities (IADLs) such as preparing meals, light housework, or more basic activities (ADLs) such as being able to dress, eat, or bathe without assistance.

The transition pattern between the $K = 4$ states is as in Table 4.27, remembering that death is an 'absorbing state'.

Grigsby and Bailey relate the transition rates between states to age (continuous measure over 70) and sex (1 = Females, 0 = Males). So there are 3.4 = 12 possible transitions to consider (missing origins or destinations are omitted from the analysis). Using the regression coefficients they supply, 1560 cases (i.e. approximately a 10 % 'sample' from the full data) are generated for transitions between two times ($T = 2$). The states are

1 = fully able,
2 = unable on 1 or more IADLS,
3 = unable on 1 or more ADLS,
4 = dead.

Table 4.27 Changes in health among persons aged over 70, US Longitudinal Study on Aging, 1984–90.

Status at beginning of interval	Status at end of Interval					
	Able	1+ IADLs	1+ ADLs	Dead	Missing	All
Able	8529	754	621	959	661	11524
Unable to perform 1 or more IADLs, able to perform all ADLs	262	457	352	312	76	1459
Unable to perform 1 or more ADLs	100	100	607	504	84	1395
Missing	195	35	56	184	747	1217
All	9086	1346	1636	1959	1568	15595

All transitions from state 4 are necessarily excluded, but 'getting better' moves (2,1), (3,1) and (3,2) are possible. Worsening health consists of the moves (1,2), (1,3) and (2,3), while mortality consists of the moves (1,4), (2,4) and (3,4).

We model the impact of the time constant covariates x_{i1} (age) and x_{i2} (sex) by making $\underset{\sim}{\pi}_{itj}$ a function of covariates $x_i = (x_{i1}, x_{i2})$ as well as of previous state:

$$\underset{\sim}{\pi}_{itj} = (\pi_{j1}(x_i), \pi_{j2}(x_i), \ldots, \pi_{jK}(x_i)).$$

Corner constraints are set on the parameters γ_{jk} (age effects) and δ_{jk} (sex effects) by assuming $\gamma_{11} = \gamma_{22} = \gamma_{33}$ and $\delta_{11} = \delta_{22} = \delta_{33}$. That is, the reference category is defined by unchanged health status.

The results in Table 4.28 (based on a single long run of 5000) are broadly consistent with those of Grigsby and Bailey. Thus transitions to mortality or worse health generally increase with age, while transitions to death are more associated with males, but transitions to worse health with being female. Thus living longer among females is counterbalanced by functional limitations. Transitions which are infrequent in the full sample of 15 595, such as (3,1), namely moving from 1+ADLs to Able, may have imprecise or wrongly signed covariate effects since in terms of the 10 % sample they are not well represented.

4.11 MULTIVARIATE CATEGORICAL OUTCOMES

The above log-linear models also broaden into circumstances where two or more categorical outcomes are jointly dependent. Such combinations of categorical outcomes, binomial or multinomial, can be studied using an extended multinomial likelihood approach, in which the response has its number of categories defined by all the separate outcomes. For example, Grizzle and Williams (1972) consider data y_{ijkl} from an international study of atherosclerosis. The categories i and j are regarded as joint responses, both binary, namely z_{1i} infarct (yes/no) and z_{2j} myocardial scar (yes/no). These partly define a

Table 4.28

	Mean	SD	2.5%	Median	97.5%
Sex effects by move type					
$\delta(1,2)$	0.865	0.588	−0.237	0.842	2.094
$\delta(1,3)$	0.652	0.485	−0.268	0.654	1.607
$\delta(1,4)$	−0.536	0.215	−0.936	−0.539	−0.101
$\delta(2,1)$	0.747	0.535	−0.286	0.730	1.803
$\delta(2,3)$	0.913	0.697	−0.341	0.886	2.382
$\delta(2,4)$	−0.395	0.521	−1.437	−0.396	0.623
$\delta(3,1)$	−0.931	0.689	−2.283	−0.931	0.418
$\delta(3,2)$	−0.193	1.018	−2.127	−0.223	1.953
$\delta(3,4)$	−0.241	0.488	−1.156	−0.252	0.739
Age effects by move type					
$\gamma(1,2)$	0.129	0.032	0.056	0.136	0.179
$\gamma(1,3)$	0.102	0.027	0.052	0.104	0.154
$\gamma(1,4)$	0.066	0.023	0.010	0.071	0.104
$\gamma(2,1)$	−0.204	0.058	−0.320	−0.202	−0.104
$\gamma(2,3)$	−0.040	0.065	−0.168	−0.035	0.092
$\gamma(2,4)$	0.037	0.043	−0.041	0.034	0.121
$\gamma(3,1)$	0.090	0.053	−0.028	0.091	0.190
$\gamma(3,2)$	−0.065	0.080	−0.241	−0.061	0.076
$\gamma(3,4)$	0.025	0.033	−0.044	0.024	0.082

multinomial outcome combining four categorical variables. The categories k and l are defined by predictor variables, with k denoting population type (New Orleans White, Oslo, New Orleans Black) and l denoting age (35–44, 45–54, 55–64, 65–69).

A major question of interest here was whether the two responses are independent within each of the 12 subtables formed by combining the levels of k and l. Thus let $\pi_{11}, \pi_{12}, \pi_{21}, \pi_{22}$ denote the four possible response categories (with $1 = Yes, 2 = No$ and $\sum_i \sum_j \pi_{ij} = 1$). Then for the subtable (k, l), infarct and scar are independent if the odds ratio

$$\omega_{kl} = \pi_{11kl}\pi_{22kl}/\pi_{12kl}\pi_{21kl} \tag{4.21}$$

equals 1. In practice we would examine the credible interval for ω_{kl} and see if it straddled the value 1. Equivalently the outcomes are independent within that subtable if the log odds $\Psi_{kl} = \log(\omega_{kl}) = 0$. The model-based estimate of ω_{kl} is likely to differ from the observed ratio $n_{11kl}n_{22kl}/n_{12kl}n_{21kl}$; so we are attempting to smooth the observed pattern of associations and obtain a picture less distorted by sampling variation.

A number of approaches to multivariate discrete outcomes are possible, see Maddala (1983, Chapter 5) and McCullagh and Nelder (1989, Chapter 6). Here we adopt the approach of Morimune (1979) and consider the application to a joint binary outcome, so that the multinomial has four categories. Suppose there is a single predictor k with K levels (e.g. grouped age) and covariate value x_{1k} at each level (such as the middle age value of each age group). For subtable k, we model each outcome via the form $\phi_{ij}(i = 1, 2; j = 1, 2)$ with $\phi_{22} = 1$, and

$$\phi_{11} = \exp(\alpha x + \beta x + \gamma x)$$
$$\phi_{12} = \exp(\alpha x)$$
$$\phi_{21} = \exp(\beta x)$$

with

$$\pi_{ij} = \phi_{ij} / \sum_l \sum_m \phi_{lm}.$$

Here $x = (x_0, x_1)$ contains continuous predictors with $x_0 = 1$ as the constant term. We test for independence (or conversely correlation) within each subtable of the predictor using ratios like (4.21). Using the just given relations among the ϕ_{ij} (which are proportional to π_{ij}) these are given by

$$\log(\phi_{11k}\phi_{22k}/\phi_{12k}\phi_{21k}) = \log(\pi_{11k}\pi_{22k}/\pi_{12k}\pi_{21k}) = \gamma x.$$

Example 4.31 Respiratory symptoms among miners Ashford and Sowden (1970) consider the joint dependent variables, both binary, of wheezing and breathlessness among coalminers who smoked but were without radiological pneumoconiosis. The predictor variable is age group, so that the covariate for each subtable can be taken as a continuous variable (the midpoint of the age band). As above we take $\phi_{22} = 1$ and $\phi_{11} = \exp(\alpha x + \beta x + \gamma x)$ with $y_{i1} = 1$ and $y_{i2} = 1$ if both breathlessness and wheeze are present (Table 4.29). Some design deficiencies of these data are discussed by McCullagh and Nelder (1989).

The covariate x_1 is the midpoint of each age interval, after a form of centering, namely

$$x_1 = (\text{age} - 42)/5.$$

Table 4.29

Age group	Breathless		Not breathless		Total	Age-point
	Wheeze	No wheeze	Wheeze	No wheeze		
20–24	9	7	95	1841	1952	22
25–29	23	9	105	1654	1791	27
30–34	54	19	177	1863	2113	32
35–39	121	48	257	2357	2783	37
40–44	169	54	273	1778	2274	42
45–49	269	88	324	1712	2393	47
50–54	404	117	245	1324	2090	52
55–59	406	152	225	967	1750	57
60–64	372	106	132	526	1136	62

Table 4.30

Odds ratio	Mean	SD	Median
ω_1	42.18	6.861	41.52
ω_2	35.57	4.873	35.17
ω_3	30.02	3.374	29.76
ω_4	25.35	2.272	25.21
ω_5	21.43	1.505	21.36
ω_6	18.13	1.046	18.08
ω_7	15.34	0.8693	15.3
ω_8	13	0.8796	12.97
ω_9	11.02	0.9505	10.98

Instead of the direct model form $\phi_{11} = \exp(\alpha x_1 + \beta x_1 + \gamma x_1)$ we take $\phi_{11} = \exp(\delta x_1)$ and then estimate γ by using the sampled values of $\delta - \alpha - \beta$. We adopt non informative priors for the constant terms and age effect terms in δ, α and β (see *Program 4.31 Respiratory Symptoms*).

We estimate $\gamma_0 = 3.058$ and $\gamma_1 = -0.165$. These values are close to those reported by McCullagh and Nelder (1989, p. 234), though more informative priors would lead to posterior estimates at variance with maximum likelihood. The fitted model for the log-odds in age group k is thus

$$\log(\pi_{11k}\pi_{22k}/\pi_{12k}\pi_{21k}) = 3.058 - 0.165x_{1k}.$$

The posterior estimates of the odds ratios ω for each age group are given in Table 4.30.

They show a clear pattern of a decline in odds ratio with age. So the association between the two outcomes falls with age – though it remains clearly pronounced even for the oldest age group.

4.12 ORDERED OUTCOMES

Categorical variables which are ordinal occur frequently in applications such as the measurement of health functioning and quality of life, socio-economic ranking, and market research. Such scales may be intrinsically categorical or arise through converting originally continuous scales into ordinal ones. For example, Best *et al.* (1996) convert continuous cognitive function scores from the Mini Mental State Evaluation instrument into a five-fold ordinal ranking, because using the scores as continuous would assume a constant effect across the whole scale, whereas a nonlinear effect is more likely. An important issue in either case is whether the number or choice of categories in the scale affects inferences. If a new category were formed by amalgamating two or more adjacent and formerly separate categories, then the effect on inferences should be assessed. These inferences may relate to substantively similar although not identical effects; for example, the effect on schizophrenic prevalence of a background in a nonmanual as against a manual socio-economic group is not necessarily just a collapsed version of the impact on schizophrenia of backgrounds more finely differentiated by socio-economic status.

The usual approach to ordinal scales assumes a latent (continuous) variable Y underlying the ordered categories. This applies even if an ordinal scale arises from grouping an originally continuous scale, in which case a new continuous scale is in a sense being identified. Suppose the states are ranked from 1 (lowest) to J (highest), with cutpoints θ_j from the continuous scale delineating the transition from one category to the next. So if $J = 4$, there are three cutpoints. One supposes there are additional start and end points to the underlying scale, namely θ_0 and θ_4, such as $\theta_0 = -\infty, \theta_4 = +\infty$. Then θ_1, θ_2, and θ_3 are free parameters to estimate subject to the constraint

$$\theta_0 < \theta_1 < \theta_2 < \theta_3 < \theta_4.$$

Other choices of endpoint are possible according to context; for example one might, again for $J = 4$, take $\theta_1 = 1, \theta_4 = 4$, and just estimate the intervening parameters subject to $\theta_1 < \theta_2 < \theta_3 < \theta_4$ (e.g. Chuang and Agresti, 1986). The case $\theta_1 = 1, \theta_2 = 2, \theta_3 = 3, \theta_4 = 4$ would correspond to the so-called fixed score model.

The probability P_{ij} that an individual i $(= 1, \ldots, n)$ will be in state j $(= 1, \ldots, J)$ is then the same as the chance that the subject's underlying score is between θ_{j-1} and θ_j. So the cumulative probability γ_{ij} that an individual i with latent score Y_i will be classified in state j or below is $\gamma_{ij} = \text{Prob}(Y_i < \theta_j)$. Hence we can write $P_{ij} = \gamma_{ij} - \gamma_{i,j-1}$ for the chance of belonging to a specific category. The indexing i is not necessarily restricted to individuals; for example, one of the earliest studies of an ordinal outcome (severity of pneumoconiosis) was among eight groups of miners classified by their length of work-time underground (Ashford, 1959).

Various link functions can be used for γ_{ij} but the most common are the logit, namely $\log\{\gamma_{ij}/(1 - \gamma_{ij})\}$ and the complementary log-log, namely $\log\{-\log(1 - \gamma_{ij})\}$ (McCullagh, 1980). The proportional-odds model uses the logit link for the cumulative probabilities with a parameterisation as follows:

$$\text{logit}(\gamma_{ij}) = \theta_j - \mu_i \tag{4.22}$$

where we can specify $\mu_i = \beta' x_i$ as incorporating individual (or group) characteristics such as treatment allocation, age, income and so on.

Consider the ratio of odds of the event $Y_i < \theta_j$ (i.e. that the individual will be in one of the states $1, 2, \ldots, j$) at different values of x, namely x_1 and x_2. Under the proportional odds model of (4.22) this ratio is

$$\gamma_{ij}(x_1)/(1 - \gamma_{ij}(x_1))/[\gamma_{ij}(x_2)/(1 - \gamma_{ij}(x_2))] = \exp[-\beta(x_1 - x_2)]$$

and is independent of category j. The negative sign on μ_i in the model ensures that larger values of $\beta' x$ lead to an increased chance of belonging to the higher categories; in a medical context, this would mean that higher levels of an adverse risk factor are associated with a more adverse outcome or more severe condition.

We may also introduce random effects specific to individuals, especially if the data are clustered in some way; for example, if there are repeated observations over time on an individual's ordinal outcome (see Chapter 8). Another possibility is nonparallel effects of covariates as expressed in a model such as

$$\text{logit}(\gamma_{ij}) = \theta_j - \mu_{ij}.$$

Example 4.32 Lung disease in miners Ashford's study referred to above of lung disease in miners specifies three ordered categories of outcome (Table 4.31).

Hence there are two cutpoints θ_j to estimate, plus an exposure effect. Work by McCullagh and Nelder (1989, Chapter 5) shows that exposure measured as a log of years worked provides a better model than a linear model in years.

Program 4.32 Lung Disease adopts this exposure form and also converts the aggregated data into equivalent data for 371 individuals. Non informative priors are set on β, while those for θ_j reflect the constraint of monotonicity. Initial values of θ_j can be set by default to the fixed score case, i.e. $\theta_1 = 1$ and $\theta_2 = 2$. So the model is

$$\text{logit}(\gamma_{ij}) = \theta_j - \beta\log(\text{exposure}_i), \quad j = 1, \ldots, 2; i = 1, \ldots, 371.$$

Also available as sampled statistics are

(a) the odds of having pneumoconiosis at 10 years exposure; and
(b) the effect of a doubling in exposure time.

McCullagh and Nelder quote central estimates of these statistics of 40.1 and 6.1.

It may be noted, however, that estimates of the main parameters and these statistics are influenced by the degree of informativeness of the priors, especially for θ_j. The full posterior summaries in the case of priors

Table 4.31

	Pneumoconiosis		
Period exposed in years	Normal	Mild	Severe
5.8	98	0	0
15	51	2	1
21.5	34	6	3
27.5	35	5	8
33.5	32	10	9
39.5	23	7	8
46	12	6	4
51.5	4	2	5

$$\theta_1 \sim \mathrm{N}(0, 1000)\mathrm{I}(, \theta_2)$$
$$\theta_2 \sim \mathrm{N}(0, 1000)\mathrm{I}(\theta_1,)$$

are given in Table 4.32 (from the second half of a run of 20 000 iterations).

There was no evidence to support a nonparallel relationship, i.e. parameters β differentiated by j, though the relevant code is included in the program.

Another approach to ordered responses which is sometimes useful, especially for a small number of categories, is to consider the successive response categories as a sequence of increasingly selective stages (the continuation ratio approach). This would be especially useful if nonparallelism were expected. In the above example, let the rows of Table 4.31 contain elements y_{i1}, y_{i2}, y_{i3}, with y_{i3} as the severe disease group. This approach would involve two binomial models, the first for the total incidence of lung disease, namely with $r_i = n_{i2} + n_{i3}$ cases from $n_i = y_{i1} + y_{i2} + y_{i3}$ at risk. The second binomial model would then be for severe cases out of all with lung disease, so that $r_i = y_{i3}$ is the response and $n_i' = y_{i2} + y_{i3}$ is the risk total. McCullagh and Nelder (1989) discuss an example for mortality due to radiation, where the first binomial is for the chance of dying vs surviving, the second models the chance of death from cancer vs death from other causes, and the third models the chance of leukaemia mortality among all cancer deaths.

4.12.1 Scores for ordered outcomes

A third possible approach to ordinally ranked data, particularly in a two-way (I rows $\times$ J columns) contingency table form, involves replacing the usual interaction term in the log-linear model with a particular multiplicative structure. Scores are attached to rows, columns or both (leading to 'row effect' models, 'row and column effect' models, etc). The row and column effect model (RC model) has been widely studied by authors such as Goodman (1979).

Suppose y_{ij} denote the original counts and p_{ij} denote the multinomial probabilities of a response $j = 1, \ldots, J$ under each of I conditions, with

$$\sum\nolimits_j p_{ij} = 1$$

Table 4.32

	Mean	SD	2.5%	97.5%
β	2.72	0.35	2.08	3.37
θ_1	9.85	1.31	7.29	12.34
θ_2	10.78	1.33	8.17	13.3
Risk at 10 years (1 in ...)	45.3	22.6	17.7	104.3
Deviance	411.3	2.25	408.8	417.1
Increased risk on doubling exposure	6.33	1.68	3.83	10.35

for all i. For example, the response might be ranked income group and the condition (the row) variable might be education type. The usual log-linear model specifies

$$\log(p_{ij}) = \mu + \alpha_i + \beta_j + \gamma_{ij}$$

with constraints $\sum \alpha_i = 0$, $\sum \beta_j = 0$ and $\sum \sum \gamma_{ij} = 0$. If rows are not ordered but columns are, one might set the interaction parameter to

$$\gamma_{ij} = j\rho_i \tag{4.23}$$

with $\sum \rho_i = 0$. This is the row effects model, treating the ordinal response as an equally spaced numerical scale with fixed scores, namely $1, 2 \ldots, J$
A generalisation of this model is to assign monotone and variable scores v_j to category j, i.e.

$$\log(p_{ij}) = \mu + \alpha_i + \beta_j + \gamma_{ij} \tag{4.24}$$

with

$$\gamma_{ij} = \rho_i v_j. \tag{4.25}$$

The scaling of v_j is arbitrary as discussed above, for example in the specification of minimum and maximum scores v_1 and v_J. If the column scores are constrained to increase with their ordering then there is a stochastic order in the column response. Thus for a pair of rows a and b, the log odds of adjacent column (response) categories j and $j + 1$ is

$$\log(p_{aj}p_{b,j+1}/p_{a,j+1}p_{bj}) = (\rho_b - \rho_a)(v_{j+1} - v_j).$$

So if $\rho_b > \rho_a$, the log-odds ratios are non-negative.

In fact in the original RC model the scores v_j are variable but not necessarily monotone, while some studies have considered the case where both v_j and ρ_i are monotone (Ritov and Gilula, 1991). One variant of the model described by (4.24) and (4.25) is to introduce an overall measure ϕ

$$\gamma_{ij} = \phi v_j \rho_i \tag{4.26}$$

where ϕ is a measure of association, restricted to non-negative values. The row and column variables are independent if and only if $\phi = 0$. So if the 95% credible interval for ϕ is entirely positive then there is strong support for dependence between row and column variables.

In a Bayesian analysis the parameters v_j and ρ_i are usually treated as fixed effects. For identifiability they may be constrained; thus for $J = 4$ in the representation described by (4.24) and (4.25) one might take $v_1 = 1, v_4 = 4$, and just estimate the intervening parameters subject to $v_1 < v_2 < v_3 < v_4$ (e.g. Chuang and Agresti, 1986). For the representation (4.26) one might adopt a particular form for the v_j and ρ_i such as $\mathrm{N}(0, 1)$ or a t density for greater robustness (Evans *et al.*, 1993). The α_i and β_j can be either fixed or random, as in a standard analysis of variance.

Example 4.33 Pain by drug type Chuang and Agresti (1986) discuss the case of an ordinal pain outcome ($j = 1, \ldots, 4$) according to drug type ($i = 1, \ldots, 4$), see Table 4.33.

They show that the homogenous model under which $p_{ij} = p_j (i = 1, \ldots, 4)$ fits poorly with the G^2 measure of fit equalling 46.74 on 9 degrees of freedom. In *Program 4.33 Drug Pain*, both multinomial and Poisson likelihoods for this model may be used. The effects α_i and β_j are specified with a corner constraint but convertible to a sum to zero constraint as in Chuang and Agresti (1986). This also applies to the unrestricted v_j and ρ_i scores of the RC model.

In the Poisson case we have

$$y_{i,j} \sim \text{Poi}(\mu_{i,j});$$

with probabilities estimated via $p_{i,j} = \mu_{ij} / \sum_j \mu_{ij}$ and under the homogeneity model

$$\log(\mu_{i,j}) = \lambda + \alpha_i + \beta_j.$$

We adopt non informative N(0, 100) priors on α_i and β_j. The deviance fit for this model, over 10 000 iterations with a 2500 iteration burn-in, has an average of 49.8, a minimum of 46.7 (exactly as in the maximum likelihood analysis of Chuang and Agresti), and maximum of 66.2. The row effects model (4.22) has an average deviance of 30.4 and minimum of 24.7 (compared with the ML deviance of 24.5).

The RC model, namely (4.24) and (4.25) but without monotonic constraints on the v_j or ρ_i has an average fit of 14.8 and a minimum of 7.2 (compared an ML deviance of 6.7). Non informative N(0, 100) priors were adopted for the scores of this model. The fitted scores (from the second half of a run of 10 000 iterations) are first converted to sum to zero (Table 4.34). They show evidence of non-normality and wide scatter (high coefficients of variation). The column scores also show a 'reversal' with a lower score for an ordinally higher category.

The second drug has the highest score ρ_2 (best analgesic effect) despite the poor definition of the scores.

Adopting the constrained model with v_1 set to 1, and θ_4 to 4, means introducing constraints on the remaining v_j to ensure monotonicity (the relevant parameters in Program 4.33 are denoted `nu.con[]`). Non informative N(0,100)

Table 4.33 Ratings of drugs in analgesic trial.

	Reduction of pain rating			
Drug	Poor	Fair	Good	Very good
Z100	5	1	10	14
EC4	5	3	3	20
C60	10	6	12	3
C15	7	12	8	2

Table 4.34 Posterior summary: RC model.

	Mean	SD	2.5%	Median	97.5%
ν_1	0.49	0.84	−0.71	0.26	2.70
ν_2	2.16	2.08	0.14	1.49	7.95
ν_3	0.46	0.84	−0.70	0.23	2.65
ν_4	−3.11	2.81	−10.70	−2.23	−0.23
ρ_1	−0.66	0.90	−3.45	−0.35	−0.06
ρ_2	−1.07	1.48	−5.27	−0.56	−0.11
ρ_3	0.58	0.88	0.02	0.28	3.19
ρ_4	1.15	1.58	0.11	0.60	5.61

priors were set on ν_2 and ν_3. This leads to a slight worsening in fit (a mean deviance of 18.0, and minimum of 10.6), but involves two fewer parameters and a more precisely defined set of scores (centred in the case of row scores), see Table 4.35.

The second drug again has the highest score, so appears to cause the best response in the sample of patients analysed. The scores $\nu_1 \equiv 1$, ν_2 and ν_3 of the three lower categories of the column ranking (poor, fair, good) appear not to be greatly differentiated so there might be a case for amalgamating them.

Example 4.34 Periodontal condition Evans *et al.* (1993) analyse survey data on periodontal condition by calcium intake, each variable being categorical with four ordered groups; see Table 4.36.

Table 4.35

	Mean	SD	2.5%	97.5%	Median
ρ_1	0.39	0.14	0.12	0.68	0.39
ρ_2	0.62	0.15	0.34	0.93	0.62
ρ_3	−0.41	0.18	−0.78	−0.087	−0.41
ρ_4	−0.60	0.21	−1.06	−0.22	−0.58
ν_2	1.23	0.20	1.01	1.74	1.17
ν_3	1.71	0.35	1.14	2.47	1.68

Table 4.36 Periodontal condition and calcium intake.

	Calcium intake			
Periodontal condition	1	2	3	4
A	5	3	10	11
B	4	5	8	6
C	26	11	3	6
D	23	11	1	2

Here A is the best condition and 1 is the lowest calcium intake. These authors adopted model (4.26), and assumed random row and column effects α_i and β_j, but a fixed effects form for v_j and ρ_i. Here we assume fixed effects throughout, with $v_j(j = 1, \ldots, 4)$ and $\rho_i(i = 1, \ldots, 4)$ not constrained to be monotonic, but taken to have an N(0, 1) prior for identifiability (see *Program 4.34 Periodontal Condition*). This will lead approximately to a unit length constraint on each of the two sets of scores. The log of the association measure has a non informative N(0, 100) prior, with the posterior estimate of ϕ based on exponentiating this. The posterior summary of the fit and centred model parameters (from the last 10 000 of a run of 15 000 iterations) are given in Table 4.37.

Scores ρ_i for condition are monotonic but those for intake show a reversal for categories 3 and 4. The posterior distribution of the ϕ association measure is, despite being skewed, clearly focused on positive values. Evans *et al.* also find a positive mean, namely 7.3, for ϕ, but a less closely defined credible interval, with the standard deviation of ϕ being 7.5. Posterior estimates of the association measure are sensitive to its prior specification; using a U(0, 100) prior for this parameter leads to a mean of 8.7 and a standard deviation of 10.5, though the median remains under 5, and the 2.5% point stands at around 1.2.

Table 4.37

Statistic	Mean	SD	2.5%	97.5%	Median
G^2	10.04	4.20	2.46	40.0	9.34
ϕ	3.11	2.13	0.86	8.6	2.37
v_1	−0.50	0.24	−1.06	−0.15	−0.46
v_2	−0.36	0.22	−0.92	−0.065	−0.32
v_3	0.54	0.28	0.15	1.23	0.49
v_4	0.32	0.21	0.04	0.88	0.27
ρ_1	0.48	0.23	0.12	1.00	0.45
ρ_2	0.41	0.24	0.09	1.00	0.37
ρ_3	−0.30	0.19	−0.77	−0.03	−0.26
ρ_4	−0.60	0.30	−1.30	−0.16	−0.56

CHAPTER 5

Ensemble Estimates: Hierarchical Priors for Pooling Strength

5.1 HIERARCHICAL PRIORS: IMPROVED PRECISION THROUGH BORROWING STRENGTH

Bayesian hierarchical random effects models facilitate the simultaneous estimation of several parameters θ_i of the same type, with the motives both of pooling strength to improve the precision of the estimate of each parameter, and of allowing for uncertainty in such estimates or associated parameters (odds ratios, treatment effects, etc.). In some of the examples of Chapter 4 and earlier we might in fact have adopted this hierarchical or random effects approach, and one of the strengths of Bayesian sampling methods is the relative tractability of quite complex random effects models. On the other hand, hierarchical modelling to pool strength depends on an exchangeability assumption between the units in the analysis, which will not hold, for example, if certain units are correlated with a subset of the other units. The notion of exchangeability also has relevance to prediction beyond the sample, i.e. to 'generalisation' of the results to broader settings. Draper (1995) defined the notion of 'uncertain exchangeability' to cover cases where exchangeability beyond the sample is in doubt – for example, if nonrandomised designs are employed.

Often the concern is with inferences within a given set of units. For example, in educational or medical league table comparisons, the parameters of interest might correspond to average exam scores by school, hospital waiting times, or surgical mortality. In disease mapping the outcomes may be death rates based on varying populations at risk. In meta-analysis the parameters are effect sizes over a set of studies. However, in other settings the notion of exchangeability is more critical to out of sample inferences. In surveys the parameters may be estimates of a mean or proportion, and inference from the survey to population subgroups will be improved by pooling information over strata, clusters or other survey defined subgroups.

Typically there are many institutions, studies or individuals but relatively little data on each separately. The statistical analysis aims at combining the information from the various sources of data (schools, hospitals, studies) exploiting the assumed similarity between the parameters in terms of their genesis. Pooling over all units thus requires exchangeability in the sense that inference is invariant to permutation of suffixes (Leonard, 1972).

In the case of rare outcomes we might expect the smoothed parameter θ_i based on pooling strength to be a more sensible estimate than the maximum likelihood estimate which treats each case i as an isolated entity. On the other hand pooling over institutions or individuals may introduce some bias in the sense that shrinkage towards the overall average occurs. Therefore the pooling method may be 'robustified' to allow for outliers or perhaps modified to allow partial exchangeability within two or more groups of the original units. In this case one is interested in shrinking towards a central value for each group (Albert and Chib, 1997).

The form of random effect variation between units may also be important for regression models. In fitting generalised linear models, data sets may show greater residual variability than expected under the exponential family (Albert and Pepple, 1989). Allowing for variability between units, for example by taking prior and sampling density which are a conjugate mixture of exponential family distributions (e.g. gamma–poisson), is one approach to overdispersion or extra-variation of this kind.

The general situation is as follows: the sequence of data points and underlying true values $(y_i, \theta_i), i = 1, \ldots, N$ are identically distributed. The density of the observations x_i, given θ_i, is $f(y_i \mid \theta_i)$. Each θ_i is distributed according to a mixture density governed by a hyperparameter λ or hyperparameters $\lambda_1, \lambda_2, \ldots$ The three-stage hierarchical model has the following components:

1. Conditionally on λ and $\theta_1, \ldots, \theta_N$, the data y_i are independent, with densities $f(y_i \mid \theta_i)$ which are independent of θ_j for $j \neq i$ and of λ.
2. Conditionally on λ the true values θ_i are drawn from the same density $\pi_1(\theta \mid \lambda)$.
3. The hyperparameter λ has its own density $\pi_2(\lambda)$.

Examples of this scenario include cases where π_1 is conjugate with the density of the observations. For example, the data may be counts x_i with Poisson means θ_i, which are themselves distributed with density π_1 which is a Gamma(α, β). The data may be discrete outcomes (Yes/No or successes in a set of trials) which are Bernoulli or Binomial with probabilities θ_i which are themselves Beta(α, β).

The advent of MCMC and other sampling techniques has, however, facilitated nonconjugate analysis. A frequent example is in the analysis of proportions y_i/n_i where the data are assumed Binomial, $y_i \sim \text{Bin}(\theta_i, n_i)$ and then the proportions are transformed to the real line via $\eta_i = \text{logit}(\theta_i)$ as in Leonard (1972) or via an arcsin transformation as in Efron and Morris (1975). The η_i are then assumed to follow a Normal or Student density.

5.1.1 Pooling strength: Poisson outcomes

Suppose we first consider a Poisson outcome in a particular (spatial health) setting. There are O_i observed health events in a small area i and E_i expected events, which are calculated by the usual demographic technique of indirect standardisation (Newell, 1988). Then our model for this small count outcome is

$$O_i \mid \theta_i \sim \text{Poi}(\theta_i E_i).$$

The maximum likelihood approach to this situation treats each observation as a separate entity and would lead to estimates $\theta_i = O_i/E_i$ of the relative risk in each small area. An alternative is to use a hierarchical prior density for the unknown parameters θ_i and assume the unknown latent rates are drawn from a larger set of rates. If the outcome O_i were Poisson, a gamma prior is often used as the density for the latent rates because it is an appropriate prior density for a necessarily positive variable (a Poisson mean), because of its flexibility in representing skewness, and for mathematical reasons (conjugacy of prior and posterior).

The full model in these circumstances can be written in three stages: at stage (1) conditional on θ_i, the O_i are independent and $O_i \mid \theta_i \sim \text{Poisson}(\theta_i E_i)$; at stage (2) conditional on the hyperparameters α and β the θ_i are independently gamma, that is $\theta_i \mid \alpha, \beta \sim \text{G}(\alpha, \beta)$; and at stage (3) the hyperparameters (α, β) of the gamma are themselves given priors. For example, George *et al.* (1993) use an exponential $\text{E}(1)$ prior on α, and a $\text{G}(b_1, b_2)$ prior on β where b_1 and b_2 are known.

Having observed the outcomes O_i, inference about the θ_i is based on the marginal posterior distribution $\pi(\theta_i \mid O_i)$. This distribution is based on integrating the product

$$\pi(\theta_i \mid O_i, \alpha, \beta)\pi(\alpha, \beta \mid O_i)$$

over the full range of the bivariate density of (α, β). The first term in the product is the posterior density of θ_i given α, β, and the data, while the second is the posterior density of the hyperparameters given the data. An empirical Bayes approximation is often made, namely

$$\hat{\pi}(\theta_i \mid O_i) = \pi[\theta_i \mid O_i, \hat{\alpha}, \hat{\beta}]$$

where $\hat{\alpha}$ and $\hat{\beta}$ are typically maximum likelihood estimates based on the observed events. However, for small sample sizes this 'data based' approach to estimating the prior may understate the impact of the uncertainty about the hyperparameters α and β.

Example 5.1 Smoothing of incidence and mortality rates An example of fully Bayes pooled estimation which combines information between separate Poisson densities is provided by a case study of childhood leukaemia deaths in two English counties in the 1950s (Knox, 1964) (Table 5.1). Death rates are classified by child age and by type of residence. The study demonstrated the higher overall mortality in urban areas and that the age distributions of urban and

Table 5.1 Deaths from childhood cancers 1951–60 (Northumberland and Durham).

Cytology	Age (yrs)	Place	Observed	Expected	Population	Max lkd rate per million child years	Max lkd standard mortality ratio
Lymphoblastic	0–5	Rural	38	24.1	103857	36.6	158
	6–14		13	36.1	155786	8.3	36
	0–5	Urban	51	31.5	135943	37.5	162
	6–14		37	47.3	203914	18.1	78
Myeloblastic	0–5	Rural	5	8.0	103857	4.8	63
	6–14		8	12.0	155786	5.1	67
	0–5	Urban	13	10.4	135943	9.6	125
	6–14		20	15.6	203914	9.8	128

rural lymphoblastic leukaemia mortality rates are different. Rural rates fall much more at later ages.

We examine how to reproduce this analysis using a fixed effects approach (analogous to conventional maximum likelihood) and provide posterior densities credible intervals for the age/area rates and SMRs. Second, we consider the impact of adopting an exchangeable gamma prior on the estimates of these rates.

To provide rates per million, the fixed effects model specifies $O_i \mid P_i \sim \text{Poisson}(\theta_i P_i)$ where P_i is child years in millions and each θ_i assigned a vague gamma prior, specifically $\theta_i \sim \text{G}(0.001, 0.001)$. The code for this (with $N = 8$) is

```
{for (i in 1:N) {theta[Cancer[i],Place[i],Age[i]] ~ dgamma (0.001,
0.001);
lambda [i] <- theta[Cancer[i],Place[i],Age[i]]* Pops[i]/100000;
O[i] ~ dpois(lambda[i])}}
```

Sampling from 5000 iterations after a burn-in also of 5000 gives the estimates of mortality rates by cancer type (L,M), place (R,U), and child age (Young, Old) shown in Table 5.2.

We now apply an exchangeable gamma prior which allows pooling of strength and adopt the priors $\alpha \sim \text{E}(1)$ and $\beta \sim \text{G}(0.1, 1)$. The code becomes

```
{for (i in 1:N) {     theta[Cancer[i],Place[I],Age[i]] ~
                      dgamma(alpha, beta);
                      lambda[i] <- theta[Cancer[i], Place[i],
                      Age[i]]* Pops[i]/100000;
                      {O[i] ~ dpois(lambda[i])}
                alpha ~ dexp(1); beta ~ dgamma(0.1, 1.0)}
```

This leads to a notable smoothing towards the overall average in terms of the central estimates of the rates (their means and medians). This is especially so for the rural myeloblastic mortality rates based on small cumulated counts.

Table 5.2 Posterior estimates: mortality rates per million.

	Mean	SD	2.5%	Median	97.5%
θ(L,R,Y)	36.55	5.89	25.98	36.25	48.94
θ(L,R,O)	8.32	2.30	4.43	8.06	13.44
θ(L,U,Y)	37.49	5.32	27.73	37.17	48.72
θ(L,U,O)	18.10	2.93	12.64	17.95	24.35
θ(M,R,Y)	4.78	2.12	1.51	4.49	9.83
θ(M,R,O)	5.12	1.84	2.16	4.88	9.32
θ(M,U,Y)	9.60	2.65	5.18	9.36	15.31
θ(M,U,O)	9.77	2.19	5.91	9.56	14.67

The posterior means of α and β are 1.6 and 0.1 respectively, and the variance in mortality rates is estimated as 326. Adopting a less informative prior on β, namely $\beta \sim G(0.001, 0.001)$ leads to the same estimates of the gamma parameters and virtually the same smoothing of death rates (Table 5.3).

With the alternative SMR outcome (see Program 5.1) we obtain the results in Table 5.4.

Table 5.3 Posterior estimates: mortality rates per million (vague prior).

	Mean	SD	2.5%	Median	97.5%
θ(L,R,Y)	34.79	5.59	24.65	34.54	46.53
θ(L,R,O)	8.81	2.31	4.84	8.61	13.87
θ(L,U,Y)	36.08	5.08	26.76	35.84	46.59
θ(L,U,O)	18.06	2.91	12.77	17.88	24.20
θ(M,R,Y)	5.81	2.30	2.15	5.51	11.10
θ(M,R,O)	5.78	1.89	2.69	5.56	10.03
θ(M,U,Y)	9.99	2.63	5.50	9.77	15.77
θ(M,U,O)	10.09	2.16	6.27	9.94	14.66

Table 5.4 Deaths from childhood cancers 1951–60 (Northumberland and Durham), SMRs.

Cytology	Age (yrs)	Place	Observed	Expected	Max lkd standard mortality ratio	Bayes smooth		
						Mean	2.5%	97.5%
Lymphoblastic	0–5	Rural	38	24.1	1.58	1.54	1.10	2.06
	10–14		13	36.1	0.36	0.39	0.22	0.62
	0–5	Urban	51	31.5	1.62	1.59	1.19	2.04
	10–14		37	47.3	0.78	0.80	0.57	1.06
Myeloblastic	0–5	Rural	5	8	0.63	0.71	0.28	1.35
	10–14		8	12	0.67	0.73	0.34	1.24
	0–5	Urban	13	10.4	1.25	1.22	0.67	1.93
	10–14		20	15.6	1.28	1.25	0.79	1.83

The smoothing towards the central average (of 1) is apparent again in the rural myeloblastic mortality outcome, and in fact only one SMR is clearly below average in terms of a 95% credible interval.

We might choose to model data of this type in various other ways, for example making the prior densities for mortality variation specific to place or cancer type, e.g.

```
theta[Cancer[i],Place[i],Age[i]] ~ dgamma(alpha [Cancer[i]],beta
[Cancer[i]].
```

Example 5.2 Bayesian graduation Various regression and smoothing methods (collectively entitled 'graduation') have been suggested for application to schedules of age-specific rates which occur in demographic and actuarial work. These may involve highly nonlinear regression models. The application considered by Carlin (1992) is a relatively simple one without an explicit regression (see *Program 5.2 Graduation*). Carlin reanalyses a schedule of deaths data from Broffitt (1988), namely deaths and exposed population d_i and e_i for $k = 30$ single years of age from ages 35 to 64.

Underlying the rate at age i is a Poisson rate θ_i (per 1000)

$$d_i \sim \text{Poi}(\theta_i e_i/1000).$$

We compare a fixed effects model and a random model where the rates θ_i are drawn from a gamma hyperdensity $\text{G}(\alpha, \beta)$. In both cases the rates are subject to an increasing condition

$$0 < \theta_1 < \theta_2 < \ldots < \theta_k < B.$$

With the random effects model adopting priors $\alpha \sim \text{E}(0.001)$ and $\beta \sim \text{G}(0.001, 0.0011)$, we obtain the smoothed rates shown in Table 5.5.

We find the random effects estimates of the death rates moves smoothed towards the average, and a predictive model check, as discussed in Chapter 2 and based on the chi-square statistic, tends to favour the random effects model predictions. The predictive fit criterion for a random effects model with monotonic θ is 0.20, whereas under a fixed effects monotonic model it is 0.06 and under a pooling model but without monotonicity it is 0.99. Values near 1 or 0 indicate failures of model fit.

5.1.2 Pooling strength: Binomial outcomes

Assume our data are in the form of numbers r_i assumed to be binomial $\text{Bin}(\pi_i, n_i)$:

$$f(r_i|n_i, \pi_i) \propto \pi_i^{r_i}(1 - \pi_i)^{n_i - r_i}.$$

In contrast to the standard binomial model which sets $\pi_i = \pi$, the proportions π_i themselves are assumed to be variable. For example, we may assume a beta density for these proportions with parameters α and β either known in advance

Table 5.5 Posterior estimates: smoothed death rates under monotonicity.

	Mean	2.5%	97.5%		Mean	2.5%	97.5%
α	1.58	0.87	2.59	$\theta 16$	4.81	3.75	6.08
β	0.25	0.12	0.44	$\theta 17$	5.20	4.01	6.55
$\theta 1$	0.66	0.23	1.21	$\theta 18$	5.62	4.29	7.15
$\theta 2$	0.85	0.38	1.44	$\theta 19$	6.22	4.72	8.00
$\theta 3$	1.07	0.56	1.68	$\theta 20$	7.28	5.56	9.37
$\theta 4$	1.26	0.71	1.90	$\theta 21$	8.24	6.33	10.41
$\theta 1$	1.49	0.88	2.16	$\theta 22$	9.23	7.22	11.53
$\theta 2$	1.78	1.11	2.51	$\theta 23$	10.12	8.02	12.49
$\theta 3$	2.08	1.36	2.87	$\theta 24$	11.02	8.72	13.48
$\theta 4$	2.40	1.65	3.21	$\theta 25$	12.14	9.73	14.90
$\theta 1$	2.67	1.87	3.53	$\theta 26$	12.98	10.41	15.84
$\theta 10$	2.96	2.08	3.87	$\theta 27$	14.03	11.22	17.33
$\theta 11$	3.38	2.49	4.36	$\theta 28$	15.02	11.98	18.97
$\theta 12$	3.76	2.86	4.76	$\theta 29$	16.23	12.70	20.87
$\theta 13$	3.99	3.08	5.07	$\theta 30$	19.31	14.09	27.34
$\theta 14$	4.21	3.25	5.34				
$\theta 15$	4.51	3.51	5.71				

or distributed according to a higher stage prior. Then conditional on α and β the π_i have density

$$f(\pi_i \mid \alpha, \beta) \propto \pi_i^{\alpha}(1 - \pi_i)^{\beta}$$

so that the posterior estimates of π_i are drawn from a beta density with parameters $\alpha + r_i$ and $\beta + n_i - r_i$.

The beta parameters may be assumed known constants *a priori.* Thus Beta(0.5, 0.5), the Jeffreys prior, and Beta(1, 1), the uniform prior, are default non-informative priors. If the binomial outcomes were of similar relative frequency then a Beta(m, m) prior with m a small positive integer may be seen as a 'sceptical prior' (Carlin and Louis, 1996, p. 51). This might be appropriate, for example, as a sensitivity check on the form of prior, since it favours values of θ_i near 0.5.

Alternatively the beta parameters may be updated by the data and assigned priors. Posterior estimates of these parameters are likely to be sensitive to prior specification if sample sizes are small.

Example 5.3 Toxoplasmosis data A subset of 10 observations (O1, ..., O10) of the toxoplasmosis data analysed by Efron and Morris (1975) consists of the 10 points as in Table 5.6.

Table 5.6

Observation	O1	O2	O3	O4	O5	O6	O7	O8	O9	O10
n_i	51	16	82	13	43	75	13	10	6	37
r_i	24	7	46	9	23	53	8	3	1	23

The numbers n_i denote people tested and r_i denote those tested positive for toxoplasmosis in each of $N = 10$ cities in El Salvador. Assume $r_i \sim \text{Bin}(p_i, n_i)$, and that α and β are known ($\alpha = \beta = 1$). This leads to posterior estimates of the proportions positive as in Table 5.7.

These represent a degree of smoothing compared with the maximum likelihood estimates; for example the MLE for observation 9 is 0.167 but the smoothed value is pulled towards the average.

To update the Beta parameters, we adopt the parameterisation $\alpha = \nu\mu$ and $\beta = \nu(1 - \mu)$ (e.g. Stroud, 1994), where μ is the prior mean and ν is the precision attached to that mean. We use an exponential prior on ν and a beta prior on μ. The results in Table 5.8 are comparable to those obtained by George *et al.* (1993) who use a prior $p(\alpha, \beta) \propto \exp(-\alpha - \beta)$. Smoothing is greatest for cities 8 and 9 with small observed counts n_i and relatively low crude rates.

Example 5.4 Stomach cancer death rates An example of a nonconjugate analysis is provided by an analysis of stomach cancer death rates in 84 Missouri cities, originally discussed by Tsutakawa *et al.* (1985). Albert and Chib (1997) present a reanalysis of these deaths data (y_i, n_i) assuming

Table 5.7 Estimated proportions testing positive.

Parameter	Mean	SD	2.5%	Median	97.5%
p_1	0.472	0.068	0.341	0.472	0.606
p_2	0.443	0.115	0.226	0.441	0.672
p_3	0.560	0.054	0.454	0.561	0.661
p_4	0.666	0.117	0.421	0.673	0.872
p_5	0.534	0.073	0.391	0.534	0.675
p_6	0.702	0.052	0.594	0.704	0.799
p_7	0.598	0.122	0.352	0.603	0.819
p_8	0.333	0.131	0.107	0.323	0.606
p_9	0.250	0.143	0.038	0.228	0.574
p_{10}	0.615	0.076	0.460	0.617	0.755

Table 5.8 Estimated proportions testing positive, beta prior updated.

Parameter	Mean	SD	2.5%	Median	97.5%
$\alpha/(\alpha + \beta)$	0.495	0.28	0.03	0.49	0.97
p_1	0.47	0.05	0.37	0.47	0.56
p_2	0.44	0.09	0.28	0.44	0.61
p_3	0.56	0.04	0.48	0.56	0.64
p_4	0.68	0.09	0.51	0.69	0.84
p_5	0.54	0.05	0.44	0.54	0.64
p_6	0.71	0.04	0.63	0.71	0.78
p_7	0.61	0.09	0.43	0.61	0.78
p_8	0.31	0.10	0.13	0.30	0.52
p_9	0.19	0.11	0.03	0.17	0.45
p_{10}	0.62	0.05	0.51	0.62	0.72

$$y_i \sim \text{Bin}(\theta_i, n_i)$$
$$\eta_i = \text{logit}(\theta_i)$$
$$\eta_i \sim \text{N}(\mu, \sigma^2).$$

They test a fixed vs random effects specification where 'fixed' refers to the assumption of a single rate θ for all areas with $\sigma^2 = 0$. (An alternative meaning of 'fixed effects' relates to the standard maximum likelihood estimates with no pooling.)

They stipulate a discrete prior on a grid of eight values with equal spacing in terms of $log(\sigma^2)$, namely $log(\sigma^2) = -3.2, -2.8, -2.4, \ldots, -0.4$. These eight values are assumed equal prior weight of 0.0625, while the value $\sigma^2 = 0$ is assigned a prior weight of 0.5. Comparing the posterior relative frequencies of these nine values will give Bayes factors to assess the most appropriate value. Albert and Chib find that $\sigma^2 = 0$ is selected with a relative frequency of 0.069 (6.9 % of the iterations in a run of 100 000 iterations). From this can be derived Bayes factors of the eight positive values against the zero value of the hypervariance.

This type of discrete prior – with widely separated sets of values (namely $\sigma^2 = 0$ vs the remainder) – is subject to the risk of an 'absorbing state'. Here we select ten possible values, the eight equally spaced as above but adding those corresponding to $log(\sigma^2) = -4$ and -3.6 (see model A in Program 5.4).

A single run of 10 000 iterations produces Bayes factors as given in Table 5.9 on the grid of values used. The frequencies are based on aggregating over iterations of the index k in Program 5.4.

By contrast a standard 'just proper' Gamma(0.001, 0.001) prior for $1/\sigma^2$ (model B in Program 5.4) gives a posterior mean for the variance of 0.074, and median 0.051, but with a wide credible interval (0.005, 0.24). The posterior mean of σ^2 exceeds the posterior median. The smoothed death rates are however broadly similar between the two approaches (Table 5.10).

Table 5.9 Stomach cancer, variance of underlying rates, Bayes factors using discrete prior.

σ^2	Frequency	%	Prior frequency (%)	Bayes factor
0.018	579	5.79	10	0.58
0.027	1116	11.16	10	1.12
0.041	1531	15.31	10	1.53
0.061	1801	18.01	10	1.80
0.091	1969	19.69	10	1.97
0.135	1612	16.12	10	1.61
0.202	878	8.78	10	0.88
0.301	386	3.86	10	0.39
0.449	119	1.19	10	0.12
0.670	9	0.09	10	0.01
Total	10000	100	100	

Table 5.10 Smoothed death rates by prior assumptions.

			Death rates per 100,000			
			Gamma prior		Discrete prior	
City	Population	Deaths	Mean	SD	Mean	SD
1	98066	99	103.7	9.7	102.8	9.6
2	53637	54	104.9	12.4	103.6	12.2
3	46394	80	154.9	19.3	158.1	18.1
4	12890	17	121.6	21.0	123.1	22.4
5	10975	11	109.0	20.1	108.1	21.0
6	7436	13	134.2	28.6	137.5	29.9
7	3814	3	107.1	24.2	106.3	25.9
8	3461	2	104.3	24.7	103.3	26.4
9	3349	1	100.2	24.7	97.5	26.4
10	3215	1	100.9	24.9	98.2	26.0
11	2708	2	108.8	25.7	108.0	28.0
12	2530	4	121.0	28.7	123.3	32.6
13	2145	4	124.7	31.7	126.6	33.5
14	1823	2	114.4	28.4	114.8	31.2
15	1668	2	115.6	28.1	115.8	31.4
16	1627	3	122.6	31.3	123.6	33.4
17	1407	1	112.1	27.8	111.5	30.3
18	1356	0	105.5	26.9	104.8	29.7
19	1339	3	125.0	33.8	126.5	36.2
20	1209	1	112.2	28.1	111.9	30.3
21	1208	1	112.7	28.3	112.7	31.4
22	1185	0	106.9	27.2	104.5	28.4
23	1104	0	107.1	27.2	105.2	29.8
24	1083	0	107.7	27.6	106.1	29.5
25	1025	1	114.0	29.4	114.5	31.8
26	917	1	115.3	30.0	115.5	32.8
27	917	0	108.1	27.4	107.3	29.8
28	877	0	108.3	28.1	107.6	29.7
29	874	0	108.7	28.6	108.3	30.7
30	857	1	115.7	30.0	116.4	33.8
31	855	0	107.9	28.1	107.8	30.8
32	854	0	108.2	28.7	107.8	31.3
33	842	0	108.4	27.8	107.5	31.3
34	799	2	122.7	34.0	125.7	37.5
35	731	5	149.4	54.5	157.9	59.5
36	721	0	109.7	28.3	108.4	31.4
37	709	1	116.7	30.4	117.6	33.6
38	706	3	132.8	40.2	135.6	45.8
39	680	1	116.0	30.3	117.4	34.6
40	676	1	116.8	32.1	117.5	33.8
41	664	0	109.8	28.5	109.9	30.8
42	657	0	109.9	28.8	109.3	33.5

Table 5.10 *cont.*

			Death rates per 100,000			
			Gamma prior		Discrete prior	
City	Population	Deaths	Mean	SD	Mean	SD
43	647	1	117.3	31.7	118.4	33.7
44	631	1	116.8	31.9	117.5	34.8
45	603	0	110.5	29.7	109.8	30.3
46	601	0	109.9	28.4	110.1	33.6
47	600	1	117.1	31.2	118.7	33.6
48	592	0	111.0	29.6	110.3	31.6
49	588	3	133.4	40.9	138.3	48.0
50	583	1	117.4	32.2	119.0	35.4
51	582	3	133.5	41.7	138.9	47.9
52	582	0	109.9	28.9	110.3	32.6
53	581	1	117.9	31.1	118.8	35.2
54	556	0	111.2	30.1	111.1	32.9
55	527	0	110.8	29.1	110.0	30.8
56	524	1	118.4	32.1	118.8	35.4
57	522	0	110.9	29.4	111.5	32.2
58	517	1	118.2	32.4	118.2	33.7
59	512	1	117.5	31.3	118.3	36.7
60	493	0	110.6	29.5	111.2	32.9
61	490	1	118.0	31.9	120.0	37.0
62	481	0	111.3	29.2	111.0	32.1
63	448	1	119.6	34.1	121.7	39.2
64	443	0	111.6	29.2	110.8	32.1
65	423	1	118.8	33.2	121.1	37.1
66	419	2	127.4	38.4	132.5	46.2
67	403	0	112.3	30.5	111.5	30.8
68	395	0	112.0	30.2	111.7	32.3
69	395	0	111.8	29.5	112.3	32.0
70	389	1	119.8	33.5	121.8	39.6
71	386	0	112.2	29.8	112.7	34.0
72	383	0	112.1	29.9	112.6	33.0
73	372	1	119.0	32.9	122.1	38.4
74	368	1	119.5	32.4	119.8	37.3
75	350	1	119.9	32.6	119.6	36.5
76	339	1	120.0	32.8	121.3	40.3
77	333	1	120.3	34.1	120.6	36.1
78	325	0	113.2	30.2	113.5	35.4
79	318	0	112.3	30.6	112.4	33.9
80	317	0	112.5	30.2	111.8	33.5
81	312	1	119.9	33.2	122.6	38.0
82	307	0	112.7	29.9	112.4	32.0
83	305	0	112.5	30.4	112.0	32.7
84	305	0	113.4	30.2	112.6	36.1

5.2 OVERDISPERSED CATEGORICAL DATA AND RANDOM EFFECTS MODELS

Outcome data in count form assumed to be generated from a Poisson model or proportions assumed to be binomial often show a residual variance larger than expected under these models, even after allowing for important predictors of the outcome. This will be evident for example in deviance statistics larger than expected under Poisson or binomial sampling. This overdispersion may arise from omitted covariates, or some form of clustering in the original units (e.g. the data are for individuals but exhibit clustered effects because individuals are grouped by household). Another generic source of overdispersion in behavioural and medical contexts arises from intersubject variability in proneness or frailty.

It is preferable to use a model accounting for such overdispersion, especially if interest focuses on the regression parameters describing the relation between the outcome and certain covariates. As Cox and Snell (1989) point out, standard errors in general linear regression models which do not account for overdispersion are likely to be too small and may result in misleading inferences.

In a regression setting overdispersion may be remedied by the inclusion of additional covariates, or special terms for modelling outliers (Baxter, 1985). Another possibility, especially if overdispersion is attributable to variations in proneness between individuals, is to assume the counts Y_i for subject i are Poisson with mean Z_i. But this mean is itself a random variable, for example with gamma form. This leads to a form of negative binomial density if Y_i and Z_i are related as follows

$$Y_i \sim \text{Poi}(Z_i)$$
$$Z_i \sim \text{G}(\mu\kappa, \kappa)$$

since then $E(Z) = \mu$ and $Var(Z) = \mu/\kappa$. We might model μ in terms of covariates; see, for example, the analysis of the data on school nonattendance below. The marginal variance of Y is then $Var(Y) = \mu + \mu/\kappa$, with smaller values of κ indicating more overdispersion in Y, relative to the Poisson assumption that $Var(Y) = \mu$. Another parameterisation of the gamma density for Z_i involves an index $1/c$ independent of μ, namely

$$Z_i \sim \text{G}(1/c, \mu) \tag{5.1}$$

leading to a quadratic variance function

$$\text{Var}(Y) = \mu + c\mu^2.$$

A specific consequence of overdispersion in terms of distorting inferences relates to tests of interaction between categorical predictor variables in log-linear models. For example, suppose counts Y_{ij} are classified by two factors i and j, but that the counts are overdispersed against the Poisson. Then tests of interaction (such as Pearson's chi-square) assuming Poisson sampling may give a misleading test of interaction between the factors. Several authors have proposed tests of interaction which allow for overdispersion, for example by adjusting the denominator of the chi-square statistic to reflect departures from the Poisson or binomial variance form (Paul and Bannerjee, 1998).

Example 5.5 Non-attendance at school Paul and Banerjee (PB, 1998) analyse Australian educational data relating to days of nonattendance Y at school for 146 children, classified by race (aboriginal or white) and age band (Primary, 1st form, 2nd form and 3rd form). The numbers of children N_{ij} in the eight possible cells range from 13 (primary aboriginal) to 26 (white 1st form), and there is considerable variability in the counts Y_{ijk}, $k = 1, \ldots, N_{ij}$ within each cell. For example, in the 2nd form aboriginal cell, days of nonattendance range from 2 to 81.

PB use an interaction model for the mean in the cells $\{i,j\}$, with $i = 1, \ldots, a; j = 1, \ldots, b$ (see Thall, 1992). This model has the form

$$\mu_{ij} = \alpha_i \beta_j (\tau + \varphi_{ij})$$

where $\alpha_a = 1, \beta_b = 1, \varphi_{ib} = \varphi_{aj} = 0$ for all i and j. The other parameters are modelled as positive fixed effects. Independence of the two factors occurs if $\underset{\sim}{\varphi} = 0$. If we then allow for overdispersion as above, the model becomes

$$Y_{ijk} \sim \text{Poi}(Z_{ijk})$$
$$Z_{ijk} \sim \text{G}(\mu_{ij}\kappa, \kappa).$$

Suppose Y_{ij0} denotes the total nonattendance count in cell $\{i,j\}$ over all N_{ij} pupils in each cell and $\bar{Y}_{ij0} = Y_{ij0}/N_{ij}$ the average count in each cell. Then PB propose the modified Pearson chi- square test

$$C^2 = \sum_i \sum_j N_{ij}(\bar{Y}_{ij0} - \mu_{ij}.\text{ind})/(\mu_{ij.\text{ind}} + \mu_{ij.\text{ind}}/\kappa)$$

with $\mu_{ij.\text{ind}}$ based on the independence hypothesis:

$$\mu_{ij.\text{ind}} = \alpha_i \beta_j \tau.$$

They actually use the alternative parameterisation (5.1) above to model the overdispersion, with $Z_{ijk} \sim \text{G}(1/c, \mu_{ij})$.

Applying this procedure (with a run of 10 000 iterations using *Program 5.5 Non-Attendance*) shows the test statistic C^2 averaging around 26, whereas its distribution is chi-square with $(a - 1)(b - 1) = 3$ degrees of freedom under the independence hypothesis. The mean nonattendance days μ_{ij} for the race–age combinations $\{i,j\}$ show the interaction between the two factors (Table 5.11).

Example 5.6 Reverse mutagenicity assay Albert and Pepple (1989) present count data, based on Ames Salmonella reverse-mutagenicity assay. The data is also analysed by Breslow (1984). The response Y_i is the number of revertant colonies observed on a plate, while the predictor is a measure x_i of dose level. Collings *et al.* (1981) propose a model which can be approximated by Poisson sampling:

Table 5.11 Parameters for non-attendance.

	Mean	SD	2.5%	Median	97.5%
C^2	26.1	9.2	13.4	24.4	48.6
α_1	1.79	0.24	1.37	1.78	2.32
β_1	0.94	0.21	0.60	0.92	1.41
β_2	0.91	0.18	0.63	0.89	1.30
β_3	1.25	0.23	0.88	1.23	1.76
μ_{11}	18.46	3.10	12.90	18.32	25.10
μ_{12}	18.30	2.65	13.57	18.13	23.86
μ_{13}	26.45	3.55	20.14	26.26	33.81
μ_{14}	18.85	2.95	13.53	18.69	24.99
μ_{21}	9.81	1.90	6.47	9.67	13.85
μ_{22}	9.52	1.47	6.94	9.41	12.61
μ_{23}	13.05	1.94	9.57	12.91	17.13
μ_{24}	10.59	1.58	7.54	10.56	13.83
ϕ_{11}	0.67	0.68	0.01	0.46	2.52
ϕ_{12}	0.90	0.88	0.02	0.63	3.26
ϕ_{13}	1.56	1.34	0.06	1.21	4.95

NB White and 3rd form are reference categories for α and β.

$$Y_i \sim \text{Poi}(\theta_i)$$
$$\log(\theta_i) = \beta_0 + \beta_1 x_i/1000 + \beta_2 \log(x_i + 10).$$

Fitting this via a standard Poison regression (see *Program 5.6 Revertant Mutagenicity*) gives a deviance averaging 46.7 (with standard deviation 2.4), which indicates overdispersion for the data set of $N = 18$ counts. A standard Poisson-gamma mixture can be performed via the parameterisation

$$Y_i \sim \text{Poi}(\theta_i)$$
$$\log(m_i) = \beta_0 + \beta_1 x_i/1000 + \beta_2 \log(x_i + 10)$$
$$\theta_i \sim \text{G}(\lambda m_i, \lambda)$$

where λ is the precision of the hyperparameter m_i. This reduces the deviance to a level in line with the available degrees of freedom.

Albert and Pepple consider a discrete prior on alternative values of $\log\lambda$. We replicate their analysis except for extending the discrete grid for $\log\lambda$ to include infinity (i.e. equivalent to the standard Poisson regression). Taking the second half of a run of 10 000 iterations gives Bayes factors as shown in Table 5.12 on seven alternative values of $\log\lambda$.

Example 5.7 Conduct disorder Johnson and Albert (1999) consider the number of times, out of four total observations on 172 schoolchildren, that a students exhibited conduct disorder. Conduct was assessed at grades 6, 8, 10 and 12. The binomial model is then appropriate with $y_i \sim \text{Bin}(\pi_i, 4)$. The available covariates are: sex $S_i(1 = F, 0 = M)$, aggressive behaviour observed

Table 5.12 Revertant colony count analysis.

(A) Gamma mixture

Parameter	Mean	SD	2.5%	97.5%
β_0	2.10	0.41	1.27	2.90
β_1	−1.12	0.46	−2.05	−0.20
β_2	0.34	0.11	0.13	0.55
λ	0.56	0.34	0.17	1.41
Deviance	18.2	6.0	8.4	31.8

(B) Discrete mixture on precision parameter

Relative frequencies of different values of precision parameter

$\log\lambda$	Frequency	%	Prior rel.freq	Bayes factor
1.0	85	0.017	0.143	1.0
0.5	379	0.076	0.143	4.5
0.0	1009	0.202	0.143	11.9
−0.5	1643	0.329	0.143	19.3
−1.0	1368	0.274	0.143	16.1
−1.5	450	0.090	0.143	5.3
−2.0	66	0.013	0.143	0.8

Parameter	Mean	SD	2.5%	97.5%
β_0	2.24	0.31	1.54	2.88
β_1	−0.98	0.36	−1.77	−0.32
β_2	0.30	0.08	0.14	0.47

at 3rd grade ($A_i = 1$ or 0) and social rejection as rated at third grade ($R_i = 1$ if rejected, $R = 0$ otherwise). Hence

$$\text{logit}(\pi_i) = \beta_0 + \beta_1 S_i + \beta_2 A_i + \beta_3 R_i.$$

A standard logit link (Program 5.7, model A) provides the coefficient values in Table 5.13.

We may monitor the usual binomial deviance

$$D = 2\sum \{y_i\log(y_i/\hat{y}_i) + (n_i - y_i)\log([n_i - y_i]/[n_i - \hat{y}_i]\}.$$

This averages around 295, with 168 degrees of freedom. This excess dispersion may mean the standard errors of the regression coefficients are understated.

Table 5.13 Conduct disorder.

Parameter	Mean	SD	2.5%	Median	97.5%
β_0	−0.37	0.13	−0.62	−0.37	−0.12
β_1	−0.17	0.16	−0.48	−0.17	0.15
β_2	0.55	0.22	0.12	0.55	0.98
β_3	−0.28	0.19	−0.64	−0.27	0.09

Table 5.14 Conduct disorder, overdispersion model.

Parameter	Mean	SD	2.5%	Median	97.5%
β_0	−0.47	0.20	−0.86	−0.46	−0.09
β_1	−0.19	0.25	−0.68	−0.19	0.31
β_2	0.69	0.35	0.01	0.69	1.38
β_3	−0.36	0.30	−0.96	−0.36	0.21
σ^2	1.30	0.38	0.67	1.25	2.18

One way to model such overdispersion is in terms of the likely correlation between conduct disorder in repeat observations from the same individual. Unobserved influences mean some individuals have a higher random effect term e_i than others. We introduce random errors at individual student level via a normal density with mean zero (see Program 5.7, model B).
Thus

$$\begin{aligned}\text{logit}(\pi_i) &= \beta_0 + \beta_1 S_i + \beta_2 A_i + \beta_3 R_i + e_i \\ e_i &\sim \text{N}(0, s^2).\end{aligned}$$

The precision (= 1/variance) of the random effects is assigned a noninformative gamma prior. The estimated coefficients from this model have standard errors adjusted upwards in a similar way as would result from a classical correction for overdispersion. The actual interest (assuming approximate posterior normality) is in the standard effect ratios $\beta_j/\text{se}(\beta_j)$, since the β_j may change in a random effects model. Thus the standard effect of aggression falls from 2.5 to 1.97, or around 21%. The conventional correction would retain $\beta_2 = 0.55$ but increase its standard error to $0.22(295/168)^{0.5}$. See Table 5.14.

5.3 MODELS FOR POLYTOMIES AND HISTOGRAMS: EXTRA-VARIATION IN MULTINOMIAL MODELS

The product multinomial approach with a form of Poisson likelihood is a widely adopted and flexible form of log-linear model. Consider groups or individuals $i = 1, \ldots, N$, with a choice between s alternatives. For example, Nelson (1984) considers personal crime victims O_{ij} grouped by city (13 cities in the US) and subject to four possible types of personal crime (robbery, aggravated assault, simple assault, and larceny with contact). The cities differ both in their overall crime rate and the distribution of crimes among the four types. One approach allows variability in the overall rate θ_i (e.g. using a gamma density to represent variability in the Poisson rate), but assumes that the multinomial choice probabilities p_{ij} are constant across i. So the rate for crime type j in city i, θ_{ij} is modelled as the product $\theta_i p_{ij}$. The proportions $p_{ij} = p_j$ may at the simplest be based on the overall proportions of victimisation of each type.

However, in this and other contexts the assumption of equal conditional probability (conditional on the overall rate) is substantively unrealistic. In the crime example, certain individual lifestyles or urban types are more likely to be associated with certain types of crime. To a large extent this type of heterogeneity may be represented by covariates specific to group (or individual) i, choice j, or both. Further, some random variability is likely to remain unexplained despite such extensions.

One option for modelling this heterogeneity is to adopt a Dirichlet prior for the conditional probabilities; this has the advantage of conjugacy when there is a multinomial likelihood.

Assume instead the independent Poisson representation of the multinomial within group i, and subject to the total $O_i = \sum_j O_{ij}$. The respective choice probabilities π_{ij}, with $\sum_j \pi_{ij} = 1$ can be written

$$\pi_{ij} = \exp(\theta_{ij}) / \sum_k \exp(\theta_{ik}).$$

Suppose the parameters of the different multinomial distributions are exchangeable, and that given μ and covariance C the vectors $\theta_i = (\theta_{i1}, \theta_{i2}, \ldots, \theta_{is})$ are independently multivariate normal with common mean μ and covariance C. This specification has in fact more generality than the Dirichlet (Leonard and Hsu, 1994). A multivariate t may be used instead for greater robustness.

Example 5.8 Grades in high schools Leonard and Hsu (1994) present mathematics test results on students by school $i = 1, \ldots, 40$ and grade j, with six grades. A method taking account of the ordering of the grades may be possible. Here we just assume the data result from 40 multinomial distributions, each with six outcomes. We assume the θ_{ij} are multivariate normal with mean μ and precision $P = C^{-1}$. We adopt a multivariate normal hyperprior for μ and a Wishart hyperprior for P with 6 degrees of freedom (see *Program 5.8 High School Grades*).

The estimate of $\pi = (\exp[\mu_1], \exp[\mu_2], \ldots, \exp[\mu_6]) / \sum_j \exp[\mu_j]$, is then shown in Table 5.15.

These are similar to the estimates of Leonard and Hsu, though grade 6 has a slightly lower probability and grade 1 a slightly higher probability. The estimated correlations between grades are as shown in Table 5.16.

Table 5.15 Estimated grade probabilities.

	Grade					
	1	2	3	4	5	6
Mean	0.095	0.217	0.264	0.252	0.073	0.099
2.5%	0.074	0.197	0.238	0.229	0.060	0.073
97.5%	0.129	0.244	0.287	0.278	0.091	0.121

Table 5.16

	1	2	3	4	5	6
1	1.00	0.93	0.77	0.23	−0.42	−0.72
2	0.93	1.00	0.92	0.52	−0.13	−0.47
3	0.77	0.92	1.00	0.76	0.18	−0.17
4	0.23	0.52	0.76	1.00	0.73	0.47
5	−0.42	−0.13	0.18	0.73	1.00	0.90
6	−0.72	−0.47	−0.17	0.47	0.90	1.00

These show high correlations for adjacent grades and negative correlations for widely separated grades.

5.3.1 Histogram smoothing

Let values of an originally continuous variable y be arranged in s histogram intervals of equal width, $\{I_{j-1}, I_j\}, j = 1, \ldots, s$ (e.g. income bands or weight intervals), with frequencies f_i in the ith interval. Often the observed histogram of frequencies is irregular because of sampling variations when *a priori* more smoothness is expected. Leonard (1973) and Leonard and Hsu (1999) propose a method to smooth an observed histogram in line with an underlying density $q(y)$. Let θ_j denote the underlying probability of an observation lying in interval j:

$$\theta_j = \int_{I_{j-1}}^{I_j} q(u)du.$$

The observed frequencies $y_1, \ldots, y_s$ are then multinomial with probability vector $(\theta_1, \ldots, \theta_s)$ and index $F = \sum_j y_j$. We can parameterise the probabilities using the multiple logit as

$$\theta_j = \exp(\gamma_j) / \sum_k \exp(\gamma_k)$$

where $(\gamma_1, \ldots, \gamma_s)$ are multivariate normal with means $g_1, \ldots, g_s$ and sxs precision matrix P. A neutral prior on the θ_j would assign them prior mass $1/s$, and this translates into the means g_j having values $-log(s)$. For the covariance matrix $V = P^{-1}$ we assume a smoothness structure

$$V_{ij} = \sigma^2 \rho^{|i-j|}$$

as in a time series autoregressive process of order 1 (see Chapter 7). This prior says that we expect adjacent points in the histogram to have similar frequencies. Let

$$\tau = \sigma^{-2}(1 - \rho^2)^{-1}. \tag{5.2}$$

The precision matrix then has the form (see Box and Jenkins, 1970)

$$P_{11} = P_{ss} = \tau$$
$$P_{jj} = \tau(1+\rho^2), j = 2, \ldots, s-1$$
$$P_{j,j+1} = P_{j+1,j} = -\rho\tau, j = 1, \ldots, s-1$$
$$P_{ij} = P_{ji} = 0 \text{ for } i = 1, \ldots, s-2; j = 2+i, \ldots, s.$$

We expect ρ to be positive though Leonard (1973) assigns it a Normal prior $\mathrm{N}(a, A)$ with sampled values constrained to be between -1 and $+1$. Leonard assigns the equivalent of a gamma prior to τ, $\tau \sim \mathrm{G}(b, bc)$ where the prior value of $1/\tau$ is c, and b is the strength of belief in this prior estimate. For example if we expected σ^2 to be 0.3 and ρ to be 0.7 then, from (5.2), the prior expectation of τ^{-1} is approximately 0.15. We might then adopt a prior $\tau \sim \mathrm{G}(1, 0.15)$ or $\tau \sim \mathrm{G}(0.5, 0.075)$. A structured covariance matrix may also be obtained using discrete priors, with (for example) most prior mass concentrated on positive values of ρ.

Example 5.9 Pig weight gain data We illustrate histogram smoothing using the distribution of gains in weight among 522 pigs as presented in Leonard and Hsu (LH, 1999) and first analysed by Snedecor and Cochran (1989). The observed frequencies are cumulated into 21 intervals with weight gains (in lbs) $19, 20, 21, \ldots, 38, 39$. They peak at 30 lbs, with $f_{12} = 72$, but show irregularities in the tails; for example, the data show equal frequencies at weight gains 25 and 26 lbs, and more pigs at gain 35 lbs than at 34 lbs.

We adopt discrete priors (see *Program 5.9 Pig Weight Gain*), both with 20 bins, on ρ and τ. For the former we take possible values $0.05, 0.1, 0.15, \ldots, 0.9, 0.95, 0.99$ and for τ we take values $0.5, 1, 1.5, \ldots, 9.5, 10$. These bin values were based on pilot analyses with broader ranges. The resulting estimates of the smoothed frequencies (`Sm[]` in Program 5.9) show less 'smoothing upwards' in the tails than the results of LH and show a value exceeding 0.9 for the autocorrelation, compared with the value of 0.7 used by LH (Table 5.17). The implied variance at around 2.2 is higher than the value of 0.3 proposed by LH.

5.4 LEAGUE TABLES AND META-ANALYSES

In recent years there has been a considerable increase in the use of quantitative comparisons of performance and service quality, especially in the areas of health, education, and social services. The indicators involved may be variously termed performance indicators, outcome indicators, or quality measures, and often form part of performance audits in public services. Similar issues occur in league table rankings of sporting performance (Morris and Christiansen, 1996). The main object in such exercises may include

(a) comparing a set of k observed outcomes against an external standard of acceptable performance, e.g. surgical mortality in a set of hospitals may be

Table 5.17 Pig weight gains, Frequency, actual and mean smoothed.

Weight gains (lbs)	Actual	Mean	SD
19	1	1.51	0.88
20	1	1.56	0.80
21	0	1.95	0.90
22	7	4.99	1.72
23	5	6.05	1.91
24	10	10.90	2.73
25	30	27.60	4.79
26	30	30.37	4.91
27	41	40.65	5.74
28	48	48.31	6.22
29	66	65.43	7.16
30	72	70.94	7.56
31	56	56.04	6.65
32	46	46.39	6.19
33	45	43.31	5.89
34	22	23.46	4.21
35	24	22.23	4.29
36	12	11.50	2.83
37	5	4.94	1.69
38	0	2.06	0.99
39	1	1.78	0.99
Parameters			
ρ		0.92	0.06
τ		3.16	1.35

compared with some acceptable maximum rate, perhaps within one or two standard deviations of a national average (Morris and Christiansen, 1996);

(b) testing a hypothesis that the rates or means being compared are equal (Berger and Deely, 1988);

(c) determining which institution or individual is best (or worst) in the existing sample or the probabilities that each institution is the best;

(d) predicting relative performance or ranking in the next observation period, on the basis of the observed data in the current period.

More refined comparisons may include finding the posterior probability that each institution has an outcome y_i which is a certain factor above or below average. For example, if the outcome were surgical deaths in hospital, then we might wish to find probabilities, for $b > 1$, that

$$P(y_i > by_j, \text{ for all } j\#i).$$

These probabilities would be measures of unacceptable performance relative to the rest of the sample (Deely and Smith, 1998). Varying the factor b would be one form of sensitivity analysis.

There are statistical issues around the uncertainty of the estimates of the outcomes for each institution, and how precisely and with what certainty institutions can be ranked. Sampling variations in rare health outcomes, possibly combined with small populations at risk, can give an impression of spatial or institutional variation when in fact the patterning is largely due to random variability. So it is increasingly recognised that the traditional (but often used) fixed effects comparison of a set of rates is less preferable than one which pools information across institutions to obtain more reliable estimates of performance (Spiegelhalter and Marshall, 1998). On the other hand if there are genuine outliers in the data it may be unwise to assume full exchangeability across institutions. For example, consider a set of normal outcomes, with known variances S_i^2,

$$y_i \sim \text{N}(\mu_i, S_i^2)$$

and suppose we assume that the means μ_i are in turn drawn exchangeably from a common normal density

$$\mu_i \sim \text{N}(\nu, \phi^2). \tag{5.3}$$

This may be modified in several ways. First we may assume that the means μ_i vary in part due to contextual influences on performance; for example, exam results may be affected by the socio-economic composition of a school's catchment area, while clinical comparisons may be affected by 'casemix' differences (some hospitals or surgeons may on average receive more clinically complex patient caseloads than others). Thus Morris and Christiansen (1996) adopt a 'severity index' z_i in a comparison of kidney graft failure rates between hospitals, so that

$$\mu_i \sim \text{N}(\beta_0 + \beta_1 z_i, \phi^2).$$

The other aspect of the exchangeable institutions approach is a common variance ϕ^2. We may modify this by adopting a density robust to outliers and to allow the observations to have different variances in the hyperdensity (5.3). For example a t density with m degrees of freedom is equivalent to

$$\mu_i \sim \text{N}(\nu, V_i\phi^2)$$
$$1/V_i \sim \text{G}(0.5m, 0.5).$$

This is the mixture interpretation (sometimes called a 'scale mixture') of the t density (Gelman *et al.*, 1995). Another procedure, useful in this and other modelling settings, is to use a contaminated or location shift normal – this allows each observation to fall in the main sample or in an outlier sample with a different mean (Verdinelli and Wasserman, 1991).

There are many other issues around the robustness of league table rankings based on performance indicators: for example, relatively slight changes in the definition of indicators used, the time span they cover and so on, may considerably affect the relative positions of the institutions, persons or areas being compared.

Example 5.10 Baseball scores Morris and Christiansen (MC, 1996) present data on runs per game Y_i for 14 American Baseball League Teams in 1993. Detroit led that season with an average of 5.549 runs per game.

The variances $V_i = V(y_{ij})/162$ over the runs scored y_{ij} in $j = 1, \ldots, 162$ games in that season are also known. The observed data (see *Program 5.10 Baseball Scoring*) are thus averages Y_i and square roots of V_i.

An exchangeable model for these data is

$$Y_i \sim \text{N}(\theta_i, V_i)$$
$$\theta_i \sim \text{N}(\nu, \tau^2), i = 1, \ldots, 14.$$

We first estimate the probabilities that each team is worst or best in terms of this model. Second we calculate predictive probabilities that each team is best or worst in a 'new' or future season, by sampling from the posterior distribution of each team's score rate. This season would have 162 games also, so the prediction is

$$Y_i.\text{new} \sim \text{N}(\theta_i, V_i).$$

The second half of a run of 50 000 iterations is used to obtain these statistics. The probability that Detroit is the best team after we take account of the sampling variation over all 14 games is 0.542 (MC obtain 0.58), and the probability that it is the best team in the 'new' season is 0.396, as accumulated in the `vector best.new[]` in Program 5.10 (MC obtain 0.438). Other predictive applications are possible, for example forecasting the entire 1994 season on the basis of the first 15 % of the games, and combining with the 1993 predictive density just discussed. We may also obtain the distribution of the ranks (for the 1993 and new 1994 seasons).

Example 5.11 Conception rates The data analysed here have been considered by Goldstein and Spiegelhalter (1996), and reanalysed by Deely and Smith (1998). They concern the conception rate among 13–15-year-old girls in a set of Scottish health boards. There are $N = 15$ such boards, with the analysis concerning the number of conceptions x_i in relation to female populations in the relevant age range n_i. Following the earlier work a Poisson model with mean rates $\lambda_i n_i$ (i.e. rates of incidence λ_i) is applied. The populations are in thousands so the incidence rates are per 1000.

To allow for variability in the underlying outcome, a gamma mixing density for the rates λ_i is selected, namely $\text{G}(\mu\beta, \beta)$ so that the mean of the gamma density is μ, the average of the underlying rates over the entire set of N areas, and its variance is μ/β. Various specifications of the priors for β and μ are possible, and any league table rankings may be sensitive to prior specification, as well as to the other broader caveats about such rankings.

Here a $\text{G}(M, 1)$ prior for μ was used, where M is a prior estimate of the mean; a simple average of the original rates x_i/n_i is a possibility, though strictly amounts to a use of the data (i.e a form of empirical Bayes). In practice the analysis is likely to be more sensitive to the choice of prior for β, than that of μ.

A non-informative gamma density G(0.1, 0.1) for β is used initially, with $M = 7.8$ (the average observed rate).

Among issues of interest are the comparison of the crude rates per 1000 x_i/n_i with the smoothed rates allowing for exchangeability between the areas being compared. Here we compare B (0.1, 0.1) and B (0.01, 0.01) priors on β, and evaluate the following:

(a) the probability that each area has the worst rate;
(b) the probability that each area has a rate less than 95% of the other rates;
(c) the predictive probability that each area has the best rate next year given the same population n_i at risk.

We obtain broadly similar conclusions under these alternative priors; for example, the probability that area 15 (Tayside) is worst is about 0.39, and areas 11–15 together have around 0.9 probability of including the worst rate. The Western Isles (area 1) has a conception rate that is less than 95% of the other rates with probability 0.30 under a B (0.1, 0.1) prior for β. This is monitored by the array `best.by.factor[]` in *Program 5.11 Conceptions*. This area has a predictive probability of 0.32 of having an observed number of conceptions per 1000 that is smaller than any other rate. Areas 2 and 9 (respectively Orkney and Shetland) also have relatively high predictive probabilities of having the lowest rate next year.

5.5 META-ANALYSIS

Statistical techniques to combine results from separate clinical trials, economic evaluations and epidemiological studies have become commonplace in the health sciences – partly in consequence of increasing size of the research literature, partly due to demand for clinical standards drawn from evidence-based medicine. Until the recent growth in quantitative review in medicine, more examples of meta-analysis occurred in the social sciences, especially in psychological and educational research (Hedges and Olkin, 1985).

The goals of meta-analysis may be seen to include the establishment of systematic protocols for study design and recording results and the combination of evidence from a range of studies which may individually be subject to sampling variation. The emphasis is on combining estimates of effect size over a number of studies, and in medical settings such effect sizes usually consist of odds ratios, risk differences, and risk ratios. They have also been applied in sociological and psychometric studies, where more attempt has been made to synthesise results based on regression or correlation coefficients. Combination of p values or test statistics (differences in means divided by a pooled standard deviation) are also possible.

In fixed effects models the assumption is made that there is a global average effect and individual studies differ from this because of sampling variation. Thus for a continuous outcome we have

$$y_i = \mu + e_i$$

where e_i is the error in the ith study, assumed to be normal or Student t. This model would need to be modified for a discrete outcome. For example, if studies differed in terms of a binomial outcome, r_i from n_i observations, then

$$r_i \sim \text{B}\,(y_i, n_i)$$
$$\text{logit}\,(y_i) = \mu + e_i.$$

An alternative fixed effects approach may be defined according to risk categories r, such that for studies $i = 1, \ldots, I$ and risk groups within studies r,

$$y_{ir} = \mu_r + e_{ir}.$$

The random effects model, by contrast, views some or all of between-study variation as due to random variations in study design, measurements, etc. These lead studies to obtain different effect parameters. For example, Hsieh and Pugh (1993), reviewing studies on the association between poverty and violent crime, find a variance in the reported correlations between studies too great to reflect sampling variability alone. In the random effects approach, variation in true effects between studies is allowed, and these are regarded as unknown parameters drawn from a population of study effects. Between-study variation is seen as a source of uncertainty that should be taken account of in estimating the overall treatment effect μ. The choice between fixed and random effects may be decided in part by a formal chi-square test of homogeneity – that is whether the between study variance component is zero or not. Alternatively a prior for this variance may be chosen which allows a wide range of values (including zero) (DuMouchel, 1990; Hedges, 1997).

A baseline assumption which requires critical evaluation is that of exchangeability between the studies: that they are not identical repeats but alike enough to be compared with a view to estimating a parameter of interest in a meta-analysis. So if there is exchangeability then there is no need to assign a different prior to any one study. On the other hand, variation in effect sizes may be related in part to features of study design (e.g. case control vs cohort studies in epidemiology, or size of area unit in sociological studies of aggregate outcomes). Then exchangeability may be more appropriately assumed within a particular study type.

Random effects models estimated by empirical Bayes methods estimate the between-study variance from the data and then condition on this estimated value. By contrast a fully Bayes method allows for the uncertainty in the estimate of the between-study variability. A typical model structure for a linear outcome would be

$$y_i \,|\, \alpha_i, \sigma_i^2 \sim \text{N}(\alpha_i, S_i^2) \tag{5.4}$$
$$\alpha_i \,|\, \mu, \sigma_\alpha^2 \sim \text{N}(\mu, \sigma_\alpha^2)$$
$$\mu \sim \text{N}(m_1, m_2)$$
$$1/\sigma_\alpha^2 \sim \text{G}\,(t_1, t_2)$$

where m_1, m_2, t_1 and t_2 are known constants. Here α_i is the true study effect, S_i^2 is the within-study variance (typically known from the original study together with the effect size y_i), σ_α^2 is the between-study variance and μ is the overall effect. We may set non informative proper priors, e.g. $m_1 = 0$, $m_2 = 1000, t_1 = 0.001, t_2 = 0.001$. Alternatively, there may be a stronger prior assumption that the overall effect (e.g. a treatment difference) is zero. For example, Silliman (1997) adopts a mixed empirical-fully Bayes approach; this involves computing a non-iterative estimate of 0.04 for σ_α^2 from 12 fluoride effectiveness studies and then establishing a 'clinical informative prior' with $m_1 = 0$ and $m_2 = 0.04$, and with $t_1 = 25, t_2 = 1$. We might also draw on a previous meta-analysis to provide our prior assumptions.

Conditional on σ_α^2, the observed effects y_i are normal about μ with variances $V_i = \sigma_\alpha^2 + S_i^2$ while the conditional distribution of μ is Normal with variance

$$\sigma_\mu^2 = (\sum_{i=1} 1/V_i)^{-1}$$

and mean

$$\sum_{i=1} y_i/V_i.$$

This is a weighted mean of the observed effect sizes, and in empirical Bayes estimation would be calculated conditional on a single estimate of σ_α^2.

We would typically be interested in whether this overall effect was positive, and specifically in the posterior probability that $\mu > 0$. Sensitivity to prior assumptions about the between-study variance may also be investigated (e.g. Thompson, 1993). This is especially important for the prior on the between-study variance when the number of studies is small (Hedges, 1997).

Example 5.12 Income inequality and homicide Hsieh and Pugh (1993) review studies linking violent crime (homicide, rape, assault, robbery) to poverty and income inequality. The analyses are for spatial aggregates (neighbourhoods, metropolitan areas, states, nations). The effect size considered here is the correlation between homicide rates and the Gini index of income inequality. There were 17 studies with evidence on this (in studies with repeated correlations on the same set of spatial units, the later year was taken). These authors used an N-weighted average coefficient to estimate the global effect but noted the variation in effect sizes between studies. Here we adopt the Fisher z transform of the original correlation coefficients and model the z_i as normal with precisions $n_i - 3$ where n_i is the number of units (counties, nations, etc.) on which the ith correlation is based.

We assume a random effects model with $m_1 = 0, m_2 = 100, t_1 = 1, t_2 = 0.01$ (see *Program 5.12 Violent Crime*). The prior on the precision favours higher values (i.e. lower values for the between-study variance) in line with a prior belief that the studies are measuring a common effect but with some variability due to study design and type of aggregate area unit. A range of other prior assumptions, especially on the between-study variance, is possible.

The pooled correlation estimate is 0.47 and the between-study standard deviation is estimated as 0.17, with 95% credible interval (0.08, 0.29). A chi-square statistic for homogeneity between studies averages 261 (on 16 degrees of freedom), with a 95% credible interval (199, 327). So a random effects approach is supported.

Example 5.13 Vitamins and infection Glasziou and Mackerras (1993) performed a fixed effects meta-analysis of vitamin A supplements in infectious disease. The simple odds ratios from the five studies, and their 95% intervals vary from 0.46 (study 3) to 0.94 (study 2), with the common odds ratio estimated as 0.70 – with a 95% interval from 0.62 to 0.79. This analysis suggests a reduced risk of infection with vitamin supplements. A random effects meta-analysis (*Program 5.13 Vitamin Supplement*) gives smooth estimates of the study odds ratios varying from 0.63 to 0.76. The overall effect is estimated as 0.70 with 95% credible interval from 0.60 to 0.87. Some idea of the uncertainty of extending this to future studies is gained by prediction (of OR.new) from the posterior density of the overall effect. This gives an average of 0.72 with 95% interval from 0.45 to 1.07. So a predictive test still gives some ground for hesitation about the effectiveness of the supplements.

5.5.1 Combining over design types and selection bias

We may be interested in combining the results from different types of study design, in a cross-design synthesis. This might include contrasts between clinical trials or observational studies, or between case control or cohort studies. Here a typical approach assumes exchangeability within a specific design category. There would be, as before, an overall population effect μ, then means λ_j for specific types of study (e.g. $j = 1$ for clinical trials, $j = 2$ for case-control and $j = 3$ for cohort studies) and finally, α_{ij} for effect parameters within study types with $i = 1, \ldots, I_j, j = 1, \ldots, J$. We could use regression to estimate the λ_j, for example by including a categorical factor z_j with levels $j = 1, \ldots, J$ as a predictor. The individual study parameters α_{ij} will be assumed to have a multivariate normal prior with means λ_j and $J \times J$ covariance matrix S.

Further aspects of meta-analysis may also be investigated by Bayesian methods. For example, the usual random effects analysis assumes the studies included are a random sample from all available studies. However, if studies are selected via literature search, there is a problem of selection bias, in that published work tends to be biased towards statistical significance. One option is to model the process of selection by assuming that the probability of a study being observed (in the sense of it being published) is a function of its effect size. The observed (published) studies are then a weighted version of the true set of study effects, with the weights being defined by the selection probabilities (Silliman, 1997).

Example 5.14 Environmental tobacco smoke Hedges (1997) presents data from a United States Environmental Protection Agency meta-analysis of 11 studies into the effect of environmental tobacco smoke on lung cancer. Two of the studies were cohort studies, the others case-control. We therefore adopt the multilevel structure outlined above to estimate the separate design effects (mean effects for case-control and cohort studies) and the grand mean pooling across designs. This type of analysis will be sensitive to prior assumptions, and since there are only two cohort studies the variance of the cohort studies will not be satisfactorily identified.

The second half of a run of 10 000 iterations (see *Program 5.14 Tobacco Smoke*, model A) gives the estimates of the individual study effects, the design-specific means, and the grand effect mean as in Table 5.18.

The probability that the latter exceeds zero is about 0.85, comparable to the analysis of Hedges (1997) who does not estimate a grand mean but separate effects for cohort and case-control studies. The device of sampling seven 'new' cohort studies is used to complete the multivariate normal sampling.

To illustrate sensitivity to selection bias, we consider the same data, but without regard to design type. A standard analysis via random effects as in (5.4) above gives an estimated central effect of 0.157 with 95% interval from −0.02 to 0.35 (see Program 5.14, model B). The probability is 0.954 that μ exceeds zero. This is close to the result of Hedges, namely a central estimate of 0.17 with an interval (0.01, 0.33).

To allow for selection bias we follow the approach of Silliman (1997) in assuming higher prior values of selection for 'more significant' studies. Here we take these as studies from among the 11 where effect y_i exceeds the survey standard deviation S_i (see Program 5.14, model C). The less significant studies are the remainder. We can model the bias by sampling two uniform U(0, 1) numbers, u_1 and u_2, and taking $w_1 = \max(u_1, u_2)$ as the weight attaching to significant studies and $w_2 = \min(u_1, u_2)$ as the weight for nonsignificant

Table 5.18 Tobacco smoke meta-analysis.

Node	Mean	SD	2.5%	97.5%
$\alpha_{1,1}$	0.1774	0.2269	−0.2651	0.6454
$\alpha_{1,2}$	0.2569	0.3293	−0.3561	0.9731
$\alpha_{2,1}$	0.07586	0.2081	−0.3788	0.4552
$\alpha_{2,2}$	0.1523	0.1469	−0.1329	0.4386
$\alpha_{3,2}$	0.1789	0.4348	−0.6601	1.022
$\alpha_{4,2}$	0.1673	0.4071	−0.5836	0.9395
$\alpha_{5,2}$	0.1654	0.4169	−0.6275	0.8979
$\alpha_{6,2}$	0.1662	0.4337	−0.6708	1.073
$\alpha_{7,2}$	0.177	0.4814	−0.6481	1.069
$\alpha_{8,2}$	0.1626	0.4367	−0.6914	0.997
$\alpha_{9,2}$	0.1682	0.4417	−0.6673	0.9953
μ	0.1604	0.1747	−0.1738	0.5094
λ_1	0.1491	0.1191	−0.08546	0.3828
λ_2	0.1718	0.1861	−0.1787	0.5509

studies. The resulting estimate of μ is 0.10 with interval $(-0.04, 0.27)$ and a 91.7 % probability that the mean effect is positive.

5.6 ESTIMATING POPULATION PARAMETERS FROM SURVEY DATA

Complex survey designs are used to estimate population parameters, especially totals, and means or rates based on them. In practice simple random sampling from an entire population is a relatively infrequent survey mode. More frequently, sampling is done independently within strata or subpopulations. The focus is therefore often on predicting parameters for population subgroups or geographical areas, perhaps crossed by survey variables. Such subgroupings may also, depending on context, be called clusters, small areas, or domains. Not only are estimates needed but also measures of their precision.

The element of prediction relates to the latent or unknown data in that part of the subpopulations which is not sampled or surveyed (Stroud, 1991). Underlying this prediction are models of the observations and the hyperparameters generating them, not just for the survey subjects, but the 'superpopulation' of observations (sampled and nonsampled together).
Thus the goal of the analysis is

(a) to obtain posterior densities of parameters such as means or proportions θ_i, for domains $i = 1, \ldots, k$, based solely on the observed totals $y_1, y_2, \ldots, y_k$ from samples of size $n_1, n_2, \ldots, n_k$ (with $y_k = \sum_{i=1}^{n_k} y_{ik}$);
(b) to predict the population totals $y'_1, \ldots, y'_k$ in the remaining non-surveyed populations of sizes $N_1 - n_1, N_2 - n_2, \ldots, N_k - n_k$. The predicted subgroup totals at population level are $Y_1 = y_1 + y'_1$, $Y_2 = y_2 + y'_2$, $\ldots$ and these form the basis for inferences about the sub-population parameters.

The hierarchical Bayes approach based on exchangeability between domains provides a way to 'borrow strength' in situations when inferences are required from relatively thinly spread multistage surveys to strata or to small areas not necessarily included in the survey. As an example of the latter, Stroud (1994) quotes the example of the Canada Youth AIDS study (King *et al.*, 1988) where the primary sampling units were school boards within two-way strata. Within regions within Canada, school boards were stratified into Catholic vs Protestant, and into rural–urban categories. In some strata no boards were sampled.

The exchangeability assumption forms the basis for random effects at primary Sampling unit (PSU) level. Such effects may be combined with regression models governing systematic variations between domains, or between relatively homogenous subgroups based on covariate combinations. For example, Stroud (1991) describes procedures used in the Canada Labour Force Survey, where geographical stratification (for strata $1, \ldots, k$) is combined with subdivision of the adult population into groups defined by age band (ages 15–24, 25–54, 55+)

and by sex. Thus the prediction of labour force numbers Y_{jk} is specific to $6 \times k$ sub-populations. This may involve a conjugate beta-binomial model as in Stroud (1991).

A model-based approach in a similar application, namely labour force participation by local area, is outlined by MacGibbon and Tomberlin (1989). The model is logit-normal in form and includes categories and continuous covariates defining sampled individuals, as well as their survey characteristics. For a binomial outcome, the probability of success is

$$\text{logit}(\pi_{jik}) = \theta_j + X_{ik}\beta + \varepsilon_i$$

for an individual lying in the jth cell in a set of categorical covariates, and being the kth sampled in the ith cell in a set of nested sampling criteria, such as PSU, or secondary sampling unit (SSU) within PSU. For example, the parameter θ_j represents the impact of fixed classifications (sex, ethnicity) on the outcome, the parameters β represent the impact of continuous covariates (age in single years, income) X_{ik}, and the ε_i represent exchangeable departures from the fixed part of the model at the PSU level. The latter effects may also be specific to the categories j.

Suppose the design is a two-stage cluster sample to assess labour force participation, with geographic areas at stage one and a random sample of individuals within the selected PSUs. Suppose the subscript j denotes sex, then for individual i in the kth PSU,

$$\text{logit } (\pi_{ik}) = \theta[J_{ik}] + X_{ik}\beta + \varepsilon_i$$

where J_{ik} is the sex of the kth individual in the ith PSU, and where the ε_i are random effects, typically Normal or Student t. For a more complex three-stage design the model might take the form

$$\text{logit } (\pi_{imk}) = \theta[J_{imk}] + \varphi_{im} + \beta X_{imk}$$

where φ_{im} is the random effect associated with the mth SSU within the ith PSU.

Alternatively (especially if there are no covariates) one may adopt a hierarchical beta-binomial model. Let y_{ic} denote 'positive' responses out of n_{ic} surveyed in stratum i and respondent category c. Category c may be based on the intersection of two or more original classifications. Then

$$y_{ic} \sim \text{ Bin } (\pi_{ic}, n_{ic})$$

with one parameterisation for the prior being

$$\pi_{ic} \sim \text{ Beta } (vm, v(1 - m))$$

where m is a prior mean proportion and v is its precision, as in Example 5.3. The mean m may be differentiated according to respondent category c, so that

$$\pi_{ic} \sim \text{ Beta } (vm_c, v(1 - m_c)).$$

Example 5.15 Simulated data on local area labour force participation MacGibbon and Tomberlin (MT, 1989) describe a simulation study in which $I = 15$

areas formed the PSUs for a study of labour activity, and within the ith PSU, $K = 25$ individuals between the ages of 20 and 40 are sampled. The assumed PSU means for the noncentred random effects $\eta_i, i = 1, \ldots, 15$ (average participation rates by area) are 0.79, 0.79, 0.96, 0.88, 0.90, 0.95, 0.86, 0.96, 0.61, 0.87, 0.781, 0.91, 0.94, 0.92 and 0.83. Activity is also affected by sex, with $\theta_1 = -0.5$ (for males) and $\theta_2 = -1$ (for females), and by age, with a coefficient of $\beta = 0.1$ on a linear age term.

MT repeated this sampling scheme for 205 replications. Here we adopt five replications, generated according to

$$\begin{aligned} &\text{Sex}[i,k] \sim \text{ Bern } (0.5) \\ &\text{Age } [i,k] \sim \text{ U}(20,40) \\ &\text{logit } (\pi_{ik}) = \theta[\text{Sex } [i,k]] + \beta \text{ Age } [i,k] + \eta_i - \bar{\eta}, \quad i = 1, \ldots, 15, k = 1, \ldots, 25. \end{aligned}$$

Having generated the data we apply a logit normal model for labour participation. Our pooled model estimate of the average participation rates by PSU, p_i, is then a simple average of the predicted probabilities p_{ikr} over cases k and replicates r, using

$$\text{logit } (p_{ikr}) = \theta[\text{Sex } [i,k,r]] + \beta \text{ Age } [i,k,r] + \varepsilon_i$$

with θ_1, θ_2 and β assigned vague priors and ε_i assumed to be Normal with zero mean and variance σ^2.

The estimates from this regression model given in Table 5.19 are pooled over replicates. In a more general analysis, we might also consider variability between replicates. The male–female differential is replicated by using centred errors and no intercept means the location of θ_1 and θ_2 is shifted.

This approach can be compared with

(a) unbiased estimates based on the observed sample proportions, with no pooling mechanism, and so less precision, and
(b) synthetic estimates which pool over areas but take account (say) only of the age–sex structure of each area's sample (Holt *et al.*, 1979). Thus the participation rate for sex and age group (a, b) is estimated via a fixed effects logistic regression over all IK sampled individuals

$$\text{logit } (p_{ab}) = \theta_{1a} + \theta_{2b}$$

and then participation rate in area i is estimated on the basis of the age–sex structure of its sampled individuals.

The comparison with (a) is illustrated in Table 5.20 using the observed proportions from the sample data for five replicates. In areas 4, 11, 12 and 15 it can be seen that the model estimate in fact come closer to the generating values.

Example 5.16 Predicting population level labour force participation Stroud (1991) presents data $\{y_{ic}, n_{ic}\}$ on labour force participation in Canada, by stratum

Table 5.19 Local area participation.

Parameter	Mean	SD	2.5%	Median	97.5%
β	0.070	0.012	0.046	0.070	0.093
σ^2	0.654	0.341	0.233	0.570	1.517
θ_1	0.465	0.397	−0.311	0.464	1.275
θ_2	−0.366	0.391	−1.125	−0.370	0.420
Area participation rates					
P_1	0.801	0.034	0.730	0.802	0.863
P_2	0.831	0.031	0.765	0.833	0.888
P_3	0.940	0.019	0.900	0.942	0.972
P_4	0.886	0.026	0.830	0.888	0.932
P_5	0.874	0.027	0.817	0.876	0.923
P_6	0.964	0.016	0.927	0.966	0.988
P_7	0.844	0.030	0.782	0.846	0.898
P_8	0.940	0.020	0.895	0.942	0.973
P_9	0.691	0.040	0.609	0.691	0.767
P_{10}	0.894	0.025	0.839	0.896	0.938
P_{11}	0.734	0.038	0.659	0.735	0.805
P_{12}	0.894	0.025	0.840	0.896	0.939
P_{13}	0.913	0.023	0.863	0.915	0.952
P_{14}	0.914	0.023	0.863	0.916	0.953
P_{15}	0.816	0.032	0.746	0.817	0.875

Table 5.20

Area	Total active	Total sampled	Sample estimates
1	99	125	0.792
2	103	125	0.824
3	119	125	0.952
4	111	125	0.888
5	109	125	0.872
6	123	125	0.984
7	105	125	0.84
8	119	125	0.952
9	84	125	0.672
10	112	125	0.896
11	90	125	0.72
12	112	125	0.896
13	115	125	0.92
14	115	125	0.92
15	101	125	0.808

$i = 1, \ldots, I(I = 8)$ and sex $(C = 2)$. The total active y in a sample of $n = 320$ was 251. The survey population in turn represents about 0.7% of the total population $N = \sum N_{ic}$. We are interested in prediction of the population-wide participation rates π_{ic} and in the rates π'_{ic} among the $[N_{ic} - n_{ic}]$ unsampled individuals. We would expect the predictive variance of the latter to exceed that of the π_{ic}, especially for relatively small sampling fractions n_{ic}/N_{ic}. We are also interested in the total predicted labour force sizes Y_i in stratum i.

For small sampling fractions, results obtained are likely to be sensitive to the form adopted for the prior and the informativeness of the prior. We compare the logit-normal and beta-binomial approaches. Under the former

$$\text{logit}(\pi_{i1}) = \theta_1 + \varepsilon_{i1} \quad \text{(males)}$$
$$\text{logit}(\pi_{i2}) = \theta_2 + \varepsilon_{i2} \quad \text{(females)}$$

for $i = 1, \ldots, 8$; $j = 1, 2$. We assume $\varepsilon_{ij} \sim \text{N}(0, \sigma^2)$ with $1/\sigma^2 \sim \text{G}(0.01, 0.01)$. Under the beta-binomial we assume $\pi_{ic} \sim \text{Beta}(vm, v(1-m))$ with $m = 0.75$ and $v = 1$. The first model (Program 5.16, model A) leads to predicted labour force sizes Y_i and participation rates π_{ic} and π'_{ic} as given in Table 5.21.

The 95% predictive interval for Y_1 is (3814, 4628), while Stroud (1991) using a normal approximation around a mean of 4183 finds an interval (3684, 4682). Slightly different and somewhat less precise results are obtained with the beta-binomial model (see Program 5.16, model B).

5.6.1 Modelling missing cells in multi-way designs

Bayesian regression may also be applied to data from a two-way or multi-way stratified design, including the case of clusters within strata. With appropriate modelling it is possible to obtain estimates of population parameters even when certain cells formed by the multi-way stratification contain no sampled data. Specifically, a nonsaturated model in terms of fixed effects on the stratifying variables may be used. This means fixed effects at the level at which the cells are empty are not included unless perhaps they are assigned informative priors. Random effects at this level may be used however. For a two-way stratification with categorical variables r with R categories and s with S categories, and with a subset of unsampled cells (r^*, s^*), the log-linear model might thus be

$$\text{logit}(p_{ikr}) = \theta_{1r} + \theta_{2s} + \varepsilon_{rs}.$$

A corner constraint on the fixed effects θ_{1r} and θ_{2s} is applied, so that $\theta_{11} = 0, \theta_{21} = 0$. The term ε_{rs} is typically a normal random effect. Stroud (1994) outlines the same approach within a beta-binomial structure.

Example 5.17 Sexual behaviour by religion and urban stratum Stroud (1994) presents survey data on frequency of sexual behaviour from a school-based study into AIDS and Youth in Canada. Thirteen schools were the PSUs and drawn from a two-way stratified design based on Catholic/Protestant denomination, and a rural/town/small city division. In the Catholic/small city stratum no schools were sampled. The estimates of the population numbers (see *Program 5.17 Youth & AIDS*) reporting frequent sexual intercourse on the basis of the sampled data are given in Table 5.22.

Table 5.21

	Mean	SD	2.5%	97.5%
Y_1	4217	206	3814	4628
Y_2	4063	200	3656	4460
Y_3	2158	144	1820	2387
Y_4	3579	185	3179	3916
Y_5	6976	334	6352	7666
Y_6	2745	166	2374	3023
Y_7	3161	157	2873	3480
Y_8	2656	132	2381	2907
$\pi_{1,1}$	0.70	0.07	0.57	0.84
$\pi_{1,2}$	0.92	0.04	0.83	0.97
$\pi_{2,1}$	0.68	0.07	0.54	0.82
$\pi_{2,2}$	0.93	0.03	0.86	0.98
$\pi_{3,1}$	0.63	0.08	0.45	0.76
$\pi_{3,2}$	0.88	0.06	0.74	0.95
$\pi_{4,1}$	0.65	0.07	0.49	0.78
$\pi_{4,2}$	0.92	0.04	0.83	0.97
$\pi_{5,1}$	0.70	0.07	0.57	0.84
$\pi_{5,2}$	0.94	0.03	0.87	0.99
$\pi_{6,1}$	0.57	0.09	0.36	0.72
$\pi_{6,2}$	0.94	0.03	0.87	0.99
$\pi_{7,1}$	0.70	0.07	0.57	0.84
$\pi_{7,2}$	0.94	0.03	0.87	0.99
$\pi_{8,1}$	0.67	0.07	0.52	0.80
$\pi_{8,2}$	0.92	0.04	0.83	0.97
$\pi'_{1,1}$	0.70	0.07	0.56	0.84
$\pi'_{1,2}$	0.92	0.04	0.83	0.97
$\pi'_{2,1}$	0.68	0.07	0.54	0.82
$\pi'_{2,2}$	0.93	0.03	0.85	0.98
$\pi'_{3,1}$	0.63	0.08	0.46	0.76
$\pi'_{3,2}$	0.88	0.06	0.74	0.96
$\pi'_{4,1}$	0.65	0.07	0.49	0.78
$\pi'_{4,2}$	0.92	0.04	0.83	0.97
$\pi'_{5,1}$	0.70	0.07	0.57	0.84
$\pi'_{5,2}$	0.94	0.03	0.87	0.99
$\pi'_{6,1}$	0.57	0.09	0.36	0.72
$\pi'_{6,2}$	0.94	0.03	0.87	0.99
$\pi'_{7,1}$	0.70	0.07	0.57	0.85
$\pi'_{7,2}$	0.94	0.03	0.87	0.99
$\pi'_{8,1}$	0.67	0.07	0.52	0.80
$\pi'_{8,2}$	0.92	0.04	0.83	0.97

The population-wide predictions are comparable to those of Stroud, obtained via a beta-binomial model, though his mean for the missing cell is 68 (in a total group of size 2324). The precision of the prediction is obviously less for this cell than for the others.

Table 5.22 Youth & AIDS study. Frequency of sexual intercourse.

		Rural	Town	Small city
Catholic	'Often' in sample	7	8	0
	Total sample	140	104	0
	Total children	2523	937	2324
Protestant	'Often' in sample	24	19	11
	Total sample	292	174	278
	Total children	4452	1391	1698
Predicted 'Often' among total children				
Catholic	Mean	129	71	98
	2.5%	61	35	4
	97.5%	224	118	514
Protestant	Mean	363	154	67
	2.5%	244	100	37
	97.5%	506	217	106

5.6.2 Two-stage sampling for proportions

For reasons of cost or lack of suitable sampling frame, it is often advantageous to sample from a selection of groupings or clusters of the original sampling units. For example, if a town contains 200 000 persons and no satisfactory sampling list exists, then a listing will have to be specially made for a survey to be undertaken. It may be better to confine this listing to a few subareas in the town. Suppose the town can be divided into 1000 subareas each containing 200 persons, then a sample of say 1% of the town's population can be obtained by first sampling 10 subareas (the 'clusters') and then sampling all the individuals in those clusters. Alternatively, two-stage sampling occurs if there is subsampling within the clusters chosen at the first stage; for example, 100 subareas could be chosen and 10% of the population sampled in each.

Suppose we are considering a categorical outcome, specifically a binary variable Y_{ik} for individual i in cluster k. Suppose the underlying average proportion in cluster k is θ_k, i.e.

$$\Pr(Y_{ik} = 1 \mid \theta_k) = \theta_k.$$

Suppose n clusters have been sampled from a possible set of N and that two-stage sampling is used with m_k units selected from the total populations M_k in cluster k. For this form of outcome we typically adopt a beta prior $\theta_k \sim \mathrm{B}\,(\beta, \tau - \beta)$, with τ a notional prior sample size.

The fundamental prediction issue here relates to the total population proportion P of individuals exhibiting a 'positive' outcome on the binary variable, where the total population is the sum over all N possible clusters of the cluster populations M_k.

Example 5.18 Visiting the doctor in the last year Nandram and Sedransk (NS, 1993) consider predicting P in a synthetic population of $N = 74$ US counties where the binary outcome is whether the individual has seen the doctor at least once in the past year. Their 'population' consists of all the sampled individuals in the 1986–87 National Health Interview Survey satisfying certain personal attributes and within the selected counties. Their two-stage 'sample' then consisted of a proportional sample from the NHIS sampled individuals in $n = 20$ of the possible 74 clusters.

They supply values of m_k and p_k in the 20 clusters where p_k is the proportion seeing the doctor. However sampling fractions m_k/M_k are not given. For illustrative purposes it is here assumed that a 10% sample is taken so that $M_k = 10m_k$; for the remaining 54 clusters, similarly sized M_k to those in the 20 sampled clusters were randomly generated.

The work of NS allows for uncertainty in the beta prior for the cluster proportions θ_k by adopting a mixture prior on the first parameter β. Thus β can take values $a_1, a_2, \ldots, a_R$ with probabilities $\Pr(\beta = a_r) = \omega_r$, where

$$a_1 < a_2 < \ldots < a_R < \tau.$$

They assume $R = 9$ possible values for β, with values of a_r and ω_r based on a previous NHIS.

Their approach is here implemented in Program 5.18 with total positive outcomes in cluster $k, m.t[k], k = 1, \ldots, 20$ pre-calculated as the products $p_k m_k$.

The results obtained for the smoothed cluster proportions and overall visiting rate (Table 5.23) are close to those obtained by Nadram and

Table 5.23 Doctor visits.

	Mean	SD	2.5%	Median	97.5%
P	0.809	0.011	0.788	0.810	0.840
$\theta(1)$	0.817	0.023	0.770	0.818	0.861
$\theta(2)$	0.832	0.025	0.781	0.833	0.879
$\theta(3)$	0.815	0.027	0.759	0.815	0.866
$\theta(4)$	0.824	0.024	0.776	0.825	0.868
$\theta(5)$	0.809	0.028	0.753	0.809	0.860
$\theta(6)$	0.809	0.029	0.750	0.810	0.861
$\theta(7)$	0.834	0.024	0.785	0.834	0.877
$\theta(8)$	0.803	0.028	0.745	0.804	0.855
$\theta(9)$	0.802	0.027	0.747	0.802	0.852
$\theta(10)$	0.809	0.023	0.762	0.810	0.853
$\theta(11)$	0.815	0.029	0.755	0.816	0.868
$\theta(12)$	0.812	0.025	0.762	0.813	0.858
$\theta(13)$	0.829	0.025	0.778	0.830	0.876
$\theta(14)$	0.816	0.028	0.758	0.817	0.870
$\theta(15)$	0.822	0.022	0.778	0.823	0.863
$\theta(16)$	0.796	0.028	0.737	0.796	0.848
$\theta(17)$	0.787	0.030	0.726	0.788	0.842
$\theta(18)$	0.787	0.028	0.730	0.788	0.840
$\theta(19)$	0.751	0.026	0.697	0.751	0.801
$\theta(20)$	0.799	0.029	0.739	0.800	0.854

Sedransk. They were obtained after a single long run of 20 000 iterations, with $\tau = 200$.

The posterior average for P is 0.8087 with a 95% credible interval (0.788, 0.839). Sensitivity analysis might involve adopting less informative ω_r values (e.g. a uniform distribution), increasing or reducing the sampling fractions, or adopting a range of values for τ.

CHAPTER 6

Latent Variables, Mixture Analysis and Models for NonResponse

6.1 INTRODUCTION

The previous chapters have considered regression modelling of associations and pooling strength in estimating outcomes over a set of units, on the assumption that an underlying common model could represent the heterogeneity between units. The question is then one of specifying the prior for this unknown common density. Typically the prior chosen has a specific parametric (e.g. conjugate) form. However, since the appropriate parametric form of prior is not known, a nonparametric specification has sometimes been proposed as an alternative (Laird, 1982). Latent variables also underlie the representation of distributions (of data or parameters) by mixtures; here one goal is to improve the robustness of inferences by approximating the true density of the sample as a mixture of classes (West, 1992; Robert, 1997). This may be appropriate for observed data which are multimodal or asymmetric and for which a parametric model assuming a single component population is implausible. In regression modelling, finite mixture models may provide additional insights about behavioural patterns as sources of heterogeneity (e.g. different impact of marketing variables on subpopulations in mixed Poisson models of purchasing behaviour) whereas in random effects models the extra-dispersion is a nuisance factor (Wedel *et al.*, 1993). In smoothing health outcomes over sets of small areas, especially when there may be different modes in subsets of areas, a nonparametric mixture may have advantages (Clayton and Kaldor, 1987). The same rationalisation is present when latent variables, whether continuous or discrete, are postulated to underlie observed associations between several categorical variables, for example in contingency tables. The associations between the manifest indicators may be related to the operation of a few, often just one, latent constructs.

Whether the main orientation is to smoothing, pooling strength, robust inference or regression mixture analysis, a finite mixture of C subpopulations or clusters is postulated, with the subpopulations representing the classes of an underlying latent variable or possibly latent variables. This idea is generalised in the Dirichlet process priors approach, where the number of possible clusters in the latent variable is not specified *a priori*. In all latent variable applications, subject matter knowledge may be important in guiding model choice, and in specifying priors which improve identifiability. The problems of identifiability of mixture models due to flat likelihoods is discussed by Bohning (2000), though from an empirical rather than fully Bayes perspective.

6.1.1 Missing data models

Estimation of all such models is implicitly or explicitly based on augmentation of the observed data to account for the missing latent data. In this way a natural link exists with models for data subject to nonresponse, where the complete data likelihood consists of the sampling model for the observed outcomes together with a model for response mechanism (Little and Rubin, 1987). The link with latent variable methodology is clearest in categorical data subject to missingness, where the missing categorical value is imputed or modelled as a function both of the observed responses and possibly of the unobserved response itself. The overlap is also illustrated in the so-called general location model for datasets including both continuous and categorical outcomes, and with values on both types of variables possibly missing. Here an overall latent class variable refers to the joint probability of all categorical variables together, where the observed data may be subject to missing values. Given this latent class category j the continuous variables are assigned a multivariate density with vector mean μ_j.

6.1.2 Latent traits vs latent classes

In the analysis of discrete outcomes tables the goal of introducing latent variables may often be to improve the understanding of complex multivariate datasets, by reducing dimensionality and making clear the main elements of the underlying structure of the data in terms of unobservable latent variables (Bartholomew and Knott, 1999). Such variables may be defined conceptually but may not be possible to measure directly. For example, quantifying human behaviour, attitudes, or physical and intellectual status may pose measurement problems in terms of a lack of an obvious single measure of the underlying construct.

An example of unobserved constructs underlying observed data is provided by physical and mental disability, definable in terms of an inability to perform the normal activities of daily living, but commonly measured in terms of multiple dimensions of functioning. Such dimensions may in turn frequently be represented by categorical, often ordinal, responses to multiple choice questions (Bandeen-Roche *et al.*, 1997). Another area where multiple scales are applied to measure an underlying variable is in mental testing; here the scales are often simply binary (right/wrong) responses and the latent variable reflects either a specific ability or general intelligence. More generally we are often faced with situations where multiple discrete indices are being used to measure a latent construct that is difficult to measure directly. In latent structure analysis the goal is to summarise the observed indicators through continuous metric latent scales. In latent class analysis by contrast, the assumption is that the associations between the observed categorical indicators represent the impact of a latent categorical variable (Wilcox, 1983).

In latent structure models with z continuous a conventional assumption improving identifiability is that there is no association between the manifest variables once the latent variable is known. Thus let the conditional probability that an individual i with a univariate latent trait value z_i exhibits a particular set of observed or manifest responses to p categorical items be denoted

$$P_{i1,i2,i3,\ldots,ip}(z_i) = \Pr(x_{1i} = i_1, x_{2i} = i_2, \ldots, x_{pi} = i_p \mid z_i). \tag{6.1}$$

Suppose we had $p = 3$ binary items, then this conditional independence assumption means the joint success probability (ones on all items), given z_i, is

$$\Pr(x_{1i} = 1, x_{2i} = 1, x_{3i} = 1 \mid z_i) = p_1(x_{1i} = 1 \mid z_i)\, p_2(x_{2i} = 1 \mid z_i)\, p_3(x_{3i} = 1 \mid z_i)$$

where $p_1(|), p_2(|)$, and $p_3(|)$ are the marginal item probabilities. A further feature of latent trait analysis is the arbitrariness of the location and scale of z: assuming (for example) that $z \sim \mathrm{N}(0, 1)$ is simply a convention.

In many applications there may be substantive reasons to assume the latent variable is categorical, or little is lost in terms of model fit by grouping the range of a continuous z into categories. Then the densities given z in (6.1) are replaced by a relatively few distinct probability values. For example, Tanner (1996) cites the example of $p = 3$ responses to dichotomous questions on abortion attitudes where the single underlying latent class variable Z_i can be simply represented as pro- or anti-abortion. We would then focus on probabilities such as

$$p_{ih} = p(x_{ih} = 1 \mid Z_i = 1), \quad r_{ih} = p(x_{ih} = 1 \mid Z_i = 0), \quad i = 1, \ldots, n; h = 1, \ldots, p.$$

In general we could have $q < p$ latent variables with $C_1, \ldots, C_q$ categories each, and the focus is on the joint probability of the categorical outcomes x_i given the categories of the latent variables. Suppose $q = 1$, and we have prior probabilities θ_j of an observation belonging to each of the C latent classes. The assumption of conditional independence here reduces to

$$\Pr(x_i \mid Z_i = j) = \sum_{k=1}^{p} \Pr(x_{ik} = k \mid Z_i = j), \qquad j = 1, \ldots, C; i = 1, \ldots, n.$$

For a randomly selected individual the marginal probability of the outcome x_i is

$$\Pr(x_i) = \sum_j \theta_j \Pr(x_i \mid Z_i = j).$$

A Bayesian analysis might also involve specifying priors to improve identifiability; for example, constraints on the prior density of a categorical latent variable with $C = 2$ classes to ensure that one class is always the more frequent.

6.1.3 Regression and transition mixtures

Discrete latent variables are also appropriate in the analysis of mixtures of distributions and regression mixtures. These may be appropriate if there are different causal processes in subpopulations of the data, or if there is multimodality or skewness. In general we may have C categories of the underlying latent variable, and then the underlying variable for case i is discrete taking value j with probability θ_j. In probabilistic terms,

$$Z_i \sim \text{Categorical}(\theta_{[1:C]})$$

with θ often assigned equal prior weights on the possible values of the latent categorisation. Then the true density $g(x_i)$ is approximated as a mixture of densities with parameters $\{\phi_j\}$ determined by the choice of Z_i

$$x_i \sim f(x_i \mid \phi_{Z_i}).$$

For example, if x were continuous but subject to multimodality or skewness because of different subpopulations in the data, then we might have

$$x_i \sim \text{N}(\mu_{Z_i}, \sigma^2_{Z_i}).$$

If our data has a time component then we may envisage changes in the categorisation of cases i and time points t. A simple form for evolution is the Markov chain, with transition probabilities stationary (homogenous) over time. Thus the transition probabilities of the $C \times C$ chain are defined by

$$\gamma_{jk} = P(Z_{it} = k \mid Z_{i,t-1} = j)$$

leading to conditional densities on the observed data

$$x_{it} \sim f(X_{it} \mid \theta_{Z_{it}}).$$

The missing data Z_{it} are the sequence of states visited by the Markov chain.

6.1.4 Latent factor models, latent class analysis and item analysis

Bartholomew (1980, 1987) considers sets of discrete manifest variables and proposes a methodology for explaining the associations among p manifest variables $x_h (h = 1, \ldots, p)$ in terms of a smaller number of q latent variables $(y_j, j = 1, \ldots, q)$. This methodology derives metrical latent variables, but for manifest variables which are categorical. Thus consider a vector of p binary or polytomous outcomes defined for $i = 1, \ldots, n$ individuals. There may also be covariates w_i defined for individual i. The goal is to derive a small dimensional set of scores y_{ij} which summarise the interrelation between the original manifest variables. Thus we are interested in posterior 'factor scores' on the latent variables y given the observed x (and possibly covariates). We are also interested in the 'factor loadings' relating the observed and manifest variables.

Suppose we have p binary outcomes. It is assumed that the responses to the p observed items are independent given the q factor scores, that is

$$\Pr(x_{i1} = i_1, x_{i2} = i_2, \ldots x_{ip} = i_p \,|\, y_{i1}, y_{i2}, \ldots, y_{iq}) =$$
$$\prod \Pr(x_{i1} = i_h \,|\, y_{i1}, y_{i2}, \ldots) \ldots \Pr(x_{ip} = i_p \,|\, y_{i1}, y_{i2}, \ldots).$$

The model therefore simplifies to separate Bernoulli sampling one for each outcome, $h = 1, \ldots, p$ and for subjects $i = 1, \ldots, n$

$$x_{ih} \sim \text{Bern}(\pi_{ih})$$

with appropriate link functions for π_{ih} and scores y_{ij}

$$G^{-1}(\pi_{ih}) = \alpha_{h0} + \sum_{j=1}^{q} \alpha_{hj} H^{-1}(y_{ij})$$

where the scores of subject i on factor j $\{y_{ij}\}$ are taken to be uniformly distributed on $(0, 1)$. These are denoted by Bartholomew (1987) as 'y scores' or factor scores, meaning patterns of joint responses on many categorical items are expressed as scores on a small number q of continuous latent variables. For example, if $q = 1$, and the items were binary tests of ability with 1 denoting a correct answer and 0 an incorrect answer, then the collection of p item responses $\{1, 1, \ldots, 1\}$ will typically represent one extreme of the underlying ability scale and the responses $\{0, 0, \ldots, 0\}$ the other extreme. The function $\pi_{ih}(y_i) = \Pr(x_{ih} = 1 \,|\, y_i)$ for the $h = 1, \ldots, p$ items is known as the response function or in educational testing as the item characteristic curve.

As developed by Bartholomew (1987) the most tractable response function (RF) model has G^{-1} as the logit transform and H^{-1} as the probit. Equivalently we may write

$$\text{logit}\{\pi_{ih}\} = \alpha_{h0} + \sum_{j=1}^{q} \alpha_{hj} z_{ij}$$

where the z_{ij} are normal. This is known as the logit/probit model for deriving factor scores from binary variables. As noted by Bartholomew (1994) the prior distribution of the z variables is essentially arbitrary, since there is no unique scale of measurement for z. For example, taking z as $\text{N}(0, 1)$ is a convention rather than something which will be confirmed by the data. The intercepts α_{h0} measure essentially the comparative frequency of positive outcomes on item h. They define the probability of a positive response for a median individual, with zero scores z_{ij} on the underlying latent variable(s).

There is also an arbitrary choice about which of two possible outcomes (for x_h binary) is regarded as 'positive'. For example, in the case of a single latent variable with two classes an identifiability constraint may be made to ensure that the first class is always associated with a positive outcome.

In the case of two or more latent variables it is necessary to fix certain α's to ensure identifiability (without such a constraint an orthogonal transform of the α_{hj} leaves the likelihood unchanged). Thus if $q = 2$, it is sufficient to set one of the regression coefficients of item h on latent variable 2, denoted α_{h2}, to equal zero (e.g. by setting $\alpha_{12} = 0$). Bock and Gibbons (1996) discuss these identifiability issues in detail, but from the alternative underlying variable (UV) approach. Bartholomew (1987) elaborates on the distinction (but ultimate equivalence) of the RF and UV approaches.

Example 6.1 Cancer knowledge sources Bartholomew (1987) considers the analysis of four binary variables on knowledge of cancer, and whether or not the following were sources of information:

(1) radio, (2) newspapers, (3) solid reading, (4) lectures.

The data can be summarised as in Table 6.1. The response patterns and their frequencies among the survey respondents are in order of factor score (from lowest to highest) as estimated by Bartholomew (1987).

We apply the logit-probit model as discussed above and take $q = 1$. Thus a single latent dimension is assumed. We conduct the analysis at the individual level even though there are no covariates to illustrate the approach. N(0, 1) priors are taken on $\alpha_{h0}, \alpha_{h1} (h = 1, \ldots, 4)$, and on the individual latent scores z_i (see *Program 6.1 Cancer Knowledge*). So the model is

$$\text{logit}\{\pi_{hi}\} = \alpha_{h0} + \alpha_{h1} z_i, \quad h = 1, \ldots, 4, i = 1, \ldots, 1729.$$

After a burn-in of 500 iterations, a further 4500 give the posterior estimates of parameters and factor scores shown in Table 6.2. These estimates are close to those of Bartholomew except that α_{21}, the regression coefficient relating the second item to the latent variable, is slightly smaller and estimated more precisely than by Bartholomew (who estimates $\alpha_{21} = 3.4$ with standard deviation 1.14). The factor scores have the same ordering except that Bartholomew finds y_9 (for the response 0100) exceeding y_8 (for the response 1011).

Table 6.1

Response pattern	Frequency	y score
0000	477	0.208
1000	63	0.280
0001	12	0.284
0010	150	0.343
1001'	7	0.358
1010	32	0.416
0011	11	0.421
1011	4	0.497
0100	231	0.558
1100	94	0.639
0101	13	0.644
0110	378	0.710
1101	12	0.727
1110	169	0.792
0111	45	0.797
1111	31	0.870
Total	1729	

Table 6.2

Parameters	Mean	SD	2.5%	50%	97.5%
α_{10}	−1.287	0.068	−1.424	−1.286	−1.158
α_{20}	0.488	0.108	0.292	0.483	0.717
α_{30}	−0.147	0.069	−0.280	−0.146	−0.015
α_{40}	−2.696	0.123	−2.946	−2.691	−2.464
α_{11}	0.727	0.095	0.545	0.726	0.912
α_{21}	2.582	0.419	1.880	2.541	3.500
α_{31}	1.484	0.173	1.189	1.470	1.860
α_{41}	0.774	0.139	0.514	0.773	1.057
π. median(1)	0.217	0.012	0.194	0.217	0.240
π.median(2)	0.620	0.026	0.572	0.619	0.674
π.median(3)	0.464	0.017	0.430	0.464	0.498
π.median(4)	0.064	0.007	0.050	0.064	0.079
y scores (factor scores)					
$y(1)$	0.130	0.201	0.000	0.030	0.726
$y(2)$	0.215	0.251	0.000	0.104	0.884
$y(3)$	0.224	0.257	0.000	0.114	0.889
$y(4)$	0.335	0.294	0.000	0.264	0.962
$y(5)$	0.329	0.291	0.001	0.252	0.959
$y(6)$	0.478	0.316	0.005	0.472	0.992
$y(7)$	0.474	0.321	0.005	0.447	0.990
$y(8)$	0.610	0.314	0.025	0.660	0.999
$y(9)$	0.530	0.323	0.009	0.548	0.996
$y(10)$	0.655	0.300	0.037	0.735	1.000
$y(11)$	0.656	0.301	0.044	0.731	1.000
$y(12)$	0.777	0.256	0.127	0.881	1.000
$y(13)$	0.779	0.259	0.127	0.886	1.000
$y(14)$	0.875	0.197	0.257	0.972	1.000
$y(15)$	0.869	0.204	0.248	0.972	1.000
$y(16)$	0.941	0.131	0.509	0.997	1.000

6.1.5 Item analysis

The RF model with G^{-1} as the probit rather than the logit and with $q = 1$ corresponds to a generalised form of the item response model, used extensively in educational and psychological testing (Albert, 1992). Item response models are frequently applied to batteries of p test or attitude items which can be scored correct ($x_j = 1$) or incorrect ($x_j = 0$), or agree/disagree, and where all items can be conceived as representing a single continuous underlying measure. There are two goals of such an analysis: first, to rank the ability, or other form of underlying trait, for each subject, and second to identify the effectiveness of different items in measuring the underlying dimension. Suppose the latent trait for subject i is denoted z_i, and the item response curve measures the probability that an individual answers correctly or affirmatively given their trait score.

This curve can be represented in general linear regression models terms, with inverse link G appropriate for a binary outcome and the mean given by a linear regression on the latent trait for subject i with coefficients for each item. So for item responses x_{ij}, $i = 1, \ldots, n, j = 1, \ldots, p$, we have a link function G^{-1}, with

$$\Pr(x_{ij} = 1) = G(\beta_j z_i - \alpha_j) \tag{6.2}$$

and item-specific regression intercepts and slopes, namely α_j and β_j. We put a negative sign on the intercepts in order that they can be interpreted as measures of difficulty of item j (assuming $x_j = 1$ is the correct answer in an ability test). For example, G^{-1} could be a logit or probit transform. The model proposed by Lord (1952) assumes G in (6.2) is the cumulative normal density

$$\Pr(x_{ij} = 1) = \Phi(\beta_j z_i - \alpha_j).$$

In the terminology of Bartholomew (1987) this is a probit/probit model.

The terms α_j measure the difficulty of the different items and the slopes β_j measure an item's power to discriminate ability or trait between subjects. For two subjects separated by a given distance from each other on the z scale, the bigger the absolute value of β_j the greater is the difference in their probability of giving a positive response. More specialised forms of this model assume all the item slopes are equal, $\beta_j = \beta$ or that all the item slopes are equal to 1. The latter model, i.e.

$$\Pr(x_{ij} = 1 \mid z_i) = \Phi(\alpha_j + z_i)$$

was considered by Rasch (1960), with z_i interpreted as an ability parameter for subject i.

Example 6.2 Introductory statistics Tanner (1996) presents binary data x_{ij}, for $p = 6$ test items (j index) and $n = 39$ students (i index) on an Introductory Statistics course. Then we have (see *Program 6.2 Introductory Statistics*)

$$\begin{aligned} x_{ij} &\sim \text{Bernoulli}(p_{ij}), j = 1, \ldots, 6; i = 1, \ldots, 39 \\ \text{probit}(p_{ij}) &= z^*_{ij} \\ z^*_{ij} &\sim \text{N}(\mu_{ij}, 1) \\ \mu_{ij} &= \beta_j z_i - \alpha_j \end{aligned}$$

where the difficulty and discrimination parameters, α_j and β_j respectively, are assigned priors $\alpha_j \sim \text{N}(0, 1)$ and $\beta_j \sim \text{N}(1, 1)$. The scores z_i are assumed to be $\text{N}(0, 1)$. The estimated parameters (from 10 000 iterations with 500 burn-in) suggest item 4 as the most difficult, with items 3 and 5 as the most discriminating in terms of identifying ability (Table 6.3).

A Rasch item analysis, with $\beta_j = \beta$, confirms that item 4 is the most difficult (see Table 6.4).

Table 6.3

Parameter	Mean	SD	2.5%	Median	97.5%
α_1	−0.15	0.30	−0.76	−0.14	0.40
α_2	−0.26	0.30	−0.86	−0.26	0.31
α_3	0.15	0.35	−0.53	0.15	0.84
α_4	0.76	0.35	0.12	0.76	1.47
α_5	0.39	0.39	−0.34	0.38	1.21
α_6	−0.20	0.35	−0.90	−0.20	0.47
β_1	0.40	0.48	−0.51	0.39	1.43
β_2	0.51	0.48	−0.35	0.48	1.59
β_3	1.34	0.75	0.17	1.23	3.20
β_4	0.89	0.52	−0.02	0.85	2.01
β_5	1.44	0.69	0.30	1.38	2.99
β_6	1.14	0.66	0.04	1.06	2.65

Table 6.4

Parameter	Mean	SD	2.5%	Median	97.5%
α_1	−0.19	0.38	−0.93	−0.19	0.59
α_2	−0.31	0.39	−1.08	−0.30	0.46
α_3	0.18	0.38	−0.56	0.18	0.95
α_4	0.95	0.41	0.19	0.94	1.77
α_5	0.43	0.39	−0.33	0.43	1.20
α_6	−0.19	0.39	−0.95	−0.18	0.57
β	0.88	0.27	0.38	0.87	1.46

6.1.6 Polytomous outcomes

The response function approach may be used for polytomous outcomes, as opposed to binary items, by adopting a multiple logit transform for G^{-1}. For items $i = 1, \ldots, p$ with $R[i]$ categories, the model has q latent scores for each individual, and parameters relating response patterns on the items to the latent scores. Intercept and slope parameters are specific to the category of each item – though as for all logit models, a reference outcome category must be chosen for identifiability. Assume $q = 1$, and denote the intercepts for item i and category k by α_{ik} and the slopes by β_{ik}. Then we may set $\alpha_{i1} = \beta_{i1} = 0$ and estimate the remaining parameters, usually as fixed effects. As noted by Bartholomew (1987) we often expect the category scores to be monotonic (ordered) if the latent variable assumption is sensible in substantive terms.

For example let the vector of responses to item 1 by subject i be denoted $x.1_i$; this contains $R(1) - 1$ zeros and a one for the category actually chosen. Let the unknown multinomial parameter for individual i on item 1 be $\pi.1_i$. Then the multiple logit specification (based on the equivalent Poisson form) for a single latent variable z is as follows

$$x.1_i \sim \text{Multinomial}(\pi.1_i, 1)$$
$$\pi.1_{i,k} = \varphi.1_{i,k}/\Sigma_h \varphi.1_{i,h}, \quad k = 1, \ldots, R(1)$$
$$\log\varphi.1_{i,k} = \alpha_{1k} + \beta_{1k} z_i.$$

Specification of suitable priors for α_{ik} and β_{ik} reflects the assumption that a latent factor model is appropriate to the data being considered.

A non informative prior is not necessarily suitable in this as in other latent trait models. Instead a range of moderately informative priors may be adopted to assess the sensitivity of posterior estimates of regression parameters, factor scores and the individual z scores to different priors. As above a standard normal prior for the z scores is frequently chosen by default.

Example 6.3 Staff assessment data Bartholomew (1987) considers data on a staff assessment exercise in a large public corporation. There are $p = 3$ items with respectively 4,3 and 3 categories; i.e. $R(1) = 4, R(2) = 3, R(3) = 3$. Higher categories on the items correspond to higher success ratings. Bartholomew presents results for the case $q = 1$ with the latent variable being interpreted as 'success in the job'. The arbitrary nature of what counts as 'positive' is apparent in the estimation of this model (see *Program 6.3 Staff Assessment*), which results in the inverse latent variable interpretable as 'failure in the job'. Thus z scores and regression coefficients may be reversed in sign, and factor scores transformed via $y' = 1 - y$ to provide comparability with Bartholomew's results.

Alternative N(0, 2) and N(0, 4) priors were adopted for the intercepts and slopes giving the results in Table 6.5 (after 5000 iterations with a 500 burn-in).

Table 6.5

	N(0, 2)				N(0, 4)			
	Mean	SD	2.5%	97.5%	Mean	SD	2.5%	97.5%
$\alpha(1,2)$	2.25	0.26	1.74	2.75	2.37	0.32	1.81	3.03
$\alpha(1,3)$	1.96	0.27	1.45	2.48	2.09	0.32	1.50	2.75
$\alpha(1,4)$	−0.54	0.50	−1.67	0.35	−0.67	0.65	−2.18	0.42
$\alpha(2,2)$	2.75	0.38	2.05	3.54	3.08	0.52	2.22	4.19
$\alpha(2,3)$	2.36	0.38	1.62	3.17	2.69	0.53	1.83	3.84
$\alpha(3,2)$	2.44	0.34	1.85	3.20	2.75	0.47	2.01	3.84
$\alpha(3,3)$	1.75	0.40	0.98	2.61	1.99	0.55	0.91	3.18
$\beta(1,2)$	−0.44	0.34	−1.06	0.28	−0.58	0.34	−1.26	0.07
$\beta(1,3)$	−1.38	0.39	−2.18	−0.64	−1.50	0.44	−2.46	−0.72
$\beta(1,4)$	−2.17	0.52	−3.30	−1.16	−2.54	0.73	−4.15	−1.26
$\beta(2,2)$	−1.30	0.37	−2.02	−0.60	−1.62	0.48	−2.60	−0.79
$\beta(2,3)$	−1.92	0.44	−2.87	−1.19	−2.23	0.50	−3.31	−1.37
$\beta(3,2)$	−0.90	0.35	−1.64	−0.23	−1.21	0.46	−2.20	−0.43
$\beta(3,3)$	−2.68	0.64	−4.10	−1.52	−3.19	0.95	−5.52	−1.85

Adopting the less informative prior may increase the discriminative power of the model (as measured by the absolute size of the slopes) but at the expense of some loss of precision.

The slopes are absolutely slightly smaller, but estimated more precisely, than under the maximum likelihood method used by Bartholomew (1987). The factor scores and z scores (common to all individuals with a given response pattern) are similar to those of Bartholomew, though do not necessarily rank the response patterns exactly the same (Table 6.6).

Thus Bartholomew scores the response pattern $\{321\}$ as lower than $\{112\}$, but the posterior mean score is here estimated as higher for the former pattern.

6.1.7 Latent class analysis (LCA)

Like latent trait analysis, latent class analysis is a form of latent variable model applied to a set of observed responses on p categorical variables. However, the

Table 6.6 Scores under $N(0, 4)$ prior for intercepts and slopes.

Response pattern	Frequency	y score	(SD)	z score	(SD)
000	1	0.99	(0.04)	2.02	(0.70)
100	7	0.98	(0.08)	1.75	(0.70)
200	1	0.94	(0.14)	1.34	(0.71)
001	2	0.96	(0.11)	1.46	(0.69)
010	3	0.93	(0.15)	1.31	(0.71)
101	13	0.92	(0.16)	1.21	(0.70)
020	1	0.88	(0.21)	1.01	(0.71)
110	10	0.88	(0.20)	1.04	(0.71)
201	5	0.83	(0.25)	0.81	(0.71)
120	5	0.81	(0.25)	0.76	(0.70)
210	3	0.77	(0.28)	0.63	(0.70)
011	12	0.81	(0.25)	0.75	(0.68)
220	1	0.66	(0.32)	0.35	(0.72)
021	4	0.72	(0.30)	0.50	(0.69)
111	64	0.72	(0.29)	0.49	(0.68)
121	36	0.61	(0.32)	0.24	(0.68)
211	38	0.54	(0.33)	0.09	(0.68)
221	31	0.40	(0.32)	−0.21	(0.68)
012	1	0.47	(0.34)	−0.06	(0.72)
311	4	0.35	(0.32)	−0.36	(0.74)
022	1	0.35	(0.32)	−0.33	(0.72)
321	3	0.24	(0.28)	−0.63	(0.75)
112	37	0.35	(0.32)	−0.32	(0.69)
122	23	0.25	(0.29)	−0.56	(0.70)
212	34	0.19	(0.25)	−0.73	(0.68)
222	41	0.12	(0.20)	−1.03	(0.71)
312	5	0.08	(0.16)	−1.20	(0.73)
322	11	0.05	(0.12)	−1.51	(0.75)

assumption of LCA is that the interaction between these manifest variables can be explained by $q < p$ latent *categorical* variables, with $C_1, C_2, \ldots, C_q$ categories (Goodman, 1974). The most common latent class models assume latent conditional independence: conditional on the level of the latent variables, the manifest variables are independent.

For example, in medical diagnostic studies the problem may be whether or not the relationships among p indicator or diagnostic marker variables support a latent variable describing degree of morbidity, e.g. see Castle *et al.* (1994). Often we have a situation similar to item analysis where all the manifest variables are binary. Thus Rindskopf and Rindskopf (1986) use four binary signifiers for the presence of heart attack risk to determine the existence of a single latent categorical variable. They consider either $C = 2$ or $C = 3$ categories in this latent diagnostic variable.

The conditional independence assumption of LCA means that within a category of the latent variable the manifest variables are independent of each other. So for a set of binary outcomes, the parameters of a latent class model consist of the proportions $\theta_1, \ldots, \theta_C$ in the C categories of the latent variable, and the 'item probabilities' of a positive response $\delta_{jk}(j = 1, \ldots, C; k = 1, \ldots, p)$ to item k within each category j of the latent variable. Suppose we have 3 binary items and a latent variable with $C = 2$ categories. Then the LCA assumptions are expressed as follows:

$$\theta_1 + \theta_2 = 1$$

$$P_k = \theta_1\delta_{1k} + \theta_2\delta_{2k}, \quad k = 1, \ldots, 3$$

$$P_{kj} = \theta_1\delta_{1k}\delta_{1j} + \theta_2\delta_{2k}\delta_{2j}, \quad k, j = 1, \ldots, 3$$

$$P_{kjm} = \theta_1\delta_{1k}\delta_{1j}\delta_{1m} + \theta_2\delta_{2k}\delta_{2j}\delta_{2m}, \quad k, j, m = 1, \ldots, 3$$

where P_k is the relative frequency of positive response to item k, P_{kj} is the relative frequency of joint positive response to items k and j and so on.

In BUGS, the choice between classes $j = 1, \ldots, C$ of a single latent construct and the probabilities of different items under the different classes may be modelled in a similar way to latent mixture regression. Thus we use the Categorical function, dcat() in BUGS, with a Dirichlet prior for the mixture proportions. While the default choice for the category proportions assumes equal prior weights in the Dirichlet prior, it may improve identifiability to adopt a slightly informative prior: an example is provided by the work of Stern *et al.* (1995) on infant temperament. Bandeen-Roche *et al.* (1997) consider modelling the latent class probabilities for each case via a regression function on covariates w_i. Then, for a single latent variable,

$$Z_i \sim \text{dcat}(P_{i,1:C})$$

where the individual and class-specific probabilities P_{ij} are modelled via binary or multinomial logit regression.

Example 6.4 Machine-design tests McHugh (1956) presents binary item data resulting from four machine-design tests given to $n = 137$ engineers. The data are an abstraction from originally continuous data with $x_j = 1$ for results above the mean on test scale j, $x_j = 0$ for results below the mean. He fits a latent class model with two classes using maximum likelihood methods. The essential coding to estimate the proportions in the $C = 2$ categories of the latent variable and the $C \times p$ probabilities of the positive response, where $p = 4$, is as follows

```
{for (i in 1:n) { Z[i] ~ dcat(theta[])
for (k in 1:p) {x[i,k] ~ dbern(delta[Z[i],k])}}
for (j in 1:C) {for (k in 1:p) {delta[j,k] ~ dbeta(1,1)}}
theta[1:C] ~ ddirch(alpha[])}
```

Beta(1, 1) priors are assumed for probabilities of a positive response δ_{jk} while the category proportions θ_j follow a Dirichlet prior.

Applied to the machine-design item data, we adopt a Dirichlet prior with weights 2 and 1 – without this device the category proportions default to equal values of 0.5. Despite this prior (see *Program 6.4 Machine Design*) we still obtain approximately equal probabilities in the latent class categories, with $\theta_1 = 0.48$, and the first class corresponding to higher ability and concentrations of positive scores on the items (Table 6.7).

Posterior probabilities of individuals of belonging to different classes $\Pr(Z_i = j | x, \theta, \delta)$ may be monitored by the statement

```
for (i in 1:n){for (j in 1:C){post.prob[i,j] <- equals(Z[i],j)}}
```

In the present case individuals with the response pattern (1, 1, 1, 1) have posterior probabilities exceeding 0.99 of belonging to the first class. This latent class analysis is therefore simply distinguishing more creative from less creative respondents.

Example 6.5 Binary indicators of myocardial infarction risk Rindskopf and Rindskopf (1986) present data on four indicators (i.e. symptoms) of cardiac risk for 94 patients. They argue that a two-class model is adequate for these data but there is some evidence that three classes can also be identified (see *Program 6.5 Diagnosis LCA*). These might be named as low, medium and high

Table 6.7

Parameter	Mean	SD	2.5%	Median	97.5%
θ_1	0.483	0.083	0.330	0.481	0.653
θ_2	0.517	0.083	0.344	0.519	0.675
δ_{11}	0.675	0.068	0.541	0.676	0.808
δ_{12}	0.703	0.067	0.571	0.703	0.834
δ_{13}	0.770	0.102	0.578	0.767	0.964
δ_{14}	0.756	0.091	0.571	0.763	0.915
δ_{21}	0.218	0.094	0.028	0.227	0.386
δ_{22}	0.274	0.092	0.076	0.290	0.428
δ_{23}	0.093	0.059	0.007	0.086	0.234
δ_{24}	0.178	0.066	0.056	0.175	0.316

risk groups of patients. Their respective probabilities $\theta_j, j = 1, \ldots, 3$ and the item probabilities are estimated as shown in Table 6.8.

The item probabilities δ under this solution show that item 4 has a high probability of being positive (86 %) even for moderate risk patients (i.e those with $j = 2$), whereas the first item has only a 29 % chance of being positive for this class. To predict the frequencies of the 16 possible response patterns on the four items under the three-class model a 'new patient' (case 95) is randomly sampled using the group probabilities θ, and the probabilities $\gamma_{95,m}$ of the $m = 1, \ldots, 16$ different patterns calculated (and multiplied by the observed total of 94). The average of these reproduces the observed patterns reasonably closely, though it includes nonzero predictions for patterns for which no patients were actually observed.

As an illustration of analysis where the latent class memberships are predicted from regression, a synthetic age variable was created for the 94 patients. This was calculated as 60, plus a normal N(0, 20) term, plus the product of the symptom total (between 0 and 4) times a coefficient distributed as N(1, 1). This produces a moderate positive age gradient in cardiac risk.

We then allocate, for the three-class model, according to $Z_i \sim \text{dcat}(P_{i1}, P_{i2}, P_{i3})$ where P_{ij} are modelled via multinomial logit regression. That is

$$P_{ij} = \eta_{ij}/(\eta_{i1} + \eta_{i2} + \eta_{i3}),$$

where

$$\log(\eta_{ij}) = a_j + b_j \text{Age}_i,$$

with non informative priors on a_j and $b_j (j = 2, \ldots, k)$, and with $a_1 = b_1 = 0$. Applying this in *Program 6.5 LCA Diagnosis (Covariate Model)* produces a

Table 6.8

Parameter	Mean	SD	2.5%	Median	97.5%
θ_1	0.47	0.08	0.31	0.48	0.61
θ_2	0.19	0.09	0.02	0.18	0.38
θ_3	0.34	0.09	0.16	0.34	0.50
$\delta_{1,1}$	0.01	0.02	0.000	0.01	0.06
$\delta_{1,2}$	0.17	0.06	0.035	0.17	0.30
$\delta_{1,3}$	0.03	0.03	0.001	0.02	0.10
$\delta_{1,4}$	0.12	0.08	0.001	0.11	0.29
$\delta_{2,1}$	0.29	0.21	0.01	0.25	0.74
$\delta_{2,2}$	0.51	0.17	0.19	0.51	0.81
$\delta_{2,3}$	0.39	0.23	0.03	0.38	0.83
$\delta_{2,4}$	0.86	0.17	0.36	0.93	1.00
$\delta_{3,1}$	0.87	0.09	0.67	0.88	1.00
$\delta_{3,2}$	0.87	0.08	0.70	0.87	1.00
$\delta_{3,3}$	0.91	0.07	0.75	0.92	1.00
$\delta_{3,4}$	0.99	0.01	0.95	0.99	1.00

similar item probability pattern as the nonregression analysis and a significant impact of age on the highest risk level.

6.1.8 Multiple categorical outcomes

LCA extends to joint distributions of several polytomous categorical outcomes. For example, consider a two-way table and let X and Y denote $p = 2$ manifest variables with levels $i = 1, \ldots, I$ and $j = 1, \ldots, J$ respectively. Let the joint probabilities in the table be denoted

$$P_{ij} = \Pr(X = i, Y = j).$$

Let Z be an unobserved categorical variable with levels $1, \ldots, C$ and let

$$\begin{aligned}\theta_k &= \Pr(Z = k)\\ \alpha_i(k) &= \Pr(X = i | Z = k)\\ \beta_j(k) &= \Pr(Y = j | Z = k).\end{aligned}$$

Under a latent conditional independence postulate the joint probabilities can be written

$$P_{ij} = \sum_{k=1}^{C} \theta_k \alpha_i(k) \beta_j(k).$$

Assume the simplest model where $C = 2$, so the latent variables is a dichotomy. Then

$$P_{ij} = \theta\alpha_i(1)\beta_j(1) + (1 - \theta)\alpha_i(2)\beta_j(2).$$

To implement the analysis we need to specify priors for the parameters θ, $\alpha_i(k)$ and $\beta_j(k)$, for $k = 1, 2$. Since

$$\sum_i \alpha_i(1) = \sum_j \beta_j(1) = \sum_i \alpha_i(2) = \sum_j \beta_j(2) = 1$$

a Dirichlet prior may be assumed for these parameters. For example,

$$\alpha(1) \sim \mathrm{D}(1, 1, \ldots, 1)$$

where the Dirichlet vector is of length I. A beta prior such as $\mathrm{B}(1, 1)$, or some other default prior, may be adopted for the mixture parameter θ when the latent variable is binary.

If individual level covariates are absent then the analysis is best conducted in terms of the aggregated counts n_{ij}. For example, if Z has two categories then, from the form of the joint probability, the probability of an observation in cell (i,j) belonging to category 1 of Z is

$$s_{ij} = \theta\alpha_i(1)\beta_j(1)/[\theta\alpha_i(1)\beta_j(1) + (1 - \theta)\alpha_i(2)\beta_j(2)]. \tag{6.3}$$

If initial values are assumed for r_{ij} where

$$r_{ij} \sim \mathrm{Bin}(s_{ij}, n_{ij}) \tag{6.4}$$

Table 6.9

	Mean	SD	2.5%	97.5%
$\alpha_1(1)$	0.090	0.064	0.004	0.241
$\alpha_1(2)$	0.793	0.084	0.632	0.951
$\alpha_2(1)$	0.348	0.071	0.217	0.495
$\alpha_2(2)$	0.089	0.050	0.007	0.196
$\alpha_3(1)$	0.562	0.078	0.411	0.716
$\alpha_3(2)$	0.118	0.064	0.010	0.248
$\beta_1(1)$	0.118	0.071	0.006	0.262
$\beta_1(2)$	0.716	0.078	0.571	0.880
$\beta_2(1)$	0.448	0.076	0.303	0.598
$\beta_2(2)$	0.241	0.071	0.088	0.378
$\beta_3(1)$	0.435	0.080	0.292	0.605
$\beta_3(2)$	0.043	0.032	0.002	0.123
$s_{1,1}$	0.018	0.023	0.000	0.083
$s_{1,2}$	0.174	0.164	0.005	0.640
$s_{1,3}$	0.495	0.289	0.024	0.974
$s_{2,1}$	0.387	0.267	0.015	0.941
$s_{2,2}$	0.846	0.119	0.541	0.993
$s_{2,3}$	0.966	0.041	0.851	1.000
$s_{3,1}$	0.421	0.274	0.018	0.952
$s_{3,2}$	0.869	0.103	0.611	0.994
$s_{3,3}$	0.972	0.033	0.882	1.000
θ_1	0.470	0.079	0.321	0.629
θ_2	0.530	0.079	0.371	0.679

then updating via full conditional densities can proceed. More complex but analogous procedures apply for three way tables and higher, and latent variables with three or more categories.

Example 6.6 Visiting frequency Evans *et al.* (1989) consider a cross-tabulation of frequency of visits ($I = 3$ categories) according to grouped length of stay in hospital ($J = 3$) for 132 long-term schizophrenic patients. They adopt estimation of an LCA model via adaptive importance sampling. Here we adopt the approach as in (6.3) and (6.4) above, with $r_{ij} \cong n_{ij}/2$ as initial values (see *Program 6.6 Latent Class Visits*). This gives the parameter estimates in Table 6.9 (after a single run of 20 000 iterations, with 5000 burn-in), with α, β and s as defined above. The mean posterior estimates are comparable to those of Evans *et al.* though the posterior parameter variances slightly exceed those reported by Evans *et al.*

6.2 MIXTURES OF DISTRIBUTIONS, REGRESSION MIXTURES AND TRANSITION MIXTURES

The structure of many data sets is too complex to be represented by a single parametric model (e.g. Poisson or normal) especially if they exhibit multimodality, outlying subsets of observations, and so on (McLachian and Basford, 1988; Lavine and West, 1992; Leonard *et al.*, 1994). Nonparametric analysis is one way of circumventing the problems raised by the complexity of observed structures. A Bayesian approach to a completely nonparametric analysis is via the Dirichlet Process Prior or DPP method (see below). Another retains the framework of parametric densities, but approximates the underlying sampling density as a finite mixture of C latent densities

$$\phi(x_i) \cong \sum_{j=1}^{C} \theta_j f_j(x_i \mid \alpha_j), i = 1, \ldots, n$$

where $f_j \mid \alpha_j)$ is a given parametric density such as the normal, exponential or Poisson. The proportions, or masses, $\theta_j, j = 1, \ldots, C$, attach to each component of the mixture, sum to one. Such a model is appropriate for a heterogenous population containing subgroups whose size is proportional to θ_j. The parameters α_j might, for example, relate to means and variances $\{\mu_j, V_j\}$ of normal subpopulations, or means μ_j of Poisson subgroups. Means can in turn be related to covariates $z_1, \ldots, z_n$ via class-specific regression parameters β_j. For example, for a mixture of Poison regressions one might have

$$x_i \sim \sum_{j=1}^{C} \theta_j \text{Poi}(\mu_{ij})$$

$$\log(\mu_{ij}) = \beta_j z_i.$$

As noted by Dempster *et al.* (1977) a mixture model can be expressed alternatively in terms of the original data and augmented data, where the latter consists of an indicator variable Z_i of group membership among the C groups, and we *assume* that Z_i is known. If w_{ij} are dummy indicators equalling one when $Z_i = j$ and zero otherwise then the likelihood of the complete data (observed and augmented data combined) is

$$\prod_{j=1}^{C} \prod_{i=1}^{n} \theta_j{}^{w_{ij}} \, [f_j(x_i \mid \alpha_j)]^{w_{ij}}.$$

In practice of course the w_{ij} are unknowns. However the complete data representation means that a mixture model can be represented hierarchically, analogous to the random effects models of Chapter 5, with at the top level the mixture density parameters α_j, then the (unobserved) class indicators

$$w_{ij} \sim g(w_{ij} \mid \alpha_j)$$

and finally the observed data

$$x_i \sim f_{Z_i}(x_i \mid \alpha_j, w_{ij}).$$

The expected posterior probabilities of class membership for individuals, i.e. $\varphi_{ij} = \Pr(w_{ij} = 1)$, have the form

$$\varphi_{ij} = \theta_j f_j(x_i \mid \alpha_j) / \sum_k \theta_k f_k(x_i \mid \alpha_k). \tag{6.5}$$

Posterior means for each case are then weighted averages of the subpopulation means with weights φ_{ij}. So in the Poisson case without covariates the expected posterior mean for case i is

$$M_i = \sum_j \varphi_{ij} \mu_j \tag{6.6}$$

and in the regression case is

$$M_i = \sum_j \varphi_{ij} \exp(\beta_j z_i). \tag{6.7}$$

The goal in the latent mixture analysis without covariates, as in (6.6), is often the same as in random effects approaches to heterogeneity: to provide a smoothed estimate (e.g. of disease morbidity) which results from pooling strength over all observations (Bohning, 2000). In the regression case (6.7) the goal might be to identify subpopulations with different economic choice patterns or health behaviours.

6.2.1 BUGS implementation

The facility to define models hierarchically in BUGS by specifying priors and likelihood, aids in the estimation of mixture models. A standard formulation is that the subpopulation proportions θ_j have a Dirichlet prior with equal prior masses on each subgroup. For example, with $C = 2$ groups, we could choose $\lambda_1 = \lambda_2 = 1$ and stipulate the prior for θ as a $\text{Dir}(\lambda_1, \lambda_2)$. The choice for subject i is then made via

$$Z_i \sim \text{Categorical}(\theta_{[1:C]}). \tag{6.8}$$

Finally the likelihood will be specified to differentiate between the parameters of subpopulations. Suppose x_i were Poisson, and $C = 2$, then

$$x_i \sim \text{Poi}(\mu_1) \text{ if } Z_i = 1,$$
$$x_i \sim \text{Poi}(\mu_2) \text{ if } Z_i = 2.$$

The means μ_j would have gamma priors such as $\mu_j \sim G(a_j, b_j)$. The prior shape and scale parameters a_j and b_j are typically set at low values such as $a_j = b_j = 0.01$ (Escobar and West, 1998).

An alternative is Gibbs sampling directly from the conditional posteriors of each parameter, possibly including the augmented data w_{ij} as extra parameters (Lavine and West, 1992). Thus, continuing the case of a Poisson without covariates, we would supply starting values for w_{ij} and then update them according to a multinomial

$$w_{ij} \sim \text{Mult}(\varphi_{ij}, 1)$$

where φ_{ij} is as in (6.5), α_j is a vector of Poisson means, $\alpha_j = \{\mu_j\}$, and

$$f_j(x_i \mid \alpha_j) = e^{-\mu_j} \mu_j^{x_i} / x_i!.$$

If $A_j = \sum_i w_{ij}$ cases are allocated to group j and the prior mass on group j is λ_j then the subpopulation proportions are updated according to a Dirichlet

$$\theta \sim D(A_1 + \lambda_1, A_2 + \lambda_2, \ldots, A_C + \lambda_C). \tag{6.9}$$

Finally the means μ_j, with respective gamma priors $G(a_j, b_j)$ would be updated according to

$$\mu_1 \sim G\Big(\sum_{Z_i=1} x_i + a_1, A_1 + b_1\Big)$$

$$\mu_2 \sim G\Big(\sum_{Z_i=2} x_i + a_2, A_2 + b_2\Big)$$

and so on. The x_i in the first gamma argument are summed only over those subjects allocated to the group concerned.

Whatever approach is used, mixture models pose problems of estimation in terms of selecting the appropriate number of categories, and obtaining well-identified solutions. Selecting a postulated value of C which is too large may mean that certain group means are very similar, or that one or more group proportions θ_j are very small (e.g. under 0.01). Choice of starting values may be important and constraints for identifiability may be imposed. Thus for $k = 4$ Poisson means, and no covariates, we might specify ordered means

$$\begin{aligned}
\mu_1 &\sim G(a_1, b_1)I(, \mu_2) \\
\mu_2 &\sim G(a_2, b_2)I(\mu_1, \mu_3) \\
\mu_3 &\sim G(a_3, b_3)I(\mu_2, \mu_4) \\
\mu_4 &\sim G(a_4, b_4)I(\mu_3,).
\end{aligned}$$

This unique labelling (Richardson and Green, 1997) is important, especially when the upper limit for the number of groups appropriate for a particular data set is being approached. Another option is to increase the prior masses λ_k on the subpopulations, increasing the prior information on their likely existence.

Example 6.7 Batting averages Laird (1982) considers data on early season batting averages for 18 major US league players in 1970. These are based on their first 45 times at bat, and the intention is to predict their final batting averages φ_i for the full season. *Program 6.7 Batting Averages* contains the number of successes `Y[1:45]` in the first 45 games, and the predictions of the eventual averages `phi[1:45]`. The mean square error

$$\sum_i (P_i - \phi_i)^2$$

based on comparing the observed proportions $P_i = Y_i/45$ with the observed eventual averages ϕ_i is 0.075. A binomial model with $C = 2$ latent classes for its parameter $\pi_{1:C}$ is taken:

$$Y_i \sim \text{Bin}(\pi[Z_i], 45)$$
$$Z_i \sim \text{Categorical}(\theta_{1:2})$$
$$\theta_{1:2} \sim \text{Dir}(\lambda_{1:2}).$$

For the binomial means we specify that one exceeds the other using the parameterisation

$$\text{logit}(\pi_1) = \beta$$
$$\text{logit}(\pi_2) = \beta + \delta$$

where both β and δ are normally distributed, but δ is confined to positive values. In this example, it is likely that the separation between two hypothesised groups of players is not pronounced, and so a mildly informative prior

$$\delta \sim \text{N}(0, 1)\text{I}(0,)$$

is taken. We find the smoothed estimates ρ_i based on long run averages of the observed $\pi[Z_i]$ are better predictors of the final averages than the unsmoothed proportions P_i. Specifically the MSE $\sum_i(\rho_i - \phi_i)^2$ is 0.025, similar to the values quoted by Laird (1982). We find two subgroups, with roughly similar masses, $\theta_1 = 0.54$, and $\theta_2 = 0.46$, with the second group having a batting average of 0.315 and the first one of 0.236.

Example 6.8 Eye tracking data Escobar and West (1998) present count data on eye tracking anomalies in 101 schizophrenic patients. The data are obviously highly overdispersed to be fit by a single Poisson, and solutions with $k = 2$, 3 and 4 groups are estimated. The full conditional posterior approach is applied. The two- and three- group solutions can be obtained without the ordered means constraint – see the first program listed in Program 6.8. This includes initialisation files, with (arbitrary) initial settings for the augmented data, Z_i as in (6.8). The two-groups solution has means $\mu_1 = 0.7$ and $\mu_2 = 11.5$ with respective subpopulation proportions 0.73 and 0.27. The three-groups solution has means 0.48, 6.7 and 19.2 with respective proportions 0.66, 0.24 and 0.10.

The four-groups solution, obtained with an ordered means prior, identifies the 46 observations with no anomalies with a subpopulation having mean of virtually zero (0.003) and a mass of 0.32. The remaining groups have means 1.3, 8.5 and 22. Smoothing, even for the 46 zero anomaly cases, is apparent in the posterior means for cases 1–46 which are estimated as 0.32. Smoothing is also apparent for higher count patients; for example, cases 92 and 93 with 12 observed anomalies have posterior means of 10.3.

For comparison, a model not involving direct sampling from conditional posteriors is also applied for the four group case (see the 3rd program in Program 6.8). This results in a similar solution except that means μ_1 and μ_2 are marginally higher (at around 0.04 and 1.42).

There may be scope for higher numbers of groups, as a DPP nonparametric mixture analysis of these data suggests later.

Example 6.9 Regression mixture of small area cardiac mortality This example again involves count response data but with a covariate also, leading to a latent mixture regression. The outcomes x_i are deaths from causes in the ICD9 range 410–414 (ischaemic heart disease) in 758 London electoral wards (small administrative areas) over 1990–1992. An offset of expected deaths E_i is included in the analysis. So if d_i denotes the covariate the model is

$$x_i \sim \sum \theta_j \text{Poi}(E_i \rho_{ij})$$
$$\log(\rho_{ij}) = \beta_{0j} + \beta_{1j} d_i.$$

Thus the relative risk ρ_{ij} (the cause-specific standard mortality ratio in area i and class j) is modelled as a function of a deprivation score d, previously transformed according to $d = \log(10 + D)$, where D is the Townsend deprivation score.

We assume $C = 2$ classes, with higher values of C leading to identifiability problems. Two alternative forms for updating the subpopulation proportions θ, denoted `theta[1:2]` in *Program 6.9 IHD SMR*, are possible. First, a Dirichlet prior on θ, Dir(1, 1), and allocation using `dcat(theta[])` and second, a Dirichlet conditional posterior, as in (6.9) above. Sampling from the Dirichlet posterior in the latter case is implemented by using Gamma(, 1) densities, with elements $A_1 + 1$ and $A_2 + 1$ where A_1 (or A_2) is the total of cases allocated in a particular iteration to class 1 (or 2). In the latter case initial classes for each small area should be supplied (and set arbitrarily) as augmented data though such initialisation is also possible under the alternative method.

A two-group solution (Table 6.10), from the second half of a run of 5000 iterations, shows the major subpopulation of small areas ($\theta_1 = 0.86$) with a clearly identified deprivation effect, namely $\beta_1 = 0.36$ with 95 % credible interval (0.29, 0.43). This subpopulation has higher cardiac mortality on average (higher intercept β_0) than the other and smaller subpopulation. In the latter the deprivation effect is not well identified, though its upper 97.5 percentile in fact exceeds that in the major ward grouping of electoral wards

Example 6.10 Zero inflated Poisson: DMFT counts in children Bohning *et al.* (1999) present two-wave data on dental problems in 797 Brazilian children,

Table 6.10

	Mean	SD	2.5%	Median	97.5%
Deprivation effects					
β_{11}	0.364	0.036	0.292	0.363	0.435
β_{12}	0.186	0.157	−0.117	0.185	0.509
Intercept					
β_{01}	0.046	0.011	0.026	0.046	0.070
β_{02}	−0.316	0.063	−0.446	−0.314	−0.209
Masses					
θ_1	0.855	0.048	0.747	0.862	0.928
θ_2	0.145	0.048	0.072	0.138	0.253

specifically numbers of teeth decayed, missing or filled (DMFT). The children were subject to a dental health prevention trial involving various treatment options. Because of an excess of children with zero faults, the Poisson density is not appropriate. To model the overdispersion they propose a zero inflated Poison (ZIP) model so that there are two subpopulations. One has mean zero by definition, the other has a mean which may depend on child covariates. Thus if y_i denotes the number of DMFT teeth for child i, the observational model is

$$f(y_i|\theta, \mu_i) = (1-\theta)\text{Poi}(y_i, 0) + \theta\text{Poi}(y_i, \mu_i)$$

where

$$\text{Poi}(y_i, \mu_i) = \exp(-\mu_i)\mu_i^{y_i}/y_i!.$$

The individual level covariates in the model for the mean μ_i are sex, ethnicity and school. School, with six categories, is in fact a health prevention treatment variable, with random assignment of treatments or combined treatments. The variables are as follows:

1. dmftb – DMFT-index at the beginning of the study
2. dmfte – DMFT-index at the end of the study (2 years later)
3. sex (0 – female; 1 – male)
4. ethnic (ethnic group: 1 – dark; 2 – white; 3 – black)
5. School (kind of prevention): 1 – oral health education;
   2 – all four methods together;
   3 – control school (no prevention measure);
   4 – enrichment of school diet with ricebran;
   5 – mouthrinse with 0.2 % NaF-solution;
   6 – oral hygiene.

The outcome is the DMFT index at the end of the project; initial DMFT scores may be included as a measure of severity of dental problems (which may impact on the final measure and so lead to effect modification). Here we estimate the probability θ of not belonging to the zero problem group, and the impact on the Poisson mean of sex, ethnicity and treatment (see *Program 6.10 DMFT ZIP Mixture*), but without a control for initial severity of teeth problems (Table 6.11). The probability θ is estimated as 0.80, compared with $\hat{\theta} = 0.78$ as estimated by Bohning *et al.* (1999).

Compared with the control school, which has fixed effect set to zero, all except school 4 (enriched diet) show a reduction in DMFT problems. Adding $\log(\text{dmftb} + 0.5)$ as a covariate for initial severity leads also to the school 6 treatment losing significance.

6.2.2 Normal mixtures

A discrete mixture of normal subpopulations also provides an example of direct sampling from conditional posteriors. Consider the example of normal

Table 6.11

Parameter	Mean	SD	2.5%	Median	97.5%
$1-\theta$	0.199	0.020	0.163	0.199	0.236
θ	0.801	0.020	0.764	0.801	0.837
β_1 (constant)	0.937	0.077	0.780	0.937	1.083
β_2 (male)	0.100	0.057	−0.010	0.098	0.214
β_3 (white)	0.082	0.064	−0.043	0.080	0.207
β_3 (black)	−0.119	0.097	−0.317	−0.118	0.068
β_4 (school 1)	−0.219	0.095	−0.409	−0.216	−0.046
β_4 (school 2)	−0.468	0.108	−0.682	−0.467	−0.261
β_4 (school 4)	−0.061	0.087	−0.227	−0.061	0.114
β_4 (school 5)	−0.220	0.094	−0.404	−0.217	−0.034
β_4 (school 6)	−0.221	0.098	−0.406	−0.222	−0.026

outcomes $y_i, i = 1, \ldots, n$, and without any covariates. Denote the parameters in latent subpopulation j as

$$\alpha_j = \{\mu_j, \sigma_j^2\}$$

for which the conjugate prior can be written

$$\mathrm{N}(\gamma_j, \sigma_j^2/\kappa_j), \quad j = 1, \ldots, C$$

as in Gelman *et al.* (1995, Section 3.3). Here γ_j is an initial guess at the mean in group j with κ_j a prior sample size reflecting strength of belief in the guess about the mean. The precision in class j

$$\tau_j = 1/\sigma_j^2$$

is drawn from a gamma density $\mathrm{Gam}(v_j t_j/2, v_j/2)$. Here $t_j = 1/s_j^2$ is a prior guess at the precision in group j, and v_j is the prior degrees of freedom for that guess.

The conditional posterior distribution of the means (Diebolt and Robert, 1994) is then a weighted mixture of the priors and the allocation of observations into groups:

$$\mu_j \sim \mathrm{N}(\{\kappa_j\gamma_j + \sum_{Z_{i=j}} y_i\}/\kappa_j + A_j, \sigma_j^2/\{\kappa_j + A_j\})$$

where A_j is the number of cases allocated to group j at a particular sampling iteration, so

$$G_j = \{\kappa_j + A_j\}$$

is a 'posterior group size'. The sum $\sum_{Z_{i=j}} y_i$ is over those cases, $i \in 1, \ldots,$ n, allocated to group j (i.e. for whom $Z_i = j$).

The posterior density of the precisions is a gamma with form $\mathrm{Gamma}(N_j/2, V_j/2)$ where $N_j = v_j + A_j$ and V_j combines prior and sample estimates of sums of squares. Let

$$M_j = \sum_{Z_i=j} y_i / A_j$$

be the current mean in group j; then

$$V_j = v_j/t_j + \sum_{Z_{i=j}} (y_i - M_j)^2 + (M_j - \gamma_j)^2 \kappa_j A_j / G_j.$$

Finally the allocation indicators w_{ij} are updated from the multinomial

$$w_{ij} \sim \text{Mult}(\varphi_{ij}, 1)$$

as in (6.5) above, where in this case $\alpha_j = \{\mu_j, \tau_j\}$ and

$$f_j(y_i \mid \alpha_j) \propto \tau_j^{0.5} \exp[-0.5\tau_j (y_i - \mu_j)^2].$$

If $A_j = \sum_i w_{ij}$ cases are allocated to group j and the prior mass on group j is λ_j then the subpopulation proportions are updated according to a Dirichlet

$$\theta \sim \text{Dir}(A_1 + \lambda_1, A_2 + \lambda_2, \ldots, A_C + \lambda_C).$$

Example 6.11 Peak sensitivity wavelenghts Spiegelhalter *et al.* (1995) present an analysis of peak sensitivity wavelengths for individual microspectrophotometry on a single monkey's eyes, as in the study of Bowmaker *et al.* (1985). They fit a normal mixture with a common variance in $C = 2$ subgroups, but a different mean. They adopt a uniform Dirichlet prior on the subgroup proportions $\theta[1{:}2]$ and

$$Y_i \sim \text{N}(\mu_{Z_i}, \sigma^2_{Z_i})$$
$$Z_i \sim \text{Categorical}(\theta).$$

Here we adopt both different means and variances and sample from the conditional posteriors as above. To obtain a well-identified solution – causing no numerical problems as we iterate to a run-length of 10 000 – prior masses $\lambda_1 = \lambda_2 = 3$ are taken. Also, as in Spiegelhalter *et al.*, one mean is constrained to exceed the other. We monitor (see *Program 6.11 Eyes*) the posterior probabilities of membership as well as the normal parameters. Different variances are clearly identified (Table 6.12).

Table 6.12

	Mean	SD	2.5%	Median	97.5%
μ_1	537.5	1.1	535.1	537.6	539.6
μ_2	550.0	1.5	545.6	550.5	551.6
π_1	0.66	0.10	0.40	0.67	0.81
π_2	0.34	0.10	0.19	0.33	0.60
σ_1	4.33	0.90	2.54	4.30	6.22
σ_2	2.28	1.23	1.13	1.83	5.92

Table 6.13

$\varphi(1,1)$	0.983	$\varphi(11,1)$	0.992	$\varphi(21,1)$	0.976	$\varphi(31,1)$	0.814	$\varphi(41,1)$	0.022
$\varphi(1,2)$	0.017	$\varphi(11,2)$	0.008	$\varphi(21,2)$	0.024	$\varphi(31,2)$	0.187	$\varphi(41,2)$	0.978
$\varphi(2,1)$	0.989	$\varphi(12,1)$	0.990	$\varphi(22,1)$	0.966	$\varphi(32,1)$	0.827	$\varphi(42,1)$	0.016
$\varphi(2,2)$	0.011	$\varphi(12,2)$	0.011	$\varphi(22,2)$	0.034	$\varphi(32,2)$	0.173	$\varphi(42,2)$	0.984
$\varphi(3,1)$	0.991	$\varphi(13,1)$	0.992	$\varphi(23,1)$	0.961	$\varphi(33,1)$	0.728	$\varphi(43,1)$	0.017
$\varphi(3,2)$	0.009	$\varphi(13,2)$	0.008	$\varphi(23,2)$	0.039	$\varphi(33,2)$	0.272	$\varphi(43,2)$	0.983
$\varphi(4,1)$	0.990	$\varphi(14,1)$	0.989	$\varphi(24,1)$	0.952	$\varphi(34,1)$	0.605	$\varphi(44,1)$	0.015
$\varphi(4,2)$	0.010	$\varphi(14,2)$	0.011	$\varphi(24,2)$	0.048	$\varphi(34,2)$	0.396	$\varphi(44,2)$	0.985
$\varphi(5,1)$	0.992	$\varphi(15,1)$	0.987	$\varphi(25,1)$	0.943	$\varphi(35,1)$	0.094	$\varphi(45,1)$	0.015
$\varphi(5,2)$	0.008	$\varphi(15,2)$	0.013	$\varphi(25,2)$	0.057	$\varphi(35,2)$	0.906	$\varphi(45,2)$	0.985
$\varphi(6,1)$	0.992	$\varphi(16,1)$	0.988	$\varphi(26,1)$	0.929	$\varphi(36,1)$	0.106	$\varphi(46,1)$	0.023
$\varphi(6,2)$	0.008	$\varphi(16,2)$	0.012	$\varphi(26,2)$	0.071	$\varphi(36,2)$	0.894	$\varphi(46,2)$	0.977
$\varphi(7,1)$	0.994	$\varphi(17,1)$	0.987	$\varphi(27,1)$	0.887	$\varphi(37,1)$	0.084	$\varphi(47,1)$	0.029
$\varphi(7,2)$	0.006	$\varphi(17,2)$	0.013	$\varphi(27,2)$	0.113	$\varphi(37,2)$	0.916	$\varphi(47,2)$	0.971
$\varphi(8,1)$	0.995	$\varphi(18,1)$	0.984	$\varphi(28,1)$	0.869	$\varphi(38,1)$	0.088	$\varphi(48,1)$	0.024
$\varphi(8,2)$	0.005	$\varphi(18,2)$	0.016	$\varphi(28,2)$	0.131	$\varphi(38,2)$			
$\varphi(9,1)$	0.991	$\varphi(19,1)$	0.978	$\varphi(29,1)$	0.851	$\varphi(39,1)$			
$\varphi(9,2)$	0.009	$\varphi(19,2)$	0.022	$\varphi(29,2)$	0.149	$\varphi(39,2)$			
$\varphi(10,1)$	0.994	$\varphi(20,1)$	0.975	$\varphi(30,1)$	0.836	$\varphi(40,1)$			
$\varphi(10,2)$	0.006	$\varphi(20,2)$	0.025	$\varphi(30,2)$	0.164	$\varphi(40,2)$			

Observations 1–26 all have a probability over 0.9 of belonging to the first group of wavelengths, and cases 35–48 all have probabilities over 0.9 of belonging to the second (Table 6.13). Only one observation (no. 34) is not clearly associated with either group, with $\varphi_{34,1} = 0.6$ and $\varphi_{34,2} = 0.4$.

6.3 LATENT TRANSITION MATRICES AND MIXTURES OF TRANSITION MATRICES

We are often interested in patterns of movement between states at successive time points. This type of stochastic process can be characterised by a Markov chain stationary over time points. Thus let the state occupied at time t be denoted $y_t (t = 0, 1, 2, \ldots, T)$ with M possible outcomes denoted $S_j (j = 1, \ldots, M)$, with transition probabilities between states being

$$\Pr(y_t = S_j \,|\, y_{t-1} = S_i) = p_{ij}(t) = p_{ij}(t+1) = p_{ij} \tag{6.10}$$

and with $\sum_j p_{ij} = 1$ for all origins $i = 1, \ldots, M$. The likelihood of a sequence of states $(y_0, y_1, \ldots, y_T)$ may under these assumptions be written

$$\Pr(y_0, \ldots, y_T) = \Pr(y_0) \prod_{t=1}^{T} \Pr(y_t \,|\, y_{t-1}) = \Pr(y_0) \prod_{t=1}^{T} p_{ij}(t) = \Pr(y_0) \prod_{ij} p_{ij}^{n_{ij}}$$

where n_{ij} is the total of moves between states i and j over all periods. One research question is whether such patterns of state change may be differentiated by observable and/or latent characteristics. Transitions governed by human behaviour or attitudes (e.g. changes of political affiliation, migration, changes in marital state) are likely to result in part from unmeasured attributes.

In other situations the data themselves constitute a time series but are not explicitly in transition form. However, they are subject to variability which may be modelled by a latent mixture variable which evolves according to a Markov chain. Here the sequence of states visited by the Markov chain constitute missing data. These hidden Markov models or Markov mixture models therefore generalise mixture distributions by introducing serial correlation in the sequence of latent state categories (Leroux and Puterman, 1992; Chib, 1996). We may have various observed covariates we think are likely to be associated with these latent behaviours, and these may be used to assist either in predicting membership of states or in modelling transition rates between the states of the latent variable

Thus for individuals i, and observed patterns of movement between states S_i with M levels, we observe histories of states $S_{it}, i = 1, \ldots, n, t = 1, \ldots, T$ and covariates x_i (fixed) or x_{it} (time varying). Underlying observed transition behaviour are latent categorisations W_i, with C levels and prior probabilities $\lambda_1, \ldots, \lambda_C$, and associated transition patterns $P[i, j, k], i = 1, \ldots, M$, $j = 1, \ldots, M$, and $k = 1, \ldots, C$. The probability that the latent categorisation of individual i is state k, π_{ik} may be related to covariates x_i via a multiple logistic model. Identifiability may require constraints on the parameters of such models. For example, order restrictions may be imposed on intercepts to ensure one latent group among two is always the most common.

In the other situation we have a discrete time series (binomial, multinomial, Poisson) for times $t = 1, \ldots, T$ with heterogeneity modelled via a latent categorisation Z_t with C levels. The category occupied at time t depends on the previous period's category, so that $Z_t \sim \text{Cat}(Q_{Z_{t-1}}), t > 1$, where Q is a vector with C cells. For example, a simple one-step dependence such as this defines a latent Markov chain, but there are other options. The first period latent category requires special treatment and can be modelled separately via $Z_1 \sim \text{Cat}(Q^*)$, with Q^* itself drawn from a Beta or Dirichlet prior. Subsequent values will be drawn according to a Markov chain.

A conceptually similar situation occurs if data for time- sequenced changes in state at micro level are not available. For example, industrial surveys might record firms by size (e.g. number of employees) but not the sequence of changes in size, while occupational or income surveys might enumerate totals in occupational or income groups but not the sequence of job changes intervening between two surveys. Assume then that the observed data consist of proportions $w_j(t)$ in different states at time t, with $j = 1, \ldots, M$ and generated by $N(t)$ trials. We may wish to estimate the underlying stationary transition matrix $P = [p_{ij}]$. Then

$$\begin{aligned} w_j(t) &= \Pr(y_t = S_j) \\ &= \sum_i \Pr(y_{t-1} = S_i)\Pr(y_t = S_j \,|\, y_{t-1} = S_i) \end{aligned}$$

$$= \sum_i \Pr(y_{t-1} = S_i) p_{ij}$$
$$= \sum_i w_i(t_{-1}) p_{ij}.$$

The observed stock totals at time t, namely $N_1(t), N_2(t), \ldots, N_M(t)$ are then multinomial with probabilities $q_i(t) = \sum_i w_i(t_{-1}) p_{ij}$ from a total $N(t)$. Lee *et al.* (1968) propose a Dirichlet prior on the transition probabilities p_{ij} in each row, and compare Bayesian and ML estimation of the underlying transition probabilities.

Example 6.12 Homicides and suicides in Cape Town MacDonald and Zucchini (MZ, 1997) present data from Cape Town (South Africa) on homicides and suicides over 313 weeks in 1986–91. The outcome $y_t = (y_{1t}, y_{2t}, \ldots, y_{5t})$ is multinomial, with categories: firearm homicide, non-firearm homicide, firearm suicide, non-firearm suicide, and legal intervention homicide (deaths following police actions). MZ assume a latent transition variable Z_t with $C = 2$ categories, so that there are ten parameters. Two are needed to identify the chain itself (since the sum of the transition probabilities in each row of a 2×2 transition matrix is 1, there are only two free parameters) and eight are needed to identify the two multinomial vectors. So the simplest model (model A) would be as follows:

$$Z_1 \sim \text{Categorical}(Q_{1:2})$$
$$Z_t \sim \text{Categorical}(P[Z_{t-1}, 1{:}2]), t = 2, \ldots, 313$$
$$y_t \sim \text{Multinomial}(\pi[Z_t, 1{:}M], n_t)$$

where n_t is the total of deaths over all five categories and π_t is the vector of multinomial parameters corresponding to latent classes $Z_t = 1$ or $Z_t = 2$. Thus the first week's data are y_{11}, fire-arm homicides, y_{21} non-firearm homicides; and so on.

In the Cape Town data, there was an increasing rate of firearm homicide over this period, and MZ actually identify a change-point at week 287. Thus we may additionally assume (model B) that the multinomial probabilities are differentiated by time period, with

$$y_t \sim \text{Multinomial}(\pi[T_t, Z_t, 1{:}M], n_t)$$

and $T_t = 1$ for $t = 1, \ldots, 287$ and $T_t = 2$ for $t = 288, \ldots, 313$. This is equivalent to using a covariate to assist in identifying the underlying Markov mixture. Under both models, the 2×2 transition matrix P is assigned a Dirichlet prior (indirectly modelled using gamma densities). There may well be reasons to assume a higher weight for probabilities on the main diagonal of the transition matrix, namely P_{11} and P_{22}. Here we adopt equal weights, $\alpha = 1$, for each probability in a row of the transition matrix, such that

$$P_{1,1:M} \sim D(\alpha, \alpha)$$
$$P_{2,1:M} \sim D(\alpha, \alpha).$$

After runs of 5000 iterations (see Program 6.12) with 500 burn-in, model A yields the estimated transition matrix and multinomial probabilities in Table 6.14.

The estimated latent transition matrices are similar in the two approaches, but the unconditional multinomial probabilities are more distinct between the 1st and 2nd subperiods under model B. Thus under model B we obtain the results shown in Table 6.15.

Table 6.14

	Mean	SD	2.5%	97.5%
$p_{1,1}$	0.83	0.06	0.69	0.92
$p_{1,2}$	0.17	0.06	0.08	0.32
$p_{2,1}$	0.64	0.13	0.38	0.88
$p_{2,2}$	0.36	0.13	0.12	0.62
1st subperiod unconditional multinomial probs Firearm homicide	0.101	0.003	0.095	0.108
Nonfirearm homicide	0.770	0.005	0.761	0.780
Firearm suicide	0.028	0.002	0.024	0.032
Nonfirearm suicide	0.079	0.003	0.073	0.085
Legal intervention homicide	0.022	0.002	0.019	0.025
2nd subperiod unconditional multinomial probs Firearm homicide	0.110	0.008	0.095	0.127
Nonfirearm homicide	0.763	0.008	0.746	0.778
Firearm suicide	0.028	0.002	0.024	0.032
Nonfirearm suicide	0.078	0.003	0.072	0.084
Legal intervention homicide	0.022	0.002	0.018	0.026

Table 6.15

	Mean	SD	2.5%	97.5%
$p_{1,1}$	0.82	0.07	0.65	0.93
$p_{1,2}$	0.18	0.07	0.07	0.35
$p_{2,1}$	0.66	0.12	0.41	0.88
$p_{2,2}$	0.34	0.12	0.12	0.59
1st subperiod unconditional multinomial probs Firearm homicide	0.099	0.003	0.092	0.106
Nonfirearm homicide	0.770	0.005	0.761	0.780
Firearm suicide	0.029	0.002	0.026	0.033
Nonfirearm suicide	0.079	0.003	0.073	0.085
Legal intervention homicide	0.022	0.002	0.019	0.026
2nd subperiod unconditional multinomial Probs Firearm homicide	0.127	0.011	0.106	0.149
Nonfirearm homicide	0.756	0.014	0.728	0.784
Firearm suicide	0.015	0.004	0.008	0.024
Nonfirearm suicide	0.079	0.009	0.061	0.097
Legal intervention homicide	0.023	0.005	0.014	0.034

The first subperiod ($t = 1, \ldots, 287$) is defined by relatively low levels of firearm homicide, and the unconditional multinomial probability of firearm homicide averages 0.099 in this period under model B. In the second subperiod, this unconditional probability increases to 0.127. In the second subperiod the state $Z_t = 1$ is relatively more common, with a rate of occurrence around 82.5% compared with 80% in the first period. Broadly similar results are obtained by MZ.

Example 6.13 Changes in attitudes to group membership Goodman (1974) analysed a cross-classification of 3398 schoolboys in terms of (a) self-perceived membership in the 'leading crowd' (with member denoted + and non-member denoted −) and (b) favourableness of attitude to the leading crowd (with favourable denoted + and unfavourable denoted −). Hence there are $M = 4$ possible attitude combinations. The data on two successive occasions (membership of and attitude to leading crowd respectively) are shown in Table 6.16.

Here a 10% sample of these data (i.e. 340 subjects) is taken, and $C = 2$ latent classes assumed (*Program 6.13 Changes in Attitudes*). The latent category probabilities are modelled using intercept terms only, but extensions to include covariates are possible (subject to identifiability). For example, we might classify cases i by initial overall attitude (at first interview), whether unambiguous (either ++ or −−) or ambiguous (+− or −+), and see whether this influences the latent categorisation. Alternatively, external covariates (age, social class, actual membership of school clubs) may be used. Thus for $i = 1, \ldots, 340$ and $k = 1, 2$ let

$$h_i \sim \text{Categorical}(\lambda_{i,1:C})$$

denote the latent class which determines which transition matrix $P[i,j,1], P[i,j,2], \ldots, P[i,j,C]$ is selected. The choice of state at $t+1$ for individual i is then represented in transition matrix form as

$$S_{i,t+1} \sim \text{Categorical}(P[S_{it}, 1{:}M, h_i]).$$

The latent category probabilities λ_{ik} are related to group-specific intercepts via multiple logistic regression. We treat the $M = 4$ states defined by the two

Table 6.16

1st Interview		2nd Interview			
		+	+	−	−
		+	−	+	−
+	+	458	140	110	49
+	−	171	182	56	87
−	+	184	75	531	281
−	−	85	97	338	554

attitudes as the focus of analysis (with states in the order ++, +−, −+ and −−), though one might construct a bivariate latent categorisation to model the separate transitions between the two attitudes.

For two groups the overall posterior probabilities of the classes, denoted the vector PI in Program 6.13 and based on the second half of a run of 15 000 iterations, are $\lambda_1 = 0.35$, $\lambda_2 = 0.65$. Their associated transition matrices, with the states 1 to 4 as defined above, are shown in Table 6.17.

In the second group, this solution emphasises continuity of attitudes for the disengaged (−− at both times) and the engaged (++ at both times). There is also a strong continuation effect for those having a favourable attitude to the leading crowd but not perceiving themselves as members. However, the second category (+−) perceiving themselves as belonging to the leading crowd but having a negative view of this crowd has much less continuity. The first group emphasises changes of attitude except for the unusual second category. The strong interchange between categories 3 and 4 is also apparent.

A three-group solution is also identified. For three groups the overall posterior probabilities of the groups are $\lambda_1 = 0.19$, $\lambda_2 = 0.25$ and $\lambda_3 = 0.56$ and their associated transition matrices are as in Table 6.18.

In these matrices only a subset of cell probabilities tend to be well identified. The first latent group is associated with changes of attitude in general, while the second seems to emphasise moves towards a negative attitude (to states 3 and 4). The third group emphasises continuity in attitudes except for state 2.

Table 6.17

	Class 1		Class 2	
Successive states	Mean transition probability	SD	Mean transition probability	SD
1,1	0.09	0.15	0.86	0.12
1,2	0.53	0.22	0.02	0.06
1,3	0.28	0.21	0.07	0.09
1,4	0.11	0.11	0.05	0.05
2,1	0.03	0.10	0.52	0.13
2,2	0.58	0.43	0.18	0.23
2,3	0.19	0.26	0.12	0.11
2,4	0.20	0.30	0.18	0.13
3,1	0.15	0.15	0.16	0.10
3,2	0.08	0.10	0.05	0.05
3,3	0.02	0.05	0.77	0.12
3,4	0.75	0.19	0.02	0.04
4,1	0.22	0.15	0.03	0.05
4,2	0.20	0.20	0.06	0.07
4,3	0.53	0.32	0.12	0.17
4,4	0.05	0.12	0.79	0.14

Table 6.18

Successive states	Class 1 Mean transition probability	Class 1 SD	Class 2 Mean transition probability	Class 2 SD	Class 3 Mean transition probability	Class 3 SD
1,1	0.14	0.28	0.38	0.41	0.80	0.17
1,2	0.68	0.39	0.13	0.25	0.05	0.10
1,3	0.07	0.18	0.36	0.38	0.12	0.13
1,4	0.12	0.21	0.13	0.17	0.03	0.05
2,1	0.48	0.43	0.24	0.30	0.26	0.24
2,2	0.27	0.41	0.41	0.44	0.45	0.34
2,3	0.19	0.34	0.04	0.11	0.15	0.12
2,4	0.06	0.16	0.31	0.33	0.13	0.15
3,1	0.63	0.35	0.00	0.02	0.07	0.11
3,2	0.14	0.16	0.07	0.11	0.03	0.05
3,3	0.03	0.09	0.01	0.04	0.89	0.12
3,4	0.21	0.32	0.92	0.12	0.00	0.01
4,1	0.04	0.11	0.10	0.14	0.08	0.08
4,2	0.15	0.29	0.31	0.27	0.02	0.05
4,3	0.79	0.30	0.39	0.35	0.06	0.10
4,4	0.02	0.08	0.19	0.29	0.84	0.15

Example 6.14 Estimating a transition matrix when only stock data is available Following Lee *et al.* (1968) we generate data for 100 individuals followed through 10 periods with $M = 4$ states and transitions between states according to the stationary matrix

$$P = \begin{bmatrix} 0.6 & 0.4 & 0.0 & 0.0 \\ 0.1 & 0.5 & 0.4 & 0.0 \\ 0.0 & 0.1 & 0.7 & 0.2 \\ 0.0 & 0.0 & 0.1 & 0.9 \end{bmatrix}$$

(see *Program 6.14 Model A*). Initial states are allocated randomly with equal probability, with the coding

```
for (i in 1:100) {y[1,i] ~ dcat(p0[])}
for (j in 1:M) {p0[j] <- 1/M}
```

We take the actual observations as the totals $N_1(t), N_2(t), N_3(t)$, and $N_4(t), t = 1, \ldots, 10$ so generated. Program 6.14 attempts to estimate P given only these stock totals. The estimated transition probabilities (from the second half of a run of 10 000) reproduce the generating matrix P reasonably well, though the transition probabilities for 'non-movers', i.e. P_{11}, P_{22}, P_{33} and P_{44} are underestimated (see Table 6.19).

Table 6.19

	Mean	SD	2.5%	Median	97.5%
$P(1,1)$	0.506	0.116	0.262	0.512	0.716
$P(1,2)$	0.233	0.131	0.020	0.226	0.497
$P(1,3)$	0.181	0.122	0.007	0.165	0.449
$P(1,4)$	0.081	0.071	0.002	0.063	0.259
$P(2,1)$	0.087	0.066	0.003	0.074	0.247
$P(2,2)$	0.409	0.178	0.067	0.415	0.742
$P(2,3)$	0.369	0.189	0.037	0.361	0.747
$P(2,4)$	0.136	0.109	0.005	0.111	0.402
$P(3,1)$	0.022	0.019	0.001	0.017	0.071
$P(3,2)$	0.126	0.084	0.005	0.115	0.312
$P(3,3)$	0.605	0.155	0.262	0.624	0.861
$P(3,4)$	0.247	0.145	0.025	0.228	0.563
$P(4,1)$	0.019	0.016	0.001	0.015	0.061
$P(4,2)$	0.089	0.066	0.004	0.077	0.245
$P(4,3)$	0.155	0.120	0.006	0.129	0.441
$P(4,4)$	0.737	0.128	0.446	0.754	0.936

6.4 MODELS FOR NONRESPONSE AND INCOMPLETE DATA

Categorical data from large-scale surveys, trials or observational studies are often collected with some records partially or completely missing on the survey or study items. Often we have a situation where a categorical response is missing for some subjects, but completely recorded covariates or survey design variables are available. For example, Phillips (1993) considers the clinical trial response of smoking (yes, no or missing) in relation to fully recorded data on centre (binary) and drug (binary). Stasny (1991) considers the binary outcome of crime victimisation; data on victimisation (Y/N) and nonresponse from the US National Crime Survey is classified according to the survey domain (urban vs rural, poverty level, and type of incorporation). Thse domains form the basis for estimating the proportion of nonrespondents who are victims. Park and Brown (1994) consider obesity in children (yes, no or unknown) in relation to their age and gender, which are completely recorded. In other instances we may have a set of mutually dependent categorical variables with missingness on all of them; Rubin *et al.* (1995) consider responses to the 1990 plebiscite in Slovenia on independence, involving answers to three questions: In favour of independence? In favour of secession from Yugoslavia? Will you attend the plebiscite?. The answers to each question were Yes, No or Don't Know

A variety of approaches have been applied to missing data problems but certain defining constraints apply to all of them. For example, simply omitting missing data from the analysis (e.g. via 'case deletion') leads to valid inferences only if the data are missing completely at random; that is, the missing data values are a simple random sample of all data values. A less restrictive assumption is that of missingness at random (abbreviated as MAR) under which the probability that an observation is missing depends on the observed data but not on the missing data. Thus let $R = \{R_{ij}\}$ be binary indicators whether response was obtained ($R_{ij} = 1$) or missing ($R_{ij} = 0$) for subjects $i = 1, \ldots, n$ and items $j = 1, \ldots, k$. These indicators are considered to constitute 'data' in addition to the observed outcomes Y_{obs}. Suppose X denotes covariates fully measured for all respondents (e.g. stratifying variables in a survey). Let $Y = \{Y_{\text{obs}}, Y_{\text{mis}}\}$ denote the full set of observed outcome data and missing outcome data.

Then the joint distribution of the response indicators R and the outcomes Y has the form

$$f(R, Y | \theta_R, \theta_Y, X) = f_1(Y | X, \theta_Y) f_2(R | Y, X, \theta_R)$$

where f_1 is the sampling density of the data with parameter θ_Y and f_2 is density for the response mechanism with parameter θ_R. Then the response mechanism, as reflected in a model with parameter θ_R for the generation of R, may be influenced by the outcomes Y, whether observed or missing. The response mechanism will be missing at random if

$$f_2(R | Y, X, \theta_R) = f(R | Y_{\text{obs}}, Y_{\text{mis}}, X, \theta_R) = f(R | Y_{\text{obs}}, X, \theta_R).$$

Alternatively stated, if responses to a set of questions involve missing responses, then the data are missing at random if the missingness on one question is conditionally independent of the outcome on that question that would have been observed, given the observed responses to the other questions. So the probability of a don't know on an item can depend on the observed outcomes on other items, but given these, it will not depend on the missing value itself.

If the MAR assumption holds and models for the outcome and response are separate (involve non-overlapping parameters) then the missingness pattern is also said to be ignorable (Rubin, 1976). That is missingness is ignorable if the parameters θ_Y and θ_R are distinct (Schafer, 1997, Chapter 2), with their joint prior factoring into independent marginal priors. Under the ignorability assumption there is no need to explicitly model the response mechanism when making inferences about θ_Y.

If, however, missingness on an item depends on the missing value of that outcome, then nonresponse is said to be non-ignorable. For example, a question on recent sexual activity may be less likely to be answered for those who were inactive (Raab and Donnelly, 1999). If nonresponse is incorrectly assumed to be random (with respect to the unobserved outcomes) then the procedures used to adjust for nonresponse may produce biased estimates of the distribution of the outcome across the full set of survey cases. Among possible options to verify which form of nonresponse model is appropriate, especially to distinguish among forms of non-ignorability, are follow-up of nonrespondents (Rubin, 1987, Chapter 12), or some form of cross-validation via random creation of additional missing data within the actually observed data.

The analysis of data subject to nonresponse is distinctive in several ways. First, there is typically no basis from the observed data to distinguish whether or not the missing data mechanism is non-ignorable (Forster and Smith, 1998), and model comparison procedures based on the observed data are not appropriate for inference. As Molenberghs *et al.* (1999) state, several models can look equally plausible in terms of fit to the observed data but their implications for the nature of the expected complete data can be considerably different; thus subject matter context, such as prior knowledge of covariate effects, is relevant to choice between models which are indistinguishable or only weakly identifiable on statistical grounds. A range of alternative models may be applied to gauge the extent of uncertainty about the true model (Fay, 1986), and the sensitivity of posterior to prior estimates. Given the paucity of information in the data about the response mechanism, different informative priors may be used in a sensitivity analysis regarding different possible assumptions about nonresponse and their implications for the complete data expected counts. A meaningful approach is to introduce parameters which measure the sensitivity to different levels of uncertainty.

For example, Rubin (1987) considers inferences on the overall mean $\bar{y}$ from a sample of n observations, when there are only r respondents. The appropriate estimator is a weighted mean

$$(r/n)\bar{y}_R + \{(n-r)/n\}\bar{y}_{\mathrm{NR}}$$

with $\bar{y}_{\mathrm{R}}$ the mean of the observed data and $\bar{y}_{\mathrm{NR}}$ the unknown mean of the missing data. For normal data, with $\sigma^2_{\mathrm{R}} = \sigma^2_{\mathrm{NR}} = \sigma^2$ and σ^2 known, the priors on the two means may be denoted

$$\bar{y}_{\mathrm{R}} \sim \mathrm{N}(\mu_{\mathrm{R}}, \sigma^2/r) \text{ and } \bar{y}_{\mathrm{NR}} \sim \mathrm{N}(\mu_{\mathrm{NR}}, \sigma^2/(n-r)).$$

The similarity between respondents and nonrespondents can be expressed in an 'uncertainty prior'

$$\mu_{\mathrm{NR}} \sim \mathrm{N}(\mu_{\mathrm{R}}, \varphi^2\mu^2_{\mathrm{R}})$$

so that the missing data mean is centred on the observed mean and there is a 95 % expectation that μ_{NR} will lie in the interval

$$(\mu_{\mathrm{R}} - 1.96\varphi\mu_{\mathrm{R}}, \mu_{\mathrm{R}} + 1.96\varphi\mu_{\mathrm{R}}).$$

The special case $\varphi = 0$ corresponds to ignorable nonresponse, and setting alternative values of φ typically small ($\varphi = 0.05$ or $\varphi = 0.1$) shows the impact of nonresponse on the posterior estimate of the overall mean.

The following development considers first patterns of missingness (monotone vs nonmonotone) and its application in multivariate imputation and the multiple imputation method. The subsequent analysis illustrates applications involving possibly non-ignorable missingness, in contingency table contexts, in settings where data are arranged in survey-defined domains, for polytomous outcomes and for mixtures of categorical and continuous outcomes.

6.5 MISSINGNESS PATTERNS: FACTORED LIKELIHOODS FOR MULTIVARIATE MISSING DATA

Many multivariate methods such as factor analysis and least squares regression are based on a reduction of the original observations on n cases, in a matrix of size $n \times p$, to the estimated mean μ and dispersion matrix Φ of the variables. Methods to estimate mean and dispersion parameters from incomplete data, for example in surveys, are therefore important given that some degree of nonresponse is the rule rather than the exception. We are also often interested in imputing missing data as a preliminary to applying simpler complete data techniques. For data with a pattern of monotone missingness, and subject to missingness at random (MAR), estimation to take account of the missing data is simplified by a factorisation of the joint likelihood. As an example of monotone missingness suppose x_1 (e.g. age) is recorded for all respondents of working age in a labour survey, occupation (x_2) is recorded for the majority of the respondents, and where occupation is recorded then income (x_3) is also reported for most subjects. However, suppose income is never reported when occupation is missing, or very rarely so – then this constitutes a monotone pattern of missingness (or one closely approximating it).

An example of monotone nonresponse in a bivariate situation is where $y_{i1}, i = 1, \ldots, n$ is always recorded but y_{i2} is recorded for only a subset of cases, namely $i = 1, \ldots, m$, and missing for the remaining cases, $i = m + 1, \ldots, n$. Thus $y_i = (y_{i1}, y_{i2})$ is complete only for cases $i = 1, \ldots, m$. If we assume bivariate normality with parameters μ and Φ, we can write the likelihood ignoring the missing data mechanism as

$$\log f(Y_{\text{obs}} \mid \mu, \Phi) \propto -0.5m \log|\Phi| - 0.5 \sum_{i=1}^{m} (y_i - \mu)\Phi^{-1}(y_i - \mu)$$
$$- 0.5(n - m)\log\phi_{11} - 0.5 \sum_{i=m+1}^{n} (y_{i1} - \mu_1)^2/\phi_{11}.$$

This likelihood is valid for complete data inference on $\{\mu, \Phi\}$ if the missingness in y_2 is not related to the unobserved values, specifically if the probability that y_{i2} is observed does not depend on the values of y_{i2} (though it may be related to the values of the fully observed y_{i1}). This likelihood may be represented directly. For example, in BUGS one might code as follows:

```
for (i in 1:m) {y[i,1:2] ~ dmnorm(mu[],T[,])}
for (i in m+1:n) {y[i,1] ~ dnorm(mu[1],T11)}
T11 <- T[1,1]
```

where `T[,]` and `T11` are precisions. However, a simpler approach to maximisation is based on factoring the likelihood into the marginal distribution of y_{i1} and the conditional distribution of y_{i2} given y_{i1}. This approach is applicable more broadly (i.e. to $p > 2$ variables) if there is a monotone missing data pattern. In the bivariate normal case with y_{i2} less observed than y_{i1},

$$f(y_{i1}, y_{i2} | \mu, \Phi) = f(y_{i1} | \mu_1, \Phi_{11}) f(y_{i2} | y_{i1}, \alpha_0, \alpha_1, \mu_{22}).$$

This follows from properties of the bivariate normal, with the second density applying to y_{i2}, $i = 1, \ldots, m$ and specifying the mean of y_{i2} in terms of a regression on the known y_{i1}. Thus

$$y_{i1} \sim \mathrm{N}(\mu_1, \Phi_{11}), \qquad i = 1, \ldots, n$$
$$y_{i2} \sim \mathrm{N}(\alpha_0 + \alpha_1 y_{i1}, \Phi_{22}), i = 1, \ldots, m.$$

Imputation of y_{i2} for the remaining cases $i = m + 1, \ldots, n$ would then be based on the regression parameters applied to the y_{i1} values observed for these cases. The regression variance is given by

$$\sigma_{22} = \Phi_{22}(1 - \rho^2)$$

and the regression parameters by

$$\alpha_1 = \rho[\Phi_{22}/\Phi_{11}]^{0.5}$$
$$\alpha_0 = \mu_2 - \alpha_1 \mu_1.$$

In practice we may estimate the regression coefficients and then predict μ_2, Φ_{22} and hence the bivariate correlation from the estimated α_0, α_1 and σ_{22}. For example, μ_2 is estimated as

$$\hat{\mu}_2 = \hat{\alpha}_0 + \hat{\alpha}_1 \hat{\mu}_1$$

and Φ_{22} as

$$\hat{\Phi}_{22} = \hat{\sigma}_{22} + \hat{\alpha}_1^2 \hat{\Phi}_{22}.$$

Example 6.15 Wormy fruit data A bivariate illustration is provided by an adaptation of data from Snedecor and Cochran (1989) relating percentage of wormy fruit (y_{i2}) to size of apple crop (y_{i1}). The objective is to estimate the mean of y_2 in the presence of six missing values on this variable (cases 13–18). These missing values are concentrated on trees with smaller crop totals. The observations suggest that smaller levels of wormy fruit occur on trees with larger crops. Thus the missingness appears to be MAR (missing at random) because it is related to the values of the fully observed crop size. So if μ_2 were estimated simply from the observed data this would underestimate the true value of μ_2. *Program 6.15 Wormy Fruit* reflects the monotone missingness pattern and the estimation of $\hat{\mu}_2$ to take account of all observed data, the essential coding being

```
for (i in 1:18) { y1[i] ~ dnorm(mu[1],tau[1])}
for (i in 1:12) { y2[i] ~ dnorm(mu2[i],tau[2])}
for (i in 13:18) { y2hat[i] ~ dnorm(mu2[i],tau[2])}
for (i in 1:18) { mu2[i] <- a+b*y1[i]}
                        mu[2] <- a+b*mu[1]
```

The mean of y_2 is estimated as 49.3 as compared with the sample mean of 45.

Example 6.16 Monotone missingness: Five-variable Hald data set Rubin (1987) introduces monotone missingness into the classic Hald engineering dataset, as in Table 6.20.

The data assumed missing are enclosed in brackets. Thus x_3 and y are completely recorded, x_1 and x_2 are missing for observations 10–13, and x_4 is missing whenever x_1 and x_3 are, but also for cases 7 to 9. Rubin derives estimates of the missing data via an E-M approach[1] based on a sweep algorithm; this estimates x_1 and x_2 given x_3 and y, and then estimates x_4 given the remaining variables. The sweeping refers to appropriate selection of covariances from the full 5×5 dispersion matrix to permit the associated multiple regressions, e.g. of x_1 given x_3 and y.

Here we reorder the variables such that $\{z_1 = x_3, z_2 = y\}$ are the fully known variables, $\{z_3 = x_1, z_4 = x_2\}$ have intermediate amounts of missingness, and $z_5 = x_4$ has the most missing data (see *Program 6.16 Hald Data*). We assume the first two z variables are bivariate normal for observations 1–13, with mean $\mu_{1:2}$ and 2×2 dispersion matrix Φ_{ij}. Then for observations 1 to 9, z_3 and z_4 are sampled from conditional univariate normals with means

$$m_{3i} = a_3 + b_3 z_{1i} + c_3 z_{2i}$$
$$m_{4i} = a_4 + b_4 z_{1i} + c_4 z_{2i}.$$

The remaining means of z_{3i} and z_{4i} (for observations 10 to 13) are predicted using the regression coefficients estimated from observations 1 to 9. Finally for observations 1–6, z_5 is sampled from a conditional univariate normal with means

$$m_{5i} = a_5 + b_5 z_{1i} + c_5 z_{2i} + d_5 z_{3i} + e_5 z_{4i}.$$

Table 6.20

Units	x_1	x_2	x_3	x_4	y
1	7	26	6	60	78.5
2	1	29	15	52	74.3
3	11	56	8	20	104.3
4	11	31	8	47	87.6
5	7	52	6	33	95.9
6	11	55	9	22	109.2
7	3	71	17	(6)	102.7
8	1	31	22	(44)	72.5
9	2	54	18	(22)	93.1
10	(21)	(47)	4	(26)	115.9
11	(1)	(40)	23	(34)	83.8
12	(11)	(66)	9	(12)	113.3
13	(10)	(68)	8	(12)	109.4
Mean	7.46	48.15	11.77	30.00	95.42

[1] E-M: expectation-maximisation, an iterative likelihood maximisation procedure (Dempster *et al.*, 1977; Mchachlan and Krishnan, 1997).

The remaining means of z_5 (for observations 7 to 13) are predicted using the regression coefficients estimated from observations 1 to 6, and using the regression predictions of m_{3i} and m_{4i} for $i = 10$ to 13. The averages μ_1 to μ_5 over observed and missing data are then averages of sample means and regression predictions.

The actual mean values for $z_3(= x_1), z_4(= x_2)$ and $z_5(= x_4)$ are 7.46, 48.15 and 30. Estimates of the remaining parts of covariance matrix of the z_j, namely $\{\Phi_{ij}, i \text{ and } j > 2\}$ may be made using relations between the (co)variances of an MVN distribution and the estimated regression coefficients. Table 6.21 gives the means (using the ordering of the z_j) obtained from the second half of a run of 20 000 iterations.

6.5.1 Multiple imputation

Multiple imputation is a specialised approach to missing data, whether ignorable or non-ignorable. The MCMC approach to multiple imputation is distinctive to the usual focus on parameter simulation via repeated sampling. In multiple imputation the MCMC process is used to create a small number M of independent draws, or 'imputations', of the missing data $Y_{\text{miss, m}}$, $m = 1, \ldots, M$, drawn from a predictive distribution. Thus one set of 'full' data consists of $\{y_{\text{obs}}, y_{\text{miss}}, 1\}$, the second of $\{y_{\text{obs}}, y_{\text{miss},2}\}$ and the Mth of $\{y_{\text{obs}}, y_{\text{miss},M}\}$. The basis for such imputation is via the usual Bayesian prediction technique treating the missing data as extra parameters, possibly under a range of nonresponse assumptions. Thus in the case of missing data at random the predictive density of y_{miss} is

$$P(y_{\text{miss}} \mid y_{\text{obs}}) = \int p(y_{\text{miss}} \mid y_{\text{obs}}, \theta_Y) p(\theta_Y \mid y_{\text{obs}}) d\theta_Y$$

so imputations can be created by first sampling from the posterior density of θ, to give θ^*, and then sampling from the predictive distribution

$$P(y_{\text{miss}} \mid y_{\text{obs}}, \theta^*).$$

These samples should be drawn after stationarity has been achieved in the MCMC sampling. Schafer (1997) recommends subsampling to better approximate independent samples of y_{miss}. The number of such samples needed is typically under $M = 10$ because Monte Carlo error is small compared with the

Table 6.21

Parameter	Mean	SD	Rubin estimate
μ_1	11.76	1.84	11.77
μ_2	95.39	4.34	95.42
μ_3	6.67	0.99	6.65
μ_4	49.87	2.64	49.97
μ_5	27.16	2.22	27.05

overall uncertainty about y_{miss} (Schafer, 1997, Chapter 4). Having generated the M 'full' data sets, they are analysed by complete data methods, yielding different estimates $\theta_1, \ldots, \theta_M$ of a single set of parameters θ. The aim is to combine these estimates for the final inferences.

Suppose the posterior variances of $\theta_1, \ldots, \theta_M$ are $U_1, \ldots, U_M$ respectively (e.g. from MCMC output). Then the within-imputation variance of the θ_i is estimated as

$$W = \sum_{i=1}^{M} U_i/M,$$

while the between-imputation variance is

$$B = \sum_{i=1}^{M} (\theta_i - \bar{\theta})^2/(M-1).$$

So the estimated total variance of the combined estimate $\bar{\theta}$ is

$$T = B(1 + 1/M) + W$$

and then

$$\bar{\theta}T^{-0.5} \sim t_v$$

where the degrees of freedom parameter of the t density is

$$v = (M-1)[1 + WB^{-1}(1 + 1/M)^{-1}].$$

If the imputations carry no information about the unknown θ then the separate MI estimates θ_i would be equal and T would be equal to W. Therefore the ratio $r = (1 + 1/M)B/W$ measures the increase in variance associated with the missing data, and $\varepsilon = r/(1+r)$ amounts to an estimated proportion of missing information. The relative efficiency of M imputations compared with an infinite number is

$$(1 + \varepsilon/M)^{-1}$$

which falls off rapidly with M for even large proportions of missing data (e.g. $\varepsilon = 0.5$, equivalent to 50% missingness).

Example 6.17 Bivariate normal simulated data, missing at random Following Schafer (1997, Chapter 2) 100 bivariate normal observations $\{y_{1i}, y_{2i}\}$ were generated with mean $\mu = (\mu_1, \mu_2) = (0, 0)$, variances $\sigma_{11}^2 = \sigma_{22}^2 = 1$ and correlation 0.9. The first variable y_1 is always observed but y_2 is subject to only about 50% response. Missing values in y_2 are assumed to be generated via a MAR model with response indicators ($R_i = 1$ for y_2 observed, $R_i = 0$ for y_2 missing) generated via a probit model with

$$R_i \sim \text{Bernoulli}(\pi_i)$$
$$\text{probit}(\pi_i) = a + by_{1i}$$

where $a = 0, b = 1$. The MAR assumption is reflected in the dependence of π_i on the fully observed y_1 but not on the y_2 which are subject to missing values. Generating data with a single run of *Program 6.17 Generate Sample Model* gave $R_i = 0$ for 49 of the 100 observations of y_2. In the second imputation stage the input data is the y_1 as just generated, but the y_2 has a missing value (NA) substituted where $R_i = 0$. In this stage a simple linear regression is used to generate $M = 5$ sets of the missing y_2 values. Thus

$$y_{2i} \sim \mathrm{N}(\mu_i, \varphi)$$

where φ is a variance parameter and $\mu_i = \alpha + \beta y_{1i}$.

This stage as in *MI Stage Model* estimates α and β at -0.07 (0.06) and 0.87 (0.06) respectively from the 51 complete observations. Note that this is a linear regression relating y_2 to y_1 and does not reproduce the Bernoulli missingness model (which is assumed ignorable). The value of β is close to the expected value of 0.9. The imputations are made at iterations 2001–2005.

In the third combined estimation stage (*Combined Inference Model*), the M full sets of observed and imputed values of $\{y_{1i}, y_{2i}\}, i = 1, \ldots, 100$, are used to make a revised estimate of α and β and of the overall mean of y_2 (which should be estimated as zero). All five sets of data are assumed to be generated by the same model as above, but with regression effects differentiated by imputation number, namely

$$y_{2ik} \sim \mathrm{N}(\mu_{ik}, \varphi_k), \; i = 1, \ldots, 100, k = 1, \ldots, 5$$

with

$$\mu_{ik} = \alpha_k + \beta_k y_{1i}.$$

The overall estimate of the coefficients $\bar{\alpha}$ and $\bar{\beta}$ is the average of the five separate effects.

One may then calculate the between-imputation variance of $\bar{\alpha} = \sum_k \alpha_k/5$ and $\bar{\beta} = \sum_k \beta_k/5$ as

$$B_1 = \sum_k (\alpha_k - \bar{\alpha})^2/M$$

and

$$B_2 = \sum_k (\beta_k - \bar{\beta})^2/M,$$

and the within-imputation variance of the two coefficients as $\bar{U}_j = \sum_k U_{jk}/M$ where $\{U_{1k}, U_{2k}\}$ are the posterior variances of α_k and β_k respectively over imputed data sets k. From the second half of a run of 15 000 iterations we obtain the results in *Table 6.22*.

The estimated total variances are then

$$T_j = \bar{U}_j + (1 + 1/M)B_j, \quad \text{for } j = 1, 2.$$

Table 6.22

	Mean	SD		Mean	SD
$\alpha_{[1]}$	−0.037	0.047	$\beta_{[1]}$	0.850	0.052
$\alpha_{[2]}$	0.036	0.047	$\beta_{[2]}$	0.899	0.051
$\alpha_{[3]}$	−0.054	0.043	$\beta_{[3]}$	0.953	0.048
$\alpha_{[4]}$	−0.062	0.035	$\beta_{[4]}$	0.845	0.040
$\alpha_{[5]}$	−0.044	0.039	$\beta_{5]}$	0.908	0.043
$\mu_{2[1]}$	0.036	0.046			
$\mu_{2[2]}$	0.116	0.046			
$\mu_{2[3]}$	0.031	0.043			
$\mu_{2[4]}$	0.012	0.035			
$\mu_{2[5]}$	0.036	0.039			

Here this gives $\bar{\alpha} = -0.032$ and $\bar{\beta} = 0.891$, $\bar{U}_1 = 0.0018$, $\bar{U}_2 = 0.0022$, and so

$$T_1 = 1.2(0.0012) + 0.0018,$$
$$T_2 = 1.2(0.0016) + 0.0022.$$

Therefore $\bar{\alpha}$ and $\bar{\beta}$ have estimated standard errors 0.057 and 0.064. The mean of the y_2 is estimated as 0.046 with a standard error of 0.057.

6.6 IGNORABLE AND NON-IGNORABLE MISSINGNESS: ESTIMATING MISSING COUNTS IN CONTINGENCY TABLES

As noted above, a missingness pattern is ignorable if models for the outcome and response are separate (involve non-overlapping parameters), so that the joint prior for parameters θ_Y and θ_R factors into independent marginal priors. There is then no need to explicitly model the response mechanism when making inferences about θ_Y. By contrast, a situation illustrating the dependence of missing values on an item on the unobserved value of that outcome involves categorical variables $X_1, X_2, X_3, \ldots$ with $I_1, I_2, I_3, \ldots$ levels which are incompletely classified on one or more of these variables. The simplest case is for two categorical variables X_1 and X_2 which are both binary with $I_1 = I_2 = 2$. Bivariate categorical data subject to missingness can be summarised as a two-way table with supplementary margins – see Baker *et al.* (1992) and Molenberghs *et al.* (1999). For example Baker *et al.* consider data from a prospective study on infant birthweight and maternal risk factors. The total observed data on 57 061 mothers consists of (a) fully observed counts $\{n_{11ij}\}$, totalling 53 047; (b) counts observed only on the smoking variable $\{n_{12i.}\}$; (c) counts observed only on the birthweight variable $\{n_{21.j}\}$; and (d) a total of 1224 mothers observed on neither, denoted $\{n_{22..}\}$. In each of the latter three circumstances there are four cells to estimate but the partially observed cases provide more initial information and will lead to more precise estimates of the complete data.

Following the general principle that the missingness can be represented as an additional form of data, the observed data are augmented to take the form z_{ijkl} where $i = 1, \ldots, I_1, j = 1, \ldots, I_2, k = 1, 2$ and $l = 1, 2$ denote present or missing response on X_1 and X_2 respectively. Thus $N = z{+}{+}{+}{+} = 57061$ where $+$ denotes summation over each subscript. Let μ_{ijkl} denote the expected counts in a log-linear model for the complete data, defined by parameters for the I_1 levels of X_1, the I_2 levels of X_2, the two levels of response on X_1 and the two levels of response on X_2. For example, let β_3 denote the main effect of presence or missingness of X_1 (smoking), and γ_{12ij} the interaction between X_{1i} and X_{2j}, which for X_1 and X_2 both binary reduces to a single parameter.

Then one possible set of parameters for a log-linear model form is

$$\log(m_{ij}) = \beta + \beta_{1i} + \beta_{2j} + \beta_3 + \beta_4 + \gamma_{12ij} + \gamma_{13i} + \gamma_{14i} + \gamma_{23j} + \gamma_{24j} + \gamma_{34}$$

where $\mu_{ij11} = m_{ij}$,

$$\log(a_{ij}) = -2[\beta_3 + \gamma_{13i} + \gamma_{23j} + \gamma_{34}]$$

where $\mu_{ij21} = m_{ij}a_{ij}$,

$$\log(b_{ij}) = -2[\beta_4 + \gamma_{14i} + \gamma_{24j} + \gamma_{34}]$$

where $\mu_{ij12} = m_{ij}b_{ij}$, and

$$\log(C) = 4\gamma_{34}$$

where $\mu_{ij22} = m_{ij}a_{ij}b_{ij}C$. When X_1 and X_2 are both binary all the parameters lose their subscripts, so we are left with eleven parameters, as compared with nine observed cell counts. For the model to be identifiable, it is necessary to set the a_{ij} or b_{ij} all equal to each other, or equal along one dimension. For example, with m_{ij} and C given, the following models are identifiable

(1) $a_{ij} = A$, $b_{ij} = B$
(2) $a_{ij} = A$, $b_{ij} = B_i$
(3) $a_{ij} = A_j$, $b_{ij} = B$
(4) $a_{ij} = A$, $b_{ij} = B_j$
(5) $a_{ij} = A_i$, $b_{ij} = B$.

Model (1) represents ignorable missingness since the chance of an observation missing does not depend on the unobserved outcome of either X_1 or X_2. When $I_1 = I_2 = 2$, this model reduces the parameter total to seven, leaving two degrees of freedom. Models 2 and 3 mean missingness on one variable is constant but missingness on the other variable depends on the outcome of the former. Models 4 and 5 also mean missingness on one variable is constant but missingness on the other variable depends on its own outcome. Another four models are identifiable but leave no degrees of freedom.

Example 6.18 Infant birthweight and maternal smoking We consider model 3 for X_1 as smoking (1 = Yes, 2 = No) and X_2 as birthweight (1 = under 2500 g,

2 = above 2500 g). This means a mother's nonresponse on the smoking variable depends on the child's birthweight. Vague normal priors are set on A_1, A_2, B and C but constrain them to be positive. The sampling of z_{ij22} is multinomial from a total of 1224 with the cell probabilities modelled as

$$\pi_{ij22} = \mu_{ij22}/\mu_{..22}$$

and where

$$\mu_{ij22} = m_{ij}A_jBC.$$

The sampling of the two types of partially missing data is binomial; for example, z_{1121} is binomial from a total of 142, with probability

$$\pi_{1121} = \mu_{1121}/\mu_{.121}$$

with $z_{2121} = 142 - z_{1121}$. The resulting posterior estimates of μ_{ij12}, μ_{ij21} and μ_{ij22} are from the second half of a run of 10 000 iterations using *Program 6.18 Birthweight and Smoking* (Table 6.23).

These estimates are close to those of Baker *et al.* The posterior value of C is sensitive to its prior but the estimates of μ_{ij22} are reasonably robust despite this. Subject matter knowledge on the extent of interaction between nonresponse on birthweight and smoking is likely to be relevant.

6.6.1 Hierarchical models for non-ignorable nonresponse

One approach to nonresponse involves hierarchical models over survey domains or population subgroups, both for the outcome of interest (e.g. whether the respondent is a smoker or obese), and for the probabilities of response or nonresponse within the sub-groups. These subgroups may be defined by known covariates (Park and Brown, 1994), or by variables used to

Table 6.23

	Mean	SD	Median
$\mu_{12(1,1)}$	185.6	12.5	186.0
$\mu_{12(1,2)}$	863.4	12.5	863.0
$\mu_{12(2,1)}$	139.8	11.4	139.0
$\mu_{12(2,2)}$	995.2	11.4	996.0
$\mu_{21(1,1)}$	81.2	5.9	81.0
$\mu_{21(1,2)}$	216.4	10.9	216.0
$\mu_{21(2,1)}$	60.8	5.9	61.0
$\mu_{21(2,2)}$	247.6	10.9	248.0
$\mu_{22(1,1)}$	152.6	146.1	105.0
$\mu_{22(1,2)}$	444.8	120.1	483.0
$\mu_{22(2,1)}$	114.8	110.2	78.0
$\mu_{22(2,2)}$	511.8	137.5	556.0
C	25.1	18.9	21.4

determine a survey design, such as urban or rural stratum of residence (Stasny, 1991). Under such models information from the entire sample or survey is used to improve estimates of the outcome and response probabilities in separate subgroups. A hierarchical model with exchangeable priors for response and the outcome may be adopted, and possibly allow differential nonresponse according to the outcome (Little and Gelman, 1998). Alternatively regression (e.g. log-linear) models may be adopted to assess whether differential nonresponse is related to stratum variables or fully observed covariates. Thus King *et al.* (1999) describe the application of hierarchical methods for estimating turnout rates for electoral precincts by race (black vs white) when all that is known are Census percents black for such precincts, and the total electoral turnout in relation to the number of voting age. They discuss the addition of relevant covariates to predict these missing variables.

In a baseline model under this hierarchical approach we may allow nonresponse probabilities to come from a single distribution, regardless of the response to the outcome of interest – this is in line with missingness at random. Alternatively different nonresponse distributions may be proposed for different types of response, leading to an analysis with missingness not at random. For example, we would assume a different chance of nonresponse according to the level of a binary outcome; in a health setting it may be that smokers are more likely not to answer a question about smoking habits than nonsmokers are.

Suppose the outcome is binary and that a population has been subdivided into $i = 1, \ldots, G$ groups defined by fully observed covariates or by stratifying variables. The groupings should be formed by variables which are expected to be associated with the response. Within subgroup i all individuals are assumed to have the same prevalence p_i of the binary outcome, but individuals in different subgroups may have different prevalences. Assume the density of p_i follows a beta distribution with parameters s and t. In a behavioural or health setting we may have informative prior knowledge about the overall level of the outcome of interest (i.e. the ratio of s to $s + t$) and possibly its variability also.

Suppose also that the chances of response are affected by the occurrence or otherwise of the binary outcome. For example, consider the outcome cigarette smoking or otherwise. Let π_{i1} denote the conditional probability that a smoker in group i is a responder, and π_{i0} denote the probability that a nonsmoker in the same group is a responder. Let R_{ij} be a dummy variable defined as 1 if the jth individual in the ith group is a responder and 0 otherwise. Let $Y_{ij} = 1$ or 0 according to whether the same individual has the behaviour, characteristic or attitude of interest. Then the total probability of nonresponse is the sum over the two possible combinations of outcome and nonresponse conditional on outcome:

$$\begin{aligned}\Pr(R_{ij} = 0) &= \Pr(R_{ij} = 0 \mid Y_{ij} = 0)\Pr(Y_{ij} = 0) + \Pr(R_{ij} = 0 \mid Y_{ij} = 1)\Pr(Y_{ij} = 1) \\ &= (1 - \pi_{i0})(1 - p_i) + (1 - \pi_{i1})p_i. \qquad (6.11)\end{aligned}$$

There may be prior information about the chance of response according to the outcome of interest. Alternatively we may adopt non informative priors such as

$$\pi_{i0} \sim \text{Beta}(\alpha_0, \beta_0), \quad \pi_{i1} \sim \text{Beta}(\alpha_1, \beta_1),$$

with the beta priors set to standard values such as $\alpha_0 = \beta_0 = \alpha_1 = \beta_1 = 0.5$. Another option is to assess sensitivity to alternative informative priors such as that smokers are more likely to not respond and conversely.

Suppose there are U_i nonrespondents in the ith group to the binary outcome, S_i respondents with the outcome present ($Y = 1$) and T_i with the outcome not present ($Y = 0$). The likelihood contributions for the latter two groups are respectively

$$\Pr(R_{ij} = 1 \mid Y_{ij} = 1)\Pr(Y_{ij} = 1) = \pi_{i1} p_i \tag{6.12}$$

and

$$\Pr(R_{ij} = 1 \mid Y_{ij} = 0)\Pr(Y_{ij} = 0) = \pi_{i0}(1 - p_i). \tag{6.13}$$

The likelihood contribution for nonresponders is the total probability in (6.11), so the total likelihood involves terms (6.11) to (6.13).

Suppose we continue with the smoking example, so that the U_i nonresponders will be made up of two latent classes, V_i nonresponders who smoke, and $U_i - V_i$ nonresponders who do not smoke. If the data are augmented to include the latent V_i then the probability that V_i of the U_i nonresponders are smokers is binomial with probability

$$w_i = (1 - \pi_{i1})p_i / \{(1 - \pi_{i0})(1 - p_i) + (1 - \pi_{i1})p_i\}.$$

That is,

$$V_i \sim \text{Bin}(w_i, U_i).$$

The conditional densities of the outcome prevalence (smoking rate) and response probabilities can then be written

$$p_i \mid V_i, \pi_{i0}, \pi_{i1} \sim \text{Beta}(V_i + S_i + s, T_i + t + U_i - V_i)$$
$$\pi_{i1} \mid p_i, V_i \sim \text{Beta}(S_i + \alpha_1, V_i + \beta_1)$$
$$\pi_{i0} \mid p_i, V_i \sim \text{Beta}(T_i + \alpha_0, U_i - V_i + \beta_0).$$

In an ignorable response model, the steps are the same, but we would take $\pi_{i0} = \pi_{i1} = \pi_i$ and so assume a common beta prior for π_{i0} and π_{i1}.

6.6.2 Parameterisation of differential nonresponse

Little and Gelman (1998) consider a reparameterisation of the differential nonresponse model where the outcome is binary. Thus for strata $i = 1, \ldots, G$ they consider the ratios

$$R_i = \pi_{i1}/(\pi_{i1} + \pi_{i0}) \tag{6.14}$$

and the overall nonresponse rate by stratum

$$\pi_i = (\pi_{i0} + \pi_{i1})/2. \tag{6.15}$$

So the parameter set $\{\pi_{i1}, \pi_{i0}, p_i\}$ is replaced by the set $\{\pi_i, R_i, p_i\}$. Choice of a common value for all i such as $R_i = 0.5$ corresponds to a missing completely at random assumption, while a prior on the R_i, such as a beta with mean 0.5, amounts to a non-ignorable response model. Following Kadane (1993), they stress the sensitivity of inferences about p_i to the assumptions made about the R_i. In fact if the R_i are assigned a prior which is too vague, such as the common default $B(1, 1)$ say, then the p_i may be over-smoothed.

They argue that in most surveys the R_i should vary less than the p_i on the basis that relative nonresponse probabilities are unlikely to vary more than the average response. They themselves consider pre-election polling data on candidate preferences ($Y = 1$ for Bush vs $Y = 0$ for Dukakis in the 1988 US Presidential election) over $G = 48$ US States. They advocate drawing on comparisons of previous actual polls and pre-election surveys with regard to assessing prior variability in the R_i. On this basis they select a B(68.4, 62.2) prior for the R_i and compare smoothed p_i with observed p_i, suggesting that Bush voters are more likely to respond and that differential nonresponse between states is small.

They also consider a survey situation where the number of nonrespondents is unknown. For example, in telephone surveys of voting intentions, if no-one answers the phone this may be for one of several reasons (e.g. number is not residential, person not at home). Thus the actual data is numbers of respondents W_i only and their split (S_i, T_i) over the G groups. In these circumstances the only identifiable parameters are the conditional probabilities of 'positive' response given any response

$$\rho_i = p_i\pi_{i1}/[p_i\pi_{i1} + (1 - p_i)\pi_{i0}] \tag{6.16}$$

with

$$S_i \sim \text{Bin}(\rho_i, W_i).$$

Example 6.19 Crime victimisation Stasny (1991) reports findings on crime victimisation within the last six months from the 1975–79 US National Crime Survey. Numbers of respondents who are victims or crime-free (S_i and T_i respectively) as well as nonrespondents U_i are classified into ten groups defined by post-stratification (domain) variables. The latter were urban vs rural neighbourhood (B vs R), city, other incorporated or not corporated place (C/I/N), and high or low neighbourhood poverty (H/L). Combinations of rural and city are excluded, so that only 10 of the 12 factor combinations actually exist.

Crude proportions of nonresponse differ by domain, and it is desirable to form an overall estimate of victimisation adjusted for nonresponse. Simple ML estimators of the victimisation rate (among respondents) and of the response rate suggest that response is higher in lower crime areas. That is, nonresponse may be nonrandom or non-ignorable. Stasny uses an empirical Bayes (EB) procedure to estimate the parameters α_0, β_0, α_1 and β_1, and thereby to estimate

p_i for the entire sample populations as well as the differential response rates π_{i1} and π_{i0}.

The EB procedure produces estimates of π_{i0} which are constant over domains ($\pi_{i0} = 93.7\%$ for all i), estimates of π_{i1} varying from 67.9% to 69.4%, and estimates of the crime rate accordingly adjusted upwards. Naive estimates of the crime victimisation rate varied from 8.7% to 27.3%; these do not reflect the higher nonresponse among victims or the variable sample sizes $n_i = S_i + T_i + U_i$ in different domains. The EB procedure of Stasny produces revised estimates of the victim rate varying from 16.6% to 30.5%, taking account of differential response and also pooling strength to improve estimates in small domains.

A fully Bayes (FB) procedure (see *Program 6.19 Crime Victimisation*) is here implemented with initial values of the augmented data V_i as well as prior values on the beta mixing parameters $\alpha_0, \beta_0, \alpha_1$ and β_1 (Table 6.24). The initial values of V_i may be set between 0 and U_i, perhaps using the values resulting from an ignorable nonresponse assumption. Thus if $S_1 = 250, T_1 = 750$ and $U_1 = 500$ then V_1 may be initially set to $0.25 \times 500 = 125$. The posterior estimates from the FB procedure have the advantage of producing credible intervals for p_i, π_{i0} and π_{i1}. They show slightly less smoothing of the victim rates than the EB procedure (ranging from 14.4% to 30.5%) and more variability between domains in the response rates among victims and crime-free.

To assess whether individual domain variables are associated with higher levels of nonresponse among crime victims, the estimated V_i and $U_i - V_i$ may be cumulated over relevant subgroups to form an odds ratio of nonresponse according to that domain variable. For example, we may wish to test whether high-poverty neighbourhoods have higher relative rates of nonresponse among victims than crime-free as compared with low-poverty neighbourhoods. The estimated V_i and $F_i = U_i - V_i$ are accumulated over the low-poverty groups, namely U/C/L, U/I/L, U/N/L, R/I/L and R/N/L; and similarly accumulated over the high-poverty groups. Then the relevant odds ratio is represented by $(V_{\text{high}}/F_{\text{high}})/(V_{\text{low}}/F_{\text{low}})$. The log odds of this ratio averages –0.04 with a standard deviation of 0.96, so there is no evidence that poverty level is associated with differential nonresponse.

Example 6.20 Coronary risk factors in children Park and Brown (1994) consider coronary risk factor study data on obesity in children (Yes, No or Unknown) in relation to their age and gender. The latter two variables are completely observed. So $G = 4$ groups are defined by the combination of sex and age group (Table 6.25).

The ignorable model is estimable via a log-linear model or via a hierarchical model for the outcome and response probabilities π_i by age and sex, assuming these probabilities are not dependent on missingness. Applying the former (see *Ignorable Missing Model* in Program 6.20) gives estimated obesity rates for younger ages very close to those obtained under the ignorable log-linear model of Park and Brown, namely 15% for males and females combined. At older ages, the estimates are slightly lower than their 28% (Table 6.26).

Table 6.24

	Mean	SD	2.5%	Median	97.5%
$p(1)$	0.255	0.042	0.184	0.256	0.328
$p(2)$	0.247	0.043	0.171	0.247	0.327
$p(3)$	0.258	0.041	0.190	0.258	0.332
$p(4)$	0.241	0.036	0.176	0.238	0.313
$p(5)$	0.280	0.053	0.187	0.278	0.378
$p(6)$	0.305	0.070	0.178	0.303	0.449
$p(7)$	0.274	0.072	0.145	0.271	0.423
$p(8)$	0.144	0.053	0.059	0.138	0.255
$p(9)$	0.150	0.034	0.091	0.149	0.219
$p(10)$	0.194	0.037	0.132	0.191	0.268
π_{10}	0.915	0.049	0.834	0.915	0.996
π_{20}	0.908	0.050	0.822	0.906	0.994
π_{30}	0.916	0.048	0.839	0.914	0.997
π_{40}	0.929	0.039	0.860	0.926	0.997
π_{50}	0.882	0.062	0.780	0.879	0.994
π_{60}	0.880	0.072	0.728	0.882	0.994
π_{70}	0.892	0.067	0.748	0.898	0.994
π_{80}	0.898	0.054	0.796	0.896	0.993
π_{90}	0.941	0.034	0.877	0.942	0.997
$\pi_{10,0}$	0.920	0.041	0.851	0.917	0.994
$\pi_{1,1}$	0.768	0.123	0.582	0.746	0.987
$\pi_{2,1}$	0.741	0.126	0.542	0.727	0.981
$\pi_{3,1}$	0.780	0.117	0.594	0.768	0.987
$\pi_{4,1}$	0.817	0.104	0.625	0.818	0.991
$\pi_{5,1}$	0.723	0.134	0.516	0.708	0.978
$\pi_{6,1}$	0.767	0.136	0.500	0.771	0.986
$\pi_{7,1}$	0.748	0.147	0.455	0.755	0.986
$\pi_{8,1}$	0.585	0.201	0.257	0.561	0.971
$\pi_{9,1}$	0.709	0.145	0.469	0.696	0.976
$\pi_{10,1}$	0.753	0.133	0.531	0.745	0.987

Table 6.25 Numbers of children by age, sex and obesity

		Obese			
Age	Sex	Y	N	DK	% Missing
Young	M	82	463	470	46
	F	81	435	418	45
Old	M	247	900	324	22
	F	272	861	303	21

Table 6.26

Parameter	Mean	SD	2.5%	Median	97.5%
$p(1)$	0.152	0.015	0.12	0.15	0.18
$p(2)$	0.159	0.016	0.13	0.16	0.19
$p(3)$	0.216	0.012	0.19	0.22	0.24
$p(4)$	0.241	0.013	0.22	0.24	0.27
$\pi(1)$	0.537	0.016	0.51	0.54	0.57
$\pi(2)$	0.552	0.016	0.52	0.55	0.58
$\pi(3)$	0.780	0.011	0.76	0.78	0.80
$\pi(4)$	0.789	0.011	0.77	0.79	0.81

Table 6.27

	Mean	SD	2.5%	Median	97.5%
Young M	71.4	10.6	52	71	93
Young F	66.4	10.0	48	66	87
Old M	70.0	8.2	54	70	87
Old F	72.9	8.4	57	73	90

The estimated number of nonrespondents who are obese are as in Table 6.27.

There is a suggestion that nonresponse to the obesity question is non-ignorable in that it is much higher among younger children, and because obesity differs by age. An explicitly non-ignorable model is here applied via hierarchical priors on the likelihoods, as in (6.11) to (6.13), of the outcome and of response according to outcome. The four groups are 1 = young males, 2 = young females, 3 = older males, 4 = older females (see *Program 6.19 Non Ignorable Missing Model*).

The resulting estimates of obesity rates are much less precisely estimated than under the ignorable model. They are also less precise for younger ages where nonresponse is greater (using the last 75 000 of a long run of 80 000 iterations). The rates at these ages, namely 26 and 28%, are higher than those obtained by the Empirical Bayes method of Park and Brown, which were 17% for young males and 18% for young females. The rates at older ages are broadly similar to those of Park and Brown, which are 26% and 30.5% (Table 6.28).

The estimates of the response probabilities suggest that response to the obesity question is greater for non-obese than obese children; i.e. obese children are less likely to respond to this question, especially at older ages. The estimated numbers of nonrespondents who are obese is not precisely estimated but averages 182 (out of 470) for younger males, and 184 (out of 418) for younger females. Among the 627 older children nonrespondents, on average 290–300 are estimated to be obese.

As an example of an explicit regression-based approach to hierarchical pooling, it may be assumed that the response rates for obese children π_{j1} have

Table 6.28 Results of non-ignorable model.

Parameter	Mean	SD	2.5%	Median	97.5%
p_1	0.26	0.13	0.09	0.23	0.53
p_2	0.28	0.13	0.09	0.27	0.52
p_3	0.28	0.07	0.17	0.27	0.39
p_4	0.29	0.06	0.19	0.28	0.40
π_{10}	0.64	0.13	0.49	0.60	0.96
π_{20}	0.67	0.14	0.50	0.64	0.97
π_{30}	0.85	0.08	0.73	0.84	0.99
π_{40}	0.84	0.08	0.74	0.83	0.99
π_{11}	0.42	0.23	0.15	0.35	0.93
π_{21}	0.40	0.22	0.16	0.32	0.93
π_{31}	0.64	0.16	0.43	0.61	0.97
π_{41}	0.70	0.15	0.47	0.68	0.98
Nonrespondents, obese					
Young M	182	136	6	154	452
Young F	184	124	5	173	407
Older M	159	96	7	155	316
Older F	137	89	6	129	295

a beta prior as above. However, it is assumed that response for non-obese has the form

$$\text{logit}(\pi_{0j}) = \text{logit}(\pi_{1j}) + \rho_j$$

where ρ_j is normal with precision τ. So non-ignorability across strata is still governed by an exchangeable model. An informative prior is adopted, with the ρ_j restricted to positive values (see *Program 6.20 Non Ignorable Regression*). τ also has a well-defined prior, namely a gamma $G(1, 1)$ density. So non-ignorability is assumed similar across strata. A more extensive outline of this type of approach is contained in Forster and Smith (1998). The proposed model leads to broadly similar inferences about obesity rates by age and sex but to more precise estimates of obese nonrespondents – as would be expected from the informative prior (Table 6.29).

An analysis in terms of the effects of the known covariates may also be performed. This entails making the response and/or outcome probabilities specific to the categories i_1 and i_2 of the known covariates X_1 and X_2. For example if the yes and no outcome totals, and the nonresponse levels, are known by categories i_1 and i_2 then we may make the response probabilities specific to these categories, via the parameterisation $\pi_{0,i1,i2}$ and $\pi_{1,i1,i2}$.

Example 6.21 Sexual activity among undergraduates Raab and Donnelly (1999, Table 4) present a simplified data set giving numbers reporting 'yes' to the question 'Have you ever had sexual intercourse?'. The proportion of responders answering yes was 73% but 2308 of the 6136 students surveyed did not answer the question. Non-ignorable nonresponse may occur if nonresponse is

Table 6.29

Parameter	Mean	SD	2.5%	Median	97.5%
p_1	0.27	0.08	0.16	0.24	0.50
p_2	0.27	0.08	0.17	0.25	0.49
p_3	0.27	0.04	0.21	0.26	0.38
p_4	0.30	0.04	0.24	0.29	0.39
π_{10}	0.58	0.19	0.25	0.56	0.97
π_{20}	0.60	0.19	0.27	0.58	0.97
π_{30}	0.81	0.11	0.56	0.82	0.99
π_{40}	0.83	0.11	0.59	0.84	0.99
π_{11}	0.33	0.09	0.16	0.33	0.50
π_{21}	0.35	0.09	0.18	0.35	0.52
π_{31}	0.63	0.09	0.44	0.64	0.78
π_{41}	0.65	0.08	0.48	0.66	0.78
ρ_1	1.21	1.18	0.04	0.90	4.28
ρ_2	1.20	1.18	0.04	0.88	4.20
ρ_3	1.19	1.18	0.04	0.89	4.17
ρ_4	1.22	1.18	0.04	0.91	4.16
Nonrespondents, obese					
Young M	188	84	87	164	425
Young F	169	74	80	148	378
Older M	152	61	76	135	305
Older F	152	57	79	137	288

more likely among the sexually inexperienced. A 'faculty effect' on nonresponse was apparent, in that medical students combined a high response rate with a low reported experience of intercourse. The authors accordingly adopted a log-linear model for non-ignorable nonresponse using known binary covariates gender and faculty, though they acknowledge (Raab and Donnelly, 1999 p. 129) that very little information is available in the observed data without further assumptions or more informative priors. We adopt an exchangeable approach as above (see *Program 6.21 Sexual Activity*), with X_1 a binary index for faculty and X_2 for gender. This is essentially the same as the saturated log-linear model for the outcome (sexual activity) and response, as estimated by Raab and Donnelly (1999, Table 5), involving faculty, gender and a gender–faculty interaction.

The resulting estimate of the population prevalence of previous sexual activity is 66% (after a single long chain of 50 000 iterations). This is lower than the 73% reported by the authors for their saturated model but happens to be close to their preferred estimate from a profile likelihood approach, namely 67% with a 95% interval from 58 to 74%. The interval here is wider, from 52 to 79%.

Estimates of π_{i0} among the subgroups defined by gender and faculty show a clear 'honesty' effect among the medical students (Table 6.30).

Table 6.30

	Mean	SD	2.5%	Median	97.5%
Prevalence (%) Response	66.2	7.4	52.4	66.3	79.3
rates Inexperienced					
π_{10} (Male non-medical)	0.526	0.203	0.277	0.473	0.954
π_{20} (Female non-medical)	0.551	0.201	0.302	0.497	0.966
π_{30} (Male medical)	0.752	0.129	0.541	0.742	0.983
π_{40} (Female medical)	0.744	0.128	0.541	0.728	0.984
Experienced					
π_{11} (Male non-medical)	0.679	0.127	0.519	0.652	0.953
π_{21} (Female non-medical)	0.727	0.124	0.562	0.706	0.970
π_{31} (Male medical)	0.782	0.104	0.622	0.763	0.985
π_{41} (Female medical)	0.808	0.094	0.652	0.799	0.986

Estimates of π_{i1} show clearly that the sexually experienced are more likely to respond to the question on activity. The medical faculty response rate is much less differentiated between the active and inactive; in fact the log-odds comparing π_1/π_0 between medical and non-medical students averages -0.4 (i.e. medical students are less likely to respond differentially) though this log-odds has low precision.

Example 6.22 Telephone survey of voting intentions Here we consider the voting data from Little and Gelman (1998) and consider a reparameterisation of the differential nonresponse model, as in (6.14) to (6.16) above, for a binary outcome. Thus for strata $i = 1, \ldots, G$ we take the ratios

$$R_i = \pi_{i1}/(\pi_{i1} + \pi_{i0})$$

and use only the expectation that $\text{var}(R_i) < \text{var}(p_i)$ to specify priors. We transform R_i and p_i via logits, and set up the prior model

$$\begin{aligned} \text{logit}(R_i) &= s_{1i} \\ \text{logit}(p_i) &= \beta_0 + s_{2i} \end{aligned}$$

where β_0 is the average 'success' rate for the binary outcome. Note that this formulation might actually permit complex bivariate associations between R_i and p_i. Here we simply take

$$s_{1i} \sim \text{N}(0, V_1)$$

where V_1 is known, and then take

$$s_{2i} \sim \text{N}(0, V_2)$$

where $V_2 > V_1$. Equivalently $T_2 < T_1$ where $T_j = 1/V_j$ are precisions. Specifically two options $V_1 = 1$ and $V_1 = 0.05$ are taken, corresponding approximately to $\text{B}(2.7, 2.7)$ and $\text{B}(41, 41)$ priors on the R_i themselves. Then we take

$$T_2 = T_1/(1+\tau)$$

where $\tau \sim G(1, 1)$. Under this approach the actual smoothed posterior means of p_i are relatively robust to changing values of V_1, but as V_1 is increased the posterior p_i become correspondingly less precisely estimated. The unsmoothed rates p_i of Bush support range from 80% in Utah (49 out of 61 surveyed) to 27% in Rhode Island (18 from 67 surveyed). The smoothed rates with $V_1 = 0.05$ range from 0.667 (with standard error 0.05) to 0.44 (0.06), again for these two states. Under the option $V_1 = 1$ they vary from 0.68 (0.15) in Indiana to 0.41 (0.16) in Rhode Island.

6.6.3 Polytomous and multivariate nonresponse patterns

The hierarchical modelling approach can be extended to situations with several categorical outcomes, either dichotomous or polytomous. In these circumstances modelling nonresponse may be simplified if there are monotonic patterns of response to categorical items or questions. This would be exemplified in a health behaviour survey, when age or sex are reported, a more controversial outcome variable (regarding drug taking say) is subject to possible missingness, but when age and sex are missing, drug taking is always missing. Moreover, for respondents answering yes or no the question on drug taking, a question on frequency of drug taking is subject to technical nonresponse by those subjects who are not drug takers. In the latter case the monotonicity arises by definition, in the former it is empirically so. Modelling nonresponse to several outcomes jointly depends both on the number of questions and their levels, and on the possible actual response patterns, involving possible response to one or more items and whether monotonicity of response obtains.

With H original questions or items, the nonresponse model is expressible in terms of a single multinomial variable combining the original items. For example, if there were $H = 3$ original binary items, the full multinomial probability vector involves a maximum of eight cells (the maximum may not be apply to all subjects if there is technical nonresponse possible on the items considered). Whether a response is obtained is likely to depend on the combination of characteristics, abilities, behaviours or attitudes. So there will potentially be response probabilities π_{im} (and nonresponse probabilities $\rho_{im} = 1 - \pi_{im}$) for subjects i and each possible joint outcome ($m = 1, \ldots, 8$ in the case of three binary items). Individual differentiation in response probabilities may be applicable if we are modelling response in terms of other fully observed variables x_i.

Suppose such covariates were absent, then the response probabilities π_{im} are likely to be differentiated by (and hence modelled in terms of) categories of the joint multinomial outcome. For example, suppose answers on age, drug taking and frequency were: young/old, yes/no and every day/every week/less frequently. Then there may be different response probabilities for young daily drug takers as opposed to young weekly drug takers, or older non-drug takers.

Simpler and more identifiable models are obtainable by deliberately reducing parameterisation, e.g. assuming the response probabilities π_{im} in the above example are independent of drug taking frequency though not of age or whether the behaviour is present or not. The response options may be reduced by monotonic response (empirical or logical); in the above example four cells are excluded because the frequency question cannot be answered when the drug taking question is not answered.

Example 6.23 Physical and mental status As an example of empirically defined monotonicity, Fay (1986) considers data on survival of subjects cross-classified by initial evaluations (poor/good) of physical and mental status. Thus there are $H = 2$ binary items. With the exception of one patient, response on the latter two conditions is monotone; among the 71 deceased, there are no cases with mental status observed for whom physical status is unobserved, and only one of the 93 survivors has mental status observed and physical status unobserved. There are, however, 18 deceased with observed physical status but mental status missing, and 15 survivors with the same response pattern. Following Fay, attention is here confined to the 71 deceased cases, of whom 18 had the partial monotone response, 10 had complete nonresponse (both physical and mental state missing), while 43 had both states observed.

Let (P, M) denote the joint outcome of physical and mental state, so that there are $2^H = 4$ possible complete outcomes: $(1, 1), (1, 0), (0, 1)$ and $(0, 0)$. These correspond to the pairings (poor physical, poor mental), (poor physical, good mental), (good physical, poor mental), (good physical, good mental). In modelling terms there are four types of observed response pattern to consider:

full response on both questions;
good physical status, mental status missing;
poor physical status, mental status missing;
both physical and mental status missing.

For nonmonotone missingness there would be six types of observed response pattern to consider, the additional two being

physical status missing, mental status good;
physical status missing, mental status poor.

The first overall pattern applies for the $i = 1, \ldots, 43$ patients with a fully observed response $y[i, j, k]$, where $j = 0$ or 1 for physical state and $k = 0$ or 1 for mental state. Suppose the paired outcomes are represented as a multinomial with four categories, $\{\delta_{i1}, \delta_{i2}, \delta_{i3}, \delta_{i4}\}$. The fully observed responses can be represented as $(1, 0, 0, 0)$ for those with both poor mental and physical state; $(0, 1, 0, 0)$ for those with poor physical but good mental state; $(0, 0, 1, 0)$ for those with good physical and poor mental state; and $(0, 0, 0, 1)$ for those with both states good. If the multinomial parameter $\theta_{im}(m = 1, \ldots, 4)$ denotes the probabilities of these outcomes, then the BUGS model for these patients is

```
delta[i,1:4] ~ dmulti(theta[i,],1)
```

with a Dirichlet prior on θ.

The next two types of pattern are the partially observed responses. Let π_{im} denote the probability of response according to the four multinomial outcomes. For the 14 deceased patients with poor physical status, but mental state missing, the 14 observed data points can be represented (for input to BUGS) as

$$(\text{NA}, \text{NA}, 0, 0).$$

That is, we know the pairings (good physical, poor mental) and (good physical, good mental) are not applicable, but we are uncertain which of the first two multinomial outcomes occurred. The total probability of nonresponse for these patients is

$$(1 - \pi_{i1})\theta_{i1} + (1 - \pi_{i2})\ \theta_{i2} = \rho_{i1}\theta_{i1} + \rho_{i2}\theta_{i2}$$

and the probability of the outcome (poor physical, poor mental), conditional on nonresponse, is

$$w_{1i} = \rho_{i1}\theta_{i1}/\{\rho_{i1}\theta_{i1} + \rho_{i2}\theta_{i2}\}.$$

The converse outcome will be the pair (poor physical, good mental).

So this outcome can be modelled by a Bernoulli, using augmented binary data $\text{PART1}_i, i = 1, \ldots, 14$, with

$$\text{PART1}_i \sim \text{Bernoulli}(w_{1i}).$$

If the probabilities s_i are constant within the 14 patients then we may also use a binomial, and adopt augmented data for the total, say $TPART$, with outcome $(1, 0, 0, 0)$, such that

$$\text{TPART} \sim \text{Bin}(w1, 14).$$

For the four patients with good physical state but mental state missing, the observed data has the form (for BUGS input)

0 0 NA NA

and the choice between outcomes $(0, 0, 1, 0)$ and $(0, 0, 0, 0)$ involves the parameters $\theta_{i3}, \theta_{i4}, \pi_{i3}$ and π_{i4}. This is modelled using augmented data $\text{PART2}_i, i = 1, \ldots, 4$.

The data strongly suggest, as Fay (1986) points out, that nonresponse to mental status occurs disproportionately among those with poor physical state, i.e. that ρ_1 and ρ_2 exceed ρ_3 and ρ_4.

The fourth response pattern is for the ten patients with both states missing. Here the total probability of nonresponse is

$$\rho_{i1}\theta_{i1} + \rho_{i2}\theta_{i2} + \rho_{i3}\theta_{i3} + \rho_{i4}\theta_{i4}$$

and the multinomial outcome can be modelled as

$$\text{COMP}_{i,1:4} \sim \text{Mult}(r_{i,1:4}, 1)$$

where

$$r_{i1} = \rho_{i1}\theta_{i1}/\{\rho_{i1}\theta_{i1} + \rho_{i2}\theta_{i2} + \rho_{i3}\theta_{i3} + \rho_{i4}\theta_{i4}\}$$

and so on.

This model (see the WINBUGS program *Program 6.23 Physical and Mental Status*) is applied to the 71 deceased patients. Starting values of the augmented data vectors *PART1*, *PART2*, and *COMP* are included in the inits file. This gives posterior estimates (with standard deviations) of the nonresponse probabilities according to combined physical and mental states:

$\rho_1 = 0.49\ (0.13)$ (poor physical and poor mental status)
$\rho_2 = 0.48\ (0.19)$ (poor physical and good mental status)
$\rho_3 = 0.32\ (0.17)$ (good physical and poor mental status)
$\rho_4 = 0.22\ (0.12)$ (good physical and good mental status).

The estimated probabilities θ_m of the four multinomial outcomes are 0.34 (0.06), 0.19 (0.05), 0.15 (0.04) and 0.32 (0.05). The estimated breakdown of the full sample, combining full respondents and partial/complete nonrespondents is 24.4, 15, 10.3 and 21.3. The estimated breakdown of the 28 full or partial nonrespondents between the outcomes is 11.8, 7.8, 2.9 and 5.5 (the vector `tnr[]` in Program 6.23).

These results are close to those of the maximum likelihood models M1 and M3 of Fay (1986) which view the disproportionately poor physical status of nonrespondents to mental status as the direct effect of physical status.

A complete analysis of this problem would consider the interaction of physical and mental state for both deceased and survivors. Then we could answer questions such as whether poor physical and/or mental state are involved in poorer survival chances. It would involve a multinomial with eight cells, providing the single patient with mental state observed and physical state missing was excluded.

Example 6.24 Survey on voting intentions in Slovenian plebiscite Rubin *et al.* (1995) present results from a 1990 survey of 2074 Slovenians regarding their views on Slovenian independence, to be assessed via a full plebiscite later in 1990. These data are also analysed in Chapter 17 of Gelman *et al.* (1995). The potential voters were asked (a) whether they were in favour of independence from Yugoslavia; (b) whether they were in favour of succession; and (c) whether they would attend the plebiscite. (These are abbreviated to I, S and A below.) All three questions are relevant to predicting the likely proportion favouring independence in the full plebiscite; only attending Yes votes would be counted as favourable in the plebiscite, and all non-attenders would be counted as No votes. There is no pattern of monotonic nonresponse to simplify the analysis.

Following the first analysis of Rubin *et al.* (1995) we assume the DKs on the three questions are missing at random. Of the 27 cells, there are eight completely classified; i.e. with answers yes or no on all three questions. There are 18 types of partially classified cells: with at least one question answered yes or no but one or both of the remaining questions being DK. There is one cell (with 96 cases in it) who are DK on all three questions.

Suppose answers to the questions are arranged in the order ISA, and Y denotes Yes, and N denotes no. Then the observed cells are arranged to correspond to YYY, YYN, YNY, YNN, NYY, NYN, NNY and NNN. The counts in the fully observed cells are $(1191, 8, 158, 7, 8, 0, 68, 14)$. Their distribution among the eight cells is governed by a multinomial parameter vector $(\theta_1, \theta_2, \ldots, \theta_8)$. We need to allocate the respondents in the 18 partially classified cells to make inferences about the Yes to Independence vote in the full plebiscite. (The completely unclassified cell adds nothing to inference on this parameter.)

A different procedure is applied according to whether one question or two questions are DK. There are 12 cells with one DK. The first of these contains 107 people and is Yes to Independence and to Secession but DK to Attendance (Y, Y, DK). Therefore persons in this cell fall in one of the first two of the eight completely classified cells, either YYY or YYN. Since (by assumption) nonresponse is not related to outcome, the choice involves the ratio $\theta_1/(\theta_1 + \theta_2)$. A binomial model is appropriate, using augmented data V_1 persons in the YYY cell to be sampled from a population of $U_1 = 107$ cases with probability

$$\theta_1/(\theta_1 + \theta_2).$$

Successes are allocated to YYY and the $U_1 - V_1$ failures to YYN. If nonresponse were related to outcome then the binomial probability would be of the form

$$\theta_1\rho_1/(\theta_1\rho_1 + \theta_2\rho_2)$$

where ρ_1 and ρ_2 are the nonresponse rates for the outcomes YYY and YYN.

The last of the 12 cells with only one DK contains 3 people with the pattern (DK, N, N). These can be allocated either to cell 4 (i.e. YNN) or cell 8 (i.e. NNN). So the appropriate binomial model for augmented data V_{12} is a binomial with $U_{12} = 3$, and probability of success

$$\theta_4/(\theta_4 + \theta_8).$$

The first of the six cells with two DKs consists of 19 people with the pattern (Y, DK, DK). These can be allocated to any one of the first four of the eight completely classified cells (namely YYY, YYN, YNY, YNN). Accordingly a multinomial model is used to allocate the 19 people to augmented variables $V_{13,1}$, $V_{13,2}$, $V_{13,3}$ and $V_{13,4}$. These variables have probabilities

$$\theta_1/(\theta_1 + \theta_2 + \theta_3 + \theta_4), \quad \theta_2/(\theta_1 + \theta_2 + \theta_3 + \theta_4),$$

and so on. The last of these six cells consists of 25 people with the pattern (DK, DK, N). These can be allocated to any one of the four completely classified cells 2, 4, 6 or 8. The multinomial choice probabilities are defined correspondingly.

Analysis is carried out with *Program 6.24 Slovenia Plebiscite.* To be implemented in WINBUGS it is necessary that the Dirichlet sampling from the posterior density of the

$$(\theta_1, \theta_2, \ldots, \theta_8)$$

is performed using eight separate gamma densities. The posterior density of the parameter of interest

$$\theta_1 + \theta_3$$

has mean 0.882 and 95% credible interval (0.867, 0.897). The actual plebiscite vote had 88.5% of the population attending and favouring independence.

Example 6.25 Incompletely coded hospital referral data This analysis relates to approximately 80 000 hospital inpatient referrals among residents of a London health district over a 12-month period. The referral record includes patient age, sex, small area of residence and ethnicity. Coding on the first three is virtually complete and the incomplete cases on these fields are omitted. However, ethnicity is extensively uncoded, and only recorded on approximately 52% of the records. An estimate of the full numbers of referrals by ethnicity as well as by age, sex and small area is required. Ethnicity is regarded as the 'response' (with four categories, namely white, black African or Caribbean, South Asian, and other), and age, sex and small area as completely observed covariates. These are variables all associated to some degree with variation in hospital use; ethnic minorities in London tend to be disproportionately concentrated in younger age groups, have higher fertility (and hence hospital referrals for maternity) and to be spatially concentrated in certain small areas.

The data are accumulated into $k = 1, \ldots, 810$ cells classified by these three fully observed variables: 45 small areas, sex, and nine age groups. The observed data consist of totals

$$n_k = S_{k1} + S_{k2} + S_{k3} + S_{k4} + U_k$$

patients in each cell, with S_{k1} known white patients, S_{k2} known black patients, S_{k3} known Asian patients and S_{k4} others, with U_k patients having unknown ethnicity.

Non-informative beta priors are assumed for the probabilities π_{km} of making a response by cell k and ethnicity m ($k = 1, \ldots, 810; m = 1, \ldots, 4$). For the multinomial outcome probabilities p_{km}, an informative Dirichlet prior is adopted based on the 1991 UK Census, these being the most recent data available giving a full population profile on ethnicity. This results in prior Dirichlet parameters

$$\alpha_{km} = 0.1 \times n_k \times \mathrm{G}(\mathrm{Area}_k, m)$$

where $G(i,m)$ denotes percentages of Census population by ethnicity in small area i ($i = 1, \ldots, 45$). Because population composition and the composition of referrals to hospital may be discrepant for a number of reasons, the prior data is downweighted by 90%.

The total probability of nonresponse in cell k is given by

$$(1 - \pi_{k1})p_{k1} + (1 - \pi_{k2})p_{k2} + (1 - \pi_{k3})p_{k3} + (1 - \pi_{k4})p_{k4}.$$

To allocate the U_k 'nonrespondents' (in fact, missing codings) to ethnicity in cell k, it is necessary to define augmented data V_{km}. These augmented counts will be drawn from a multinomial with size U_k and probabilities

$$w_{k1} = (1 - \pi_{k1})p_{k1}/\{(1 - \pi_{k1})p_{k1} + (1 - \pi_{k2})p_{k2} + (1 - \pi_{k3})p_{k3} + (1 - \pi_{k4})p_{k4}\}$$
$$w_{k2} = (1 - \pi_{k2})p_{k2}/\{(1 - \pi_{k1})p_{k1} + (1 - \pi_{k2})p_{k2} + (1 - \pi_{k3})p_{k3} + (1 - \pi_{k4})p_{k4}\}$$
$$w_{k3} = (1 - \pi_{k3})p_{k3}/\{(1 - \pi_{k1})p_{k1} + (1 - \pi_{k2})p_{k2} + (1 - \pi_{k3})p_{k3} + (1 - \pi_{k4})p_{k4}\}$$
$$w_{k4} = (1 - \pi_{k4})p_{k4}/\{(1 - \pi_{k1})p_{k1} + (1 - \pi_{k2})p_{k2} + (1 - \pi_{k3})p_{k3} + (1 - \pi_{k4})p_{k4}\}.$$

The outcome numbers are E_{km}, predicted patient referrals by cell k and ethnicity m. They will then be updated by separate Dirichlet densities for each of the $K = 810$ cells with elements

$$P.E_{k1} = S_{k1} + V_{k1} + \alpha_{k1},$$
$$P.E_{k2} = S_{k2} + V_{k2} + \alpha_{k2},$$
$$P.E_{k3} = S_{k3} + V_{k3} + \alpha_{k3},$$
$$P.E_{k4} = S_{k4} + V_{k4} + \alpha_{k4},$$

where S_{km} are the referrals in cell k where ethnicity is coded, and α_{km} are the priors based on Census data.

To implement the model in WINBUGS, it is necessary to use separate gamma densities $G(P.E_{k1}, 1), G(P.E_{k2}, 1), G(P.E_{k3}, 1)$ and $G(P.E_{k4}, 1)$ to sample from the relevant Dirichlet. Starting values of the augmented data V_{km} must be supplied. Among the results obtained (see *Program 6.25 Hospitalisations by Ethnicity*) are that the percentage white in the full set of referrals is estimated at 94.45%, with 95% credible interval from 94.2 to 94.7%. Total probabilities of response (complete coding) by ethnicity are formed by aggregating over the 810 cells; they show whites as having an estimated 53.2% coding rate, compared with 49.7% among black patients, 50.6% among South Asian patients, and 48% among patients with other ethnicity.

6.6.4 Mixtures of continuous and categorical data

Suppose we have data consisting of a mixture of P continuous and M categorical variables, combined in vectors X_i and Y_i respectively for cases $i = 1, \ldots, n$, and with some or all variables containing missing values for some subjects. This type of data structure occurs frequently in certain methodological contexts (e.g. analysis of variance and discriminant analysis), and sample survey data often contains a mixture of the two types of data. Then a general location model for the joint distribution $\{X_i, Y_i\}$ has been proposed for such data; it consists of a marginal distribution of the categorical variables Y_i, and the conditional distribution of the continuous variables X_i given Y_i. Specifically, suppose the categorical variables have levels $J_1, \ldots, J_M$ respectively and we form the multinomial variable Z with $C = \prod_m J_m$ cells. Thus for $M = 2$ binary variables Y_1 and Y_2, Z would have cells $\{1,1\}, \{1,2\}, \{2,1\}$ and $\{2,2\}$ formed by crossing Y_1 and Y_2.

Given the classification of case i in one of the C cells of Z, the density of X_i is multivariate normal or Student t, with mean determined by the cell of Z_i. Thus

$$\Pr(Z_i = j) = \pi_j, \quad j = 1, \ldots, C \text{ with} \sum_j \pi_j = 1$$

and either

$$X_i | Z_i \sim \mathrm{N}_P(\mu_j, \Phi)$$

or possibly

$$X_i | Z_i \sim \mathrm{T}_P(\mu_j, \Phi, \delta)$$

where μ_j is a vector of dimension P, and δ in the multivariate T is a degrees of freedom parameter. The dispersion matrix Φ is usually assumed to be constant across cells of Z (a similar assumption to that of classical discriminant analysis), though this may be subject to modification.

This model was applied to missing data problems by Little and Schluchter (1985) and its use in this context is considered further by Little and Rubin (1987, Chapter 10) and Schafer (1997). As noted by Little and Rubin (1987) this model is essentially equivalent to logistic or multiple logit regressions of each Y on the remaining variables and of multiple linear regressions of each X_i on the remaining variables. Specifically a regression of the each of $X_1, \ldots, X_P$ on $Y_1, \ldots, Y_M$ allowing for main effects and interactions between all the Y is equivalent to a multivariate analysis of variance (Schafer, 1997).

Given the wide range of possible regression models for typically extensive sets of variables, and the additional complications if there is missing data (e.g. whether to assume missingness at random or otherwise), the inferences from modelling and imputation may be strongly dependent on prior assumptions. In particular, a simplification of the dependence of the means of the X_i on the Y is likely to be better identified than the full main effects and interactions model. For example, we may just allow for main effects of $Y_1, \ldots, Y_M$ on the means of $X_1, \ldots, X_P$ (Schafer, 1997). This is likely to be appropriate if the number of cases n is not large in relation to the possible combinations C of the categorical variables.

Example 6.26 St Louis study of psychological symptoms in children Both Little and Schluchter (1985) and Little and Rubin (LR, 1987) consider data on psychological disorders in children in $i = 1, \ldots, 69$ families characterised in terms of metrical child ability scores and family risk $\{g_i\}$ of disorder ($G = 3$ categories, low, medium and high according to household circumstances and parental psychological histories). For each of two children in each family data was available on two continuous variables, reading and verbal comprehension scores. Here we take $X_i = \{X_{1i}, X_{2i}, X_{3i}, X_{4i}\}$, where X_1 =reading score of child 1, $X_2 =$ comprehension score of child 1, X_3 =reading score of child 2, X_4 =comprehension score of child 2. Also observed for each child was a binary (child) psychological symptom indicator, D_1 for child 1 and D_2 for child 2, with $D_{1i} = 1$ and $D_{2i} = 1$ if a symptom of disorder was present in the 1st and 2nd child in family i.

From a substantive point of view the interest is likely to be in ability scores given psychological symptoms, or the impact of parental history on child symptoms. The data are subject to extensive missingness (with only risk group g_i being recorded for all 69 children), and can be modelled in several ways, for example including intra-family correlations, and allowing for non-ignorable missingness. Thus the chance that D_1 and/or D_2 are missing may differ according to whether one or both outcomes shows symptoms of disorder.

Here we consider (for illustrative purposes) one possible model choice, namely one allowing for non-ignorable missingness of D_1 and D_2, and then modelling X_i as multivariate normal given g_i, D_{1i} and D_{2i}. We form the multinomial variable Z with four categories based on crossing D_1 and D_2, such that

$$
\begin{aligned}
Z_{i1} &= 1 \text{ if both } D_{1i} = D_{2i} = 1, Z_{ij} = 0 \text{ for } j \neq 1 \text{ giving a vector } Z = (1,0,0,0) \\
Z_{i2} &= 1 \text{ if } D_{1i} = 1,\ D_{2i} = 0, \quad Z_{ij} = 0 \text{ for } j \neq 2 \text{ giving a vector } (0,1,0,0) \\
Z_{i3} &= 1 \text{ if } D_{1i} = 0, D_{2i} = 1, \quad Z_{ij} = 0 \text{ for } j \neq 3 \text{ giving a vector } (0,0,1,0) \\
Z_{i4} &= 1 \text{ if } D_{1i} = D_{2i} = 0, \quad Z_{ij} = 0 \text{ for } j \neq 4 \text{ giving a vector } (0,0,0,1).
\end{aligned}
$$

The means of $X_k, k = 1, \ldots, 4$ are then specific for combinations of risk group g_i and Z_{ij}. Following the example of physical and mental status considered above, there are 29 children with both D_1 and D_2 observed. For these children the BUGS model is based on observed frequencies in the four categories of Z, with multinomial sampling:

```
Z[i,1:4] ~ dmulti(π[i,1:4],1).
```

The next four types of pattern are partially observed responses on D_1 and D_2. Let r_m denote the probability of response according to the four multinomial outcomes of Z, with $\rho_m = 1 - r_m$ the probability of nonresponse. For the five children with $D_{1i} = 1$ but D_{2i} not known then the child may belong to cells 1 or 2 of Z and the data points can be represented (for input to BUGS) as

NA NA 0 0

That is, we know cells 3 and 4 are not possible because for them $D_{1i} = 0$, but we are uncertain which of the first two multinomial outcomes occurred. The total probability of nonresponse for these children is

$$(1 - r_1)\pi_1 + (1 - r_2)\pi_2 = \rho_1\pi_1 + \rho_2\pi_2$$

and the probability of the outcome $(D_{1i} = 1, D_{2i} = 1)$, conditional on nonresponse, is

$$w_1 = \rho_1\pi_1/(\rho_1\pi_1 + \rho_2\pi_2).$$

The converse outcome will be $(D_{1i} = 1, D_{2i} = 0)$. So the outcome $(D_{1i} = 1, D_{2i} = 1)$ can be modelled by a Bernoulli, with $P_{1i} \sim \text{Bern}(w_1)$ among these five children. These principles carry through to other partial responses.

Complete nonresponse on symptoms results in inputs on Z

NA NA NA NA

with the choices modelled via a multinomial (see the physical and mental status study in Example 6.23 above).

The cell means of the four continuous ability variables X_k are then specific to a 12-fold categorisation based on crossing the symptom category l of Z (= 1 to 4) and family risk category m (= 1 to 3). We adopt a simplified parameterisation in which

$$\mu_{klm} = \nu_k + \lambda_{lk} + \kappa_{mk}$$

so that interactions between symptom and risk category are excluded in modelling the continuous ability scores. There is a single 4×4 dispersion matrix applying to all 12 cells formed by combining symptom and risk group. Noninformative priors are taken on all parameters.

The resulting posterior means, based on the second half of a run of 2500 iterations, show the four ability variables are generally highest in the low risk group, a pattern also detected by Little and Rubin (1987). Also similar to the findings of LR are the standard deviations and correlations of the multivariate density of the continuous variables, with the highest correlations being between X_1 and X_2 and between X_3 and X_4. Little and Rubin found the highest correlation to be between the two comprehension scores X_2 and X_4.

The probabilities of nonresponse $\rho_j (j = 1, \ldots, 4)$ according to the four cells of Z show the highest nonresponse to occur for the intermediate outcomes $D_{1i} = 1, D_{2i} = 0$ and $D_{1i} = 0, D_{2i} = 1$. The totals in the four cells of Z (aggregating over risk groups g_i) are estimated as $(20, 13, 22, 14)$, compared with the LR model B estimates of $(25, 10, 23, 12)$. These are the vector `tot.Z[]` in Program 6.26.

6.7 NONPARAMETRIC MIXTURE MODELLING VIA DIRICHLET PROCESS PRIORS

In applications of parametric hierarchical models, including parametric mixture models, there are questions of sensitivity of inferences to the assumed forms (e.g. normal, gamma) for the priors. The distributions of parameters, including higher stage hyperparameters, are often uncertain, and not acknowledging this uncertainty may unwarrantedly raise the precision attached to posterior inferences.

Dirichlet process priors (DPP) have been proposed as a way of avoiding the limitations involved in assuming a standard parametric form either for the sample data, or for random effects and fixed effects within a model that assumes sampling from a parametric density. An instance in classical statistics would be the Kaplan–Meier estimate of the survival curve, and in fact Susarla and Ryzin (1976) show how the Kaplan–Meier estimate can be viewed as a form of DPP.

Suppose we wish to make inferences about an underlying location parameter from a sample $\{x_i\}$ which is subject to asymmetry or multimodality. For example, suppose the data were $x = \{1, 1, 2, 4, 4, 4, 5, 9, 9, 10, 17, 20\}$, then how might we assess the probability that a new value of x would lie in the interval 11–15, or that the mean exceeds 7? A parametric mixture might be fitted, but this would be difficult for the small sample and subject to uncertainty about the appropriate number of subgroups. A DPP process offers an alternative approach which deals with the clumping in the data without tying down the analysis to a particular number of clusters.

Let $x_i, i = 1, \ldots, n$ be drawn from a distribution with unknown parameters θ_i, φ_i with

$$f(x_i | \theta_i, \varphi_i).$$

Following Escobar and West (1998), suppose the form of prior for certain parameters θ_i is subject to greater uncertainty than other parameters φ_i. We therefore adopt a Dirichlet process prior for the θ_i but a conventional parametric prior for φ_i. In certain examples θ_i might be constant over i and consist of a single parameter (e.g. an overall mean), while in other examples there may be no parameters φ_i with conventional priors.

The idea of a Dirichlet process is to set up a baseline prior G_0 from which candidate values for θ_i are drawn. Suppose that θ_i denote unknown means for each case $i = 1, \ldots, n$. Then we may anticipate a degree of clustering in these values, so that data for similar groups of cases suggests the same value of θ_i would be appropriate for them. In certain cases such as the eye-tracking anomaly data considered earlier we would anticipate the maximum number of clusters to be less than the number of distinct observations; in that example, there were only 19 distinct values of the count of anomalies, even though there were 104 observations.

In other cases heterogeneity in the data might be such that every single case might potentially be a cluster. Thus if every x_i were distinct in value, or even though some x_i were matched they had different covariates, then the maximum number of clusters could be n. In general therefore we draw from G_0 $m = 1, \ldots, M$ values of θ^*, denoted θ^*_m, corresponding to the anticipated maximum possible number of clusters. This maximum may be n or it may be less if there are repeat observations and no regression is involved. In practice, we expect that only $M^* \leq M \leq n$ distinct values of the M sampled will be allocated to one or more of the n cases.

We then select the most appropriate value θ_m^* for case i using a Dirichlet vector of length M and with a precision parameter α; thus the group (cluster) indicator for case i is chosen according to

$$Z_i \sim \text{Categorical}(P)$$

where P is of length M, and has uniform elements determined by a precision parameter α. Some of the M clusters may not be appropriate for any of the observations (i.e. there may be empty clusters) and therefore we may monitor the actual number of clusters selected. The actual nonparametric prior F may be denoted in full as

$$F \sim \text{D}(\alpha G_0).$$

There are several ways to implement a DPP prior. Suppose, for example, we are considering a set of locations θ_i for cases i with baseline density G_0 taken as $\text{N}(\mu, \sigma^2)$. Then following Sethuraman (1994) one way to generate the DPP prior is to regard the θ_i as iid (independently and identically distributed) with density function $q()$ such that

$$q(\theta_i) = \sum_{j=1}^{\infty} p_j \text{N}(\theta_i \,|\, \mu_j, \sigma_j^2)$$

where in practice the mixture is truncated at M components with $\sum_{j=1}^{M} p_j = 1$. The cluster parameters θ_i result from independent draws from baseline priors on $\tau = \sigma^{-2}$ and $\mu\,|\,\tau$. For example, the potential precisions τ_j would be drawn from a common density, such as a gamma $\text{G}(v/2, vs^2/2)$ where s^2 is a prior guess at the average variance, and then the means are drawn from a normal density $\text{N}(m, b\tau^{-1})$ where m and b are assumed to be known constants.

The mixture weights p_j are constructed by 'stick-breaking'; let $r_1, r_2, \ldots$ be a sequence of $\text{Beta}(1, \alpha)$ random variables. Then draw M beta variables

$$r_j \sim B(1, \alpha), \quad j = 1, \ldots, M$$

and set

$$\begin{aligned} p_1 &= r_1 \\ p_2 &= r_2(1 - r_1) \\ p_3 &= r_3(1 - r_2)(1 - r_1) \end{aligned}$$

and so on. This is also known as the 'constructive definition' of the Dirichlet process (Walker *et al.*, 1999).

Note that the DPP prior procedure has some apparent resemblance to standard parametric mixture modelling. The differences lie in the fact that empty clusters present no problem, that the number of clusters is random, and that the average number of clusters emerging from a particular data set depends crucially on the value or prior assumed or α. The goal is rather to use the clusters to achieve a nonparametric smoothing of the data or effects. Inferences about the underlying population may then be based on sampling new values which may be drawn from different clusters than the observed data (Turner and West, 1993). The parameter α then determines the chances that a new observation will be drawn from existing or new clusters.

As an example, suppose we had a model for a Poisson outcome, $x_i \sim \text{Poi}(\mu_i), i = 1, \ldots, n$, and anticipate heterogeneity in the Poisson means such that

$$\log(\mu_i) = \beta + \varepsilon_i$$

with $\varepsilon_i \sim \text{N}(0, 1)$. To insert a DPP stage, we regard the N(0, 1) as the baseline prior G_0 and sample $M \leq n$ candidate values from it. We then allocate the cases to one of these candidate values according to the Dirichlet process prior. This process is repeated at each iteration in an MCMC chain. So if case i is allocated to cluster j (i.e. if $Z_i = j$) with candidate value ε_j^*, then $\varepsilon_i = \varepsilon_j^*$ and $x_i \sim \text{Poi}(\mu_j^*)$, where $\log(\mu_j^*) = \beta + \varepsilon_j^*$. The average error attaching to case i will be taken over the candidate values assigned at each iteration in the chain.

For large values of α the allocation will be such that most candidate values will be selected and the actual density of ε will be close to the baseline. Selecting a large α leads to more clusters and may result in 'overfitting' or densities that seem implausibly undersmoothed in terms of prior beliefs about the appropriate number of subgroups (Hirano, 1998). For small α, the allocation is likely to be concentrated on a small number of the candidate values. In this case the DPP model comes to resemble a finite (parametric) mixture model.

Appropriate priors, for example with gamma form, may be set on the precision parameter α. These in turn induce a prior on the actual number of clusters present at any iteration (Antoniak, 1974; Escobar and West, 1998). It is often sufficient, however, to select a few trial values of α and assess the impact on the average number of actual clusters (Ibrahim and Kleinman, 1998). Some possible problems with the identifiability of this parameter are considered by Leonard (1996), especially in data without any ties in the outcome variable.

Example 6.27 Eye-tracking data We continue the analysis of the eye-tracking data above, and adopt a gamma-Poisson mixture to model the heterogeneity. The standard approach to this type of data assumes Poisson sampling, with $x_i \sim \text{Poi}(\mu_i)$ and gamma priors on the Poisson means, $\mu_i \sim \text{G}(a, b)$ where a and b are preset or themselves assigned priors. Following Escobar and West (1998) we initially choose a baseline gamma prior for the μ_i with a and b having preset values, $a = b = 1$ (see *Program 6.27 Eye Tracking*). The insertion of a DPP stage means we select $M \leq n$ candidate values μ_j^* from $\text{G}(1, 1)$ and then allocate the actual 104 cases to one of these values. Because there are only 19 distinct cases we may take $M = 19$ as the maximum possible number of clusters. Here we take $M = 10$.

This example shows the ability of a nonparametric analysis to detect discrepancies between prior and data. For example, if we also set $\alpha = 1$ then a bimodal posterior distribution is obtained for larger values of O_i because the $\text{G}(1, 1)$ prior is too inflexible to accommodate them. Thus case 92 with $x_i = 12$

has posterior mean of 13.5 (SD = 2.7) but the posterior density shows the conflict between prior and data (Figure 6.1). For this value of the precision parameter α, the average number of clusters chosen is 7.

Other options are possible. We could let the gamma prior for the μ_i no longer have preset values. In this case, still with the Dirichlet precision parameter set at $\alpha = 1$, the DPP analysis yields $a = 0.7, b = 0.08$. The posterior for μ_{92} is no longer bimodal but still has some skewness. The median number of clusters is 7.

Example 6.28 Kaplan–Meier survival data This example discusses survival models, as considered more extensively in Chapter 9. Nonparametric assumptions have figured large in both the modelling of hazards themselves and underlying frailties. Walker *et al.* (1999) reanalyse the survival data from Kaplan and Meier (1958). This consists of four complete failure times $T_i (i = 1, \ldots, 4)$ at 0.8, 3.1, 5.4 and 9.2 months, and censored cases at 1, 2.7, 7 and 12.1 months $(i = 5, \ldots, 8)$. They use a generalisation of the Dirichlet process to model the parameter $F \doteq 1 - \exp(-\lambda)$, the probability of failure before 1 month. The exponential model for the failure times, with rate λ, is

$$T_i \sim \exp(\lambda)\mathrm{I}(W_i,)$$

where W_i are the censored failure times (for $i = 5, \ldots, 8$), and $W_i = 0$ for $i = 1, \ldots, 4$. We adopt a gamma baseline prior for λ, namely $\mathrm{G}(1, 1)$. For the precision parameter α, we initially assume a fixed value 0.5 and then adopt the constructive prior with $C = 10$. From a run of 5000 iterations (and 500 burn-in) we obtain estimates $\lambda = 0.132$ (s.d. = 0.054) and $F = 0.122$ (s.d. = 0.047). Adopting a higher value of α ($\alpha = 2.5$) gives a correspondingly higher values

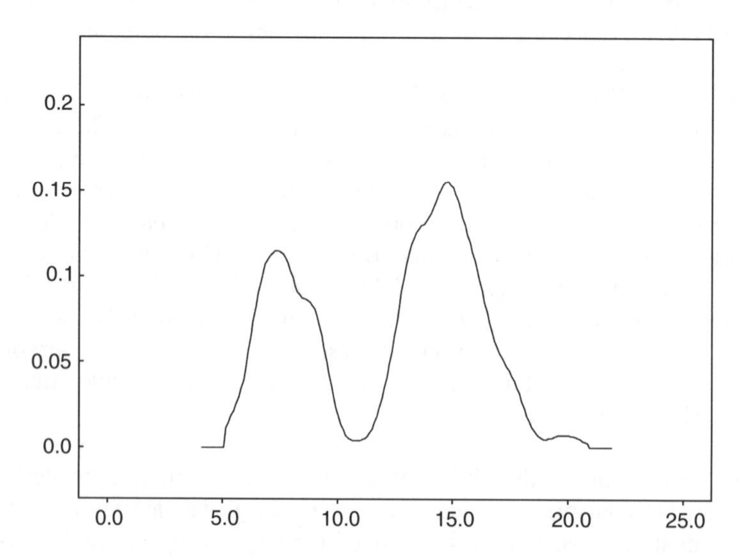

Figure 6.1 Posterior density of case 92 under fixed parameter baseline gamma prior.

of λ, namely $\lambda = 0.18$ (s.d. $= 0.062$), but thereafter values of a stabilise: taking $\alpha = 10$ gives $\lambda = 0.192$ (s.d. $= 0.059$).

The simple parametric analysis, sampling directly from G(1, 1) (set `A = A.PAR` in Program 6.28) gives $\lambda = 0.118$ and $F = 0.110$. Thus a larger probability of failure before 1 month results from a non-parametric model, as Walker *et al.* (1999) also found.

Example 6.29 SIDS deaths in N Carolina counties These data has been analysed by Cressie (1993) and relate to deaths x_i from Sudden Infant Death Syndrome (SIDS) over a five-year period, 1974–78, in 100 counties in North Carolina. We assume the SIDS deaths are Poisson with $x_i \sim \text{Poi}(\mu_i), \mu_i = E_i\gamma_i$, with E_i being expected deaths, and $100\gamma i$ being standard mortality ratios for SIDS. We adopt the stick-breaking algorithm and take trial values of α as a sensitivity analysis. It is possible to assign a prior for α but there may be numerical problems if small values of α are drawn; the cumulation of products $(1 - r_{j-1})(1 - r_{j-2}) \ldots (1 - r_1)$ may generate small numbers. This may occur if the original data are not markedly heterogeneous.

In the case of the SIDS data, sensitivity in the posterior estimates of γ_i is not pronounced under different values of α or variations in M, the number of potential clusters. We consider $\alpha = 5$, with $M = 5$, and $\alpha = 1$ with $M = 3$. We obtain similar smoothing under these and under a 'standard' pooling model, when a normal error term is included in the specification for $\log(\gamma_i)$:

$$\log(\gamma_i) = u_i$$
$$u_i \sim \text{N}(0, \Phi).$$

For $\alpha = 5, M = 5$, there are five counties with standard mortality ratios for SIDS clearly above or below 100. In most counties the 90 and 95 % credible intervals straddle 100: the 5 % point is below 100 and the 95 % point is above it. Table 6.31 compares the smoothed relative risks ($\times 100$) under the two DPP priors and an exchangeable normal prior. The lowest RRs are in two counties where SIDS counts are low in relation to a relatively large total of observed births. The more constrained DPP option with $\alpha = 1$ leads to slightly narrower interval around the mean.

Table 6.31 Smoothed SIDS relative risks, 95 % interval exceeding 100 or under 100.

	Observed		$\alpha = 5, M = 5$			$\alpha = 1, M = 3$			Exchangeable normal prior		
County	Deaths	Births	Mean	2.5%	97.5%	Mean	2.5%	97.5%	Mean	2.5%	97.5%
Anson	15	1570	296	145	506	283	165	440	275	152	458
Forsyth	10	11858	63	37	91	66	50	85	57	34	86
Halifax	18	3608	192	111	299	199	127	284	192	118	286
Robeson	31	7889	171	115	239	176	128	231	172	120	236
Wake	16	14484	69	44	96	69	52	87	65	42	93

6.7.1 Data smoothing

Turner and West (1993) consider modelling complex biomedical datasets $\{x_1, \ldots, x_n\}$ via Normal mixtures

$$x_i \sim \sum_{i=1}^{M} p_j \mathrm{N}(w_j, V_j)$$

where M is the maximum anticipated number of components (clusters) in the data. A Dirichlet mixture allows M to be infinite or to equal the sample size, but in practice the number of clusters is less than the sample size and depends on the extent of multimodality. So we envisage x drawn from a specific Normal density $\mathrm{N}(w, V)$ with mean w, in turn drawn from a (discrete) mixing distribution $\mathrm{G}(w)$. As usual in the DPP approach we supply a prior estimate $G_0(w)$ of $G(w)$ and a precision parameter α, such that large values of α correspond to the belief that the true density is close to G_0. We then assess how the $x_1, \ldots, x_n$ should be allocated to one of the potential clusters using a Dirichlet process; thus we have (for the case $V_j = V$)

$$\begin{aligned}
x_i &\sim \sum_{i=1}^{M} p_j \mathrm{N}(w[Z_i], V), && i = 1, \ldots, n \\
w_j &\sim \mathrm{N}(m, dV), && j = 1, \ldots, M \\
Z_i &\sim \text{categorical}(A[1{:}M])
\end{aligned}$$

where $A = (\alpha, \alpha, \ldots, \alpha)$ and the parameter d determines the spread of distinct cluster means w_j. If d is large the means will be dispersed, as may be suitable for asymmetric or skewed data.

Turner and West investigate sensitivity to the priors for V, α and d. For α they argue that a gamma prior $\alpha \sim \mathrm{G}(a_1, a_2)$ with a relatively large shape parameter a_1 is appropriate, but combined with a scale parameter a_2 such that the prior places weight on relatively low means for α; typical values might then be $a_1 = 10, a_2 = 100$. The actual number of clusters selected M^* will then be approximately $\alpha \ln(1 + n/\alpha)$, so that for $\alpha = 1$, and n between 300 and 1000, k is likely to be between 2 and 10. For V they suggest a prior based on the observed noise variance s^2, but with a prior degrees of freedom v which allows some flexibility, $1/V \sim \mathrm{G}(0.5v, 0.5vs^2)$.

Example 6.30 Galaxy velocities This approach is illustrated using the data on velocities (km/second) for 82 galaxies from Roeder (1990) and also analysed by Chib (1995). These are drawn from six well-separated conic sections of the Corona Borealis region. We adopt a $\mathrm{G}(50, 100)$ prior for α – an alternative $\mathrm{G}(50, 200)$ prior produced similar results in terms of the predictive density (for new cases) – and take the maximum number of clusters as $M = 10$. New cases from the overall density are sampled by selecting a new cluster and then sampling from the density for the appropriate cluster mean; we use this predictive density to assess whether the mean velocity exceeds 25 000 km/second. A possible coding (see *Program 6.30 Galaxy Mixture*) is

```
# new case
C.new ~ dcat(pi[1:M])
x.new ~ dnorm(t[C.new],Pr.X)
# Histogram ordinate of predictive densities at intervals 00.999,1
1.999,22.999,etc.
Ord.new <- round(x.new + 0.5)
# Test for mean velocity exceeding 25000 km/sec
Test <- step(x.new-25)
```

For the parameter d a gamma prior is used, namely $d \sim G(2.5, 0.1)$. The resulting smooth density shows the subpopulations (at approximate 9–10 000 and around 30 000 km/sec) as apparent in the original data (Table 6.32).

The probability that the mean exceeds 25 000 km/sec is estimated at around 8–9%, and the parameter d at around 37. The posterior for α has mean 0.6 prior and the average number of nonempty clusters selected is around 6.1 (with s.d. 1.4), with 95% of nonempty clusters being between 3 and 9.

6.7.2 Other nonparametric priors

Recently generalisations of the Dirichlet process prior have been proposed, such as stochastic process priors and partition priors (Walker *et al*., 1999). The latter are also denoted Polya tree priors (Walker and Mallick, 1997), and consist of a set of binary tree partitions to allocate a case to its appropriate cluster value selected from the baseline prior G_0 (subsequently denoted G for simplicity). Suppose we were modelling small area health outcomes x_i for $i = 1, \ldots, N$:

$$x_i \sim \mathrm{Poi}(E_i \mu_i)$$
$$\log(\mu_i) = \beta_0 + e_i$$

and adopt an $N(0, 1)$ density for e_i as the baseline G. The simplest Polya tree would have one level only and select candidate values e_i^* from two possibilities. The choice would be between candidate values selected from the partition of the real line, either from $B_0 = (-\infty, G^{-1}(0.5))$, or from $B_1 = (G^{-1}(0.5), \infty)$. Thus the partitions of the parameter space at level 1 is based on the 50th percentiles of the base prior G. The partition at the median (B_0, B_1) constitutes the breaks at level 1 of the tree.

The next binary partition would involve subdivisions of B_0 and B_1, so that $(B_{00}, B_{01}, B_{10}, B_{11})$ are the breaks at level 2. The choice would then be between candidate values selected from $B_{00} = \{-\infty, G^{-1}(0.25)\}, B_{01} = \{G^{-1}(0.25), G^{-1}(0.5)\}, B_{10} = \{G^{-1}(0.5), G^{-1}(0.75)\}$ or $B_{11} = \{G^{-1}(0.75), \infty\}$. The number of sets, namely ranges of bands from which candidate values (for parameter values or cluster means) are chosen, is thus 2^m at level m. Candidate values in the lowest and uppermost bands are selected from truncated densities, with a form defined by G_0. For intervening bands j, they may be selected from a uniform density with $G^{-1}[(j-1)/2^m]$ and $G^{-1}(j/2^m)$ as the endpoints.

Table 6.32 Actual and predictive distributions: galaxy velocities (in intervals of thousands, ordinate is upper limit of bands 1000 km wide).

Predictive density (sample 3000)			Actual density ($n = 82$)		
Ordinate	Frequency	% Frequency	Ordinate	Frequency	Rel frequency (%)
5	1	0.03	10	5	6.10
6	6	0.20	11	2	2.44
7	8	0.27	17	2	2.44
8	27	0.90	19	4	4.88
9	77	2.57	20	18	21.95
10	88	2.93	21	13	15.85
11	47	1.57	22	6	7.32
12	14	0.47	23	11	13.41
13	6	0.20	24	9	10.98
14	9	0.30	25	6	7.32
15	11	0.37	26	1	1.22
16	19	0.63	27	2	2.44
17	41	1.37	33	2	2.44
18	124	4.13	35	1	1.22
19	351	11.70			
20	453	15.10			
21	433	14.43			
22	314	10.47			
23	377	12.57			
24	254	8.47			
25	122	4.07			
26	62	2.07			
27	27	0.90			
28	13	0.43			
29	6	0.20			
30	5	0.17			
31	16	0.53			
32	30	1.00			
33	22	0.73			
34	22	0.73			
35	8	0.27			
36	3	0.10			
37	1	0.03			
43	2	0.07			
48	1	0.03			

Walker and Mallick (1997) liken the choice of an appropriate candidate value to a cascading particle. The choice between B_0 and B_1 is a Bernoulli choice governed by probabilities C_0 and $1 - C_0$. The probability C_0 may be selected from a prior beta density but Walker and Mallick (1997, p. 851) suggest $C_0 = 0.5$ on the basis that the first partition is centred at the median.

In general, if the option B_ε is selected then the particle may move to either $B_{\varepsilon 0}$ or $B_{\varepsilon 1}$ at the next step. The respective probabilities of these options may be denoted $C_{\varepsilon 0}$ and $C_{\varepsilon 1} = 1 - C_{\varepsilon 0}$. These probabilities are random beta variables with

$$(C_{\varepsilon 0}, C_{\varepsilon 1}) \sim \text{Beta}(\alpha_{\varepsilon 0}, \alpha_{\varepsilon 1})$$

where $\alpha_{\varepsilon 0}$ and $\alpha_{\varepsilon 1}$ are both non-negative. The choice of values for $\alpha_{\varepsilon 0}$ and $\alpha_{\varepsilon 1}$ should reflect prior beliefs about the underlying smoothness of F. For m large, we would then set $\alpha_{e0} = \alpha_{e1} = c_m$ in such a way that $F(B_{\varepsilon 0})$ and $F(B_{\varepsilon 1})$ are close. This may be done by setting

$$c_m = cm^d \quad \text{for } c > 0, d > 1,$$

so that c_m increases with m (in line with prior expectations that some degree of pooling should be appropriate, based on the smoothness). For example $c_m = cm^2$ or $c_m = cm^3$ may be used with $c = 0.5$ or $c = 0.1$. Larger values of c mean the posterior will resemble the baseline prior G more closely. The Dirichlet process prior corresponds to $c_m = 1/2^m$.

In the small area health example, suppose we take $C = 3$ as the maximum number of levels. With N parameters e_i for each area to choose values for, and taking $c_m = cm^3$, we might therefore use the coding

```
c <- 0.5
for (m in 2:C) { cm[m] <- c*pow(m,3)}
for (i in 1:N){    L[1,i] ~ dbern(0.5)
for (m in 2:C) {p[m,i] ~ dbeta(cm[m],cm[m])
                L[m,i] ~ dbern(p[m,i])}
# level 1 choice (convert L = 0, 1 to B = 1,2)
  B[1,i] <- L[1,i] + 1
# choices at level 2 and above
  for (m in 2:C) { B[m,i] <- sum(BC[m,i,1:m-1]) + L[m,i] + 1
  for (n in 1:m-1) {BC[m,i,n] <- L[m-n,i]*pow(2,n)}}
# select from baseline
  e[i] <- Z[B[C,i]];
  x[i] ~ dpois(mu[i]);
  log(mu[i]) <- log(E[i]) + e[i]}
```

The options for the baseline `Z[]` would then be based on the selected prior G, e.g. an N(0, 1), and the 12.5th, 25th,...87.5th percentiles of G^{-1}.

Example 6.31 Seeds and extracts Walker and Mallick (WM, 1997) reanalyse the factorial layout data from Crowder (1978). The original model of Crowder proposed variation of expected proportions within cell means

$$y_{ij} \sim \text{Bin}(\pi_{ij}, n_{ij}), \qquad i = 1, \ldots, 4; j = 1, \ldots, n_i$$

with π_{ij} then distributed according to four beta densities $\text{B}(a_i, b_i)$, with mean $\pi_i = a_i/(a_i + b_i)$. WM instead propose a Polya tree non-parametric prior for the overdispersion under a logit transform of the π_{ij}. Following their analysis, we take

$$\text{logit}(\pi_{ij}) = \beta x_i + \varepsilon_{ij}$$

where the base probability ε_{ij} for a Polya Tree prior with $C = 3$ levels is taken to be $N(0, 1)$. The beta weights $c_m = 0.1m^2$ were taken for $m = 2, 3$. The results, from the second half of a run of 50 000 iterations for the three factorial effect parameters and for the cell probabilities are similar to those of WM, with only β_3 clearly different from zero (Table 6.33). *Program 6.31 Seeds and Extracts Polya Tree* was used for this analysis.

Example 6.32 Smoothing SIDS deaths incidence rates via Polya tree This procedure is applied with 2^3 partitions (i.e. $C = 3$) to the SIDS data analysed above – see *Program 6.32 SIDS Deaths Polya Tree.* This results in somewhat less shrinkage to the overall mean than the constructive DPP prior used above.

Table 6.33

Parameter	Mean	SD	2.5%	Median	97.5%
$\beta1$	−0.25	0.36	−0.96	−0.25	0.45
$\beta2$	−0.50	0.42	−1.32	−0.50	0.32
$\beta3$	0.85	0.42	0.01	0.85	1.68
$\varepsilon11$	−0.74	0.47	−1.65	−0.74	0.16
$\varepsilon12$	−0.27	0.42	−1.08	−0.27	0.56
$\varepsilon13$	−0.65	0.41	−1.46	−0.65	0.16
$\varepsilon14$	0.27	0.43	−0.56	0.26	1.13
$\varepsilon15$	−0.01	0.45	−0.89	−0.02	0.87
$\varepsilon21$	0.55	0.72	−0.84	0.53	2.01
$\varepsilon22$	0.32	0.43	−0.53	0.32	1.16
$\varepsilon23$	0.55	0.44	−0.30	0.55	1.40
$\varepsilon24$	−0.06	0.44	−0.94	−0.06	0.80
$\varepsilon25$	−0.25	0.42	−1.08	−0.25	0.57
$\varepsilon26$	0.48	0.61	−0.70	0.47	1.69
$\varepsilon31$	0.60	0.55	−0.48	0.59	1.67
$\varepsilon32$	0.03	0.50	−0.95	0.03	1.01
$\varepsilon33$	−0.17	0.51	−1.18	−0.16	0.82
$\varepsilon34$	0.73	0.46	−0.17	0.73	1.63
$\varepsilon35$	−0.80	0.82	−2.45	−0.79	0.75
$\varepsilon41$	−0.89	0.61	−2.10	−0.88	0.29
$\varepsilon42$	0.05	0.47	−0.87	0.04	0.97
$\varepsilon43$	−0.09	0.50	−1.07	−0.09	0.88
$\varepsilon44$	0.40	0.46	−0.50	0.39	1.31
$\varepsilon45$	−0.25	0.67	−1.57	−0.25	1.06
$\pi1$	0.38	0.03	0.32	0.38	0.44
$\pi2$	0.69	0.03	0.62	0.69	0.75
$\pi3$	0.36	0.04	0.28	0.36	0.45
$\pi4$	0.49	0.05	0.40	0.49	0.58

There are 11 of the 100 counties (italicised below) with 95% intervals for standard incidence ratios either completely above 100 or below 100. The relative risks for counties in Table 6.34 are based on the second half of a single long run of 10 000 iterations.

Table 6.34

County	Mean RR	SD	2.5%	Median	97.5%
County 1	133.3	37.9	72.0	138.3	213.5
County 2	49.0	27.6	19.0	44.5	114.4
County 3	74.6	55.1	22.1	60.2	215.3
County 4	***350.1***	***56.5***	***212.0***	***339.4***	***490.1***
County 5	72.8	42.3	23.6	64.8	198.7
County 6	61.4	40.2	20.6	49.0	159.7
County 7	123.6	45.0	51.3	111.7	218.1
County 8	202.4	82.1	77.2	200.5	372.7
County 9	203.7	75.4	90.1	201.4	367.2
County 10	110.1	45.0	45.3	102.8	213.7
County 11	63.7	19.8	29.7	63.2	107.3
County 12	75.8	28.8	29.5	71.1	144.2
County 13	48.9	21.1	22.2	47.0	102.7
County 14	85.8	30.5	37.5	83.5	151.7
County 15	91.2	70.8	22.8	69.8	326.0
County 16	103.5	41.3	43.2	100.4	206.7
County 17	101.0	56.5	28.3	91.6	222.1
County 18	***50.6***	***19.1***	***24.1***	***48.3***	***95.4***
County 19	75.2	38.5	25.9	68.5	162.8
County 20	103.0	56.9	28.7	93.0	222.2
County 21	90.1	57.1	26.1	75.2	217.2
County 22	90.9	70.5	23.0	69.9	326.6
County 23	100.0	28.8	49.8	98.9	159.2
County 24	***208.6***	***56.6***	***112.3***	***204.2***	***345.9***
County 25	106.7	28.5	60.6	103.6	164.0
County 26	96.6	14.7	67.7	99.8	121.2
County 27	111.1	74.0	26.6	93.8	332.4
County 28	73.7	53.4	21.2	59.3	211.3
County 29	76.4	24.1	33.3	72.1	127.0
County 30	75.2	23.6	31.6	71.3	123.9
County 31	85.3	36.0	29.7	81.9	163.4
County 32	99.4	23.5	60.4	100.3	149.5
County 33	131.4	41.5	66.5	124.8	215.4
County 34	***45.9***	***13.6***	***25.2***	***46.8***	***74.2***
County 35	83.5	44.6	26.9	72.6	202.2
County 36	63.6	17.9	31.0	63.5	103.4
County 37	77.1	57.0	21.7	62.8	217.8
County 38	80.4	59.7	22.3	64.7	220.3
County 39	115.9	52.1	43.8	105.5	220.9

(*Continued*)

Table 6.34 *cont.*

County	Mean RR	SD	2.5%	Median	97.5%
County 40	193.3	90.1	61.0	198.0	378.2
County 41	71.9	15.1	46.5	69.9	103.7
County 42	***232.6***	***63.2***	***140.5***	***210.5***	***364.2***
County 43	84.1	30.3	30.9	81.5	149.7
County 44	63.5	31.2	23.9	58.4	143.1
County 45	96.9	39.2	39.8	92.0	203.1
County 46	216.5	84.1	86.6	204.3	379.3
County 47	211.7	82.1	86.4	202.7	378.5
County 48	85.2	64.9	22.9	67.9	261.0
County 49	57.1	23.4	24.8	50.1	109.0
County 50	95.4	52.8	27.7	86.1	217.8
County 51	78.1	27.8	30.8	73.7	145.1
County 52	104.9	69.6	26.9	89.7	325.3
County 53	107.4	44.0	44.2	101.3	211.4
County 54	133.8	40.7	67.9	138.3	217.4
County 55	167.2	57.7	78.4	155.1	326.6
County 56	121.2	51.5	46.7	108.2	222.2
County 57	61.4	39.6	20.0	49.6	160.1
County 58	123.2	72.2	29.6	104.8	329.4
County 59	77.8	40.7	25.9	69.7	188.9
County 60	***110.8***	***9.8***	***100.4***	***108.3***	***140.4***
County 61	66.3	45.7	20.8	50.4	200.5
County 62	115.9	57.7	31.6	104.2	232.8
County 63	95.3	37.7	40.1	91.3	200.7
County 64	98.9	31.8	47.6	97.2	166.4
County 65	105.8	29.1	57.9	103.4	162.8
County 66	***281.1***	***85.6***	***138.6***	***318.3***	***432.8***
County 67	126.1	25.0	83.8	121.6	173.2
County 68	70.0	28.8	27.6	67.0	142.6
County 69	106.6	70.9	26.1	89.8	325.4
County 70	95.8	45.6	29.7	89.5	208.3
County 71	147.0	70.1	48.8	140.0	337.4
County 72	113.3	74.9	26.9	95.3	333.1
County 73	122.3	56.4	44.9	108.5	232.6
County 74	133.3	35.9	73.0	138.4	212.1
County 75	108.8	71.6	26.9	93.0	328.4
County 76	81.6	27.8	36.5	79.2	145.1
County 77	78.7	32.9	28.5	72.4	152.7
County 78	***188.6***	***31.8***	***138.2***	***200.5***	***237.1***
County 79	***171.9***	***41.1***	***100.8***	***162.5***	***237.6***
County 80	***44.7***	***18.9***	***21.0***	***43.9***	***92.8***
County 81	187.5	53.5	98.3	199.3	331.2
County 82	73.4	30.8	27.7	69.3	144.8
County 83	163.9	57.3	73.7	151.8	324.7
County 84	104.7	41.9	43.4	100.9	206.8

Table 6.34 *cont.*

County	Mean RR	SD	2.5%	Median	97.5%
County 85	58.6	31.4	21.8	49.6	140.4
County 86	82.4	31.7	30.1	78.9	151.7
County 87	180.2	93.9	49.2	155.3	379.9
County 88	122.9	63.9	31.7	107.0	320.2
County 89	94.4	73.0	22.7	71.3	325.3
County 90	60.0	24.3	25.5	56.0	113.9
County 91	94.5	40.2	30.8	90.0	201.9
County 92	***57.5***	***13.9***	***30.6***	***55.5***	***87.7***
County 93	179.1	84.9	57.4	157.2	366.0
County 94	213.3	90.3	77.0	202.7	389.4
County 95	65.2	36.5	22.3	57.1	152.6
County 96	132.9	31.4	82.3	138.6	207.6
County 97	71.4	30.0	27.4	68.0	143.0
County 98	142.8	40.9	74.0	142.3	219.4
County 99	67.3	39.3	22.4	59.2	159.5
County 100	62.5	41.2	20.5	49.3	165.2

CHAPTER 7

Correlated Data Models

7.1 INTRODUCTION

Estimation of the parameters in the multiple regression model, $t = 1, \ldots, N$

$$y_t = x_t\beta + u_t$$

depends on the assumptions made about interdependencies between the errors u_t themselves and about the relationship between x and u. The standard assumptions are that u is uncorrelated with x, and that the disturbances are pair-wise uncorrelated, so that $V = E(uu') = \sigma^2 I$. These assumptions still pertain in the case of variance heterogeneity, where the diagonal of V consists of distinct variances $\sigma_1^2, \sigma_2^2, \ldots, \sigma_N^2$ but off-diagonal elements are zero. Among the conditions where these assumptions are likely to be violated are when we have (a) autocorrelations in y or u over time and space; (b) systems of equations where errors are correlated across equations; and (c) dependencies between disturbances and predictors, as in simultaneous equation and measurement error models.

7.1.1 Time series analysis

With data arising in sequence in time, successive observations are likely to be dependent. Time series analysis aims to explain the stochastic mechanism that gave rise to an observed series, to identify shifts in series due to interventions or other changes in conditions, and to predict future values of the series. All these aspects of modelling make use of the assumed dependencies between consecutive measurements.

Stochastic dependence in consecutive observations themselves is observed in a wide range of applications (Cox *et al.*, 1996). For example, Helfenstein (1991) cites time dependencies in environmental medicine; a high concentration of a pollutant on one day has a tendency to be followed by a high reading the following day, so demonstrating a positive autocorrelation in the series of observations. Time series of economic indicators such as prices and output levels also usually show autocorrelation over time. Another sort of dependency takes the form of regular seasonal or cyclical fluctuations, as in many climatic or biomedical series.

In other cases the autocorrelation is in terms of regression disturbances. For example, if we consider an economic outcome, wage settlements y_t in relation to time-varying predictors x_t such as profit levels, unemployment, and so on, sequenced by times $t = 1, \ldots, N$. If unusually large settlements occur at time t (as compared with those predicted by the x_t) this may generate excess pressure in subsequent periods $t+1, t+2, \ldots$, so that a run of positive disturbances occurs. This might be modelled by first order serial correlation in the disturbances:

$$y_t = x_t\beta + u_t$$
$$u_t = \rho u_{t-1} + e_t$$

where e_t are uncorrelated white noise. The disturbances u_t and u_{t-s} $(s = 1, 2, \ldots)$ are now correlated such that $E(u_t u_{t-s}) = \rho^s\sigma^2$ and $E(uu')$ is given by

$$\begin{vmatrix} 1 & \rho & \rho^2 & \rho^{N-1} \\ \rho & 1 & \rho & \rho^{N-2} \\ \ldots & \ldots & & \ldots \\ \rho^{N-1} & \rho^{N-2} & & 1 \end{vmatrix}.$$

Such continuity in time is the basis for extrapolation into the future, for example via autoregression on previous values of the series y_t itself, or on forecast (or known) values of x_t in future periods $N+1, N+2, \ldots$ Another goal of time series analysis is the detection of changes in structure in the observed series – either as a result of an 'intervention' such as a change in economic policy, pollution incident, or the administration of a medical treatment. For example, Gordon and Smith (1990) cite examples involving biochemical series on individual patients, such as white blood cell counts in dialysis patients, where rapid on-line detection of shifts in the structure of time series is important.

One major class of models for time series data are the autoregressive moving average models of Box and Jenkins (1970). The implementation of these rests on the assumption of second-order stationarity, after removing trends, and cyclical or seasonal regularities. Thus the mean, variance and autocovariances of the data series actually modelled are assumed fixed in time. Because the data often require differencing they need to be 'integrated' back into the original undifferenced form: hence the notation ARIMA for Autoregressive Integrated Moving Average. The notation ARMA refers to the general set of techniques involving autoregressive or moving average assumptions (or priors in Bayesian terms). These models provide a flexible approach to a range of time series problems, and have been heavily used in forecasting. However, they are prone to the risk of overfitting – for example, extending an AR(2) model to an AR(3) may lead to an apparently significant extra coefficient, but without a notable improvement in overall fit and at the loss of precision of the preceding AR coefficients. An overparameterised model may also not yield sensible forecasts (Harvey and Todd, 1983). Many observed series exhibit clear upward or downward trends (i.e. are not stationary), and application of ARIMA methods then requires transformation of the original data series and/or differencing, though maybe only first or second differencing.

An alternative approach is to focus on the observed components of many series, such as trends, seasonal cycles, or changing impacts of covariates and model their stochastic evolution directly, without differencing to gain stationarity. Thus a typical time series may consist of up to four components:

$$\text{observed series} = \text{trend} + \text{seasonal effects} + \text{regression term} + \text{irregular effects}.$$

One option for modelling these effects is by a set of fixed coefficients, e.g. a polynomial in time to describe the trend in the level of the series, and seasonal dummies to represent seasonal factors. Such a model places equal weight on all observations when predicting the future.

A more flexible approach is to retain the regression framework, but allow time-varying coefficients such that forecasts place more weight on recent observations. Harvey (1990) defines such structural time series models as regression models in which the explanatory variables are functions of, or vary over, time and the parameters also are time varying. The similar and closely related fully Bayesian methodology for state space time series modelling has been denoted dynamic linear modelling. Recently, West and Harrison (1997) have explored the potential for mixtures of ARMA and DLM components to model the same data, for example an AR process for the stationary component of y_t and another model for the nonstationary trends.

Whatever approach is adopted to time series methods, the usual wider modelling issues are relevant. These include possible outliers and robust alternatives to the Normal (in the case of continuous y_t) to analyse them. For example, Hoek *et al.* (1995) discuss the problem of analysing time series with a deterministic or stochastic trend. Such a trend corresponds to forms of nonstationarity (e.g. a unit root) in the autoregressive model of the series unless there is predifferencing. They show the use of a robust Student t sampling model with small degrees of freedom reduces the probability of such nonstationarity in the lag coefficient(s) of AR models.

7.1.2 Spatial modelling

The role and impact of spatial dependence have similarities to those of temporal dependence, and there is an increasing convergence in the methods used to model them (Diggle, 1997). For example, just as temporal residual dependence may be related to omitted predictors so may spatial dependence. Thus poor health outcomes and early mortality are generally higher in socially deprived areas, and if we construct a model for such outcomes that omits area deprivation, then the residuals from this model will tend to be positively correlated (clustered) in space, since social deprivation tends to be spatially clustered. Similarly in an analysis of crop yields following different treatments, there may be underlying variations in soil fertility which are spatially correlated and, unless allowed for, will distort the assessment of treatment effects

(Besag and Higdon, 1999). Spatial dependence in regression in a set of N discrete areas

$$y_i = x_i\beta + u_i$$

means that the errors are dependent across the areas, so that u has a multinormal distribution with $N \times N$ dispersion matrix Σ. This is a nonspherical error matrix, in contrast to the standard form $\sigma^2 I_N$. Among a range of models for this form of correlation is the simultaneous scheme, discussed further below, where a set of weights $W = [w_{ij}]$ relates areas i and j:

$$u = \rho Wu + e \tag{7.1}$$

and where $e \sim \mathrm{N}(0, \sigma^2)$ denotes uncorrelated white noise. These weights are often defined in terms of spatial contiguity. Suppose we had five areas, with area 1 having neighbours (a common boundary with) areas 2 and 3. Suppose the neighbours of four other areas were:

area 2, neighbours 1, 3, 4
area 3, neighbours 1, 2
area 4, neighbours 2, 5
area 5, neighbour 4.

If we set $C_{ij} = 1$ if areas i and j are adjacent, then the relevant weights w_{ij} are defined by standardising the C_{ij} such that $w_{ij} = C_{ij}/\sum_j C_{ij}$. For the above example,

$$w = \begin{bmatrix} 0 & 1/2 & 1/2 & 0 & 0 \\ 1/3 & 0 & 1/3 & 1/3 & 0 \\ 1/2 & 1/2 & 0 & 0 & 0 \\ 0 & 1/2 & 0 & 0 & 1/2 \\ 0 & 0 & 0 & 1 & 0 \end{bmatrix}.$$

If further we set $D = (I - \rho W)$, then the covariances between the disturbances are, from (7.1), seen to be

$$E(uu') = \sigma^2 DD'.$$

The modelling of spatial dependencies has included recent advances from a Bayesian perspective (Mollié, 1996) and an expanding range of applications (e.g. disease mapping and image analysis). In a similar way to temporal correlation, an observed pattern of spatially dependent residuals in a regression for an outcome y_i across a set of areas may have several possible sources. It may represent a genuine pattern of spatial dependence, but a set of specification issues also occur. Measurement errors, nonconstant variance (heteroscedasticity) as well as omitted covariates may produce apparent spatial effects (Anselin and Griffith, 1988). Cressie and Read (1989) emphasise the potential for spatial prewhitening and transformation to eliminate spatial dependence, and achieve spatially stationary errors, in a similar manner to that carried out in ARIMA time series analysis.

A more fundamental criticism of the spatial correlation models for sets of discrete areas concerns the arbitrariness of the weights matrix for distance decay. These may lead to overlaps in spatial modelling between what is known in terms of measurable systematic factors and what is unknown. Arora and Brown (1977) argue that common representations of these weights, such as the inverse decay model $w_{ij} = d_{ij}^{-\kappa} (\kappa > 0)$ in terms of interarea distances d_{ij}, amount to measurable accessibility proxies. As such they should arguably be used as explanatory variables and not assigned to the error term.

The application of spatial models to sets of discrete units also raises possible issues of nongeneralisability to higher or lower levels of aggregation. This does not apply in all cases; for example, in agricultural plot trials, the plot to which different treatments have been applied is the only level of interest. However, in many regional applications the presence of the 'modifiable area unit problem' suggests a conceptual advantage in modelling in continuous space, so that the analysis does not have to be repeated at various different scales of aggregation.

7.1.3 Contemporaneous correlation in systems of equations and in measurement error models

Social, economic and epidemiological models are often based on an idealised view of the data generating mechanism. Thus Rothenburg (1974) typifies many economic models as implicitly assuming a form of experimental process generating the observed data, with the outcome envisaged as random variable whose probability distribution is a smooth function of the variables describing the conditions of the experiment. However, if the available data are in fact not generated from controlled experiments then standard statistical modelling assumptions may need to be evaluated in the same way as priors and data models are subject to sensitivity evaluations. The explanatory variables in economic and social relationships are themselves the product of complex stochastic mechanisms and just like the outcome may be subject to errors of measurement or operationalisation. Also the clear distinction between inputs and outputs is no longer likely to hold, inducing simultaneous dependencies and correlated errors across equations with interdependent outcomes.

A relatively simple instance of such stochastic dependencies occurs in sets of equations describing the same process but across different sets of units. These occur in various applications, especially in econometrics and also in areas such as regional science (White and Hewings, 1982). For example, suppose the demand D_j for a set of goods $j = 1, \ldots, M$ is a function of their relative prices $P_1, \ldots, P_M$ and of money incomes Y. The demand equations are then

$$D_j = f_j(P_1, \ldots, P_M, Y).$$

Suppose that a time series is available on the quantities in a set of such equations. Then as the equation estimates and their errors are traced through time, the disturbances at any particular time are likely to reflect omitted variables that are common to all the equations – economic examples might be general economic conditions, purchase tax changes and so on. This is also likely in situations where the equations relate to investment by firms within a particular industry, or to production functions in different regions. Correlation between disturbances from different equations at one time is referred to as contemporaneous correlation as distinct from autocorrelation over time or space. When contemporaneous correlation is present, joint estimation of all equations is likely to be appropriate, rather than independent estimation of each.

Measurement errors extending to predictors as well as outcomes are another source of stochastic variability which invalidates the standard regression assumptions. For example, Maddala (1981) states there 'is no doubt that almost all economic variables are measured with error'. Consider the linear regression model with a single covariate

$$y_i = x_i\beta + u_i$$

where the errors u_i have constant variance and are uncorrelated. The standard assumption is that x_i is known and fixed, and the distributional properties of the least squares estimator $b_{yx} = (\underset{\sim}{x}'\underset{\sim}{x})^{-1}\underset{\sim}{x}'\underset{\sim}{y}$ depend on this assumption. However, suppose the predictor x is measured with error, such that the observed x can be written

$$x = X + \delta$$

where δ is random and X is the true value. This is in fact the 'classical' measurement error model. Errors of measurement arise from genuine inaccuracies in the measurement device, but also from random fluctuations around the true average (e.g. in biological series), or from the operationalisation of unmeasurable concepts (e.g. ambition, ability) by proxy variables. They may also occur because a precise measuring instrument X is too expensive to administer routinely and may only be available (for comparison with the surrogate x) in a small validation subsample. The presence of stochastic predictors leads to contemporaneous correlation between these predictors and the error term u, so that b_{yx} underestimates the true strength of the effect of the true X on y. Measurement error in one predictor can also distort the association between y and other predictors, even if the latter are measured without error (Richardson, 1996). Of course such measurement errors also may occur when the u_i are correlated or when the regression model is nonlinear or relates to discrete outcomes.

Stochastic predictors also occur when the values of a number of dependent variables are jointly determined. Instead of assuming only one-way causation or association from x to y, we allow for feedback or reverse impacts such that jointly dependent outcomes are determined by a system of equations. Suppose we have y_1 and y_2 denoting two jointly dependent or endogenous variables, and x_1 and x_2 are exogenous variables determined outside the system. Then a two-equation system may involve the relationships such as

$$\begin{aligned} y_{1i} &= \gamma_1 y_{2i} + u_{1i} \\ y_{2i} &= \gamma_2 y_{1i} + \beta_1 x_{1i} + \beta_2 x_{2i} + u_{2i}. \end{aligned}$$

Because of the joint determination of y_1 and y_2 and hence of u_1 and u_2, the latter are no longer uncorrelated with the predictors or with each other.

The examples below consider the potential of fully Bayesian estimation in models with contemperaneous correlation, and spatio-temporal correlations. The emphasis is on the associated issues in a Bayesian approach to data analysis and does not pretend to cover completely the full range of Bayesian methodology applied in these areas. To take two examples, there is an extensive and complex Bayesian methodology for simultaneous equations, going back to the 1960s, and which has yet to be completely tackled by MCMC methods. Similarly the diversity of developments in dynamic linear models within a state space framework can only be hinted at.

7.2 TEMPORAL CORRELATION DESCRIBED BY AUTOREGRESSIVE PROCESSES

The first order autoregressive AR(1) process is the simplest model to describe dependence in the values of an outcome variable over successive time points. This model describes the evolution of an outcome in relation to its own previous values and stochastic noise, without reference to further external data (e.g. covariates). It is the basis for an extensive class of univariate and multivariate models for time series analysis and forecasting.

This model, for data $y_1, y_2, \ldots, y_N$ in time order, has the form

$$y_t = \beta_0 + \beta_1 y_{t-1} + u_t, \quad t = 1, 2, \ldots, N$$

where β_0 and β_1 are parameters modelling respectively the overall level of the process, and the dependence between successive observations. The common assumption is that the u_t are independently and normally distributed, with mean zero and common precision $\tau = 1/\sigma^2$ across all time points t. Equivalently y_t is normally distributed with mean $\mu_t = \beta_0 + \beta_1 y_{t-1}$ and precision (around the regression) τ. This is the centred form adopted in BUGS. The above is a first order autoregressive process, denoted AR(1). Lags in $y_{t-2}, y_{t-3}, \ldots, y_{t-p}$ with associated parameters $\beta_2, \beta_3, \ldots, \beta_p$ would then lead to AR(2), AR(3), ..., AR(p) processes.

In the classical approach, both the level of the process and the variance (precision) of the u_t are assumed constant over time (the stationarity assumption). However, many time series in practice are nonstationary, showing strong upward or downward trends. Nonstationary time series can be transformed by differencing. Usually first or second differencing is sufficient to ensure that the transformed series

$$z_t = y_t - y_{t-1} \text{ (first differences)}$$

or

$$w_t = z_t - z_{t-1} = (y_t - y_{t-1}) - (y_{t-1} - y_{t-2}) \text{ (second differences)}$$

are stationary. We can denote such differencing operations using a backward shift function B, so that $B(y_t) = y_{t-1}$. It follows that $z_t = (1 - B)y_t$, $w_t = (1 - B)z_t$ and so $w_t = (1 - B)^2 y_t$. The number of times that a series must be differenced to achieve stationarity is called the order d of the process.

The backward difference operator can be used to represent the AR(p) process as a polynomial in B. Thus, assuming y_t has been centred around its mean, the AR(p) process can be written

$$y_t - \beta_1 y_{t-1} - \beta_2 y_{t-2} - \ldots - \beta_p y_{t-p} = \beta(B) y_t = y_t(1 - \beta_1 B - \beta_2 B^2 - \ldots - \beta_p B^p).$$

The stationarity condition implies that the coefficients $\beta_1, \ldots, \beta_p$ are confined to a region C_p which satisfies the condition that the roots of $\beta(B)$ lie outside the unit circle. For example, if $p = 1$ then C_1 consists of the interval -1 to $+1$, while if $p = 2$, C_2 is a triangle.

7.2.1 Priors on autoregressive coefficients

In contrast to classical methods, the Bayesian approach to estimation does not necessarily restrict β_1 in the AR(1) process to be between -1 and $+1$, and so applies to both explosive and nonexplosive cases (Zellner, 1996). By monitoring the proportion of values of β_1 which exceed the stationarity bound, one may test for stationarity without necessarily imposing it *a priori* (Broemeling and Cook, 1993). This means we can assess whether a process is explosive or nonexplosive, without necessarily predifferencing to eliminate a trend. Similarly for an AR(p) process there need be no restriction of $\underset{\sim}{\beta} = (\beta_1, \ldots, \beta_p)$ to the region C_p defined by the roots of $\beta(B)$. We can assess the probability of stationarity by monitoring the proportion of estimates satisfying it. For general lag p models, the roots of the polynomial in the lag operator $\beta(B) = (1 - \beta_1 B - \beta_2 B^2 - \beta_3 B^3 - \ldots - \beta_p B^p)$ can be evaluated at each sample of the $\beta_1, \ldots, \beta_p$ and the probability that the roots lie outside the unit circle monitored.

For $p > 1$, a program derived from Schur's theorem (Henrici, 1974) may be used to check on stationarity (e.g. within an MCMC run) without solving the characteristic equation. Nonstationarity with estimated `AR[]` parameters occurs if any of the `NS[]` in the following program are unity rather than zero.

```
model {     a [1,1] <- -1
            for ( kk in 1:p) { a [ kk + 1,1] <- AR[kk]
   for (j in 1: p1-kk) { b[j,kk] <- a[1,kk]* a[j,kk]-a[p+2-kk,kk]*
     a[p+3-kk-j,kk]
            a [j,kk + 1] <- b [j,kk]}
            NS[kk] <- step (-b [1,kk])}}
```

Thus for $(\beta_1, \beta_2) = (1.4, -0.6)$ there is stationarity, but not for $(\beta_1, \beta_2) = (-0.5, 0.5)$.

In the absence of accumulated knowledge about stationarity, noninformative priors on σ, and unconstrained priors on the elements of $\underset{\sim}{\beta} = (\beta_1, \ldots, \beta_p)$ are appropriate. For example, the following noninformative prior in an AR(1) model

$$p(\beta_1, \sigma) \propto 1/\sigma$$

lead to posterior densities of standard form on β_1 and σ^2 (Broemeling and Cook, 1993; Zellner, 1996). So it is possible to sample directly from the full conditional densities of the parameters.

Of course one may assume *a priori* that the process is nonexplosive; an expectation of a stationary rather than explosive process in an AR(1) model would involve a prior constraint that $|\beta_1| < 1$. This could be imposed by taking a prior on the real line (e.g. a Normal) and then using rejection sampling, for example by using the $I(,)$ function in BUGS. It could also involve assuming β_1 uniform between -1 and $+1$, $U(-1, 1)$, or adopting a reparameterisation $\zeta_1 = \log(1 + \beta_1) - \log(1 - \beta_1)$ so that the new parameter ζ_1 covers the whole real line (Naylor and Smith, 1988). Berger and Yang (1994) consider the problems in devising a prior for the AR(1) model which ascribes equal prior weight to the stationary and explosive options.

7.2.2 Stationarity priors in the $AR(p)$ model

An alternative Bayesian methodology, at least for a univariate outcome, adopts the stationarity condition at the outset. It is based on the reparameterisation of the β_j in terms of the partial correlations r_j of the AR(p) process (Barndorff-Nielsen and Schou, 1973; Jones, 1987; Marriott and Smith, 1992; Marriott *et al.*, 1996). In an AR(p) model let

$$\underset{\sim}{\beta} = (\beta_1^{(p)}, \beta_2^{(p)}, \ldots, \beta_p^{(p)})$$

with $\beta_j^{(p)}$ the jth coefficient in an AR(p) model. Then the stationarity conditions that β lies within C_p become equivalent to restrictions that $|r_k| < 1$ for $k = 1, 2, \ldots, p$. The transformations relating $\underset{\sim}{r}$ and $\underset{\sim}{\beta}$ for $k = 1, \ldots, p$ and $i = 1, \ldots, k - 1$ are

$$\begin{aligned} \beta_k^{(k)} &= r_k \\ \beta_i^{(k)} &= \beta_i^{(k-1)} - r_k \beta_{k-i}^{(k-1)} \end{aligned}$$

For example, for $p = 3$ the transformations would be

$$\begin{aligned} \beta_3^{(3)} &= r_3 \\ \beta_1^{(3)} &= \beta_1^{(2)} - r_3\beta_2^{(2)} = \beta_1^{(2)} - r_3 r_2 \quad (\text{for } k = 3, i = 1) \\ \beta_2^{(3)} &= \beta_2^{(2)} - r_3\beta_1^{(2)} = r_2 - r_3\beta_1^{(2)} \quad (\text{for } k = 3, i = 2) \\ \beta_1^{(2)} &= \beta_1^{(1)} - r_2\beta_1^{(1)} = r_1 - r_2 r_1 \quad (\text{for } k = 2, i = 1). \end{aligned}$$

It may be noted that these partial correlations r_j play a central role in identifying the order of an AR process, and we could apply Bayesian procedures to test their significance at various lags (see Box and Jenkins, 1970, Chapter 6).

Barnett *et al.* (1996) outline procedures for selecting the order of an AR(p) model, using the methods of George and McCulloch (1993). These procedures may be applied to either the r_j or the β_j.

As in Marriott and Smith (1992), the usual Fisher transformations for correlations may be used such that r_j^* is a normal or uniform draw on the real line. Then the r_j are obtained from $r_j^* = \log([1 + r_j]/[1 - r_j])$. Alternatively, Jones (1987) proposes the partial correlations be generated using beta variables $r_1^*, r_2^*, r_3^*, \ldots, r_k^*$, with beta priors B(1, 1), B(1, 2), B(2, 2) and B($\{(k+1)/2\}$, $\{k/2\} + 1$), where $\{x\}$ here denotes the integer part of x. These are then transformed to the interval $[-1, 1]$ via $r_1 = 2r_1^* - 1, r_2 = 2r_2^* - 1$, etc. An alternative prior structure recently proposed for AR models applies to the real and complex roots of the characteristic equation and leads on to time series decompositions (Huerta and West, 1999).

7.2.3 Initial conditions as latent data

A remaining complication in the analysis of the AR(p) process is the implicit reference to latent (unobserved) quantities before the observed system started. For example, in an AR(1) process the first observation is modelled as

$$y_1 = \beta_1 y_0 + u_1$$

where y_0 is unknown, and in an AR(2) process the first two observations are modelled as

$$y_1 = \beta_1 y_0 + \beta_2 y_{-1} + u_1$$
$$y_2 = \beta_1 y_1 + \beta_2 y_0 + u_2$$

where y_0 and y_{-1} are unknown. In general for an AR(p) process with $t = 1$ defining the first actual observation, the latent variables are $y_0, y_{-1}, \ldots, y_{1-p}$.

Different assumptions may be adopted regarding the latent data $y_0, y_{-1}, \ldots$ (Zellner, 1996). In an AR(2) model we know y_2 and y_1 but not y_0 or y_{-1} so we cannot specify the model for the points y_2 or y_1 in the same way as the rest of the data. The Bayes method allows us to treat the unobserved or 'missing' data points such as y_0 in an AR(1) model as extra parameters. One option is to write the composite unknowns, such as $\beta_1 y_0$ in the AR(1) model, and $\beta_1 y_0 + \beta_2 y_{-1}$ and $\beta_2 y_0$ in the AR(2) model, as new parameters. These would be modelled as fixed effects. For example, in the AR(1) case y_1 would be normal with mean $M_1 (\equiv \beta_1 y_0)$ and variance σ^2. If the stationarity assumption is made then y_0, y_{-1} etc. may be modelled formally within the exact likelihood for an AR(p) process (Newbold, 1974; Marriott *et al.*, 1996). Another option, discussed below, is 'backcasting' to estimate the latent starting data.

Alternatively we can condition on the initial observations; that is, we specify the likelihood for values of t only when we know both y_t and y_{t-1} (for an AR(1) process), or all of y_t, y_{t-1}, and y_{t-2} (for an AR(2) process), and so on. This amounts to treating the initial observations as known and fixed (i.e. having zero variance). The conditional likelihood approach makes it easier to deal with models involving higher order lag dependence, but involves a loss of data in the likelihood.

The importance of assumptions about initial observations diminishes with longer series of observed points. Note that if we took (say) first differences then the rule for initial observations in an AR(1) model for differences is the same as the rule for initial observations in an AR(2) model for the original data. This is because $z_1 = y_1 - y_0$ and since y_0 is unknown, z_1 is unknown. Therefore the model for z_2 will be the first we need to consider (unless we condition on initial observations).

7.2.4 Predictions

Both in classical and Bayesian time series analysis, out-of-sample predictions are a major goal. Two major kinds of regression-based forecasts can be distinguished: pure autoregression, of the Box–Jenkins kind, on previous values of the series itself, and those deriving from explanatory regressions on predictors x_t also changing in time. If forecasts are based on an explanatory model, then values x_t for $t = N+1, N+2, \ldots$ must be known in the future (e.g. they may be functions of time, or seasonal dummies), or forecasts will also be required of these predictors.

Suppose we are predicting using autoregression to one period ahead beyond the end of the sample, namely to $N+1$, and that we have applied an AR(1) process to y, or to differences in y. On the assumption that the AR(1) process is correct, and with y now generically representing a possibly differenced version of an original time series, then the prediction of y_{N+1}, namely $\hat{y}_{N+1}$, will have mean $\hat{\beta}_0 + \hat{\beta}_1 y_N$ and sampling of this future value will be from the same density assumed for the observed y_t. The forecast error is then just $y_{N+1} - \hat{y}_{N+1}$. The forecast for y_{N+2} will have mean $\hat{\beta}_0 + \hat{\beta}_1 \hat{y}_{N+1}$, and so will accumulate errors both from the model fitted up to time N and from the prediction error of y_{N+1}. Forecasts for successive periods follow recursively using the same principle. If we had differenced the original series we would need to transform back by 'integrating' the differenced form of the variable.

One common practice in time series modelling is to validate the forecast within the sample; one option is to fit the selected model to periods $t = 1, \ldots, F$, where F is less than N. Periods $F+1, F+2, \ldots, N$ are used to validate competing models, since y_{F+1}, y_{F+2}, etc. are known and can be compared with the predictions. Another option is to make short-term predictions ahead within the sample (one-step-ahead predictions), and then choose the model minimising such prediction errors. If f_t denotes the forecast for period t (and y_t is also known) then mean square error or mean absolute error statistics are often used to summarise the fit of such predictions. One is based on the relative deviations $F_t = (f_t - y_{t-1})/y_{t-1}$ and $Y_t = (y_t - y_{t-1})/y_t$ and can be written

$$\text{MSE} = \sum (F_t - Y_t)^2/N = (\bar{f} - \bar{y})^2 + s_p^2$$

where s_p^2 is the variance of the prediction errors. Another measure is the mean of the absolute deviations $|y_t - f_t|$.

Example 7.1 US unemployment Fuller (1976) considers an AR(2) model for the quarterly US unemployment rate y_t over the 25 years 1948–1972, and then carries out predictions to the four quarters of 1973. The data model is then (applying a normal approximation to the binomial)

$$y_t \sim \mathrm{N}(m_t, \tau), t = 1, \ldots, 100.$$

The means for $t = 1, 2$ involve extra compound parameters modelling the impact of parameters and initial data, so that

$$\begin{aligned} m_1 &= \beta_0 + M_1 \qquad \text{(where } M_1 = \beta_1 y_0 + \beta_2 y_{-1}) \\ m_2 &= \beta_0 + \beta_1 y_1 + M_2 \quad \text{(where } M_2 = \beta_2 y_0) \\ m_t &= \beta_0 + \beta_1 y_{t-1} + \beta_2 y_{t-2}, \quad t = 3, \ldots, 100. \end{aligned}$$

The predictions for 1973 quarters are generated recursively as follows:

$$\begin{aligned} y_{100+j} &\sim \mathrm{N}(m_{\text{new}.100+j}, \tau), \quad j = 1, \ldots, 4 \\ m_{\text{new}.100+j} &= \beta_0 + \beta_1 y_{100+j-1} + \beta_2 y_{100+j-2}. \end{aligned}$$

As in Fuller (1976), the predictions are for a falling rate in 1973, though with increased uncertainty (larger prediction variance) for the later forecasts (Table 7.1).

Example 7.2 AIDS cases via dependent Poisson model McKenzie (1988) proposes a class of dependent Poisson models analogous to AR and MA processes for continuous data. These are based on the 'binomial thinning' operation for Poisson data; thus a Poisson count X is envisaged to be the outcome of X trials with success probability α, and define the random variable $\alpha^* X = \sum_{k=1}^{X} B_k(\alpha)$ where $B_k(\alpha)$ is a Bernoulli with success probability α. Then let the first observation of a time series of counts X_1 be distributed as Poisson with mean θ, and let subsequent observations X_t have mean

$$\mu_t = \alpha^* X_{t-1} + (1 - \alpha)\theta.$$

Table 7.1

Parameter	Mean	SD	2.5%	Median	97.5%
β_0	0.62	0.15	0.32	0.62	0.91
β_1	1.56	0.08	1.40	1.56	1.71
β_2	−0.69	0.08	−0.83	−0.69	−0.53
Y_{101}	5.08	0.34	4.43	5.08	5.76
Y_{102}	4.90	0.63	3.65	4.90	6.16
Y_{103}	4.75	0.88	3.00	4.75	6.48
Y_{104}	4.65	1.05	2.60	4.62	6.73

Table 7.2

	Mean	SD	2.5%	97.5%
μ_2	0.12	0.22	0.00	0.76
μ_3	1.11	0.21	1.00	1.70
μ_4	2.07	0.27	1.17	2.70
μ_5	2.82	0.51	1.51	3.47
μ_6	1.12	0.22	1.00	1.75
μ_7	4.06	0.29	3.09	4.65
μ_8	8.94	0.44	8.01	9.57
μ_9	17.36	0.89	15.07	18.27
μ_{10}	22.23	1.02	19.75	23.26
μ_{11}	28.79	1.96	24.12	31.06
μ_{12}	19.28	0.98	17.01	20.24
μ_{13}	24.27	0.99	22.02	25.27
μ_{14}	35.46	1.52	31.90	37.11
α	0.94	0.04	0.84	0.99
θ	1.59	1.66	0.04	6.17

This is analogous to an AR(1) process. Extensions of this modelling approach to ARMA models in general, and to panel data have been investigated by Bockenholt (1999) and others.

We illustrate an AR(1) model of this kind with the data of Dobson (1990), relating to a series of 14 quarterly death totals from AIDS in Australia in the early 1980s. The total rises from 0 in early 1983 to 45 in mid-1986. Dobson proposes the growth model $X_t \sim \text{Poi}(v_t)$. Applying the above model (see *Program 7.2 INAR Dependent Poisson*) with vague priors on α and θ we find the predicted series and parameters given in Table 7.2.

7.3 AUTOREGRESSIVE MOVING AVERAGE MODELS

In the AR(p) model the observed value of an outcome is related to its past values and to a random error. The error term u_t can be seen as a random shock (also called an innovation process), the impact of which may not necessarily be absorbed in the same period. The moving average model for the error term allows for dependence on past values of the error as well as of the observed series $y_1, y_2, \ldots, y_N$. For example the model, for centred data, might take the form

$$y_t - \beta_1 y_{t-1} = u_t - \gamma_1 u_{t-1}, \quad t = 1, 2, \ldots, N.$$

This defines a first order moving average MA(1) in u_t as well as AR(1) dependence in the data y_t. A second order moving average MA(2) would involve a term $\gamma_2 u_{t-2}$. Taken together, the number of lags p in the autoregression of the y_t, the number of lags q in the moving average, and the order of differencing d, determine an ARIMA(p, d, q) model. If the data are undifferenced then a process with autoregressive lag p and moving average lag q is denoted ARMA(p, q).

Thus for an ARMA(3,3) model for y_t with mean zero, we would have

$$y_t - \beta_1 y_{t-1} - \beta_2 y_{t-2} - \beta_3 y_{t-3} = u_t - \gamma_1 u_{t-1} - \gamma_2 u_{t-2} - \gamma_3 u_{t-3}$$

or as a polynomial in the backward shift parameter

$$\beta(B) y_t = \gamma(B) u_t \tag{7.2}$$

with $\beta(B) = 1 - \beta_1 B - \beta_2 B^2 - \beta_3 B^3, \gamma(B) = 1 - \gamma_1 B - \gamma_2 B^2 - \gamma_3 B^3$.

The BUGS centred mean formulation for a Normal or Student t outcome refers to errors with 'unstructured' white noise form. As a device to allow for moving average error effects (i.e. for a form of structured dependence in the errors) we introduce an extra measurement error term, so partitioning the original variance about the regression. Thus the observations have the form

$$y_t \sim \mathrm{N}(\mu_t, 1/\tau_\varepsilon)$$

where $y_t = \mu_t + \varepsilon_t$, with $\varepsilon_t \sim \mathrm{N}(0, 1/\tau_\varepsilon)$. Then μ_t is itself specified in terms of structured errors u_t with mean 0 and variance τ_u. For an MA(1) model, for example,

$$\mu_t = u_t - \gamma_1 u_{t-1}.$$

Similar questions of re-expression are involved in putting an MA(q) model into state-space terms (Harvey, 1990, Chapter 3).

Moving average parameters are often estimated by minimising a non-linear least squares function. Suppose the model were an MA(1), namely $y_t = u_t - \gamma_1 u_{t-1}$; then the estimate of γ_1 would be based on minimising $\sum u_t^2 = \sum (y_t + \gamma_1 u_{t-1})^2$. So in effect the least squares MA parameter estimates are conditional on a single point estimate (the minimum) of the variance term. By contrast, Bayesian estimation would allow for uncertainty in the estimated variability of the u_t.

The constraint of invertibility for an MA(q) model may be formally specified in the prior in a similar way as for the AR(p) model. This again involves reparameterising, in terms of the partial autocorrelations r_j, the MA coefficients

$$\gamma^{(q)} = (\gamma_1^{(q)}, \gamma_2^{(q)}, \ldots, \gamma_p^{(q)})$$

with $\gamma_j^{(q)}$ the jth MA coefficient in an MA(q) process. Then the invertibility conditions requiring that $\gamma^{(q)}$ lie within a region C_q become equivalent to restrictions that $|r_k| < 1$ for $k = 1, 2, \ldots, q$. The transformations for $k = 1, \ldots, p$ and $i = 1, \ldots, k-1$ are

$$\gamma_k^{(k)} = r_k$$
$$\gamma_i^{(k)} = \gamma_i^{(k-1)} - r_k \gamma_{k-i}^{(k-1)}.$$

If both lag and moving average terms are included in an ARMA(p, q) model then both sets of coefficients would be modelled via this parameterisation.

Example 7.3 Randomly generated moving average process 500 values are drawn from the series $y_t = u_t + 0.5u_{t-1} + 0.3u_{t-2}$, with $u_t \sim \mathrm{N}(0, 1)$, leading to 498 values of an observed outcome. The model to estimate is therefore

$$y_t = u_t - \gamma_1 u_{t-1} - \gamma_2 u_{t-2}, \quad t = 1, \ldots, 498$$

and $u_t \sim \mathrm{N}(0, 1/\tau_u)$. To fit an MA model in BUGS we partition the variance of y_t around $\mu_t = \mathrm{E}(y_t)$, introducing an unstructured error term (or measurement error) η_t such that

$$y_t = u_t - \gamma_1 u_{t-1} - \gamma_2 u_{t-2} + \eta_t.$$

The first two means (for y_1 and y_2) draw on the shocks u_0 and u_{-1} which we assume are drawn from the same density as the remaining errors u_t. The priors on γ_1 and γ_2 are constrained to invertibility, as in Marriott *et al.* (1996) (see *Program 7.3 MA2 Data*).

Iterations 10–50 thousand of a single long run give an estimated residual variance, namely $[1/\tau_\eta + 1/\tau_u]$, to be close to the value used to generate the data. The first MA lag parameter has a 95% credible interval which includes the value 0.5 used to generate the data, but the second MA lag parameter is overestimated (in absolute terms) (Table 7.3).

7.3.1 Initial conditions in the ARMA model

The full ARMA model involves the latent data $y_0, y_{-1}, \ldots, y_{1-p}$ and $u_0, u_{-1}, \ldots, u_{1-q}$ which initiate the process. Marriott *et al.* (1996) outline Gibb's sampling procedures for the exact likelihood of the ARMA model, including these latent data. Alternatively, their values may be predicted via a 'backcasting' approach, which is based on the duality between the backward and forward versions of the ARMA model. In a stationary time series model, the direction of time is irrelevant in that we could equally think of time as running backward as well as forward. For example, an ARMA(1,1) model for centred observations

$$y_t = \beta y_{t-1} + u_t - \gamma u_{t-1}$$

can also be generated by the corresponding backward model

Table 7.3

	Mean	SD	2.5%	Median	97.5%
γ_1	−0.55	0.03	−0.47	−0.55	−0.61
γ_2	−0.37	0.03	−0.32	−0.37	−0.44
σ^2	0.99	0.06	0.88	0.99	1.13

$$y_t = \beta y_{t+1} + b_t - \gamma b_{t+1}$$

where b_t has the same distribution as u_t. Starting with $b_T = 0$, these equations can be used to generate $b_{T-1}, \ldots, b_1$, and then y_0 and u_0, these being the latent quantities needed for an unconditional ARMA(1,1) model.

More generally, let $F = B^{-1}$ refer to the operation $u_{t+1} = Fu_t$ and $y_{t+1} = Fy_t$. The ordinary forward version of the ARMA model for a centred (zero mean) dependent variable

$$\beta(B)y_t = \gamma(B)u_t$$

can be written equivalently in a backward form

$$\beta(F)y_t = \gamma(F)b_t.$$

Such an approach has been found to be a suitable approximation to the full likelihood method (Pai *et al.*, 1994; Ravishanker and Ray, 1997). Box and Jenkins (1970) outline a recursive procedure which switches between the processes in order to identify initial conditions.

In BUGS one possible device is to declare the same set of outcome data y twice over two 'different' sets y_1 and y_2, but with the same variance parameter. The full backcasting procedure involves selecting a lead time parameter Q, exceeding both p and q, such that y_{n+t} and b_{n+t} are negligible for $t > Q$. Similarly, the antecedent latent data u_{1-t} and y_{1-t} are effectively zero for $t > Q$. The full set of actual ($t = 1, \ldots, N$), antecedent ($t < 1$) and subsequent ($t > N$) data, is of length $2Q + N$. Box and Jenkins (1970) illustrate this procedure with an extract from their IBM share series with $p = q = 1$ and $Q = 5$.

7.3.2 Normality assumptions

As well as differencing, the performance of time series models will depend on the closeness to assumed normality. One option is to use a robust density (such as the t distribution) as the data model for the y_t (Chib, 1993). Another is to transform the series (e.g. via logs or square roots) to stabilise the variance. This may be coupled with 1st; or 2nd differencing in the transformed values to achieve stationarity. Thus the Box–Cox transform has been proposed for time series (Granger and Newbold, 1976):

$$v_t = \{(y_t + d)^\theta - 1\}/\theta, \quad \theta \neq 0$$
$$v_t = \log(y_t + d), \quad \theta = 0,$$

the goal being to create a variable more closely approximating normality than the y variable.

Power transformations have less effect on data with a limited range (i.e. of variability in relation to mean), because under these conditions, the transforms are virtually linear. It may be noted that if v_t is assumed normal then y_t is no longer necessarily normal. For instance, if $v_t = y_t^{0.5}$ is a normally distributed autoregressive process with order 1 (i.e. first differences in v are taken giving an

ARIMA(1,1,0) process), then the original series y_t is approximately an ARIMA(2,1,2) process with non-normal errors.

Example 7.4 Simulated ARMA(1,1) model forecasts Fuller (1976) considers forecasts to N_1+1, N_1+2, N_1+3 of $N_1 = 7$ observations generated from the ARMA(1,1) time series

$$y_{1t} = 0.9y_{1,\,t-1} + 0.7u_{1,\,t-1} + u_{1t}$$

where $u_{1t} \sim \mathrm{N}(0, 1)$. Fuller's derivation of the forecasts is based on the theoretical autocorrelations deriving from the known AR(1) and MA(1) parameter values of 0.9 and 0.7.

To permit stochastic estimation and predictions we supplement these seven values by a second series of $N_2 = 493$ values, denoted y_{2t}, and generated by the same model. We specify separate Normal models for the $y_{1t}, y_{2t}, u_{1t}, u_{2t}$ with

$$\begin{aligned} y_{1t} &\sim \mathrm{N}(\mu_{1t}, \zeta_y) \\ y_{2t} &\sim \mathrm{N}(\mu_{2t}, \zeta_y) \\ u_{1t} &\sim \mathrm{N}(0, \zeta_u) \\ u_{2t} &\sim \mathrm{N}(0, \zeta_u) \end{aligned}$$

and for $t > 1$

$$\begin{aligned} \mu_{1t} &= b_1 y_{1,\,t-1} + u_{1t} + b_2 u_{1,\,t-1} \\ \mu_{2t} &= b_1 y_{2,\,t-1} + u_{2t} + b_2 u_{2,\,t-1}. \end{aligned}$$

The parameters b_1 and b_2 are constrained to stationarity and invertibility respectively. The initial mean terms μ_{11} and μ_{21} are for simplicity set equal to an additional parameter M. The total variance around the regression is 'partitioned' between ζ_y and ζ_u.

The posterior estimates of the precision $\tau = (1/\zeta_y + 1/\zeta_u)$, from the second half of a run of 25 000 iterations is 1.1 (see *Program 7.4 ARMA(1,1) Joint Series*). Note that the burn-in period for convergence from null starting values of $b_1 = b_2 = 0$ takes over 1000 iterations. The AR lag parameter is closely reproduced but the MA lag parameter is slightly underestimated. The three predicted values of the first series have posterior averages close to those of Fuller, which were (with standard errors in parentheses): 3.04 (1.00), 2.73 (1.89), 2.46 (2.37). See Table 7.4.

Example 7.5 Luteinizing hormone data Venables and Ripley (VR, 1994) analyse a series of 48 readings y_t, centred around the mean (i.e. $y_t = h_t - \bar{h}$), of luteinizing hormone readings h_t in a human female. They consider especially AR(1), ARMA(1,1) and AR(3) models. The series has some discontinuities that make modelling more complex; there is also some evidence of an upward trend so a complete model might allow for changes in level or take first differences. However, we follow VR in using undifferenced data. In these circumstances, one-step-ahead forecasts are a useful measure of predictive fit.

Table 7.4

	Mean	SD	2.5%	97.5%
$y_{1,8}$	3.02	1.01	1.11	5.02
$y_{1,9}$	2.72	1.82	−0.84	6.27
$y_{1,10}$	2.45	2.25	−1.97	6.77
b_1	0.90	0.02	0.86	0.93
b_2	0.65	0.04	0.55	0.72

The ARMA(1,1) option is

$$y_t = \beta y_{t-1} + u_t - \gamma u_{t-1}$$

where u_t is white noise. In this instance the parameters γ and β are constrained to lie in regions C_q and C_p defined by the interval $[-1, 1]$ in both the AR and MA cases. Therefore β and γ may be modelled by assigning both r^*_β and r^*_γ a $B(1, 1)$ prior and then setting

$$\beta = 2r^*_\beta - 1,$$
$$\gamma = 2r^*_\gamma - 1.$$

VR find the first lag coefficient β in the ARMA(1,1) model to be significant, but the moving average parameter θ to be not well identified; they estimate

$$y_t = 0.463(0.218)y_{t-1} + u_t + 0.200(0.241)u_{t-1}$$

and a residual variance of 0.205.

Here we partition the variance as in the preceding two examples. The other specification issue relates to the missing initial conditions; thus the model for the first term is $y_1 - \beta y_0 = u_1 - \gamma u_0$, where u_0 and y_0 are additional unknowns. We model the missing data by using a backcasting approach, which involves copying the data into two series $y.a[i]$ and $y.b[i]$, and fitting the forward model to $y.a[i]$ and the backward model to $y.b[i]$. The backward model is

$$y_t = \beta y_{t+1} + b_t - \gamma b_{t+1}$$

where u_t and b_t are from the same density. Then we model $y.a[i], i = 2, \ldots, N$ and $y.b[i], i = 1, \ldots, N - 1$ in terms of known data and the parameters γ and β. The backcast mean for $t = 1$ can then serve as the mean for y_1 in the forward model.

This model is implemented in the listing *Program 7.5 Model A*. The equation parameter posterior means and standard deviations from the second half of a run of 50 000 iterations, are

$$y_t = 0.473(0.131)y_{t-1} + u_t + 0.359(0.343)u_{t-1}.$$

The estimate of β is similar to that of VR, but that of the MA(1) term, namely $\gamma = -0.36$ (0.34), is not well identified. The residual variance is estimated as 0.197. This problem shows signs of having a common factor, i.e. the likelihood is relatively flat along the line $\beta = -\gamma$ with the AR and MA lag parameters close to summing to zero (Cryer, 1986, Chapter 8). To improve identification we might, however, constrain γ to be negative.

The mean absolute deviation of the one-step-ahead forecasts is 0.46. Note that this is the average comparing predictions and actual data over all iterations, rather than a comparison of posterior one-step-ahead means with observed y_t. An examination of the posterior estimates of the one-step-ahead errors shows a relatively low lag 1 correlation of 0.25, an indication of adequate predictive fit.

An AR(3) model is estimated with the stationarity constraints outlined above – see *Program 7.5 Model B*. The latent data y_0, y_{-1} and y_{-2} involved in the regression means for $t = 1, 2, 3$ are modelled as drawn from a normal density with mean zero and the same variance as the observed $y_t, t = 1, \ldots, 48$ (this is one among several possibilities of modelling the initial conditions). The second half of 10 000 iterations yields similar estimates as follows (these are similar to those cited by VR)

$$y_t = 0.658(0.145)y_{t-1} - 0.066(0.174)y_{t-2} - 0.234(0.145)y_{t-3}.$$

However the MAD of the one-step-ahead forecasts from the AR(3) option is higher than for the ARMA(1,1) model just discussed. The highest absolute deviations are for the points (such as y_6 and y_{15}) involving discontinuities in the series.

Example 7.6 Forecasting birth rates Miller (1984) considers the modelling and forecasting of birth rates specific to age group of mother using ARIMA methods. We consider a subset of the Miller data involving forecasts of the US birth rate y_t based on 33 observations in the years 1948–80 for older teenage mothers (i.e. aged 15–19). The short series available do not support the development of complex (e.g. high lag dependence) models, nor do they allow great precision in parameter estimation.

We compare a selected range of values of θ in the transformation

$$v_t = \begin{cases} (y_t^\theta - 1)/\theta, & \theta \neq 0 \\ \log(y_t), & \theta = 0. \end{cases}$$

These are applied to improve the normality approximation, and for comparing fit with the undifferenced or 'linear' transformation $v_t = y_t$. Specifically, we compare the values $\theta = 1$ (linear), $\theta = 0.5$ (square root), $\theta = 0$ (logarithmic) and $\theta = -1$ (inverse). For the inverse transformation use of an additional constant $c = 1000$ giving $v_t = 1000(1 - 1/y_t)$ was found beneficial to forecasting.

We then apply an AR(1) model to the first differences z_t of the transformed series v_t, i.e. assume an ARIMA(1,1,0) model for the undifferenced series (see *Program 7.6 Teenage Birthrate*). Thus the model is

$$z_t = \beta_0 + \beta_1 z_{t-1} + u_t.$$

We adopt a composite parameter to model the unknowns in the mean of the initial observation z_2. We compare forecasting within the sample period by using the 28 observations up to 1975, and then measuring forecast accuracy (via the root of the mean square error) of different transformations using the actual and predicted values for 1976–80. In this period, a long downward movement in fertility was ended and the series for this and other ages of mother levelled off or even turned up again. So the forecasting ability of the models is put under a heavy test.

In terms of the square root of the mean square error, the best performing AR(1) model involves the inverse transformation. The same conclusion was reached by Miller (1984). This transformation choice also has the best identified parameters. The estimates and forecast birth rates under this transformation are given in Table 7.5.

However, in terms of matching the slowdown in the fall in fertility, the logarithmic transformation does best, with the forecast for the last year, 1980, actually showing a slight predicted upturn. In the case of this, but not the other transforms, there is a contrast between the mean and median predictions for 1976–80.

The probability of stationarity in the AR(1) model (i.e. that $-1 < \beta_1 < 1$) can be calculated for each transformation. For the log transform, this is 0.858 based on using the BUGS step function. For the inverse transformation (i.e. $\theta = -1$), this probability is 0.997.

Example 7.7 Trapped lynx, 1821–1934 One of the classic time series data sets consists of the number of lynx y_t trapped each year in the Mackenzie River district of North-West Canada between 1821 and 1934. Despite the erratic fluctuations apparent in these data, an ARMA(3,3) process has been found to give a suitable fit. The model is therefore

$$y_t - \beta_1 y_{t-1} - \beta_2 y_{t-2} - \beta_3 y_{t-3} = u_t - \gamma_1 u_{t-1} - \gamma_2 u_{t-2} - \gamma_3 u_{t-3}.$$

Here we adopt the backcasting approach to estimating latent starting values of y and u, and the reparameterisation suggested by Jones (1987) to ensure

Table 7.5

Parameter	Mean	SD	2.5%	Median	97.5%
RMSE	30.6	3.6	23.2	30.7	37.7
β_0	−0.12	0.10	−0.32	−0.12	0.06
β_1	0.43	0.20	0.04	0.43	0.81
Forecasts					
Y_{1976}	55.5	1.5	52.7	55.5	58.6
Y_{1977}	54.7	2.6	49.8	54.6	59.7
Y_{1978}	54.0	3.5	47.2	53.8	61.4
Y_{1979}	53.3	4.4	44.9	53.2	62.8
Y_{1980}	52.8	5.2	43.1	52.5	64.0

Table 7.6

	Mean	SD	2.5%	Median	97.5%
β_1	1.8390	0.1556	1.544	1.843	2.079
β_2	−1.4080	0.2429	−1.780	−1.413	−0.952
β_3	0.2803	0.1426	0.019	0.278	0.499
γ_1	0.6939	0.1958	0.318	0.727	0.986
γ_2	−0.0414	0.1287	−0.280	−0.029	0.178
γ_3	−0.5330	0.0687	−0.674	−0.545	−0.385

stationarity and invertibility (see *Program 7.7 Lynx Series*). We adopt a lead time parameter for backcasting set at $Q = 10$.

After a single long run of 5000 iterations (with 500 burn-in), the autoregressive parameters β and MA parameters γ have posterior means close to those quoted by Marriott *et al.* (1996), though a natural log rather than log base 10 transformation was here applied to the original counts (Table 7.6).

7.4 OUTLIER AND INTERVENTION MODELS

As in other areas, time series model estimation and selection may be affected by outliers. In a time series setting the detection of outliers and of shifts in the series are closely interrelated. McCulloch and Tsay (1994) and Barnett *et al.* (1996) discuss outlier models which allow for additive outliers (to be added to an outlier outcome y_t) and innovation outliers (to be added to outlying random shocks u_t). For example, consider an ARMA(1,1) model

$$y_t - \beta y_{t-1} = u_t - \gamma u_{t-1}.$$

To allow for additive and innovation outliers, an additional error term O_t is introduced such that

$$y_t - \beta y_{t-1} = u_t - \gamma u_{t-1} + o_t.$$

Then the two errors u_t and o_t are modelled as finite mixtures of normals. Thus $o_t \sim \text{N}(0, K_{1t}\sigma^2)$ and $u_t \sim \text{N}(0, K_{2t}\sigma^2)$ and $K_t = (K_{1t}, K_{2t})$ has a bivariate multinomial distribution with $K_{1t} \geq 0$ and $K_{2t} \geq 1$. If K_{1t} exceeds 0 then there is an additive outlier at point t in the series, while if K_{2t} exceeds 1 there is an innovation outlier. We may restrict the prior choices on the possible pairings of $(K_{1t}.K_{2t})$ so that only an additive or an innovation outlier is allowed.

We may use a discrete prior on a small number of possible pairs, designed to encompass the likely ranges. Since BUGS models normal data with precisions rather than variances, in practice we divide the precision τ (the inverse of σ^2) by K_{1t} and K_{2t}. Thus Barnett *et al.* propose a seven-point discrete prior on (K_1, K_2), namely (0,1), (3.3,1), (10,1), (32,1), (0,3.3), (0,10), (0,32). Hence we replace zero values of K_1 by 0.001 to avoid divisions of the precision by zero.

Another generalisation of the standard stationary model allows for a shift in the level of the process, as a consequence of 'interventions' such as government policy shifts, changes in sales strategies or natural disasters. Thus McCulloch and Tsay (1994) consider autoregressive models allowing random level shifts, as in

$$\begin{aligned} y_t &= \mu_t + v_t \\ \mu_t &= \mu_{t-1} + \delta_t \eta_t \\ v_t &= \beta_1 v_{t-1} + \beta_2 v_{t-2} + \ldots + \beta_p v_{t-p} + u_t \end{aligned} \tag{7.3}$$

where δ_t is the probability of a shift at time t $(t \geq 2)$, drawn from a Bernoulli density with parameter ε. The shifts themselves are distributed as $N(0, \sigma_\eta^2)$. See Notes. The same principle may be applied to random shifts in regression coefficients.

Example 7.8 Luteinizing hormone data We first analyse the luteinizing hormone data with an ARMA(1,1) model, and use the backcasting method (see *Program 7.8 Model A*) with lead time parameter of $Q = 5$. This yields a similar estimate of β, namely 0.48 (with standard deviation 0.13), to that reported above. However γ is estimated to have a higher absolute value, though still imprecisely estimated, namely -0.37 (with standard deviation 0.36).

Adopting the outlier procedure above and the seven-point discrete prior mentioned there (model B), we find little evidence of innovation outliers, but points 5, 13 and 45 have significant additive outlier effects. The effect is to reduce the size of β, with its estimate now being 0.32 (with standard deviation 0.11), but γ is now estimated at -0.76 (with standard deviation 0.17). The outlier model, by smoothing out deviations from the underlying time series pattern, seems to have made the MA process clearer.

Example 7.9 Quarterly road accidents We analyse the UK road accidents data (for quarters) for 1968–84 as considered by Harvey and Durbin (1986) and also by McCulloch and Tsay (MT, 1994). There are 64 quarterly observations, whereas MT considered 192 monthly observations. They took log transforms of the count of fatal road accidents and subtracted monthly means to eliminate seasonal variation. Here we subtract quarterly means and adopt an AR(3) model as they did. We include a backcasting procedure to model latent initial data, with $Q = 5$ (see *Program 7.9 Quarterly Accidents*). We also follow MT in adopting an informative beta prior for the probability of shifts, and assume (adopting the notation of model (7.3) just described) that $\sigma_\eta^2 = 10\sigma_u^2$. This is because we expect the shift variance to exceed the more stable white noise variability.

MT detect two major shifts, one in late 1974, attributable to a petrol price increase by OPEC, and one in early 1983, related to the introduction of a seat belt law in the UK. Monitoring the means $\mu_t = \mu_{2t}$ of the 'forward series' (the

Table 7.7

Parameter	Mean	SD	2.5%	Median	97.5%
β_1	0.89	0.09	0.71	0.90	1.07
β_2	−0.20	0.12	−0.44	−0.21	0.03
β_3	0.21	0.09	0.04	0.21	0.39

term `Model[]` in *Program 7.9*) suggests a shift at points 24 and 57 (Quarter 4, 1974 and Quarter 1, 1983). The difference between μ_{56} and μ_{58} is about −0.25. This implies a 22% fall, i.e. 100exp(−0.25), in the accident level in early 1983. The estimated AR parameters are given in Table 7.7.

7.5 REGRESSION WITH AUTOCORRELATED ERRORS IN TIME

In the standard linear model for N observations, $y_t = x_t\beta + e_t$, the error term e_t is assumed uncorrelated between different observations, so that $\text{cov}(e) = \sigma^2 I$ where I is an $N \times N$ identity matrix. If correlation exists then the covariance matrix is no longer diagonal. Thus suppose $\rho(B)e_t = u_t$ where u_t defines an unstructured exchangeable error with a constant variance σ^2 and where $\rho(B) = 1 - \rho_1 B - \rho_2 B^2 - \ldots \rho_p B^p$ defines the lagged autocorrelation in the actual regression errors. The observations may be normal, Student t, or with suitable link functions, Poisson or binomial. More generally we can have regression models with errors following an ARMA(p, q) form

$$y_t = x_t\beta + e_t, \quad t = 1, \ldots, N$$
$$e_t - \rho_1 e_{t-1} - \rho_2 e_{t-2} - \ldots - \rho_p e_{t-p} = u_t - \gamma_1 u_{t-1} - \gamma_2 u_{t-2} - \ldots - \gamma_q u_{t-q}.$$

As an example first order autocorrelation, i.e. AR(1) dependence, in the errors e_t would imply

$$e_t = \rho_1 e_{t-1} + u_t. \tag{7.4}$$

Then $\text{var}(e_t) = \rho_1^2 \text{var}(e_{t-1}) + \sigma^2 + 2\rho_1 \text{cov}(e_{t-1}, u_t) = \rho_1^2 \text{var}(e_t) + \sigma^2$, and

$$\text{var}(e_t) = \sigma^2/(1 - \rho_1^2).$$

We can also show that $\text{corr}(e_t, e_{t-1}) = \rho_1$, so that ρ_1 describes correlation between observations one period apart. It follows that $\text{corr}(e_t, e_{t-k}) = \rho_1^k$. From (7.4) it can be seen that the variation in the errors e_t will be understated if the model does not explicitly allow autocorrelation – this in turn means credible (confidence) intervals for the components of β that are too narrow.

Bayesian estimation of the autoregressive AR(p) error model is simplified by conditioning on the first p observations when there is a p-order autodependence in the e_t (Chib, 1993). This avoids specifying a prior for the latent data values $e_0, e_{-1}, \ldots, e_{1-p}$. Another option is to use composite parameters for terms involving the latent data. Thus for a first order autoregressive process the points $t = 2, \ldots, T$ are modelled as

$$y_t = x_t\beta + e_t$$
$$e_t = \rho_1 e_{t-1} + u_t.$$

The first point is modelled as

$$y_1 = x_1\beta + M + u_1$$

where $M = \rho_1 e_0$ and involves the latent datapoint e_0 (Zellner and Tiao, 1964). So y_1 has mean $x_1\beta + M$ and variance σ^2, with M regarded as an unknown parameter (typically modelled as a fixed effect). A full likelihood approach to the ARMA(p, q) errors regression model is developed by Chib and Greenberg (1994). As in the ARMA(p, q) model involving the lagged observations, back-casting provides another way of estimating the initial latent data, so avoiding a conditional approach which loses some observed data.

Note that the prior assumptions for the Bayesian regression with autoregressive errors do not restrict $\rho_1, \ldots, \rho_p$ to satisfy the stationarity constraint, where the roots of $\rho(B)$ lie outside the unit circle. Thus for an AR(1) model, there is no necessary restriction on ρ_1 and explosive processes can be accommodated. However, a model with this form of prior involving regression on a covariate may lead to identifiability problems if the changing level of y_t could be due equally to changes in the level of x_t as to nonstationary errors. Zellner and Tiao (1964) illustrate the dependence which may occur between a nonstationary error process and the posterior density of the regression parameter β in an AR(1) error model with a single covariate.

Example 7.10 Regression with autocorrelated errors: Electricity consumption Chib (1993) applies the AR(p) model in the error term with conditioning on the initial observations and without necessarily assuming stationarity in the AR process *a priori*. It is also assumed that the parameters $(\underset{\sim}{\beta}, \sigma^2)$ are independent of the lag coefficient vector $\rho = (\rho_1, \rho_2, \ldots, \rho_p)$, and that x_t does not contain any lagged values of y. Under this specification the complete conditional distributions on the variance, the regression term $\underset{\sim}{\beta}$ and the autoregressive parameters ρ are available to be sampled directly from standard normal and gamma conditional densities.

Here we specify priors and normal data likelihoods for the data used in Chib (1993). The data consists of the log of electricity consumption in kilowatt hours per customer, and the covariates are:

PCI (per capital income, denoted `xtran[,2]` in Program 7.10),
the log of the real electricity price (PE),
the log of the real gas price (PG),
cooling degree days (CDD), and
heating degrees days (HDD).

After a single long run of 10 000 (with 5000 burn-in) the posterior estimates, and the comparable results of Chib (1993) are as shown in Table 7.8.

Table 7.8

	Mean	SD	2.5%	Median	97.5%	Chib Mean	Chib SD
CONST	−7.146	35.51	−95.13	−6.463	71.4	−8.75	8.11
PCI	0.6364	0.1659	0.29	0.6418	0.9492	0.669	0.144
PE	−0.1744	0.06499	−0.3062	−0.1731	−0.05105	−0.188	0.065
PG	−0.1085	0.07485	−0.2587	−0.1074	0.03711	−0.106	0.069
CDD	3.05E-05	3.08E-05	−2.21E-05	2.86E-05	9.09E-05	2.44E-5	2.26E-5
HDD	3.33E-04	3.28E-05	2.67E-04	3.34E-04	3.97E-04	3.36E-4	2.67E-5

Example 7.11 Cobb–Douglas production function Judge *et al.* (1988) analyse $N = 20$ observations from a series $\{y_t, x_{1t}, x_{2t}\}$ denoting the logarithms of output Q_t, labour L_t and capital K_t respectively. The Cobb–Douglas production relation is multiplicative

$$Q_t = \alpha L_t^{\beta_1} K_t^{\beta_2} \eta_t$$

with $e_t = \log \eta_t$ Normally distributed. Expressed in log-linear terms, the Cobb–Douglas model is therefore

$$y_t = \beta_0 + \beta_1 x_{lt} + \beta_2 x_{2t} + e_t.$$

Economic theory suggests parameter constraints

$$0 < \beta_1 < 1, \ 0 < \beta_2 < 1$$

though values outside this range are not absolutely excluded. For this small data set prior assumptions on parameters, and other factors such as choice of scale in y and x (centred or not), and the treatment of the first observation will be important. A full model might also allow for changes in the level of y beyond those produced by changes in x_2 and x_3.

A model allowing for AR(1) dependence in the errors is an initial approach to reproducing the dynamic features of these data, so that

$$e_t = \rho e_{t-1} + u_t.$$

Judge *et al.* compare Bayes estimates of this model with generalised least squares, nonlinear least squares, and maximum likelihood. They point out that Monte Carlo evidence of finite sample estimation with the latter three methods suggests that ρ will be underestimated. They obtain a parameter value (and SD) of $\rho = 0.67$ (0.19), as compared with maximum likelihood of 0.56 (0.19). The ML estimates of the other parameters are $\beta_0 = 4.06$ (5.77), $\beta_1 = 1.67$ (0.28), $\beta_2 = 0.76$ (0.14) and $\sigma^2 = 6.07$ (1.9). The AR(1) errors model can be rewritten as

$$y_t = \beta_0 + \beta_1 x_{1t} + \beta_2 x_{2t} + \rho(y_{t-1} - \beta_0 - \beta_1 x_{1,t-1} - \beta_2 x_{2,t-1}) + u_t$$

for $t = 1, \ldots, N$. In the BUGS centred form

$$\begin{aligned} y_t &\sim \mathrm{N}(\mu_t, \sigma^2) \\ \mu_t &= \beta_0 + \beta_1 x_{1t} + \beta_2 x_{2t} + \rho(y_{t-1} - \beta_0 - \beta_1 x_{1,t-1} - \beta_2 x_{2,t-1}). \end{aligned} \tag{7.5}$$

The values y_0, x_{10} and x_{20} are latent data (additional unknowns), and one way to proceed is to condition on the first observation (*Program 7.11 Cobb Douglas Function, Model A*). We initially apply this method with vague priors on all β_j and a uniform prior on ρ constraining it to $[-1, 1]$. This gives (from the second half of 50 000 iterations) estimates as shown in Table 7.9.

The posterior mean of the autocorrelation parameter exceeds the estimate cited by Judge *et al.*, as well as those from classical methods.

If stationarity is assumed with $-1 < \rho < 1$ (see model B, Program 7.11) then the conditional variance of the first observation is $\sigma^2/(1 - \rho^2)$ with mean $\mu_1 = \beta x$, while subsequent observations have mean μ_t as in (7.5) above. This leads to an estimate of $\rho = 0.64$ (and 95% interval 0.25, 0.97) with $\beta_2 = 1.79$ $(1.47, 2.11)$ and $\beta_3 = 0.80$ $(0.47, 1.10)$.

Another approach, not restricted to stationarity, follows Zellner (1971) in modelling all the data together, i.e.

$$y_t \sim \mathrm{N}(\mu_t, \sigma^2)$$

for $t = 1, \ldots, N$ but with the terms in μ_1 involving latent data represented by a composite parameter M, such that

$$\mu_1 = \beta_0 + \beta_1 x_{11} + \beta_2 x_{21} + M - \rho\beta_0.$$

M is assigned a vague prior in *Program 7.11 Model C.*

This yields similar results to the conditional method, and suggests a 20% probability of nonstationarity. A variant on this theme is to treat the first data point separately, so that

$$y_t \sim \mathrm{N}(\mu_t, \sigma^2)$$

for $t = 2, \ldots, N$ with μ_t as in (7.5), and with y_1 is a separate fixed effect. So the model for μ_1 does not need to involve latent terms, and $\mu_1 = \beta_0 + \beta_1 x_{11} + \beta_2 x_{21}$. This option comes closer to Judge *et al.* (1988), with results as shown in Table 7.10.

Finally, we might try to model the latent data directly. For example, we could assume x_{10} is drawn from the same density as the known x_{1t} $(t = 1, \ldots, N)$, and similarly for x_{20}; they would then be regarded as missing data. We might

Table 7.9

Parameter	Mean	SD	2.5%	Median	97.5%
β_1	−23	108.9	−309	−5.57	170.5
β_2	2.01	0.79	0.51	1.96	3.71
β_3	0.79	0.16	0.47	0.79	1.10
ρ	0.83	0.18	0.38	0.91	1.00
σ^2	9.53	3.84	4.57	8.73	19.15

Table 7.10

Parameter	Mean	SD	2.5%	Median	97.5%
β_1	−0.97	4.66	−11.95	−0.65	8.74
β_2	1.82	0.58	0.67	1.81	3.03
β_3	0.78	0.16	0.45	0.78	1.09
ρ	0.73	0.19	0.30	0.76	0.98
σ	9.03	3.66	4.37	8.26	18.28

take all data as centred ($\underset{\sim}{z} = \underset{\sim}{x} - \bar{\underset{\sim}{x}}, w = y - \bar{y}$) and then set a prior for the predictors including time 0:

$$z_{1t} \sim \mathrm{N}(0, \xi_1), t = 0, \ldots, N;$$
$$z_{2t} \sim \mathrm{N}(0, \xi_2), t = 0, \ldots, N.$$

The latent value of the dependent variable at time 0 can then be modelled in terms of z_{10} and z_{20}, so that $w_t \sim \mathrm{N}(\mu_t, \sigma^2), t = 1, \ldots, N$ and

$$w_0 \sim \mathrm{N}(\mu_0, \sigma^2)$$
$$\mu_0 = \beta_0 + \beta_1 z_{10} + \beta_2 z_{20} + M$$

where M is a composite parameter as above. We apply this approach to the centred data in this example and the second half of a run of 50 000 iterations yields a posterior mean for ρ of 0.76. It may be noted that in all these analyses the density of ρ shows skew (median exceeds mean) and in fact suggests values of $\rho > 1$ may be feasible.

Example 7.12 Non-explosive series Zellner and Tiao (ZT, 1964) present a series of artificial data y_t ($t = 1, \ldots, 16$) generated from known investment expenditures x_t taken from Haavelmo (1947). The data follow the non-explosive model

$$y_t = \beta x_t + e_t$$
$$e_t = \rho e_{t-1} + u_t$$

where $\rho = 0.5, \beta = 3$ and $u_t \sim \mathrm{N}(0, 1)$. The aim is to 're-estimate' these parameters from the resulting data (see *Program 7.12 Non-explosive Series*).

We assume a vague prior for β, a uniform $\mathrm{U}(-1, 1)$ prior for ρ, and use the composite parameter $M = \rho e_0$ to model the first mean. The results from the second half of a 50 000 iteration run reproduce the generating model closely, except that the variance of the u_t is overstated (Table 7.11).

ZT also present an explosive series generated with the same model except that $\rho = 1.25$. This poses identifiability problems unless a more informative prior on ρ and β is adopted. Alternatively, sampling directly from the conditional posterior densities may be used, or other methods (e.g. DLM techniques) allowing a changing level.

Table 7.11

Parameter	Mean	SD	2.5%	Median	97.5%
β	2.87	0.19	2.43	2.88	3.19
ρ	0.47	0.26	−0.07	0.48	0.93
σ^2	2.10	0.96	0.95	1.89	4.60

Example 7.13 Transfer function for soil and air temperature We consider the interrelation between an 'output' variable soil temperature y_t and an input variable air temperature x_t measured at hourly intervals over 5 days (so $N = 120$) (Lai, 1979). In the single-input single-output transfer function the model is

$$y_t = \frac{A(L)}{B(L)} x_{t-b} + u_t$$

where $A(L)$ is a polynomial function of order s, with $A(B) = (A_0 - A_1B - A_2B^2 - \ldots - A_sB^s)$, and $C(B)$ is a function of order r with $C(B) = (1 - C_1B - C_2B^2 - \ldots - C_rB^r)$. The output variable thus depends both on the past r values of itself and on the s values of the input, starting at lag b (i.e. at times $t - b, t - b - 1, \ldots, t - b - s$).

The noise series u_t is also of ARMA (p, q) form as in (7.2), with

$$u_t = \Theta(B)[\Phi(B)]^{-1}a_t.$$

Thus an ARMA(1,1) error model would lead to

$$u_t = \phi_1 u_{t-1} + a_t - \theta_1 a_{t-1},$$

and an AR(1) model would give (setting $\rho = \phi_1$)

$$u_t = \rho u_{t-1} + a_t$$

where $a_t \sim \mathrm{N}(0, \tau_a)$. Then in terms of the lag operator

$$u_t = a_t/(1 - \rho B).$$

The identification of possibly appropriate values of r, s, b, p and q depends on examination of autocorrelations between transformations of x_t and y_t (after both are subject to the same 'pre-whitening' transformation) (e.g. see Box and Jenkins, 1970).

In the present case the values $r = s = b = p = 1$ and $q = 0$ were identified using these procedures. As usual we may initially difference the original variables y_t and x_t and actually work with $w_t = y_t - y_{t-1}$ and $z_t = x_t - x_{t-1}$. We can then rewrite the model (with $b = 1$ so that $z_{t-1} = x_{t-1} - x_{t-2}$) as follows

$$w_t(1 - C_1B)(1 - \rho B) = (A_0 - A_1B)(1 - \rho B)z_{t-1} + a_t(1 - C_1B)(1 - \rho B).$$

This may be written in full as

$$w_t = C_1 w_{t-1} + \rho w_{t-1} - C_1 \rho w_{t-2} + A_0 z_{t-1} - (A_0 \rho + A_1) z_{t-2} + A_1 \rho z_{t-3} + a_t - C_1 a_{t-1} - \rho a_{t-1} + \rho C_1 a_{t-2}.$$

Lai (1979) obtains via nonlinear least squares, the following estimates (SEs in brackets): $C_1 = 0.639$ (0.036), $A_0 = 0.085$ (0.0064), $A_1 = -0.0214$ (0.0088) and $\rho = -0.26$ (0.073). Therefore rising values of air temperature are associated with a significant later rise in soil temperature.

Here a non informative multivariate normal prior is used for the parameters C_1, A_0 and A_1 (see *Program 7.13 Soil Temperature*). Both the differenced series w_t and the innovations a_t are assumed normal, with non informative gamma priors used for the precisions τ_w and τ_a.

After 10 000 iterations (with 500 burn-in), the estimates obtained, with their 90% credible intervals, were $C = 0.626$ (0.51, 0.73), $A_0 = 0.087$ (0.066, 0.107), $A_1 = -0.0219$ (-0.053, 0.007) and $\rho = -0.15$ (-0.46, 0.27). Allowing for uncertainty in the estimate of σ_a^2 therefore increases the uncertainty in the noise lag ρ, and to a lesser extent the parameters C_1, A_0 and A_1. The dependence of w_t on z_{t-2} (in the parameter A_1) is predominantly negative, but shows some lack of clarity. The first lag effect is clearly positive.

The above model may be converted back to the original undifferenced y_t and x_t and this gives a model with

$$y_t = \alpha_1 y_{t-1} + \alpha_2 y_{t-2} + \alpha_3 y_{t-3} + \beta_1 x_{t-1} + \beta_2 x_{t-2} + \beta_3 x_{t-3} + \beta_4 x_{t-4} + a_t - C_1 a_{t-1}.$$

This re-expression is the basis for future prediction (i.e. of soil temperature for hours 121, 122, etc.). In estimation of this model, values of C_1 can be sampled from the density N(0.62, 0.07) obtained above, or C_1 may be re-estimated as a moving average parameter with invertibility specified by the prior. This reformulation seems to be overparameterised in that 90% credible intervals for $\alpha_2, \alpha_3, \beta_2, \beta_3$ and β_4 all straddle zero (i.e. the posterior density between the 5th and 95th percentiles contains both negative and positive values). So estimations of both forms of the model suggest that a simpler transfer function may be appropriate.

7.5.1 Cointegration and unit root testing

One problem with least squares regression of one time series y_t on another x_t is that spurious regression effects may occur if the variables are undifferenced – and if the regression effect is taken as fixed. Thus Granger and Newbold (1974) demonstrated spurious effects in a regression

$$y_t = \alpha + \beta x_t + \varepsilon_t \tag{7.6}$$

when the time series were independently generated random walk series

$$y_t = y_{t-1} + e_{1t}$$
$$x_t = x_{t-1} + e_{2t}$$

with e_{1t} and e_{2t} uncorrelated over time and with each other. The residuals $\hat{\varepsilon}_t$ from such a regression are likely to be highly autocorrelated over time, and usual tests for β in (7.6) are misleading. One remedy for this is differencing or detrending of the original series, and analysis in terms of differences or detrended series rather than the original levels.

However, long-run or equilubrium relationships between the levels (rather than differences) of series are often of theoretical economic importance. At equilubrium, successive values of the levels of variables are equal, so $y_t = y_{t-1} = \ldots = y^*$ and similarly for x. So a dynamic relationship such as

$$y_t = \alpha + \beta x_t + \gamma x_{t-1} + \delta y_{t-1} + u_t$$

becomes in steady state equilibrium

$$y^* = \alpha + \beta x^* + \gamma x^* + \delta y^*$$

with solution

$$y^* = \alpha/(1 - \delta) + (\beta + \gamma)x^*/(1 - \delta).$$

Granger (1981) shows that analysis in terms of levels may be admissible if series are cointegrated, meaning that the trends in y_t and x_t cancel out when the residual

$$\varepsilon_t = y_t - (\alpha + \beta x_t)$$

is formed, and ε_t is then stationary. More generally, a pair of series are cointegrated if there is a linear combination z_t of them such that $z_t = b_1 y_t + b_2 x_t$ is stationary.

Much of the econometric literature on whether a time series variable y_t is stationary focuses on the presence of a unit root in the regression

$$y_t = \rho_0 + \rho_1 y_{t-1} + \varepsilon_t$$

with the hypothesis of interest being whether $\rho_1 = 1$. An alternative stochastic unit root (STUR) model was proposed by Granger and Swanson (1997), such that

$$\begin{aligned} y_t &= \rho_t y_{t-1} + \varepsilon_t \\ \rho_t &= \exp(\omega_t) \end{aligned} \tag{7.7}$$

where ω_t is itself autoregressive AR(p) with mean $\mu.\omega$. So for ω_t following an AR(1) process

$$\begin{aligned} \omega_t &= \varphi_0 + \varphi_1 \omega_{t-1} + \eta_t \\ &= \mu.\omega + \varphi_1(\omega_{t-1} - \mu.\omega) + \eta_t \end{aligned} \tag{7.8}$$

where

$$\mu.\omega = \varphi_0/(1 - \varphi_1),$$

$\eta_t \sim \text{N}(0, \sigma_\eta^2)$ and $\varepsilon_t \sim \text{N}(0, \sigma_\varepsilon^2)$. Jones and Marriott (JM, 1999) consider Bayesian estimation of this model. The analysis focuses especially on the value of $\mu.\omega$. Negative values of $\mu.\omega$ correspond to $E(\rho_t)$ being under 1, while a unit root process has $\mu.\omega = 0$ and so $E(\rho_t) = 1$. Positive values of $\mu.\omega$

Table 7.12

Parameter	Mean	SD	2.5%	Median	97.5%
$\mu.\omega$	−0.46	0.12	−0.70	−0.45	−0.24
σ^2_ε	53.7	5.3	44.1	53.7	64.1
σ^2_η	0.45	0.16	0.18	0.44	0.76

indicate either an 'explosive' nonstationary time series or the need for an alternative model other than that of (7.7) and (7.8).

Example 7.14 Cointegrated macro-economic series Here we consider the application of a STUR model to assess whether the residuals from a regression of one time series on another indicate that the series are cointegrated. Griffiths *et al.* (1993) consider US Quarterly Consumption C_t and disposable income Y_t at constant prices, between Winter 1947 and Fall 1980 ($N = 136$ values). A simple least squares regression of C_t on Y_t gives

$$\hat{C}_t = \alpha + \beta Y_t = 326 + 0.862 Y_t$$

with the marginal propensity to consume ($\beta = 0.862$) having an apparent t ratio of 189. The question is whether the two series are cointegrated to an extent that we can regard the estimate of β as a long-run (steady state) marginal propensity.

In *Program 7.14 US Consumption and Income* we form the errors ε_t as $C_t - (326 + 0.862 Y_t)$ and then assess the unit root hypothesis for the ε_t using the STUR model. For the ω_t series we take $p = 1$, and assign vague priors to $1/\sigma^2_\eta$ and $1/\sigma^2_\varepsilon$. Following JM we assume a vague prior on $\mu.\omega$ but take a more informative prior $\varphi_1 \sim$ N (0, 1) on the AR(1) parameter in (7.8). We focus on the probability that $\mu.\omega$ is less than zero to assess the unit root hypothesis. In fact the mean of the ω_t series is estimated as −0.46 with $s.d. = 0.12$ and there is zero probability that the mean exceeds 0. This finding corresponds to that of Griffiths *et al.* using a modified Dickey–Fuller test. The parameters obtained are given in Table 7.12.

7.6 LINEAR DYNAMIC MODELS AND TIME VARYING COEFFICIENTS

Classical time series regression models and forecasting methods resting on stationarity assume fixed parameters and observation variances. However, a model with good fit in the sample period may have poor performance after that if the underlying parameters are evolving through time. When the forecasting accuracy of fixed coefficient models is low or their forecast errors are strongly autocorrelated, we may consider allowing time-varying regression coefficients, and possibly varying precisions governing the evolution of these coefficients and of the observations also. Forecasts based on such a dynamic framework may have better forecasting performance both within the sample and to future periods, since forecasts are based on the most up-to-date coefficient values, revised in line with changing structural dependencies between successive values of the outcome or between the outcome and predictors. On the other hand, introducing time variability may be less advisable if long-term forecasts are to be made, since DLM forecasts tend to become diffuse in the long term (Harrison and West, 1997, Chapter 9).

There have been extensive developments in varying coefficient models for time series data, with a specifically Bayesian methodology called dynamic linear modelling being developed in the 1970s and 1980s (West and Harrison, 1989). An introduction to dynamic coefficient models in forecasting is provided by Pole *et al.* (1994). In fact, however, random coefficients are quite feasibly included within the 'traditional' ARMA model (e.g. McCulloch and Tsay, 1994), and there is a wide interface between the two approaches, with the possibility of mixed models combining both approaches (West and Harrison, 1997).

The estimation of varying coefficients is based on the assumption of their belonging to a common density, and Bayesian methods are often appropriate to 'pooled strength' estimation of coefficients from a common density and evolving in a stochastic fashion. They also have advantages in dealing with outliers and shifts in time series (through outlier detection methods or robust regression approaches). On the other hand, varying coefficient models are highly parameterised and may be less well identified than more 'conventional' fixed coefficient time series models.

As an example of the typical framework which may be adopted consider a univariate Normal outcome $y_t, t = 1, \ldots, N$, with changing variance V_t through time. The mean of y_t is to be modelled in terms of two varying predictors x_t and z_t, so that

$$y_t \sim \mathrm{N}(\mu_t, V_t)$$

$$\mu_t = \beta_{1t} + \beta_{2t}x_t + \beta_{3t}z_t.$$

The evolution of the varying coefficients, or 'state' parameters, β_{1t}, β_{2t} and β_{3t} may be specified via separate (univariate) random walk priors, again assumed Normal, with (for $t > 1$)

$$\begin{aligned}\beta_{1t} &\sim \mathrm{N}(\beta_{1,t-1}, W_1),\\ \beta_{2t} &\sim \mathrm{N}(\beta_{2,t-1}, W_2),\\ \beta_{3t} &\sim \mathrm{N}(\beta_{3,t-1}, W_3).\end{aligned} \tag{7.9}$$

So conditional on β_t the y_t are assumed independent of y_s and β_s for all past and future values of s. The state vector β_t evolves in time according to a Markovian process with a particular evolution density; here the evolution is governed by Normal densities. Given β_{t-1}, β_t is conditionally independent of preceding values of β_s, for $s < t - 1$. The notation in (7.9) uses the centred method for expressing a normal prior, and we might also have written

$$\beta_{1t} = \beta_{1,t-1} + w_{1t}$$
$$w_{1t} \sim \mathrm{N}(0, W_1)$$

and so on. The variances W_1, W_2 and W_3 of the regression coefficients are, for illustration, taken as fixed but dynamic forms are possible. The first period coefficients are modelled separately as fixed effects with known prior parameters, e.g. $\beta_{11} \sim \mathrm{N}(a_{11}, E_{11})$ where a_{11} and E_{11} are known, with a typical prior taking $a_{11} = 0$ and E_{11} large.

An alternative to univariate random walk priors on these coefficients might have been a multivariate evolution, for example if B_t denotes the vector $(\beta_{1t}, \beta_{2t}, \beta_{3t})$, then

$$B_t \sim \mathrm{N}_3(B_{t-1}, W)$$

where W is a dispersion matrix. With appropriate links from outcome y_t to the regression mean μ_t, this framework adapts to discrete outcomes.

7.6.1 The general DLM framework

This approach is an illustration of a more general evolution mechanism for uinivariate or multivariate observations, and for the parameters in the regression function. The standard form for such dynamic models involves one step and linear dependence in the regression or state parameters – hence the terminology Markovian transition dependence. A transition matrix M_t defines this transition between states. Consider the usual linear regression model $y_t = x_t\beta + \varepsilon_t$ (with x containing p predictors, the first being an intercept) but with a changing coefficient vector β_t:

$$y_t = x_t\beta_t + \varepsilon_t \tag{7.10}$$

$$\beta_t = M_t\beta_{t-1} + \omega_t \tag{7.11}$$

with the first equation defined for $t = 1, \ldots, N$ periods, and the second for $t = 2, \ldots, N$.

Equation (7.10) defines the observational model, for an outcome y_t and its relation to p covariates x_t. The first of the covariates is often a constant or intercept term, representing the level of the series. If y_t is univariate then $\underset{\sim}{\beta_t}$ is a vector of the form $(\beta_{1t}, \beta_{2t}, \ldots, \beta_{pt})$ while if y_t is multivariate (with m variables) it will be an $m \times p$ matrix at each time point. The most general form of the 'state-space' model has errors ε_t and ω_t with time-dependent variances V_t and W_t, denoting observation and system variances respectively.

The second equation, known as the state or system equation, specifies the evolution of the varying parameters through time. If the model includes p changing regression coefficients, then the evolution is governed by a $p \times p$ transition matrix M_t. With M_t usually taken as known and fixed in time (e.g.

$M_t = I$ the identity matrix), the state equation often reduces to simple forms of autoregressive prior, such as the random walk prior in the example above. The state equation can be seen to define a prior distribution on the sequence $(\underset{\sim}{\beta}_1, \underset{\sim}{\beta}_2, \ldots, \underset{\sim}{\beta}_N)$ in a way analogous to a random effects prior in a mixed cross-sectional model (Knorr-Held, 1997).

Note that if $M_t = I$ and $W_{jt} = 0, j = 1, \ldots, p$, then we have the usual fixed coefficient multiple regression. V_t is a time specific observation variance or covariance matrix depending on whether y_t is univariate or multivariate. At its most general W_t is a time-specific $p \times p$ covariance matrix for the components of the regression vector, but for tractable estimation the state parameters may be assumed to vary independently of each other and W_t reduces to a vector of variances, $W_t = (W_{1t}, W_{2t}, \ldots, W_{pt})$. A further frequently made assumption (Gamerman, 1997, Chapter 5) is that the variances V_t and W_t are constant over time, though the DLM still involves evolution in levels or regression effects.

This model is heavily parameterised, and practical application often involves simplifications to the model structure, informative priors to produce identifiability, or both. There is also often a clear empirical Bayes flavour in applications of DLMs, with alternative priors closely related to the data, and often being used in deriving alternative 'what-if' forecasting scenarios (West and Harrison, 1989; Pole, 1994)

The system is typically set in motion by assigning priors to the regression parameters in the first period, $\beta_{11}, \beta_{21}, \ldots, \beta_{p1}$, usually with known levels and variances, i.e. $\beta_{11} \sim \mathrm{N}(a_{11}, E_{11}), \beta_{21} \sim \mathrm{N}(a_{21}, E_{21}), \ldots, \beta_{p1} \sim \mathrm{N}(a_{p1}, E_{p1})$. Regression parameters for $t > 1$ are determined by preceding values (at $t-1$ or possibly before also), drawing on the prior for the first period value and the evolution error terms ω_t Sometimes detailed contextual knowledge would be available on the level and variances of the series and regression parameters, as illustrated in the examples of Pole *et al.* (1994) and West and Harrison (1989).

7.6.2 On-line vs retrospective analysis

Updating parameters through times $t = 1, \ldots, N$ based on data D_t for current and preceding periods constitutes an 'on-line' analysis. The increased uncertainty through time about the state parameters in such an analysis is apparent for a univariate normal outcome, y_t. Thus, suppose we assume the first period regression parameter has a prior

$$\underset{\sim}{\beta}_1 \sim \mathrm{N}(a_1, E_1)$$

where $a_1 = (a_{11}, a_{21}, \ldots, a_{p1})$ and $E_1 = (E_{11}, E_{21}, \ldots, E_{p1})$ are both known. Then the priors for successive $(t > 1)$ regression vectors are updated on the basis of the posterior densities of the preceding regression terms, namely $\underset{\sim}{\beta}_{t-1} \sim \mathrm{N}(\underset{\sim}{m}_{t-1}, \underset{\sim}{C}_{t-1})$, so that

$$\underset{\sim}{\beta}_t \sim \mathrm{N}(\underset{\sim}{b}_t, R_t)$$

where $\underset{\sim}{b}_t = \underset{\sim}{m}_{t-1} M_t$ and $R_t = M_t C_{t-1} M_t + W_t$. So the state parameters accumulate uncertainty through time. Updating or 'filtering' equations, which for normal outcomes take a Kalman filter form, are discussed in detail in West and Harrison (1989) and West *et al.* (1985). They form the basis, in the normal outcome case, for direct sampling of the parameters and forecasts of the outcome variable from full conditional densities.

Another option in state-space models is retrospective smoothing of the on-line regression coefficients β_{jt} and overall means μ_t. The initial 'on-line' estimates are revised to take account of the full span of observations, and their precision accordingly increased. This may be applied in a BUGS setting by a secondary analysis of posterior estimates and standard deviations. Suppose we have a univariate and centred Normal outcome y_t modelled in terms of a level and a regression on a single predictor x_t. From an initial sampling run, we have estimated posterior means b_t and variances R_t for the regression coefficient β_t (i.e. $\beta_t \sim \mathrm{N}(b_t, R_t)$ approximately), and an estimate of the fixed variance W governing the evolution of this coefficient. Assume also that the evolution of the β_t followed a random walk. Suppose we denote the information contained in the full span of observations $\{y_t, x_t\}$ from $t = 1, \ldots, N$ as D_N and the full regression parameter is set as φ. Then we seek to revise the on-line estimates according to the retrospective predictions

$$p(\beta_t | D_N) = \int p(\beta_t | D_N, \varphi) p(\varphi | D_N)\, d\varphi, \quad t = 1, \ldots, N.$$

The second term in the integral is the posterior distribution of φ, and the first term can be shown to equal $p(\beta_t | D_t, \varphi)$. The retrospectively smoothed estimate of β_t is then a weighted average of the initial on-line estimates of $\beta_t, \beta_{t+1}, \ldots, \beta_N$, with weights calculated using the estimates of b_t, R_t and W. For example, one-step retrospective smoothing is governed by

$$\beta_{t-1} | D_t \sim \mathrm{N}(e_{t-1}, f_{t-1})$$

where $e_{t-1} = b_{t-1} - B_{t-1}(b_t - b_{t-1})$, $f_{t-1} = R_{t-1} - B_{t-1}(R_{t-1} + W - R_t)B_{t-1}$ and $B_{t-1} = R_{t-1}(R_t + W)^{-1}$.

7.6.3 Random walk and other autoregressive priors

As examples of typical simplifications suppose in (7.10) and (7.11) above that $M_t = I$ and $p = 1$, namely an intercept-only model. Then we have a random walk for the level β_{1t} of the series

$$y_t = \beta_{1,t} + \varepsilon_t$$

with the evolving level specified by (for $t > 1$)

$$\beta_{1,t} = \beta_{1,t-1} + \omega_{1t}.$$

In the case of a normal outcome $\varepsilon_t \sim \mathrm{N}(0, V_t)$ and $\omega_{1t} \sim \mathrm{N}(0, W_{1t})$, with W_{1t} a scalar parameter. The first period prior may be modelled as a separate fixed effect, with $\beta_{11} \sim \mathrm{N}(a, E)$, where a and E are known. If also $W_{1t} = W_1$ is fixed through time, then in centred form we can write the data model and prior for the evolving state parameters as

$$y_t \sim \mathrm{N}(\beta_{1t}, V_t), \quad t = 1, \ldots, N$$
$$\beta_{1t} \sim \mathrm{N}(\beta_{1,t-1}, W_1), \quad t > 1.$$

Such a random walk may be assumed for any regression parameter also. Another simple prior for the level or regression parameters is a second order random walk,

$$\beta_{1t} = 2\beta_{1,t-1} - \beta_{1,t-2} + \omega_{1t}, \quad t > 2.$$

If we have evidence for a trend in the level then an additional time-varying parameter (sometimes called a slope) τ_{1t} may be introduced to represent this growth:

$$y_t = \beta_{1,\tau} + \varepsilon_\tau$$
$$\beta_{1t} = \beta_{1,\tau-1} + \tau_{1,t-1} + \omega_{lt}$$
$$\tau_{lt} = \tau_{1,t-1} + \omega_{2t}$$

with $\omega_{2t} \sim \mathrm{N}(0, W_{2t})$. A range of options is possible if there are several covariates. For example, Fahrmeir (1992) considers evolution only in the level, while keeping the other coefficients constant.

7.6.4 Discounted priors for time-specific variances

If the system variances are specified to be changing through time, Ameen and Harrison (1985) suggest a discounting process in the prior information about the variances, carried over from one time to the next. This avoids estimation of each time-specific variance but allows some flexibility through time. For example, we might specify prior information on the first period error terms ω_{j1} $(j = 1, \ldots, p)$, and downweight this information in successive periods. If the system variances are varying over time, then there is wide scope for modelling like this rather than trying to estimate each of the $N \times p$ variances.

Suppose we have an intercept $X_1 = 1$ and a single covariate X_2, then we may specify priors on the precisions $P_{11} = 1/W_{11}$ and $P_{21} = 1/W_{21}$ at time 1 and then discount precisions in future periods using a discount factor δ. Thus

$$P_{jt} = \delta P_{jt-1}, \quad j = 1, 2; \quad t > 2. \tag{7.12}$$

The discount factor is between 0 and 1, and usually exceeds 0.9. A discount factor of 0.95 is approximately equivalent to a 5% increase in uncertainty in each time period. The idea is that as information ages it becomes less reliable or useful. If we define a diagonal matrix, constant in time, with diagonal elements $D_{jjt} = 1/\delta$ for $j = 1, 2$, then the prior covariance R_t for the state vector at time t becomes

$$R_t = D_t M_t C_{t-1} M_t D_t + \omega_t.$$

Pole *et al.* (1994) suggest a few standard values (0.9, 0.95, 0.99) are tried and their predictions and fit compared. They consider that the likelihood is often flat in terms of distinguishing between such values, and so a prior might favour values near 1.

7.6.5 Priors under interventions

Often instability will be caused by external events or 'interventions' (e.g. a competitor opening a new product line). We may introduce an extra error term at the time of the intervention to accommodate the anticipated series shift. This may be illustrated with an economic example. Following Pole *et al.* (1994) assume sales (S) of a commodity at time t are assumed to depend only on prices (P) at t. Assume also that evolution of the level (L) and sales effect (β) is confined to a random walk autoregressive prior with a fixed variance. Then the observation model is

$$S_t = L_t + \beta_t P_t + \varepsilon_t$$

with state priors for all periods except the first

$$L_t = L_{t-1} + \omega_{1t},\ \omega_{1t} \sim \mathrm{N}(0, W1), \quad t = 2, \dots N$$
$$\beta_t = \beta_{t-1} + \omega_{2t},\ \omega_{2t} \sim \mathrm{N}(0, W2), \quad t = 2, \dots, N.$$

The first period priors may be based on known means and variances, typically minimally informative, for example

$$\beta_1 \sim \mathrm{N}(0, 100);\ L_1 \sim \mathrm{N}(0, 100).$$

If the intervention is at time I and affects only the level of sales then we might modify the autoregressive prior for the level with an additional effect η_{1t} operating only from time I. Thus

$$\begin{aligned}
L_t &= L_{t-1} + \omega_{1t}, & t &= 1, \dots, I-1\\
L_t &= L_{t-1} + \omega_{1t} + \eta_{1t}, & t &= I, \dots, N\\
\omega_{1t} &\sim \mathrm{N}(0, W_1), & t &= 1, \dots, N\\
\eta_{1t} &\sim \mathrm{N}(0, H_1), & t &= I, \dots, N.
\end{aligned}$$

If the intervention at time I affected the price–sales relationship (e.g. a government price control) then a similar modification could be made to the prior for β_t to reflect the greater uncertainty about the parameter's future evolution. If it is not assumed that the variances of ω_{1t} and ω_{2t} are constant then discontinuities may also be modelled via the discounting mechanism; to allow for greater uncertainty (i.e. about the smoothness of the process) around a particular time point a larger than usual discount factor is adopted (West *et al.*, 1985).

Another way to model various shifts in a series is via a latent class process. Gordon and Smith (1990) illustrate how discontinuities in medical time series can be modelled by extending the changing trend (changing slope) model above

$$y_t = \beta_t + \varepsilon_t^{[j]}$$
$$\beta_t = \beta_{t-1} + \tau_t + \omega_t^{[j]}$$
$$\tau_t = \tau_{t-1} + \eta_t^{[j]}$$

with $J = 4$ possible latent classes $j = 1, \ldots, J$. Here y_t denotes the measured biochemical variable, β_t its 'actual' or true level, and τ_t is the trend or slope of the series. Thus for state $j = 3$, say, the η_t have a large variance, the ω_t have virtually no variance, and the ε_t have a 'typical' variance: $\varepsilon_t^{[3]} \sim \mathrm{N}(0, 1)$, $\omega_t^{[3]} \sim \mathrm{N}(0, 0.01), \eta_t^{[3]} \sim \mathrm{N}(0, 100)$. Choice of this state corresponds to a marked change in the slope of the series. Choice of other categories of j at a particular time t may refer to typical changes in observed level only (with no discontinuity in either slope or actual level), or marked changes in actual level but not in slope or measured level, or marked change in measured level.

Example 7.15 Milk production West and Harrison (1989, Section 3.3) consider the interrelationship over 1970–1982 between y_t (annual milk production in lbs $\times 10^9$) and x_t (milk cow totals $\times 10^6$). See Table 7.13.

They suggest a baseline observation model without an intercept for $t = 1, \ldots, 13$, namely

$$y_t = \beta_t x_t + v_t$$

and a system model

$$\beta_t = \beta_{t-1} + w_t.$$

They suggest a fixed variance $W = 0.05$ for the evolution equation errors w_t . In the centred notation therefore

$$\beta_t \sim \mathrm{N}(\beta_{t-1}, W).$$

If additionally the precision of the y_t is set to 1, then the implementation via BUGS (see *Program 7.15 Milk Production*) shows β_t increasing in the second half of the 13-year period as a result of increasing productivity (Table 7.14).

At $t = 11$ the forecast for year 12 has posterior mean 129.3 and credible interval (123.8, 134.7), whereas the actual observation was 130.0. At $t = 12$, the forecast for the next year has mean 131.6 and credible interval (126.1, 137.1), whereas the actual observation is 135.7. The forecasts from a locally dynamic model improve over those from the standard static model where $W = 0$.

Table 7.13

	1970	1971	1972	1973	1974	1975	1976	1977	1978	1979	1980	1981	1982
y	117	117.6	120	115.5	115.6	115.4	120.2	122.7	121.5	123.4	127.5	130	135.8
x	12	11.8	11.7	11.4	11.2	11.1	11	11	10.8	10.7	10.8	10.9	11

Table 7.14

Parameter	Mean	SD	2.5%	Median	97.5%
$\beta[1]$	9.785	0.079	9.63	9.79	9.94
$\beta[2]$	10.04	0.076	9.89	10.04	10.19
$\beta[3]$	10.22	0.076	10.07	10.22	10.37
$\beta[4]$	10.16	0.078	10.01	10.16	10.32
$\beta[5]$	10.32	0.078	10.16	10.32	10.47
$\beta[6]$	10.45	0.080	10.29	10.45	10.61
$\beta[7]$	10.89	0.080	10.74	10.89	11.05
$\beta[8]$	11.14	0.080	10.98	11.14	11.30
$\beta[9]$	11.27	0.082	11.11	11.27	11.43
$\beta[10]$	11.54	0.082	11.38	11.54	11.70
$\beta[11]$	11.86	0.081	11.70	11.86	12.02
$\beta[12]$	11.97	0.080	11.81	11.97	12.13
$\beta[13]$	12.29	0.085	12.13	12.29	12.46
One-step-ahead forecasts					
$\hat{y}_2$	115.4	3.0	109.7	115.4	121.3
$\hat{y}_3$	117.5	2.9	111.7	117.5	123.2
$\hat{y}_4$	116.5	2.9	110.8	116.5	122.1
$\hat{y}_5$	113.8	2.8	108.3	113.8	119.4
$\hat{y}_6$	114.5	2.8	109.1	114.5	120.0
$\hat{y}_7$	114.9	2.8	109.4	114.9	120.5
$\hat{y}_8$	119.8	2.8	114.3	119.8	125.3
$\hat{y}_9$	120.3	2.8	114.8	120.3	125.6
$\hat{y}_{10}$	120.6	2.7	115.2	120.6	126.0
$\hat{y}_{11}$	124.6	2.8	119.3	124.6	130.0
$\hat{y}_{12}$	129.3	2.8	123.8	129.2	134.7
$\hat{y}_{13}$	131.6	2.8	126.1	131.6	137.1

Example 7.16 Market share, promotion and prices Variance discounting may be illustrated by the sales model of Pole *et al.* (1994). They consider a weekly time series over two years (1990 and 1991) of the market share S_t of a consumer product (so the total observations are $N = 104$). Fluctuations in market share are related to (a) the price of the product relative to the average for such products, denoted P_t; (b) an index of the promotion level for the product, OWNPROM_t; and (c) an index of promotions of alternative competing products, CPROM_t. On economic grounds, the impact on the product's market share of increased competitor promotion activity or raised price should be negative, while promotion of the brand itself should enhance market share. The share variable is a percentage but varies within a narrow range (about 40–45%) and can be approximated by a normal. The predictors are provided in standardised form by Pole *et al.* (1994).

First a static model, with regression coefficients and their variances fixed through time (and the observation variance also fixed), is applied

$$S_t = \beta_0 + \beta_1 P_t + \beta_2 \text{OWNPROM}_t + \beta_3 \text{CPROM}_t + \varepsilon_t.$$

This model fit uses the priors suggested by Pole *et al.*, namely

$$\beta_0 \sim \text{N}(42, 25),$$
$$\beta_1 \sim \text{N}(0, 4),$$
$$\beta_2 \sim \text{N}(0, 4)$$

and

$$\beta_3 \sim \text{N}(0, 4).$$

Forecasting market share one week ahead with this model gives evidence of autocorrelation in the forecast residuals. The forecasts tend to be high in weeks 10–20 of 1990 and the last few weeks of 1990, but lower through 1991. This may indicate insufficient temporal flexibility in the parameters describing the level of market share and the impact on market share of the three predictors. The lag one correlation in the forecast residuals is estimated to be 0.68.

The DLM version of this model (*Model B in Program 7.16*) has the form

$$S_t = \beta_{0t} + \beta_{1t} P_t + \beta_{2t} \text{OWNPROM}_t + \beta_{3t} \text{CPROM}_t + \varepsilon_t.$$

The first period regression effects are modelled as fixed effects with priors as in the static model above. Succeeding regression components ($t > 1$) follow a random walk prior,

$$\beta_{jt} \sim \text{N}(\beta_{j,\,t-1}, \kappa_{jt}).$$

Informative priors are used in the sense that the β_{1t} and β_{3t} are constrained to be negative, and the β_{2t} are constrained to be positive.

We allow both the variance of the observations and the evolving parameters to vary through time, with the variation being produced by time-specific discount factors, as in (7.12). Thus if $\pi_{jt} = 1/\kappa_{jt}$ are the precisions of the regression parameters and P_j denotes the initial precisions ($j = 0, \ldots, 3$), then

$$\pi_{jt} = (\delta_j)^{t-1} P_j.$$

The initial precision of the observations is Q and subsequent precisions τ_t are similarly discounted with a factor δ_s. We assign gamma priors, $\text{G}(0.5, 0.5)$ to Q, and $\text{G}(1, 1)$ to $P_j (j = 0, \ldots, 3)$. Following Pole *et al.* δ_s is set to 0.99.

We use the mean absolute deviation, MAD, as a measure of predictive fit for the one-step forecasts of S_t. As an example of the possibilities of varying the discounts to improve predictions, the MAD is compared for two models:

(a) a fixed precision on the predictors ($\delta_1 = \delta_2 = \delta_3 = 1$), but variable precision on the level ($\delta_0 = 0.99$);
(b) a fixed precision on the level ($\delta_0 = 1$), but variable precision on the predictors ($\delta_1 = \delta_2 = \delta_3 = 0.99$).

The first option has a lower average and median MAD than the latter: over the second half of single runs of 40 000 iterations, the median MAD of model (a) is 0.314 but on model (b) is 0.320. The lag one correlation in the forecast residuals under model (a) is estimated at under 0.15.

As illustrations of the regression fit are posterior estimates in Table 7.15 of $\beta_{0t}, \beta_{1t}, \beta_{2t}$ and β_{3t} for $t = 1, t = 26, t = 52, t = 78$ and $t = 102$ under model (a).

A run of 5000 iterations was used to derive a full set over all weeks of estimates of the price coefficient β_{1t}. This varies from -1.5 to -0.8, with a major fall in the impact of price in the first quarter of 1991 (Figure 7.1). Pole *et al.* attribute this to increased promotion activity on the brand (i.e. a rise in OWNPROM_t) in this period, and also to their being a relative price advantage for the product around this time.

Example 7.17 Growth in physician expenditures A standard form of dynamic linear model for time series assumes normal errors in both the observation and state equation, and assumes the value of the transition matrix M_t is known. For an illustration of the modification of these assumptions we consider a univariate series y_t of annual physician expenditures in dollars, as in Carlin *et al.* (1992). The regression model is confined to the level x_t of the series, so x_t can be considered as representing the true annual expenditures. This then amounts to a form of measurement error model, with x_t as the 'signal'.

Table 7.15

	Mean	SD	2.5%	Median	97.5%
$\beta_{0,1}$	42.36	0.38	41.63	42.36	43.11
$\beta_{1,1}$	−1.77	0.51	−2.74	−1.79	−0.74
$\beta_{2,1}$	0.44	0.26	0.04	0.40	1.02
$\beta_{3,1}$	−1.06	0.53	−2.14	−1.05	−0.14
$\beta_{0,26}$	41.85	0.47	40.84	41.88	42.70
$\beta_{1,26}$	−1.59	0.29	−2.13	−1.59	−1.02
$\beta_{2,26}$	0.51	0.24	0.10	0.49	1.02
$\beta_{3,26}$	−1.26	0.49	−2.30	−1.25	−0.39
$\beta_{0,52}$	41.99	0.34	41.39	41.98	42.69
$\beta_{1,52}$	−1.50	0.44	−2.35	−1.50	−0.59
$\beta_{2,52}$	0.45	0.22	0.08	0.43	0.92
$\beta_{3,52}$	−0.78	0.27	−1.33	−0.77	−0.28
$\beta_{0,78}$	41.08	0.65	39.85	41.06	42.40
$\beta_{1,78}$	−1.53	0.45	−2.40	−1.54	−0.66
$\beta_{2,78}$	0.66	0.29	0.16	0.63	1.28
$\beta_{3,78}$	−0.67	0.28	−1.28	−0.65	−0.17
$\beta_{0,102}$	41.64	0.77	40.13	41.62	43.19
$\beta_{1,102}$	−2.14	0.82	−3.94	−2.11	−0.63
$\beta_{2,102}$	0.65	0.32	0.12	0.62	1.33
$\beta_{3,102}$	−0.76	0.38	−1.60	−0.71	−0.14

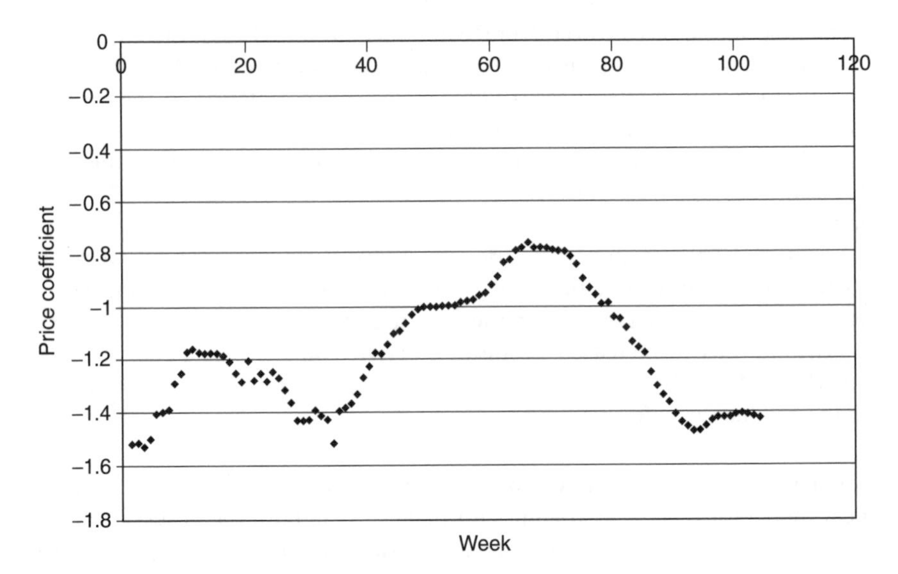

Figure 7.1 Changing price effect, DLM model (a).

We assume the scalar M_t is a parameter rather than known, though we set $M_t = M$, and adopt a uniform prior with values between 1 and 1.2 to model the exponential growth parameter (see *Program 7.17 Physician Expenditures*). Assuming normal errors, we have

$$y_t \sim \mathrm{N}(x_t, V_y), t = 1, \ldots, N$$
$$x_t = Mx_{t-1} + u_t, t > 1$$
$$u_t \sim \mathrm{N}(0, V_x).$$

For the initial value x_0 we assume, as in Carlin *et al.* (1992), a prior normal density,

$$x_0 \sim \mathrm{N}(2500, 10000)$$

so that x_0 is expected to lie within the range 2300–2700 with approximately 95 % probability. The resulting posterior mean for M is 1.093. The full summaries of F and x (from the second half of a run of 20 000 iterations) are given in Table 7.16.

To allow for departures from normality in the state evolution, scale mixtures can be used. One form of such mixture leads to a Student t density (Gelman *et al.*, 1995, Chapter 12). We can model the t density directly, or by scaling the common variance V_x by factors C_t, for each time point, $t = 1, \ldots, N$. These factors are drawn from an inverse Gamma density with parameters $v/2$ and 0.5 (v is the degrees of freedom of the t density).

Here we adopt this approach assuming the degrees of freedom parameter v to be drawn from a uniform density with limits (3, 500). The posterior mean of the latter in a run of 10 000 (and 5000 burn-in) is around 129 with median 91, so

Table 7.16

	Mean	SD	2.5%	Median	97.5%
F	1.093	0.007	1.080	1.093	1.104
x_1	2487	92	2307	2487	2663
x_2	2631	34	2547	2633	2698
x_3	2746	34	2665	2747	2822
x_4	2869	35	2795	2868	2952
x_5	3043	34	2971	3042	3126
x_6	3278	33	3205	3278	3351
x_7	3570	36	3469	3574	3631
x_8	3694	36	3634	3689	3794
x_9	4067	35	3990	4067	4148
x_{10}	4422	34	4356	4419	4506
x_{11}	4909	35	4831	4910	4982
x_{12}	5471	44	5341	5481	5520
x_{13}	5683	33	5609	5684	5756
x_{14}	5902	38	5852	5895	6019
x_{15}	6496	34	6411	6498	6565
x_{16}	6904	49	6863	6891	7064
x_{17}	8054	44	7917	8065	8100
x_{18}	8738	40	8613	8745	8790
x_{19}	9168	48	9124	9156	9322
x_{20}	10280	36	10190	10290	10350
x_{21}	11110	53	11070	11100	11270
x_{22}	12630	34	12570	12630	12720
x_{23}	14300	43	14170	14310	14360
x_{24}	15820	56	15640	15830	15870
x_{25}	16920	34	16840	16920	16990
x_{26}	18210	44	18150	18200	18340

there is no evidence of non-normality for the evolution of these parameters. The posterior of F is slightly more variable than when normality is assumed, but has a mean close to that above.

Example 7.18 Simulated univariate nonstationary model An example of a univariate nonstationary model was discussed by Kitagawa (1987) and reanalysed by Carlin *et al.* (CPS, 1992). It involves generating outputs y_t from an underlying state variable x_t in turn generated randomly as follows (for $t = 1, \ldots, 100$, and preset values $x_0 = 0, A = 0.5, B = 25, C = 8, D = 1.2, E = 0.05, \tau_u = 10, v = 10, \tau_v = 1$)

$$x_t = Ax_{t-1} + Bx_{t-1}/(1 + x_{t-1}^2) + C\cos(Dt - D) + u_t$$
$$y_t = Ex_t^2 + v_t \tag{7.13}$$

where $u_t \sim \mathrm{T}(0, \tau_u, v)$, and $v_t \sim \mathrm{N}(0, \tau_v)$. The analysis of Carlin *et al.* involves 're-estimation' of the coefficients A, B, C and variances τ_u and τ_v from the simulated data (see the data simulation program listed in Program 7.18). The $x_t (t = 1, \ldots, 100)$ are assumed unknowns in their analysis and the initialising x_0 also becomes a parameter. The other features of the generating model are assumed known.

In a first possible analysis with the generated data (*Program 7.18 Non-linear Model*), we assume both y and x are known data. We treat the x as generated from a Student t density with unknown degrees of freedom. The analysis with known x is then a nonlinear regression through time. (The data file contains the full set of generated y and x from a single iteration of the simulation program). We adopt priors on the parameters parallelling those of Carlin *et al.*, except that we take A, B and C as additional unknowns. We also generalise the regression for the generated output variable to $y_t = E + Fx_t + Gx_t^2 + v_t$ and take E, F and G as unknown parameters. The posterior estimates (Table 7.17) of the parameters of the state and observation equations are close to those of the generating model (these being from the second half of a run of 10 000 iterations). The observation equation parameters reproduce the dominant effect G of the quadratic term in x_t with the posterior density of F by contrast straddling zero. The Student t degrees of freedom is overestimated – suggesting a closer approximation to normality than in the generating model, where $\nu = 10$.

In a second analysis (*Program 7.18 Latent x Model*), similar to that of Carlin *et al.*, we take the underlying x_t as unknowns, and take only A, B and C as unknown regression parameters with the same informative priors as were adopted by them. We adopt the same regression form (7.13) for y_t as originally used by CPS. The posterior estimates of A, B and C (from the second half of a run of 10 000 iterations) are close to the generating values (0.5, 25, 8), and are precisely estimated. The Student t degrees of freedom is, however, again overestimated. Carlin *et al.* simulate the next latent value, x_{101}, and derive a bimodal posterior density. This recurs in the present analysis (Figure 7.2), even though the generated data are different from those of Carlin *et al.*

Example 7.19 Binomial data on consumer awareness Fahrmeir (1992), West *et al.* (1985) and West and Harrison (1989) analyse binomial data on consumer awareness in relation to sales expenditure. The full data consist of 168 weekly

Table 7.17

	Mean	SD	2.5%	97.5%
A	0.4262	0.03842	0.3513	0.5015
B	26.360	1.608	23.240	29.550
C	8.384	0.474	7.447	9.323
E	0.147	0.126	−0.100	0.394
F	−0.003	0.009	−0.022	0.015
G	0.049	0.001	0.047	0.050
ν	47.6	30.8	4.8	95.7

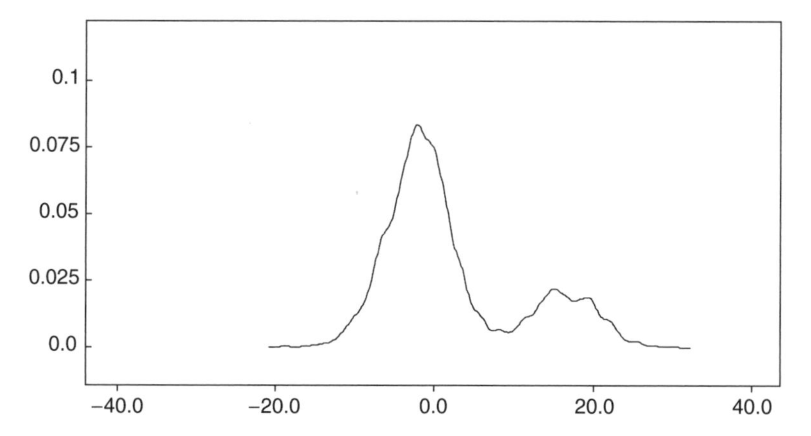

Figure 7.2 Posterior density of next period forecast.

observations on the numbers r_t responding as aware of a particular advertising campaign, out of a constant total $n_t = 66$. Different analyses of these data relate to the impact on awareness of variations in the level of TV advertising. Here we consider a subset of the full data involving 75 weeks observations, with the TV advertising variable in the form of a TVR measure x_t (TV ratings). Alternative models for these data involve a cumulative 'adstock' measure based on x_t (West *et al.*, 1985). A nonlinear model involving minimum and maximum advertising awareness levels, and a diminishing returns effect of the TVR variable is developed by West and Harrison (1989). Here we simply examine the effect of the untransformed x_t on the awareness ratings; the TVR measure peaks in weeks 4–6, 17–18, 21–26, 29–32 and 35–37.

The outcome, numbers aware, is binomial

$$r_t \sim \text{Bin}(\pi_t, n_t)$$

with

$$\text{logit}(\pi_t) = \alpha_t + \beta_t x_t.$$

The intercepts are allowed to adjust readily through time by adopting a first period precision $1/W_{\alpha 1} \sim G(0.01, 0.01)$ and then a discount factor of 0.9 for subsequent precisions $1/W_{\alpha t}, t \geq 2$. Following West *et al.* (1985) we may expect the TVR effect β_t to be fairly stable, and so adopt a fixed variance model for β_t with $W_\beta = \pi W_{\alpha 1}$, where $0 < \pi < 1$ and π is uniform.

Figure 7.3 shows the resulting path for the advertising effect (from the second half of a run of 20 000 iterations), which shows a peak at around week 43.

We can monitor change in the α_t and β_t by comparing averages for the first and last 25 weeks, i.e. 1:25 and 50:75 respectively in Program 7.19. These show a fall in the level of awareness in the latter third of the period, and a mean increase in the impact of TVR though with such imprecision as to be consistent with no advertising impact (Table 7.18).

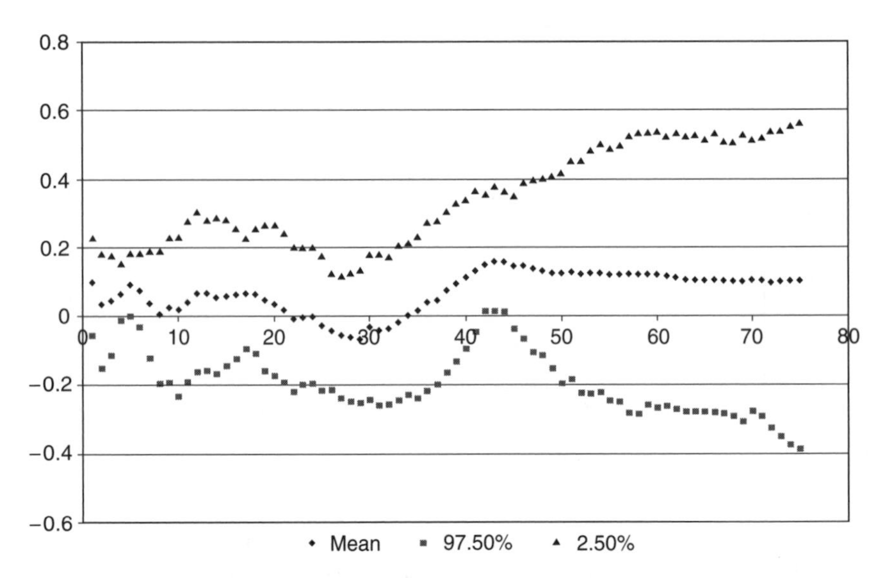

Figure 7.3 Changing impact of advertising.

Table 7.18

Period averages	Mean	SD	2.5%	Median	97.5%
$\alpha_{1,25}$	−0.671	0.075	−0.820	−0.673	−0.528
$\alpha_{50,75}$	−1.283	0.423	−2.071	−1.285	−0.429
$\beta_{1,25}$	0.048	0.045	−0.044	0.048	0.135
$\beta_{50,75}$	0.122	0.298	−0.447	0.124	0.738
Change in α	−0.562	0.399	−1.299	−0.566	0.278
Change in β	0.110	0.285	−0.413	0.108	0.699

Example 7.20 Aids forecasting Gamerman and Migon (GM, 1991) present a forecasting model based on DLM principles and applied to cumulative notified cases of AIDS (alternatively one could analyse new rather than cumulative cases). The series relates to Brazil and is by month from September 1985 to December 1988. The overall trend is upwards but irregular, and a nonlinear model is indicated. Of particular importance is whether the growth of the disease is 'explosive', and increasing without limit, or whether instead the growth indicates the cumulative total is tending to an upper asymptotic limit. Although the data are counts, they are large enough to justify a normal approximation, such that

$$Y_t \sim \mathrm{N}(\mu_t, \sigma_t)$$

with $\sigma_t = \mu_t/\omega_t$. GM use the multiplier ω_t to account for 'overdispersed' Poisson data, while $\omega_t = 1$ is equivalent to the variance law of the Poisson. The mean of the counts, μ_t, may be expressed in various ways as a function of time to represent different possible forms of growth path. Logistic growth implies an inverse link, with

$$1/\mu_t = \beta_1 + \beta_2\beta_3^t.$$

This is a particular case of a larger model set with link $g(\mu)$ on μ. Thus if

$$g(\mu) = \mu^\lambda, \quad \lambda \neq 0$$

and

$$g(\mu) = \log(\mu), \quad \lambda = 0$$

then $\lambda = -1$ gives the logistic. Exponential growth is given by $\lambda = 1$, while the Gompertz curve is obtained under the log transform, $\lambda = 0$. Suppose we write $g(\mu_t) = \phi_{1t}$ and define the recursions

$$\phi_{1t} = \phi_{1,t-1} + \phi_{2,t-1}$$
$$\phi_{2t} = \phi_{2,t-1}\phi_{3,t-1}$$
$$\phi_{3t} = \phi_{3,t-1}.$$

Then $\phi_{1,t+k} = \beta_{1t} + \beta_{2t}\beta_{3t}^k$ with $\beta_{1t} = \phi_{1t} + \phi_{2t}/(1-\phi_{3t})$, $\beta_{2t} = -\phi_{2t}/(1-\phi_{3t})$ and $\beta_{3t} = \phi_{3t}$.

This model is estimated for the exponential cases ($\lambda = 1$) in *Program 7.20 AIDS Forecasts*. The guideline means and variances for the first period parameters ϕ_{11}, ϕ_{21} and ϕ_{31} are as in Gamerman and Migon (1991). Discounts of 0.975 were used for the priors on the precisions of subsequent ϕ_{1t}, ϕ_{2t} and ϕ_{3t}. Priors on the parameters ϕ_{1t}, ϕ_{2t} and ϕ_{3t} themselves may be restricted to 'realistic' values; thus ϕ_{3t} is necessarily positive and likely to be close to 1. To model overdispersion, greater stability was obtained with a constant multiplicative parameter γ, following a Gamma(1, 1) prior, such that

$$\sigma_t = (\mu_t)^\gamma.$$

As in GM, predictive fit can be assessed by analysing average one-step-ahead prediction errors (e.g. via the mean absolute error parameter). An on-line analysis without interventions is adopted; other options include retrospective smoothing and interventions to reflect discontinuities in the series in mid-1987. Of major interest are short-term forecasts beyond the end of the series in December 1988. These can be compared with actual values as known after December 1988 but not used in estimating the model.

For forecasting into 1989 the crucial parameters are those in the last two periods of the observed data ($t = 39, 40$), and it may be noted that the mean of ϕ_{3t} slightly exceeds 1 in month 40 – though with a wide range of uncertainty. In the present analysis (based on the second half of a run of 25 000 iterations), the forecasts for the exponential growth case, namely `Y.new[]` in Program 7.20, are slightly lower than those obtained by MG (Table 7.19).

Table 7.19

Parameter	Mean	SD	2.5%	Median	97.5%
MAE	125.3	5.9	113.6	125.2	137.1
γ	0.86	0.06	0.72	0.86	0.98
$\phi_{3,35}$	1.052	0.156	0.759	1.050	1.370
$\phi_{3,36}$	1.039	0.162	0.733	1.035	1.364
$\phi_{3,37}$	0.991	0.165	0.679	0.984	1.339
$\phi_{3,38}$	1.022	0.201	0.630	1.024	1.416
$\phi_{3,39}$	1.011	0.247	0.554	1.014	1.453
$\phi_{3,40}$	1.006	0.264	0.540	1.011	1.462
Forecasts	Predicted	SE (predicted)	Actual		
Jan 89	5487	155	5477		
Feb 89	5758	279	5712		
Mar 89	6029	405	6202		
Apr 89	6301	533	6421		
May 89	6571	661	6857		
Jun 89	6842	790	7182		
Jul 89	7114	919	7538		
Aug 89	7385	1046	7787		

7.7 SPECTRAL TIME SERIES ANALYSIS

An alternative approach to time series analysis is based on the representation of a stationary series with zero mean, $y_t(t = 1, \ldots, N)$, in terms of harmonic components at different frequencies. Many stationary time series show cyclical fluctuations that can be represented by a mixture of sine waves. Spectral methods based on such cycles find wide applications in engineering and medicine; examples include spectral components in heart rate, blood pressure and the EEG (Campbell, 1996). Suppose we consider a time series y_t with fluctuations at a known periodicity (or wavelength) L. Periodicity refers to the full time span of a cycle, encompassing upturn and downturn, till the start of the next cycle.

Then if u_t denotes random variation with mean zero,

$$y_t = A\cos(\omega t + \theta) + u_t \tag{7.14}$$

where $\omega = 2\pi/L$ is the frequency of the cycle, A is its amplitude and θ is its phase. The angle $\omega t + \theta$ is in radians. The inverse of L, namely $1/L = 2\pi/\omega$ is then the portion of the cycle occurring in a unit time. Since $\cos(\omega t + \theta) = \cos(\omega t)\cos\theta - \sin(\omega t)\sin\theta$, the model (7.14) for y_t is expressible as

$$y_t = \alpha\cos(\omega t) + \beta\sin(\omega t) + u_t$$

where $\alpha = A\cos\theta, \beta = -A\sin\theta$. Since also $\cos^2\theta + \sin^2\theta = 1$, the amplitude can be written $A = (\alpha^2 + \beta^2)^{0.5}$.

In practice, many series will contain a mixture of cycles (this is true of the EEG and heart rate for example). The Fourier series representation of a time series is then in terms of a discrete number of sine waves with frequencies $\omega_j = 2\pi j/N$ for $j = 1, \ldots, m$, where the maximum number of frequencies is $m < N/2$. Thus assuming normal variation about the Fourier series means

$$y_t \sim \mathrm{N}(\mu_t, \sigma^2)$$
$$\mu_t = \sum_{k=1}^{m} [\alpha_k \cos(\omega_k t) + \beta_k \sin(\omega_k t)]. \tag{7.15}$$

By increasing m, we gain an orthogonal decomposition accounting for increasingly more of the variance of the original time series. The least squares solutions to the coefficients in (7.15) provide the periodogram, which expresses the proportion of variability in $\{y_t\}$ explained as the frequencies ω_j change. The periodogram ordinate for frequency j is defined by

$$I(\omega_j) = \left[\sum_{t=1}^{N} y_t \cos(\omega_j t)\right]^2 \Big/ N + \left[\sum_{t=1}^{N} y_t \sin(\omega_j t)\right]^2 \Big/ N. \tag{7.16}$$

Underlying the periodogram is a smooth relation between variability and frequency known as the spectrum $f(\omega)$. This is a continuous density such that $f(\omega)d\omega$ is the contribution to the total variance of y_t due to frequencies within $(\omega, \omega + d\omega)$, and can be written in terms of covariances γ_k in y_t at lag k:

$$f(\omega) = \gamma_0 + 2\sum \sum_{k=1}^{\infty} \gamma_k \cos(k\omega).$$

If σ^2 is the variance of y_t, then the normalised spectrum $g(\omega) = f(\omega)/\sigma^2$ represents the proportion of variance within $(\omega, \omega + d\omega)$. Mathematically there is no loss of generality in restricting frequencies ω in the spectrum to the range $(0, \pi)$ (Diggle, 1990).

For a purely random process (white noise) with variance σ^2, we have $\gamma_0 = \sigma^2$ and all other covariances zero. So the spectrum for such a process is constant. There are a range of procedures to estimate the spectrum from the empirical values of the periodogram, and to test whether an observed series is approximately of white noise form via its empirical periodogram (Diggle, 1990, Chapter 4).

Of interest here are the parametric representations of the spectrum for various time series models (e.g. ARMA models), so that we can consider estimating the parameters of such models using the distributional properties of the periodogram ordinates. Specifically we can write the periodogram (7.16) as

$$NI(\omega_j) = [A(\omega_j)]^2 + [B(\omega_j)]^2.$$

If we assume the y_t are Gaussian white noise with zero mean and variance σ^2, then $A(\omega_j)$ and $B(\omega_j)$ are linear combinations of the y_t also having zero mean, and variance $\text{var}[A(\omega)] = \text{var}[B(\omega)] = \sigma^2 N/2$. The covariance of $A(\omega)$ and $B(\omega)$ is zero, as it involves the cross product $\sum_{t=1}^{N} \sin(\omega t)\cos(\omega_t)$. So $2I(\omega)/\sigma^2 = 2[\{A(\omega)\}^2 + \{B(\omega)\}^2]/(N\sigma^2)$ is chi-square on two degrees of freedom, and hence in turn, $I(\omega) \sim \sigma^2/2\chi_2^2$. If additionally y_t is stationary, then $I(\omega) \sim f(\omega)\chi_2^2/2$. We can then write this relationship in a log-linear regression form as

$$\log(I[\omega_j]) = \log(f[\omega_j]) + \varepsilon_j,$$
$$\varepsilon_j \sim \log(\chi_2^2/2).$$

This result has lead to proposals for general linear modelling approaches relating the periodogram ordinates, $I(\omega_j)$ to the parameters of spectrum, with log link and gamma errors. Thus Bloomfield (1976) proposed a class of models for stationary data of the form

$$\log\{f(\omega_j)\} = 2\sum_{k=0}^{p} \theta_k \cos(k\omega_j)$$

where p is the 'order' of the model. Consider more generally, the ARMA models of order (p, q) for y_t, as discussed above, with

$$y_t = \alpha_1 y_{t-1} + \alpha_2 y_{t-2} + \ldots + \alpha_p y_{t-p} + z_t + \beta_1 z_{t-1} + \beta_2 z_{t-2} + \ldots + \beta_q z_{t-q}$$

where the errors z_t have zero mean and variance σ^2. This model has a spectral form

$$f(\omega_j) = \sigma^2\left[\left\{1 + \sum_{k=1}^{q} \beta_k \cos(k\omega_j)\right\}^2 + \left\{\sum_{k=1}^{q} \beta_k \sin(k\omega_j)\right\}^2\right] \Big/ \left[\left\{1 - \sum_{k=1}^{p} \alpha_k \cos(k\omega_j)\right\}^2 + \left\{\sum_{k=1}^{p} \alpha_k \sin(k\omega_j)\right\}^2\right]. \tag{7.17}$$

Suppose we consider estimating a Bloomfield model using the periodogram. Then writing $v_j = 1/f(\omega_j)$ and supposing we have a single replication of a series,

$$I(\omega_j) \sim \text{G}(1, v_j)$$
$$\log(v_j) = -2\sum_{k=0}^{p} \theta_k \cos(k\omega_j) = -2[\theta_0 + \sum_{k=1}^{p} \theta_k \cos(k\omega_j)]. \tag{7.18}$$

If we had r replications of a series then $I(\omega_j) \sim \text{G}(r, rv_j)$ (e.g. Cameron and Turner, 1987).

Example 7.21 Cyclical and trend components in TB cases Kafadar and Andrews (KA, 1996) present an analysis of tuberculosis cases in the US in the 1980s notified in 13 segments of 4 weeks each over a period of 10 years, 1980 to 1989 (i.e. 130 segments). They detect both segment effects (e.g. due to season) and a longer term cycle that repeats itself approximately twice during the period. There is also evidence of an overall trend in TB cases, which is modelled by a simple linear term in time. We are therefore interested in the residuals by segment and by year (`res.seg[]` and `res.year[]` in Program 7.21). The data are counts but a normal approximation is applicable given their average of around 1500 per segment. Using data generated from their results we thus fit three models, with a normal outcome $y_i \sim \mathrm{N}(\mu_i, \sigma^2)$:

(a) time trend in month $(t = 0, \ldots, 129)$

$$\mu_i = \beta_1 + \beta_2 t;$$

(b) time trend and segment

$$\mu_i = \beta_1 + \beta_2 t + \gamma[S[i]]$$

(c) trend, segment and cyclical trend

$$\mu_i = \beta_1 + \beta_2 t + \gamma[S[i]] + \delta_1 \cos(\omega t) + \delta_2 \sin(\omega t).$$

The final model implies a period of $2\pi/\omega$ segments. The results from the first model (using the second half of a run of 10 000 iterations) show a downward trend but complicated by segment effects: the first segment is lower than average, the last higher. Fitting the segment effects as in model (b) shows some evidence of a cycle (see the residuals for years in Figure 7.4), reaching a low point in the mid-1980s but rising to a shallow peak in the later 1980s. Adding a cycle as in model (c) suggests a period of around 72.5 segments (i.e. around 5.6 years).

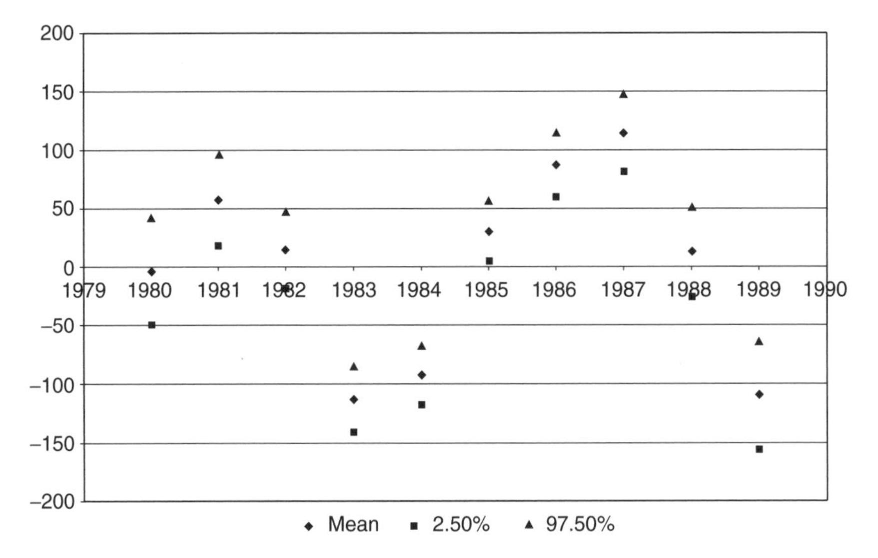

Figure 7.4 Year residuals from Model (b).

Example 7.22 Enrolments at Yale Lindsey (1989) presents a series of annual enrolments E_t at Yale University from 1796–1975. He shows first differences in the series approximate a random walk, and applies the Bloomfield model to $y_t = E_{t+1} - E_t, t = 1, \ldots, 179$. Hence the Fourier series frequencies are defined by

$$\omega_j = 2\pi j/N, \quad j = 1, \ldots, m$$

for $N = 179$ and $m = 89$. He fits Bloomfield models with a range of orders up to $p = 5$. Here we illustrate for $p = 2$ (see *Program 7.22 Yale Enrolments*) obtaining parameter estimates as given in Table 7.20.

The variance is estimated as $\exp(2 \times 6.13) \approx 210,000$. By comparison we can we fit a simple AR(2) model to the original series y_t (conditioning on the first two years)

$$y_t = \beta_0 + \beta_1 y_{t-1} + \beta_2 y_{t-2} + u_t$$

with $u_t \sim N(0, \sigma^2)$. We obtain a variance estimated at 213 000 and $\beta_2 = -0.128$ with standard deviation 0.074.

Example 7.23 Lynx capture data We reanalyse the lynx data using the spectral approach, and assuming an ARMA(3,3) model. The spectrum is as above and priors on α and β are set to ensure stationarity and invertibility. This leads to a nonlinear GLM regression with gamma errors (see *Program 7.23 Lynx Spectral*). Note that the usual MA parameters are obtained as minus the β parameters. The last 1000 of 5000 iterations yield results as shown in Table 7.21.

These are close to those obtained earlier by estimation in the time domain, with the second MA parameter being better identified.

Table 7.20

Parameter	Mean	SD	2.5%	Median	97.5%
θ_1	−0.047	0.076	−0.193	−0.048	0.103
θ_2	−0.145	0.083	−0.308	−0.146	0.018
θ_0	6.13	0.05	6.02	6.12	6.23

Table 7.21

Parameter	Mean	SD	2.5%	Median	97.5%
α_1	1.832	0.153	1.484	1.838	2.098
α_2	−1.408	0.239	−1.826	−1.417	−0.879
α_3	0.278	0.145	−0.039	0.286	0.535
β_1	−0.906	0.186	−1.164	−0.949	−0.459
β_2	0.239	0.181	−0.110	0.259	0.566
β_3	0.595	0.102	0.380	0.599	0.790

7.8 TIME SERIES FORMULATIONS OF NONPARAMETRIC REGRESSION

Suppose values (y_i, x_i) of response and p covariates $x_i = (x_{i1}, x_{i2}, \ldots, x_{ip})$ are observed for N subjects. The response is related to the covariates via unknown functions $g_1, g_2, \ldots, g_p$. Thus for a continuous outcome variable y_i, assumed normal,

$$y_i \sim \mathrm{N}(\mu, \sigma^2)$$
$$\mu_i = g_1(x_{i1}) + g_2(x_{i2}) + \ldots + g_p(x_{ip}).$$

In standard parametric multiple regression, the forms $g_1, g_2, \ldots$ are assumed known except for a few parameters. However, the assumption that these forms are of known form (e.g. linear throughout the range of both x_{ij} and y_i) may be unwarranted and discount a changing relationship between y_i and each x_{ij} as they vary over their range. For example, there are likely to be local nonlinearities or 'bath-tub' effects which result in linear correlations of zero. So it may be preferable, at least initially, to assume that the $g(x_j)(j = 1, \ldots, p)$ are smoothly changing functions of their arguments. For example, Erkanli and Gopalan (1996) discuss the prediction of oxygen uptake rate, OXY, in a sample of students in terms of (a) weight, WT; (b) time to run 1.5 miles in minutes, TIME; and (c) heart rate at rest, HEART. So

$$\mathrm{OXY} \,|\, \mathrm{WT}, \mathrm{TIME}, \mathrm{HEART} = g_1(\mathrm{WT}) + g_2(\mathrm{TIME}) + g_3(\mathrm{HEART}) + e.$$

If g_1, g_2 and g_3 are linear functions of their arguments then we have the standard normal regression. However, in biostatistical applications, smooth nonlinear functions of uptake in relation to these predictors may be more appropriate.

Following Wood and Kohn (1998) and Wecker and Ansley (1983) we aim to select a prior for the smooth functions $g_j(j = 1, \ldots, p)$ such that the posterior estimate of y_i (e.g. its posterior mean) is also smooth. The prior should also be flexible in the face of wide variability in the form of observed regression relationships. Suppose $p = 1$, so that x_i consists of a single covariate. Assume also that values of x in a sample of N are ranked from lowest to highest such that $x_1 \leq x_2 \leq x_3 \leq \ldots \leq x_N$. Often also the x_i are transformed to be positive or so that they lie within the range $(0, 1)$ (Ansley and Kohn, 1987).

7.8.1 Continuous time prior

One form of prior for the model generating such data assumes an underlying continuous process (Wahba, 1978). For a univariate regressor $x_i(i = 1, \ldots, N)$ with

$$0 < x_1 < \ldots < x_N < H,$$

this prior assumes the observations are generated by a signal plus noise model

$$Y_i = g(x_i) + e_i$$

where the $e_i \sim N(0, \sigma^2)$ are white noise and the signal $g(x)$ is generated by the stochastic differential equation

$$d^2g(x)/dx^2 = \tau dW(x)/dx. \tag{7.19}$$

In the latter $W(x)$ denotes a Weiner process which consists of an accumulation of independently distributed stochastic increments, with starting value $W(0) = 0$ and variance $\text{var}[W(x)] = 1$. The parameter τ plays a central role in governing the degree of smoothing. Large values of τ mean the smooth is very close to reproducing the actual data; i.e. there is little smoothing and the smooth in fact just interpolates the existing observations. A value of $\tau = 0$ corresponds to complete smoothing (i.e. the posterior mean is linear). Integrating (7.19) gives

$$g(x) = \alpha_0 + \alpha_1 x + \tau \int_0^x W(u)du. \tag{7.20}$$

The ratio $\varphi = \tau^2/\sigma^2$ is the signal-to-noise ratio of variances, and for a metric normal outcome its inverse $\lambda = 1/\varphi$ corresponds to the smoothing parameter in a cubic smoothing spline $g()$ with knots at each of the x_i (note that higher order Weiner processes lead to polynomial splines of other degrees).

Let the nonlinear part of g in (7.20) be $f(x) = \lambda \int_0^x W(u)du$, then the state-space evolution for this type of prior is based on f and its first derivative, $\{f(x_i), f'(x_i)\}, i = 1, \ldots, N$. Denote this pair by $f_i = \{f_{i1}, f_{i2}\}$, and define the differences $\delta_i = x_i - x_{i-1}$, with $\delta_1 = x_1$. The initial terms f_{11}, and f_{12} are treated as unknown fixed effects. Then the evolution through increasing values of x is defined by

$$f_i = F_i f_{i-1} + e_i, \quad i \geq 2$$

where

$$F_i = \begin{bmatrix} 1 & \delta_i \\ 0 & 1 \end{bmatrix},$$

and with

$$e_i \sim N(0, \tau^2 U_i), \tag{7.21}$$

where U_i is

$$U_i = \begin{bmatrix} \delta_i^3/3 & \delta_i^2/2 \\ \delta_i^2/2 & \delta_i \end{bmatrix}. \tag{7.22}$$

Because τ defines the dispersion matrix of the state-space evolution, its prior in BUGS requires a special form, such as a discrete prior. Alternatively one may sample its values (or more specifically those of τ^2) directly from its conditional density (see Carter and Kohn, 1994, pp. 545–546).

The posterior mean of g is a cubic smoothing spline, cubic in the subintervals (x_{i-1}, x_i), and linear for $x \leq x_1$ and $x \geq x_N$. The cubic smoothing function g minimises

$$\sum_1^N [y_i - g(x_i)]^2 + \lambda \int_0^H \{f''(u)\}^2 du. \tag{7.23}$$

The function g is related to the simple moving average smoother, and a reduction in the value of τ (i.e. increase in λ) amounts to a wider bandwidth of a weighted moving average smoother (Diggle, 1990).[1]

7.8.2 Fourier series prior

A spectral (Fourier series) approach to the continuous time prior is discussed by Lenk (1999) and Kitagawa and Gersch (1996). Thus for the outcome

$$y_i = g(x_i) + e_i$$

with x defined on the interval $[a, b]$, the nonparametric component is represented by the series

$$g(x) = \sum_{k=1}^{\infty} \theta_k \varphi_k(x) \tag{7.24}$$

where

$$\varphi_k(x) = \left(\frac{2}{b-a}\right)^{0.5} \cos\left\{\pi k\left(\frac{x-a}{b-a}\right)\right\}.$$

Since a smooth g will not have high frequency components, the θ_k are subject to decay as k increases. For example, we may take a 'geometric smoother' prior

$$\theta_k \sim \mathrm{N}(0, \tau^2 \exp(-\psi k), \quad \psi > 0.$$

The parameter ψ determines the rate of decay of the Fourier coefficients, and thus the smoothness of g, and will have an appropriate prior for a positive parameter (e.g. an exponential density). In practice the Fourier series is truncated above at $K < N$, so

$$g(x) = \sum_{k=1}^{K} \theta_k \varphi_k(x).$$

7.8.3 Discrete time priors

Discrete time priors for additive nonparametric regression are discussed by Fahrmeier and Lang (1999). Thus consider a formulation of the above model

$$y(t) = g(t) + e(t)$$

with arguments t being equally spaced design points (e.g. successive years in a time series), and with $e(t) \sim \mathrm{N}(0, \sigma^2)$ as above. Then the second order random walk (or 'smoothness prior')

$$g(t) = 2g(t-1) - g(t-2) + u(t) \tag{7.25}$$

where $u(t) \sim \mathrm{N}(0, \tau^2)$ may be proposed as a prior for the changing mean of $y(t)$. Providing $y(t)$ is normal, the posterior means $\hat{g}(1), \hat{g}(2), \ldots, \hat{g}(N)$ are equivalent to the estimated posterior modes of $g(t)$ derived by minimising

$$\sum_{t=1}^{N} [y(t) - g(t)]^2 + \frac{\sigma^2}{\tau^2} \sum_{1}^{N} (g(t) - 2g(t-1) - g(t-2))^2.$$

This is a discretised version of the cubic smoothing spline minimisation, though for non-Gaussian outcomes, the equivalence no longer pertains.

For y_i observed on ordered values of a covariate $x_i, x_1 < \ldots < x_N$, but non-equally spaced observations, then the variance of the random walk prior must be modified to take account of the step sizes $\delta_i = x_i - x_{i-1}$. Thus if y_i is Gaussian with

$$y_i = g(x_i) + e_i$$

then a first order random walk prior would have the form

$$g(x_i) = g(x_{i-1}) + u(x_i)$$

with $u(x_i) \sim \mathrm{N}(0, \delta_i \tau^2)$. Random walks of the second order for non-equally spaced covariates are discussed by Fahrmeier and Lang (1999), and involve weights based on the relative size of δ_i and δ_{i-1}.

7.8.4 Implementation of time series priors

In many cases we would be interested in just a response and its relation to a single covariate, and possible nonlinearities in this relation. Or we might fit conventional linear forms to all but one covariate, and apply a nonparametric approach to the remaining one. Thus a standard form for generalised linear models is the semi-parametric GLM form (Hastie and Tibshirani, 1990) with a conventional GLM except for a single nuisance covariate to which a smooth is applied. For example, the analysis of insecticide treatment effects on grain yield may be confounded by a fertility trend in the physical plot order. A smooth is therefore applied to plot number but otherwise treatment effects are treated conventionally.

A full model for p covariates involving linear terms in each, as in conventional regression, as well as p smooths is heavily parameterised. Just as in dynamic linear models based on time-varying coefficients, various simplifications are probably needed in practice, or informative priors may be indicated for identifiability (Erkanli and Gopalan, 1996). As emphasised by Hastie and Tibshirani (1990, Chapter 6), there is a risk of overfitting the data and interpreting spurious patterns in the fitted curves. Precisely identified parameters with narrow credible intervals will identify the most important features of the fit. With smooths on two or more covariates, one approach is to consider different versions of the data according to the ranking from lowest to highest on each covariate.[2]

Example 7.24 Weight – height ratios in infants An example of a one-to-one relation is provided by the analysis of Wecker and Ansley (1983) of the weight-to-height ratio among boys aged up to six. The ratio is available at age in months from 0.5 to 71.5 in intervals of 0.5, giving $n = 72$ cases. Both the nonparametric analysis of Wecker and Ansley and the earlier analysis by Gallant (using nonlinear least squares) show a steep rise in the ratio at younger ages, but a slower increase after the 12th month.

In *Program 7.24 Weight-height Ratio* the ratio is treated as a normal outcome, though a lognormal analysis is also possible. Using a second order random walk prior (7.25) for the regularly spaced time variable shows a clear nonlinear pattern very approximately described by two separate linear effects (Figure 7.5). The 'signal-to-noise ratio' φ averages 0.035.

Example 7.25 Wool prices Diggle (1990) presents data from an Australian wool price series. The data, in the form of logs of ratios comparing actual prices with a floor price at weekly markets, exhibit serial dependence. The predictor is time in weeks from 1/1/1976 with gaps occurring because wool markets are not held. With data of this type, Diggle argues that the usual methods to choose the smoothing parameter may lead to undersmoothing, because they attribute to $g()$ the serially dependent fluctuations in y. The prior used ensures the signal-to-noise variance ratio (i.e. τ^2/σ^2) is under 1. As compared with the original

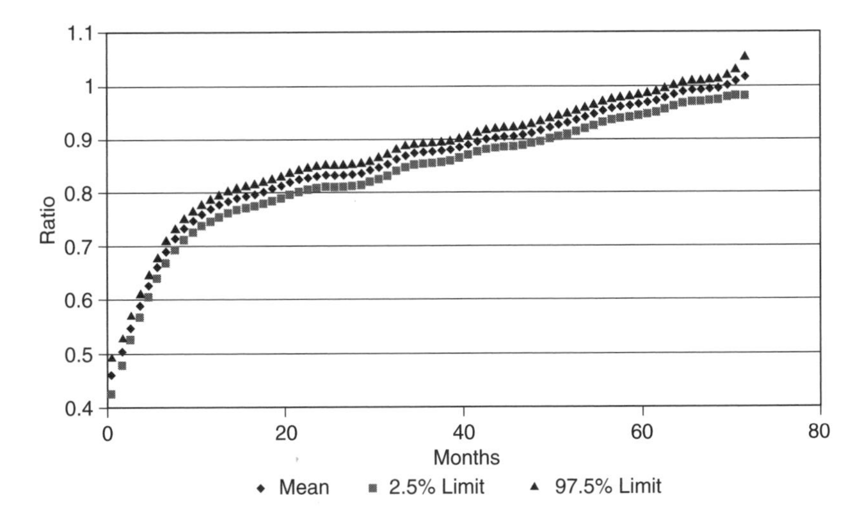

Figure 7.5 Weight-height ratio and age in months.

data, this smooth (the array `g[]` in *Program 7.25 Wool Prices*) reduces the variances of first differences by about 2.5 % and of the second differences by 4.3 %.

By contrast, taking τ^2 to be 0.66×10^{-4} (i.e. scaling τ down by a factor of 100) leads to a reduction in the variance of first differences of around 60 % and in the variance of second differences of approaching 90 % (see Figures 7.6 and 7.7). The need here to adjust standard methods to achieve a smooth suggests there may still be an element of choice in the degree of smoothing chosen, even though data-driven methods to estimating τ^2 do reduce the element of choice in smoothing.

Example 7.26 O-ring failures Data on O-ring failures in relation to temperature T provide an example of nonparametric regression for a binary outcome. As in Wood and Kohn (1998), assume an unknown continuous variable y underlies the observed binary response, w_i. Let the latent variables be related to a smooth on temperature as follows

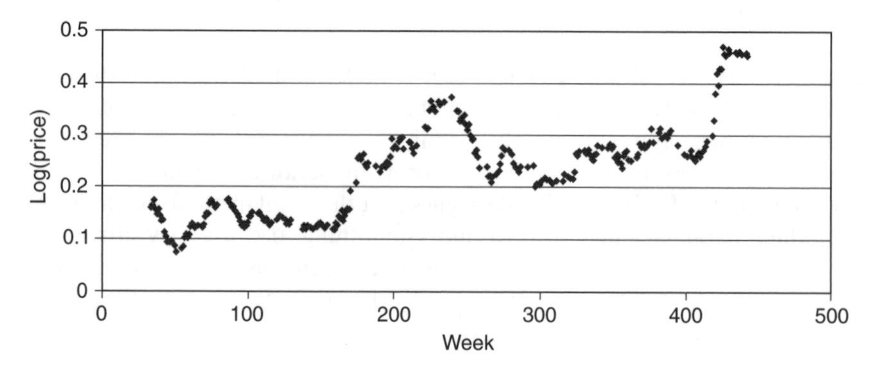

Figure 7.6 Wool prices: Standard smooth.

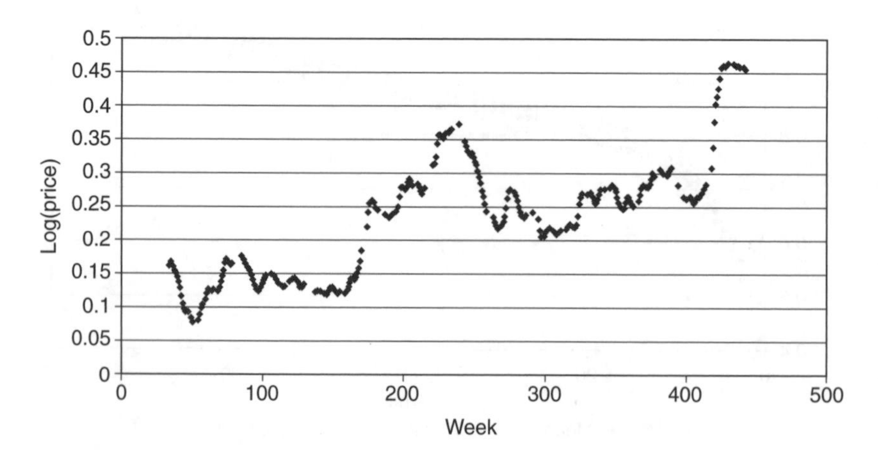

Figure 7.7 Wool prices: Enhanced smooth.

$$y_i = g\{T_i\} + e_i = \alpha_0 + s(T_i) + e_i$$

with $e_i \sim N(0, 1)$. Let $w_i = 1$ if $y_i > 0$, and let $w_i = 0$ otherwise. Then

$$\Pr\{w_i = 1 \mid g(T_i)\} = \Pr\{y_i > 0 \mid g(T_i)\} = \Pr\{e_i > -g(T_i)\} = \Phi\{g(T_i)\}$$

which reproduces the standard probit link. First order random walk priors and continuous time priors are implemented in Program 7.26. The code allows for the grouping of the covariate values; there are 23 observations but only 16 distinct covariate values.

The latent variables g[] are supplied with initial values in the Inits file and updated values generated with the code:

```
for (i in 1:n) {y[i] ~ dnorm(g[i],1) I(a[i],b[i])
                a[i] <- -10*equals(w[i],0)
                b[i] <- 10*equals(w[i],1)
                g[i] <- alpha0 + f[i,1]}
```

Thus if $w_i = 0$, y_i is sampled from $N(0, 1)$ while restricted to negative values in the interval $(-10, 0)$ while if $w_i = 1$, y_i is constrained to be sampled from positive values in the interval $(0, 10)$. The smooths in temperature show a reasonable match under the two methods though both the binary nature of Y_i and the small sample will add to uncertainty. The effect of temperature in enhancing the risk of O-ring failure is concentrated at lower values of temperature (Figure 7.8).

Example 7.27 US meat consumption Pollock (1993) applies cubic spline smooths to data on meat consumption in the US between 1919 and 1941. He

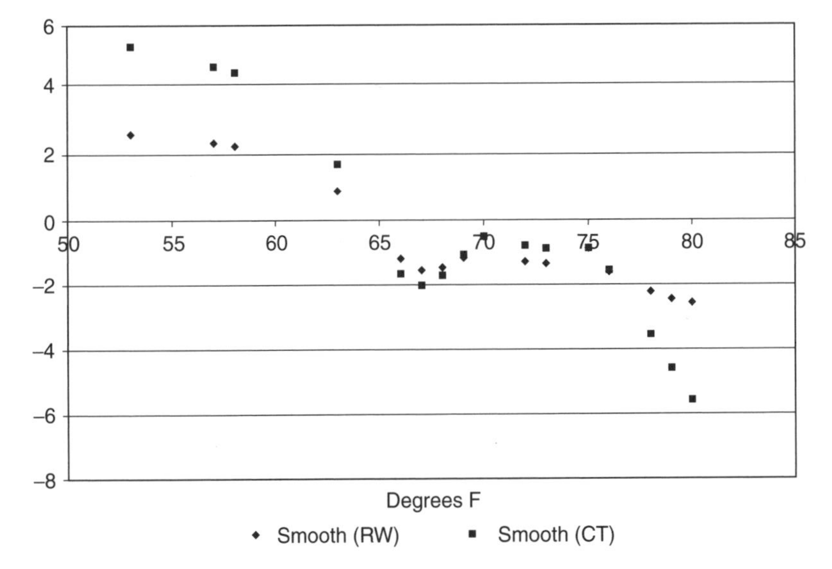

Figure 7.8 O-ring failures by temperature.

uses alternative values $\varphi = 0.125$ and $\varphi = 0.75$. Here we adopt a second order random walk prior and a continuous time prior to estimate the smooth. In the latter (see CT Model in Program 7.27) we adopt a discrete prior for the signal variance τ^2, which in view of Pollock's analysis and the nature of the data is distributed over possible values between 0.5 and 10. The posterior estimate of τ^2 with this prior is around 6.

In the discrete time random walk model (see RW model in Program 7.27) the value of τ^2 has a median around 5.6 as compared with a median noise variance of 20. Figure 7.9 shows a close match in the smooths obtained, both demonstrating the fall in consumption from 1924 to the mid-1930s.

Example 7.28 Michigan road accidents Lenk (1999) analyses data on road accidents in Michigan from the start of 1979 to the end of 1987. The monthly accident counts are large so that their logs are taken to be approximately normal. One influence on such accidents may be economic prosperity, which is proxied by the unemployment rate z. Seasonal effects are also present in the data, together with a slow upward trend over the entire period. The aim is then to assess nonlinearity in g after allowing for seasonal effects and other covariates, as summarised in the systematic regression term βz:

$$y_i = \alpha_0 + \alpha_1 x_i + \beta z + g(x_i) + e_i.$$

The linear growth over time in months x_i is measured by α_1.

Lenk considers the smooths (a) after adjusting for seasonal effects only and (b) after adjusting for both seasonal effects and unemployment. Using the same Fourier series approach as Lenk (see *Program 7.28 Accidents and*

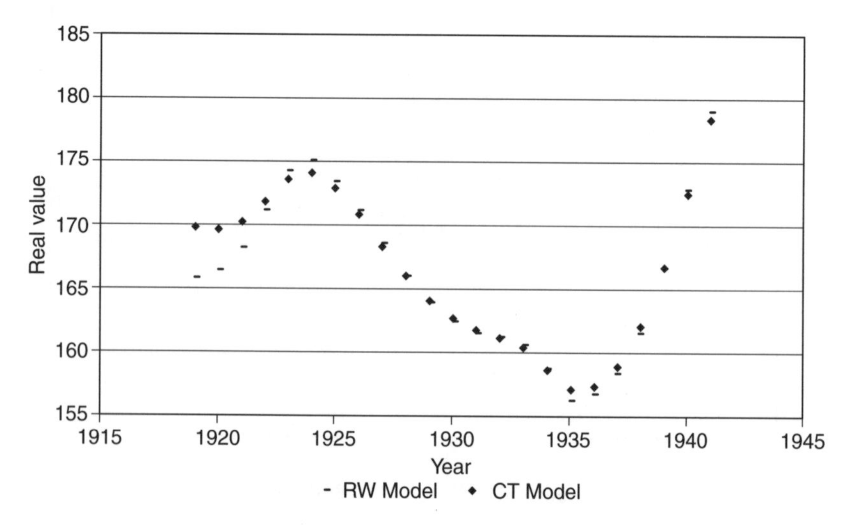

Figure 7.9 US meat consumption, alternative smooths.

Table 7.22

Parameter	Mean	SD	2.5%	Median	97.5%
Constant	4.745	0.203	4.380	4.799	5.045
Month	−0.0031	0.0052	−0.0117	−0.0001	0.0025
Spring	−0.194	0.047	−0.284	−0.193	−0.100
Summer	−0.232	0.047	−0.323	−0.232	−0.134
Autumn	−0.106	0.046	−0.199	−0.106	−0.021
Unemployment	−0.463	0.095	−0.637	−0.458	−0.282

Unemployment), we confirm that a model without log(unemployment) as a covariate shows a clear nonlinearity over time, even after removing the seasonal effects. Including unemployment much attenuates the nonlinearity in the smooth though it is still apparent. The regression coefficients β in the full model predicting accident levels (from the second half of a run of 5000 iterations in Winbugs12) are given in Table 7.22.

7.9 SPATIAL CORRELATION

Tests of hypotheses and models for regression typically assume independent identically distributed variates or regression errors. This independence assumption is as unlikely to hold for spatially located data as for data arranged in time sequence. Spatial data typically show systematic variation, with Tobler (1970) stating the first law of geography as 'everything is related to everything else, but near things are more related than distant things'. If in fact observations are spatially dependent, then standard models assuming independent observations may lead to misleading inferences. For example, Cliff and Ord (1975) applied a Monte Carlo experiment to show different type I errors in t tests comparing samples means with (a) two independent samples of size N, and (b) two sets of spatially autocorrelated observations. Similarly in regression models, if errors show positive spatial correlation, then the standard errors of the regression coefficients are understated and the variation explained is overstated.

The analysis of spatial patterning is frequently specified in terms of attribute measures in a finite set of areas or spatial zones. The attributes associated with these areas are not necessarily assumed to vary continuously over space. A different modelling approach, used in areas such as geostatistics, is in terms of variables $y(s_i)$ varying continuously over two-dimensional spatial locations $s_i = (s_{1i}, s_{2i})$ where s_{1i} and s_{2i} are grid reference locations. Here the objects of analysis include the modelling of trend over space, and the prediction of $y(s)$ for a new, as yet unobserved, point – this is, spatial interpolation.

Below we consider the modelling of associations between variables defined over finite sets of areas, before considering continuous spatial processes.

7.9.1 Hypothesis tests

Suppose we consider first an iid sample $x_1, x_2, \ldots, x_N$ from a population following $N(\mu, \sigma^2)$ and $y_1, y_2, \ldots, y_N$ are another iid sample from $N(\lambda, \sigma^2)$. Then under the prior assumption $\pi(\lambda, \mu, \sigma^2) \propto 1/\sigma^2$ the posterior density of $\delta = \lambda - \mu$ is a t density with $2N - 2$ degrees of freedom. This density has mean $\bar{x} - \bar{y}$ and variance $2s^2/N$, where $s^2 = \sum_{i=1}^N \{(x_i - \bar{x})^2 + (y_i - \bar{y})^2\}/(n-2)$.

Suppose instead we wish to construct a set of values x_i and y_i for N areas, with ρ_x and ρ_y specifying the degree of spatial correlation within the respective series. We use an $N \times N$ contiguity matrix to express the spatial interdependencies within the N areas, with $C_{ij} = 1$ if areas i and j are adjacent and $C_{ij} = 0$ otherwise, and with diagonal elements C_{ii} set to zero. The W_{ij} are then scaled by their row sums to give $W_{ij} = C_{ij}/\sum_{j=1}^N C_{ij}$ such that $\sum_{j=1}^N W_{ij} = 1$. We additionally specify two variables u and v without spatial dependence, namely $u \sim N(\mu, \sigma^2)$ and $v \sim N(\lambda, \sigma^2)$. Then spatially correlated samples are generated by the matrix equations

$$x = (I - \rho_X W)^{-1} u$$

$$y = (I - \rho_Y W)^{-1} v.$$

Cliff and Ord (1975) showed that positive autocorrelation in both sampled data sets results in an overstatement of significance of a difference in sample means. That is, such spatial effects mean the type I error exceeds its nominal level.

Example 7.29 Lattice experiment For a 5×5 lattice (a table of squares as in a chess board, for example) of $N = 25$ cells we generate x and y as above assuming $u \sim N(0, 1)$ and $v \sim N(0, 1)$. As above we set $C_{ij} = 1$ if cells i and j are adjacent and $C_{ij} = 0$ otherwise. We make two alternative assumptions: first that $\rho_Y = \rho_X = 0$ and second that $\rho_Y = \rho_X = 0.5$. After 5000 iterations with x and y generated from an uncorrelated areas assumption (see *Program 7.29 Lattice Matrix)* we find a probability of 0.051 that the t test statistic $(\bar{x} - \bar{y})/(s\sqrt{2}/\sqrt{N})$ exceeds its tabulated percentile of 1.67 (on $2N - 2 = 48$ d.f.). This is an empirical probability obtained by the usual step() function accumulation of iterations where the observed value of the t statistic exceeds 1.67. However, with $\rho_Y = \rho_X = 0.5$ the type I error rate is over 18 %.

7.9.2 Regression models for area outcomes

Inferences may also be affected in broader hypothesis testing situations, such as regression analysis. In the usual linear model for normal continuous observations $y_i, i = 1, \ldots, N$ we have

$$y_i = x_i \beta + u_i$$

with x_i an $N \times p$ matrix of explanatory variables and u_i distributed with mean zero and constant variance over all cases i. Hence the variance-covariance matrix is diagonal with form $V = \sigma^2 I$. This framework may well not apply for spatial processes where the outcomes y_i for areas $i = 1, \ldots, N$ are interdependent. The variance–covariance matrix $V = E(uu^T)$ will then have nonzero off-diagonal elements reflecting dependence between the outcomes of close neighbours. This constitutes a generalised least squares problem and spatial models may attempt to estimate V (this is possible for example with panel data on areas). More commonly parametric analysis, summarising the correlation over space, is applied either in specifying the generation of u_i or parameterising the V matrix (Richardson, 1992).

In a simultaneous autoregressive (SAR) model we specify the covariation using weights S_{ij} which model the spatial correlation between areas i and j. Thus for y_i normal,

$$y_i = \mu_i + \sum_{j=1} S_{ij}(y_j - \mu_j) + e_i$$

with S_{ii} defined as zero. The terms e_i are exchangeable (not intercorrelated), usually with a common variance σ^2.

A specific form of this model (an 'order 1' version) is when $S = \rho W$, with ρ a measure of spatial correlation, such that $-1 \leq \rho \leq 1$. The matrix W is a known interaction matrix, based on simple contiguity, inter-area distances, or shared boundary lengths. It is assumed that the original elements C on which W is based have been standardised so that the row sums of W (the elements defining interactions for a given origin) sum to one. While C may well be symmetric, W no longer is. The choice of weights represents a specification issue in spatial regression (Arora and Brown, 1977), though Besag and Higdon (1999) specify the possibility of priors on the elements W_{ij} (or C_{ij}) so that the data can inform the best choice of weights.

So if $\mu_i = x_i\beta$ then the order 1 SAR model may be written

$$y = x\beta + u$$
$$u = \rho W u + e$$

or as

$$y = x\beta + \rho W(y - x\beta) + e. \tag{7.26}$$

Because $u = (I - \rho W)^{-1} e$, the variance – covariance matrix is now $V = E(uu^T)$ $= \sigma^2 (I - \rho W)^{-1} (I - \rho W^T)^{-1}$. Consider the case where the original elements of C were $C_{ij} = 1$ for neighbouring areas and zero otherwise, and with the standardised matrix $W_{ij} = C_{ij} / \sum_j C_{ij}$. Then (7.26) becomes

$$y_i = x_i\beta + \rho \sum_{j \in N[i]} W_{ij}(y_j - x_j\beta) + e_i$$

where $N[i]$ denotes the subset of all areas which are adjacent to area i (its 'neighbourhood'). Ord (1975) presents an equivalent convenient form of this model

$$y = \rho y_L + x\beta - \gamma x_L + e$$

where $x_L = Wx, y_L = Wy$ and $\gamma = \rho\beta$.

Simultaneous schemes for spatial outcomes are expressed in terms of joint density functions for area i and its neighbours considered together. In conditional schemes for such outcomes (denoted CAR models for conditional autoregression), the conditional density of y_i given values in the remaining areas is specified (Sun *et al.*, 1999):

$$\Pr[y_i \,|\, (y_j, j \neq i)].$$

If C denotes an interaction matrix (with $C_{ii} = 0$) then for a Gaussian outcome, conditional means and variances may be expressed as follows:

$$E(y_i \,|\, y_j, j \neq i) = \mu_i + \sum_{j=1} C_{ij}(y_j - \mu_j)$$

$$\text{var}(y_i \,|\, y_j, j \neq i) = \sigma_i^2.$$

The distribution of the Y_i is multivariate normal with mean $(\mu_1, \mu_2, \ldots, \mu_N)$ and variance

$$(I - C)^{-1}\Sigma$$

where $\Sigma = \text{diag}(\sigma_1^2, \sigma_2^2, \ldots, \sigma_N^2)$. The symmetry condition of the multivariate normal outcome implies that $C_{ij}\sigma_j^2 = C_{ji}\sigma_i^2$.

Suppose we assume a constant variance over cases, with $\Sigma = \sigma^2 I$. If we also set $B = I - C$, then the joint density of the Y_i is

$$|B|^{0.5}(2\pi\sigma^2)^{-N/2}\exp[-1/2\sigma^2(y - \mu)B(y - \mu)^T].$$

Alternatively setting $C = \rho D$ with $-1 \leq \rho \leq 1$, and D symmetric, gives $\text{var}(Y) = \sigma^2(I - \rho S)^{-1}$.

Besag *et al.* (1991) proposed a special 'intrinsic' or ICAR form of the CAR model for the case where $C_{ij} = 0$ unless areas i and j are adjacent. If they are contiguous then the $C_{ij} = C_{ji}$ are positive weights. Many presentations of this model are for the case where the Y_i are Poisson counts, for example of deaths or new disease cases, with expected values (in the demographic standardisation sense) E_i. The E_i are based on the age-sex distribution of area i and applying standard (e.g. national) rates for the event or condition. Then

$$Y_i \sim \text{Poi}(E_i\alpha_i)$$

where the relative risk α_i (e.g. standard mortality ratio) in area i is linked with covariates or random effects via a log transform.

Thus, in the absence of any covariates for each area, let θ_i denote spatially interdependent log relative risks

$$\log(\alpha_i) = \theta_i.$$

The intrinsic CAR model is specified by an improper density based only on pairwise differences among the θ_i:

$$p(\theta_i) \sim \exp[-\sum_{j<i} C_{ij}\varphi(\theta_i - \theta_j)]$$

with $\varphi(z)$ an increasing function of $|z|$. The improper nature of the density arises because it specifies only differences in risks not their level. The conditional density of θ_i is

$$p(\theta_i \,|\, \theta_j, j \neq i) \propto \exp[-\sum_{j \in N[i]} C_{ij}\varphi(\theta_i - \theta_j)].$$

Identifiability of ICAR spatial effects therefore requires either (a) a sum to zero constraint on the θ_i; (b) setting one effect to a known value, e.g. $\theta_1 = 0$; or (c) setting the overall intercept to zero (Best, 1999).

The most common form of ICAR prior takes $\varphi(z) = z^2/2\sigma^2$, with $\tau = 1/\sigma^2$ the corresponding precision parameter, so that the joint density of the risks is

$$p(\underset{\sim}{\theta}) \propto \tau^{-N/2}\exp[-0.5\tau \sum_i \sum_{j<i} C_{ij}(\theta_i - \theta_j)^2]$$

and the conditional mean of a particular risk is

$$E(\theta_i \,|\, \theta_j, j \neq i) = \sum_j C_{ij}\theta_j / \sum_j C_{ij}.$$

If the C_{ij} are simply contiguity dummies, then

$$p(\underset{\sim}{\theta}) \propto \tau^{-N/2}\exp[-0.5\tau \sum_i \sum_{j\sim i} (\theta_i - \theta_j)^2]$$

where $j \sim i$ denotes that areas i and j are contiguous. The conditional mean is then

$$\sum_{j \in N[i]} \theta_j / H_i = \theta_{[i]}$$

with $\theta_{[i]}$ denoting the average of the H_i neighbours of area i. The conditional variance in this case depends on the number of neighbours, since

$$\text{var}(\theta_i \,|\, \theta_j, j \neq i) = \sigma^2/H_i.$$

This model implies a high degree of spatial interdependence and may be modified to allow for a mixed or compromise scheme where some variation is explained by a white noise (exchangeable) error term. Thus a mixed spatial model has a form starting with the form

$$\log(\alpha_i) = \varepsilon_i + \theta_i \tag{7.27}$$

where the ε_i have no spatial structure; for example, we might have $\varepsilon_i \sim \text{N}(0, \sigma_\varepsilon^2)$. Depending on the relative strength of unstructured as against spatially structured variation, individual area rates or risks will be smoothed towards the global or neighbourhood averages.

Table 7.23

	Mean	SD	2.5%	Median	97.5%
α_1	−0.748	0.062	−0.873	−0.747	−0.628
α_2	−0.0094	0.0032	−0.016	−0.009	−0.004
ρ	0.779	0.260	0.050	0.850	1.000

Table 7.24

	Mean	SD	2.5%	Median	97.5%
α_1	−0.728	0.064	−0.855	−0.730	−0.604
α_2	−0.0090	0.0033	−0.016	−0.009	−0.003
ρ	0.857	0.165	0.400	0.900	1.000

Example 7.30 Blood group distribution Upton and Fingleton (UF, 1985) present data earlier analysed by Cliff and Ord (1981) on the distribution of the proportion of blood group A in 26 Irish counties. The covariate used here is their variable X_2, namely county ranks based on 'towns' per unit area (i.e. a measure of urbanisation). There are around 55 000 people in the sample, and the data are in the form of counts $\{y_i, n_i\}$ of group A and total populations respectively.

The mixed CAR formulation just outlined, as in equation (7.27) is adopted initially (see Program 7.30 Model A). Drawing on the second half of a run of 30 000 iterations yields comparative estimates of σ_θ and σ_ε of 0.36 and 0.08. The approximate formula for the proportion of variation accounted for by the spatial effect averages 0.79. The effect of the covariate is negative but the 95% interval $(-0.021, 0.004)$ straddles zero. A lack of significance was found also by UF using a SAR model estimated by maximum likelihood.

We also apply a SAR model (model B) model with a uniform $U(-1, 1)$ prior on the spatial correlation parameter ρ. In the first SAR application we just use contiguity dummies, $C_{ij} = 1$ or 0 to define the spatial similarity matrix W in equation (7.26). By contrast to the CAR model, we find (from a single run of 20 000 iterations) a clear negative effect for rank of towns per unit area (Table 7.23).

An alternative SAR analysis (model C) is based on the proximities $s_{ij} = w^*_{ij} / \sum_j w^*_{ij}$, where $w^*_{ij} = b_{ij}/d_{ij}$, d_{ij} is intercounty distance and b_{ij} is the proportion of county i's total boundary length with contiguous counties that is shared with county j. This yields similar effects of the covariate but a slightly enhanced correlation effect (Table 7.24).

7.10 CONTINUOUS SPATIAL PROCESSES

In many applications spatial modelling is concerned with elaborating the distribution of an outcome in continuous space, including spatial smoothing, and interpolation to new points within that space. Continuous space models have a conceptual advantage over discrete neighbour-based models in that they do not need to be revised at each different level of spatial aggregation. In fact, discrete spatial models are frequently applied as an approximation to data which exist in continuous space.

Most generally in a continuous spatial process, the values $y(s)$ of the process have variance $\sigma^2(s)$ and their level has two components: an underlying mean 'trend' over space, $\mu(s)$, and a random noise variable, $\varepsilon(s)$. Fluctuations in the mean level $\mu(s)$ of the process over space are described by the term first order variation, while second order variation refers to spatial dependencies between readings $y(s_1)$ and $y(s_2)$ at points $s_1 = (s_{11}, s_{21})$ and $s_2 = (s_{12}, s_{22})$. The covariance between two particular points, $y(s_i)$ and $y(s_j)$ is defined as

$$C(s_i, s_j) = E[(y(s_i) - \mu(s_i))(y(s_j) - \mu(s_j))]$$

so the correlation is

$$\rho(s_i, s_j) = C(s_i, s_j)/[\sigma(s_i)\sigma(s_j)].$$

Spatial dependencies only exist at the scale of the observed range of distances between sample locations. Spatial variations at a smaller scale that cannot be resolved (i.e. distinguished from random noise) are denoted as 'nugget' variation in geostatistics.

There are a variety of techniques for spatial smoothing (such as spatial moving averages and Kernel smoothers) which are closely related to those used in time sequenced data, and are used to produce smoothed estimates of the mean of the process over the study region. The goal of such smoothing is frequently prediction of $y(t)$ at a new location on the basis of existing readings (e.g. the prediction might be of mineral concentrations at an as yet unmined site).

Progress in analysing continuous spatial data rests on stationarity assumptions which parallel those in time series analysis, and specify similarity in terms of spatial lags analogous to lags in time. Often we evaluate distances d_{ij} between points s_i and s_j and define lags in terms of distance bands appropriate to the data at hand (Cressie and Read, 1989; Bailey and Gattrell, 1995). Questions of second order variation then focus on the relationship between variation in the differences $y(s_i) - y(s_j)$ at varying distances d_{ij} or distance bands. A correlation effect will be apparent in smaller variations in differences in lower distance bands. The direction between s_i and s_j is also potentially relevant (e.g whether i is north of j or vice versa), though often the assumption is made that the process is 'isotropic', so that the variance of the differences in y is independent of direction.

Geostatistical smoothing or 'kriging' methods for spatial interpolation originate in the assumption of intrinsic stationarity, proposed by Matheron (1963). Under this assumption $\mu(s)$ is assumed constant, at least constant locally, so that $y(s) = \mu + \varepsilon(s)$ and the variance of the difference in y for given distance lag h is constant

$$\text{var}[y(s) - y(s + h)] = 2\gamma(h).$$

The function $\gamma(h)$, as it varies over h, is known as the semi-variogram or simply variogram. A spatial process is stationary if additionally the variance of the $\varepsilon(s)$ is constant over space, i.e. $\sigma^2(s) = \sigma^2$.

The variogram is typically estimated empirically as the average squared difference between paired data values in a given lag band. For example, denote $y_i = y(s_i)$ for simplicity, and suppose the analysis relates to small area health outcomes as in Oliver *et al.* (1992). For illustration, suppose lag 1 covers all distances between locations of under 15 km, lag 2 those between 15 and 25 km, lag 3 those between 25 and 35 km, and so on. Then the lag 1 value $\gamma(1)$ is the average of the squared differences $(y_i - y_j)^2$ for all $n(1)$ differences between pairs whose separation is under 15 km. For N observations there will then be $n = N(N-1)/2$ pairs of values to consider, with $n = n(1) + n(2) + \ldots + n(G)$, where G is the maximum distance (band) between points present in the data. in the case of a spatially continuous (i.e. correlated) variable, differences between y_i and y_j are expected to be smaller at closer distances, so the variogram function will be small at first and tend to increase as lag h increases. This increase with h may tail off so that a more or less constant variation exists above a certain lag; this ceiling variability is sometimes known in geostatistics as the 'sill' variance. If a variable shows little spatial dependence its variogram will tend to be patternless, i.e. non-increasing.

The empirical variogram is the basis for smooth variogram models in h, which are used for spatial interpolation. These models are typically based on the aggregated average estimates $\hat{\gamma}(h)$ from within distance bands, but for small sets of points one may model the n actual differences $(y_i - y_j)^2$ in terms of distance. Among the models proposed are spherical and exponential variogram models (Bailey and Gattrell, 1995), with the exponential model being

$$\gamma(h) = \sigma^2(1 - \mathrm{e}^{-3h/r}).$$

In this model r is the range (the distance at which spatial dependence ceases to be evident) and the stationary variance σ^2 is the sill variance. An extension allows for a discontinuity δ at the origin (a nugget effect) leading to

$$\gamma(h) = \delta + (\sigma^2 - c)(1 - \mathrm{e}^{-3h/r}).$$

The variogram model is then the foundation for specifying spatial dispersion matrices,

$$C(h) = \sigma^2 - \gamma(h)$$

which can be used to model the distribution of observations y_i in space and to predict y at a new spatial location (Bailey and Gattrell, 1995). It may be noted that estimation of the variogram is subject to possible distortion from outliers (Cressie and Read, 1989; Gunst and Hartfield, 1997).

As well as directly estimating the variogram of the data itself, spatial modelling of this kind also allows for large-scale trends, for example, via a linear or quadratic surface based on the grid locations in the region. So the model for $y(s)$ is more generally

$$y(s) = \beta x(s) + \varepsilon(s)$$

where $x(s)$ consist of functions of the grid coordinates s_{1i} and s_{2i}. A nonparametric option sometimes adopted is 'median polish', which if the observations are on irregular spatial units may involve an overlay of a regular grid (Cressie and Read, 1989; Bailey and Gattrell, 1995). The aim is to decompose the means ('trend') of the observed value y_i into 'row' and 'column' effects deriving from the rows and columns of the grid overlay. Once the large-scale trend is controlled for, the interest is in second order spatial dependencies, as shown by the covariogram structure of the residuals $e_i = y_i - \hat{\mu}_i$. The estimation of the variogram then relies on differences $Q_{ij} = (e_i - e_j)^2$ at different spatial lags (Diggle *et al.*, 1998).

Example 7.31 Variogram of data generated from spatially continuous process We consider the generation of points in a square bounded by points $(0, 0), (0, 100), (100, 100)$ and $(100, 0)$. Fifty points $s_i = (s_{1i}, s_{2i})$ are generated by taking s_{1i} and s_{2i} separately from $\mathrm{U}(0, 100)$ densities, with separations given by the Euclidean distances h_{ij} between s_i and s_j. The outcomes $y_i(s_i)$, $i = 1, \ldots, 50$ are assumed generated from a process with mean 160 and covariances between points

$$C(h) = 36\mathrm{e}^{-3h/30}.$$

This involves a multinormal model $y[1{:}50] \sim \mathrm{dmnorm}(mu[1{:}50], P[,])$ where the $mu[]$ are identically 160, and the precision matrix is the inverse of a covariance matrix with diagonal values 36 and off- diagonal values determined by the covariance function $C(h)$ (see Program 7.31).

The variogram is then estimated with normal sampling from the resulting generated set of locations s_i and outcomes $y_i(s_i)$. A trend is estimated even though not present in the generating model. This involves a 10×10 grid determined by row and column effects. The resulting variogram (from the second half of a run of 10 000 iterations) is defined by distance bands or spatial lags of 5 units, 0–5, 5–10, etc., with a maximum of 20 lags (see Figure 7.10; Table 7.25).

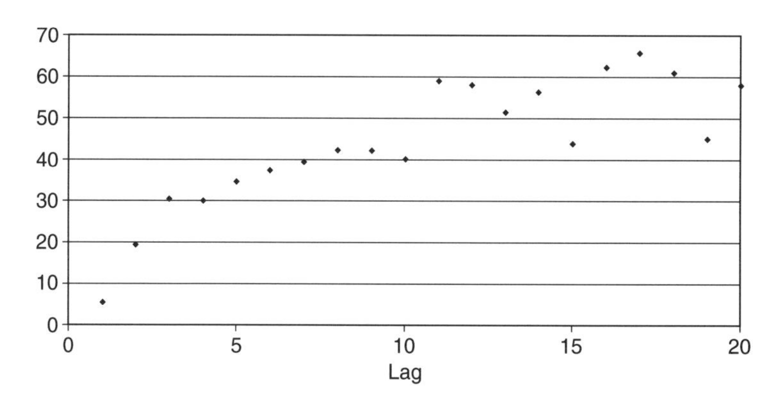

Figure 7.10 Estimated variogram (means).

Table 7.25 Variogram estimates from original generated data.

Variogram at lag	Mean	SD	2.5%	Median	97.5%
1	5.3	5.5	0.2	3.7	20.2
2	19.4	9.1	6.7	17.7	42.1
3	30.3	10.9	13.3	28.6	55.5
4	30.1	11.0	13.3	28.5	55.9
5	34.5	11.0	17.4	33.0	60.2
6	37.2	11.2	19.4	35.9	62.9
7	39.3	13.8	18.5	37.3	72.4
8	42.1	12.5	22.3	40.6	71.0
9	42.1	12.1	22.6	40.7	70.1
10	40.0	12.2	20.7	38.5	68.0
11	59.1	17.8	30.9	56.9	99.9
12	58.3	16.7	31.2	56.3	96.0
13	51.2	14.6	27.9	49.7	84.9
14	56.2	18.2	27.7	53.9	98.0
15	43.7	14.1	21.9	41.9	76.7
16	62.2	19.7	30.4	59.8	106.8
17	65.6	21.6	31.2	62.8	115.2
18	60.8	21.5	27.4	57.9	112.4
19	44.9	19.0	17.1	41.7	90.5
20	57.7	27.9	18.6	52.5	126.2

The variogram shows reasonably monotonic increase at lower lags, though with uncertainty around the estimates and some skewness in their distributions. Adopting a Student *t* with 5 d.f. as the sampling density made little difference to the variogram.

However, if the value of $y_1 = 190$ is set to replace the generated value of 161.2 then the sampling density does affect the variogram estimates (Table 7.26). Gunst and Hartfield (1997) undertake a comparable analysis with an artificial outlier, and propose alternative robust variogram estimators. A robust sampling model is an alternative approach.

Example 7.32 SIDS deaths in North Carolina Cressie and Read (1989) presents an analysis of SIDS deaths S_i in relation to total live births in 100 counties of North Carolina. A mixed ICAR form as described above is initially adopted with the deaths taken as Poisson and total births as an exposure. Thus $S_i \sim \text{Poi}(\mu_i B_i)$ where B_i are total births and μ_i is the predicted SIDS death rate in county *i*.

One possible covariate believed to influence SIDS occurrences is the proportion of births to nonwhite mothers. We investigate the impact of this variable before and after allowing for spatial structure in the errors. In the case of these data spatial dependence is pronounced (with $\sigma_\theta = 0.60$ compared with $\sigma_\varepsilon = 0.14$) and introducing it amplifies rather than lessens the effect of the covariate (see *Program 7.32 SIDS in N. Carolina*).

Table 7.26 Variogram estimates for perturbed data.

	Normal sampling		Student t sampling	
Variogram at lag	Mean	SD	Mean	SD
1	9.9	10.9	5.5	6.1
2	33.9	17.7	22.3	11.6
3	51.1	20.9	33.0	12.4
4	53.0	21.6	33.3	13.4
5	62.1	22.9	37.4	14.9
6	70.2	25.3	40.7	14.9
7	61.4	24.5	41.6	15.4
8	72.9	26.1	44.9	15.3
9	71.8	25.6	46.4	15.6
10	68.4	24.8	43.5	15.3
11	86.8	32.1	61.5	22.1
12	84.0	29.3	60.6	19.9
13	74.5	26.5	54.0	16.8
14	82.8	32.4	59.5	21.7
15	69.1	27.4	46.7	15.4
16	90.7	36.0	65.0	23.6
17	88.9	37.6	69.9	24.6
18	83.7	36.8	62.0	24.7
19	78.0	38.8	44.5	19.7
20	87.7	52.7	57.4	30.5

Cressie and Read (CR, 1989) present an alternative approach to these data drawing on continuous spatial modelling concepts. They adopt a binomial model $S_i \sim \text{Bin}(p_i, B_i)$ and consider the transformations of the dependent variable to remove, or attenuate, the dependence of the (binomial) variance on the mean. Thus defining W_i to equal $(S_i + 1)/B_i$, they search for a trend surface model μ_i and transformation g such that

$$g(W_i) = \mu_i + \varepsilon_i$$

with ε_i purged of spatial dependence. They advocate the Freeman–Tukey transform of W_i, namely $Z_i = \sqrt{(}S_i/B_i) + \sqrt{W_i}$ and show that the quantities $Z_i\sqrt{n_i}$ have approximately equal variance.

They then overlay a 9×24 rectangular grid on the map of irregular county boundaries and use a nonparametric median polish technique to fit row and column effects to $Z_i = g(W_i)$ such that spatial trends are eliminated, and a stationary set of residuals obtained. The median polish technique is intended to be robust (resistant) to unusual observations. CR use covariogram analysis of the weighted residuals $\hat{\varepsilon}_i\sqrt{n_i}$ to demonstrate this stationarity.

The spirit of their analysis is followed here with the adoption of a binomial model, but here the estimated probabilities p_i are transformed rather than the observed rates S_i/B_i. We adopt a simple square root transform of the p_i, and overlay a 48×19 grid on the original irregular lattice. A nonparametric

Table 7.27

		Mean	SD	2.5%	Median	97.5%
Unweighted variogram	Lag 1	0.0022	0.0005	0.0012	0.0021	0.0031
	Lag 2	0.0022	0.0004	0.0013	0.0022	0.0030
	Lag 3	0.0019	0.0004	0.0011	0.0019	0.0027
	Lag 4	0.0023	0.0005	0.0013	0.0024	0.0034
	Lag 5	0.0023	0.0005	0.0013	0.0023	0.0033
	Lag 6	0.0026	0.0006	0.0014	0.0026	0.0037
	Lag 7	0.0027	0.0006	0.0015	0.0027	0.0039
	Lag 8	0.0028	0.0006	0.0017	0.0028	0.0039
	Lag 9	0.0026	0.0005	0.0016	0.0026	0.0035
	Lag 10	0.0025	0.0004	0.0016	0.0025	0.0033
Weighted variogram	Lag 1	3.62	0.79	2.21	3.55	5.31
	Lag 2	3.77	0.82	2.18	3.75	5.54
	Lag 3	3.63	0.87	2.06	3.57	5.68
	Lag 4	4.09	0.95	2.44	3.99	6.40
	Lag 5	4.14	0.83	2.67	4.07	6.02
	Lag 6	4.38	1.00	2.57	4.27	6.77
	Lag 7	5.00	1.14	3.07	4.91	7.63
	Lag 8	5.23	1.21	3.26	5.11	8.07
	Lag 9	4.74	1.10	2.92	4.62	7.31
	Lag 10	4.42	1.05	2.68	4.32	6.86

analysis of the resulting row and column effects is made using a Dirichlet process (Polya tree) model. As in the analysis of CR, we define the variogram in terms of 20-metre distance bands between counties. So spatial lags $h = 1, 2, 3, \ldots$ are for intercounty distances $d_{ij} \in (20h - 10, 20h + 10)$. Then the variogram $2\hat{\gamma}(h)$ is the average in each band of the weighted differences $Q^w_{ij} = (\hat{\varepsilon}_i\sqrt{n_i} - \hat{\varepsilon}_j\sqrt{n_j})^2$. Table 7.27 presents the variograms, based on both the weighted and unweighted differences $Q^u_{ij} = (\hat{\varepsilon}_i - \hat{\varepsilon}_j)^2$. Two features to note are the relatively flat variograms, for both weighted and unweighted analysis, and the skewness in the weighted covariogram estimates. The analysis of CR is actually subsequent to the exclusion of an outlier, but this would seem to be an optional extra in terms of inducing spatial stationarity.

7.11 MEASUREMENT ERROR, SEEMINGLY UNRELATED REGRESSIONS, AND SIMULTANEOUS EQUATIONS

Models in many areas of scientific enquiry include abstract constructs that are often not directly measurable or are scales that cannot be measured precisely. For these variables we have to use proxies which are imperfect measures of the true variable, or recognise that no instrument is available to measure a certain quantity with exact accuracy. Examples of latent constructs from education and sociology include ability, attitudes or social status, while in medical studies the underlying variable may be health status. Example in medical applications of variables which cannot be measured precisely include continuous variables such as pulse rate, blood pressure and temperature, and discrete outcomes such as disease diagnosis.

More generally, measurement error regression models apply when one or more of the true predictors, denoted $X_1, \ldots, X_q$, are measured with error. They can therefore be regarded as a generalisation of standard regression models. Thus in sociological or epidemiological applications, relating area crime or disease rates to average incomes, models typically allow for sampling variability in the outcome (crime, disease) but not in the predictor (income). Measurement errors may apply to several covariates. As a specific example, income attainment is likely to depend on educational attainment, which is proxied by years of schooling, and on ability, which is proxied by IQ or other psychometric scales. Returns to education in income function models may be biased if ability is omitted or if it is not recognised to be measured with error (Berndt, 1990).

There are various formulations of measurement error models, posing potentially different problems of identifiability and estimation. The error term in the standard regression model encompasses both measurement error in the dependent variable and errors from omitted covariates. Thus while the observed dependent variable y is in effect treated as a proxy for an unobserved true dependent variable Y, the true values of predictors $X_1, \ldots, X_q$ are assumed to have been obtained. That is, it is assumed that $x_j = X_j$ for all $j = 1, \ldots, q$. Applying this standard procedure when in fact there are measurement errors leads to point estimates of regression coefficients on the true X_i which are understated in absolute terms, and their precision overstated (because the uncertainty in the relationship between x and X is not allowed for).

One version of the measurement error model assumes instead an underlying deterministic relation between the true values of response and the predictor(s). For $q = 1$ and assuming linear dependence, this implies

$$Y_i = \beta_0 + \beta_1 X_i \tag{7.28}$$

where the true values are linked to the observed values $\{y_i, x_i\}$ with additive errors:

$$y_i = Y_i + \varepsilon_i, x_i = X_i + \delta_i.$$

This is known as the classical errors assumption. We generally assume the errors δ_i and ε_i have zero means and that they are either (a) uncorrelated and independent of each other, or (b) drawn from a bivariate density. For example, under (a) we might have

$$\begin{gathered}\varepsilon_i \sim \mathrm{N}(0, \sigma_\varepsilon^2)\\ \delta_i \sim \mathrm{N}(0, \sigma_\delta^2)\\ \mathrm{cov}(\delta_i, \delta_j) = \mathrm{cov}(\varepsilon_i, \varepsilon_j) = 0 \text{ for } i \neq j\\ \mathrm{cov}(\delta_i, \varepsilon_j) = 0, \text{ all } i, j\end{gathered}$$

It is possible to regard the X_i themselves as unknown constants in a fixed effects ('functional') measurement error model. More tractable is a random effects approach with the X_i drawn from a common density with expectation μ_X and variance σ_X^2 but assumed independent of the two error terms, so that

$$
\begin{aligned}
Var(x) &= Var(X) + var(\delta) \\
&= \sigma_X^2 + \sigma_\delta^2.
\end{aligned}
$$

Our information about the distribution of X_i may consist of knowledge about such distributional features, e.g. approximate parameters for a density such as $X_i \sim \mathrm{N}(\mu_X, \sigma_X^2)$. However, it often includes a relationship to measures on ancillary causal influences Z_i, which may be regarded as measured without error. For example, in the income–consumption model considered below, we might expect the latent 'permanent consumption' X to be related to accurately known investment and government expenditure totals Z_1 and Z_2. We would include this relationship in the model by setting up a joint distribution between X and $\underset{\sim}{Z}$, or by forming a regression of x_i on Z_i. In epidemiological examples, one risk factor for a disease X, may be measured with error and another Z measured accurately (e.g. a demographic attribute). If x_i is the actual measure of X_i then information about the true X_i may be provided in two ways: first by a measurement model, describing how x_i relates to X_i, as in the classical errors model above; second by an exposure model, showing how 'exposure' to the risk factor X is related to accurately measured attributes, or other risk factors, Z.

A slightly different specification to (7.28) is the 'errors in equation' model in which the relation between the true variables is subject to error

$$Y_i = \beta_0 + \beta_1 X_i + u_i \tag{7.29}$$

where the random effects u_i are independent of the X_i and have mean zero and variance σ_u^2. So the errors in equation model may also be written

$$
\begin{aligned}
y_i = Y_i + \varepsilon_i = (\beta_0 + \beta_1 X_i + u_i) + \varepsilon_i &= \beta_0 + \beta_1 X_i + v_i \\
x_i &= X_i + \delta_i
\end{aligned}
$$

where v_i combines measurement error in the dependent variable and the equation error, $v_i = \varepsilon_i + u_i$, and ε_i and δ_i are measurement errors, with mean zero and variances σ_ε^2 and σ_δ^2. The latter two sets of errors are taken to be uncorrelated with each other and with the latent variables, Y_i and X_i. So $\mathrm{cov}(\varepsilon, \delta) = \mathrm{cov}(\varepsilon, X) = \mathrm{cov}(\varepsilon, Y) = \mathrm{cov}(\delta, X) = \mathrm{cov}(\delta, Y) = 0$.

An example from economics illustrating measurement of underlying constructs is Friedman's model for the relation between permanent income Y_i and permanent consumption X_i. The true regression between these latent variables is one of simple proportionality:

$$Y_i = \beta_0 + \beta_1 X_i + u_i$$

where the errors u_i are mutually uncorrelated, and also uncorrelated with the X_i. However, instead of Y_i and X_i, we observe totals of consumption and income x_i and y_i which include randomly distributed or 'transitory' components, ε_i and δ_i respectively.

The true regression (7.29) can then be rewritten in terms of observable data (in centred form) as follows

$$y_i = \beta_0 + \beta_1 x_i + u_i + \varepsilon_i - \beta_1 \delta_i.$$

That is

$$y_i = \beta_0 + \beta_1 x_i + w_i$$

where $w_i = u_i + \varepsilon_i - \beta_1 \delta_i$. This relation cannot then be consistently estimated by usual least squares principle because x and w are correlated: $\text{cov}(x, w) = \text{cov}(X + \delta, u + \varepsilon - \beta_1 \delta) = \text{cov}(X + \delta, \; - \beta_1 \delta) = -\beta_1 \sigma_\delta^2$.

The normal measurement error model with X_i as a random latent variable is unidentifiable by conventional least squares because it has an additional parameter σ_δ^2 that the standard regression model does not have. There are several possible identifiability assumptions:

(a) that either σ_ε^2 or σ_δ^2 is known or has a known density;
(b) that the ratio $\lambda = \sigma_\varepsilon^2 / \sigma_\delta^2$ is known or has a known informative density;
(c) that β_0, the intercept, is known;
(d) that the 'reliability coefficient' $\kappa = \sigma_X^2 / [\sigma_X^2 + \sigma_\delta^2] = [\text{var}(x) - \sigma_\delta^2] / \text{var}(x)$ is known (or can be defined within limits).

Restriction on the reliability coefficient occurs in a multiple regression situation where some predictors are measured without error. Suppose $Y = b_0 + b_1 X + b_2 Z + u$, with Z accurately measured but X measured with error, so that the observation is $x = X + \delta$. Then the regression in terms of observed data is $y = b_0 + b_1 x + b_2 z + w$, where $z = Z$. It can be shown that the maximum possible value of $\sigma_\delta^2 / \text{var}(x)$ is $1 - r^2$ where r is the correlation between observations x and z. So in cases of multicollinearity between predictors, stringent limits on σ_δ^2 as a proportion of $\text{var}(x)$ are needed. Maddala (1981) gives an extreme example where y is imports, x is gross domestic production (measured with error) and z is consumption (accurately measured), and where $\text{corr}(x, z) = 0.99789$. This implies that the variance of the measurement error in x cannot be greater than 0.211 % of the variance of x.

In some circumstances the relationship between x and X may be estimable from a calibration or validation subsample in which both measures (the 'true' measure and its proxy) are obtained. If X is latent then information on it may be improved by applying several instruments all assumed to reflect the same underlying X. An example from Richardson (1996) below illustrates this, and is implemented below.

The other major possible formulation of measurement errors (Berkson errors) may be appropriate in certain experimental situations where various preset levels of a dose or other treatment variable are administered. The administered quantities of an injected drug, say, are at levels at $x = 1, 2, 3, \ldots \text{cm}^3$ but actual concentrations in a patient will depend on the patient's physiology. Thus x is no longer random and a more appropriate model for the latent variable X is that its values are centred on the fixed series of experimental values of x.

Example 7.33 Underlying X from an asymmetric density Cheng and van Ness (CV, 1998) present simulated data for a univariate regression in which the underlying true predictor X is generated as a chi-square with $\kappa = 4$ degrees of freedom. Then the 'observed' predictors are generated according to $x_i = X_i + \delta_i$ where $\delta_i \sim N(0, \sigma_\delta^2)$, and $i = 1, \ldots, 36$. The 'observed' responses are generated according to $y_i = Y_i + \varepsilon_i$ where $\varepsilon_i \sim N(0, \sigma_\varepsilon^2)$ and with a deterministic regression model relating Y to X

$$Y_i = \beta_0 + \beta_1 X_i.$$

CV set $\sigma_\varepsilon^2 = \sigma_\delta^2 = 1$ and $\beta_0 = 0, \beta_1 = 1$. CV actually present the generated values of x and y only, not the original X. We analyse these x and y values in Program 7.33.

We can model this non-normal generation of the predictors using a gamma density $G(n, 0.5)$ where the n is assigned a discrete prior with equal weights on three possible integer values, 1, 3 and 10 (corresponding to chi-squared densities with 2, 6 and 20 degrees of freedom).

Similarly, although we know that $\lambda = \sigma_\varepsilon^2/\sigma_\delta^2 = 1$, we can allow for uncertainty in this ratio by adopting a discrete prior with equal weights on three values 0.1, 1 and 10. We assign standard a vague priors for $1/\sigma_\varepsilon^2$. Because BUGS works with precisions we actually sample $1/\sigma_\varepsilon^2$ and then derive $1/\sigma_\delta^2$ by multiplying $1/\sigma_\varepsilon^2$ by λ.

We adopt standard vague priors for the regression parameters, though in substantive examples it would be useful to employ any likely constraints implied by subject matter knowledge (e.g. $\beta_1 > 0$ in the relation between permanent income and permanent consumption). In the present simulated data example we reproduce reasonably well the original regression parameters, the precision of y around $Y(1/\sigma_\varepsilon^2)$ and the chi-square d.f. parameter underlying X (Table 7.28).

7.11.1 Double regressions

Consider the simple case where only y_i and x_i are observed, and the errors δ and ε are assumed independent and uncorrelated, where

$$y_i = Y_i + \varepsilon_i$$
$$x_i = X_i + \delta_i.$$

Moreover $Y_i = \beta_0 + \beta_1 X_i + u_i$ as in (7.29) above.

In this situation, bounds on β_1 can be assessed by combining regression of y on x with regression of x on y. The slope from the latter regression is $1/\beta_1$ and the intercept is $-\beta_0/\beta_1$ so the regression can then be solved for y to provide two

Table 7.28

Parameter	Mean	SD	2.5%	Median	97.5%
$1/\sigma_\varepsilon^2$	0.81	0.24	0.40	0.79	1.34
β_0	0.38	0.61	−0.78	0.38	1.45
β_1	0.95	0.09	0.78	0.94	1.14
$\kappa/2$	2.17	0.98	1.00	3.00	3.00
Latent X					
X_1	9.14	0.78	7.59	9.15	10.65
X_2	2.63	0.89	0.95	2.63	4.42
X_3	7.24	0.77	5.66	7.25	8.78
X_4	4.32	0.85	2.67	4.30	6.01
X_5	6.23	0.77	4.64	6.22	7.73
X_6	0.94	0.65	0.04	0.86	2.32
X_7	1.75	0.80	0.28	1.71	3.39
X_8	2.21	0.84	0.60	2.17	3.95
X_9	3.44	0.78	1.80	3.46	4.98
X_{10}	4.55	0.81	3.01	4.51	6.24
X_{11}	1.91	0.80	0.33	1.93	3.42
X_{12}	2.00	0.80	0.42	1.97	3.61
X_{13}	5.35	0.76	3.77	5.38	6.77
X_{14}	2.39	0.85	0.76	2.36	4.11
X_{15}	7.26	0.80	5.77	7.22	8.88
X_{16}	5.43	0.78	3.82	5.47	6.91
X_{17}	4.80	0.81	3.08	4.84	6.28
X_{18}	1.77	0.77	0.34	1.76	3.34
X_{19}	1.69	0.76	0.24	1.70	3.17
X_{20}	1.24	0.70	0.09	1.18	2.74
X_{21}	13.68	0.90	11.83	13.72	15.33
X_{22}	2.44	0.77	0.87	2.47	3.91
X_{23}	1.06	0.70	0.05	0.99	2.61
X_{24}	1.47	0.85	0.10	1.45	3.21
X_{25}	0.93	0.64	0.03	0.84	2.37
X_{26}	1.52	0.78	0.17	1.49	3.08
X_{27}	6.99	0.78	5.43	7.01	8.49
X_{28}	8.18	0.83	6.52	8.22	9.73
X_{29}	5.34	0.79	3.72	5.33	6.90
X_{30}	1.98	0.78	0.39	2.00	3.52
X_{31}	11.98	0.81	10.35	11.98	13.58
X_{32}	2.45	0.86	0.71	2.46	4.03
X_{33}	2.02	0.77	0.47	2.02	3.54
X_{34}	2.12	0.84	0.47	2.11	3.79
X_{35}	9.60	0.79	8.05	9.58	11.19
X_{36}	1.86	0.90	0.29	1.83	3.66

estimates of the slope (Maddala, 1981, Chapter 11). If additional and accurately measured predictors $Z_1, Z_2, \ldots$ of x are available, then additional steps of regressing x on $Z_1, Z_2, \ldots,$ and of y on $Z_1, Z_2, \ldots$ provide instruments $\hat{x}_i$ and $\hat{y}_i$. If the underlying true regression remains of the type (7.28) then regressions of y on $\hat{x}$ and x on $\hat{y}$ provide estimates of β_1 and $1/\beta_1$ respectively (Judge *et al.*, 1988). Goldberger (1972) provides an equivalent maximum likelihood procedure.

One way of implementing this via Bayesian sampling involves introducing two 'copies' of the data, one involving y_1 and x_1 in which y_1 is the outcome, and the other involving y_2 and x_2 in which x_2 is the outcome. This is illustrated using a data set relating true national consumption Y and true income X, taken from Judge *et al.* (1988).

Example 7.34 Income–consumption data. These data are produced via a response model

$$Y_t = \alpha + \beta X_t + v_t$$

with $\alpha = 10$, $\beta = 0.8$ and $v_t \sim \mathrm{N}(0, 0.5)$. The measurement model is

$$x_t = X_t + \delta_t$$

with $\delta_t \sim \mathrm{N}(0, 0.2)$. The ancillary variable model is specified deterministically as

$$x_t = \pi_0 + \pi_1 Z_{1t} + \pi_2 Z_{2t}$$

with $\pi_0 = 2, \pi_1 = 3$ and $\pi_2 = 5$. The simulation assumes assigned values to Z_{1t} and Z_{2t} corresponding approximately to quantities such as investment and government spending. The resulting simulated observations y_t and x_t correspond to income and consumption respectively.

Judge *et al* form instruments $\hat{x}$ and $\hat{y}$ by regressing x on Z_1 and Z_2, and y on Z_1 and Z_2 respectively. The regression of y on $\hat{x}$ then provides estimates $\hat{\alpha} = 9.097$ and $\hat{\beta} = 0.8368$ while the regression of x on $\hat{y}$ provides estimates of $-\hat{\alpha}/\hat{\beta} = -10.83$ and $1/\hat{\beta} = 1.1938$.

In the Bayesian version we assign non informative priors to $\alpha, \beta, \pi_0, \pi_1, \pi_2$ and to $\tau_v = 1/\sigma_v^2$ and $\tau_\delta = 1/\sigma_\delta^2$. We impose a prior restriction that the precision in the measurement model is greater than in the response model, i.e. that $\lambda = \sigma_v^2/\sigma_\delta^2$ exceeds one. An alternative would be to select a set of possible values of λ and assess sensitivity of estimates of the response model parameters. It is also assumed that $\beta < 1$ in line with economic expectations. Running *Program 7.34 Income and Consumption* for 30 000 iterations (and 500 burn-in) leads to posterior means for α and β of 8.46 (with 95 % interval from 4.8 to 14) and 0.86 (0.67, 0.99). These are denoted `b[1:2]` in Program 7.34. These estimates have narrower credible intervals than those based on the 'regression on instruments' procedure.

7.11.2 Epidemiological applications

In epidemiological applications, the relationship between accurately measured risk factors or demographic attributes for a disease, namely Z_i, and latent risk variables X_i, may be termed the exposure model, characterised by a parameter π (Richardson, 1996), $P[X_i | Z_i, \pi]$. The relationship between proxies x_i for X_i, and X_i itself, is named the measurement model, $P[x_i | X_i, \delta]$. The actual model for the disease outcome Y_i, specifies the impact of X_i, or of X_i and Z_i jointly, on frequencies, rates or probabilities of the disease, $P[Y_i | X_i, Z_i, \beta]$.

Thus in the research of Bernardinelli *et al.* (1997) the outcome Y_i consists of counts of new cases (that is the 'incidence') of insulin-dependent diabetes mellitus (IDDM) in 366 Sicilian communes over 1989–92. The interest is in the relation of IDDM incidence to population resistance to IDDM based on historic exposure to malaria. Resistance (X_i) cannot be measured directly and is proxied by an accurate measured variable, Z_i; namely, geographic patterns of malaria cases in a year prior to the Second World War (1938), when malaria was last widespread. The exposure and measurement parts of the model are represented by assuming (a) that the case counts Z_i are binomial in terms of known populations N_i and rates θ_i:

$$Z_i \sim \text{Bin}(\theta_i, N_i)$$

and (b) that the malaria incidence rates are measured around the latent resistances X_i as follows,

$$\text{logit}(\theta_i) \sim \text{N}(X_i, \rho)$$

where ρ is an unknown measurement error variance. Bernardinelli *et al* (1997) present arguments to justify a certain choice of the parameter ρ, as the data supplies no information on this parameter. The underlying true risks X_i are assumed spatially correlated according to a Markov random field prior (i.e. a discrete spatial correlation model).

The disease model, for IDDM counts Y_i, is Poisson with expected counts E_i based on known age/sex structures of populations in 1989–92. Thus

$$Y_i \sim \text{Poi}(E_i \varphi_i)$$
$$\log(\varphi_i) = \beta_{0i} + \beta_1 X_i + s_i$$

where X is long-term resistance to IDDM based on endemic malaria in the past. Note that Z could also be treated as Poisson but equivalent age/sex information is not available for 1938 and the simpler binomial model is adopted. The geographic variation in IDDM incidence, beyond that due to historic resistance X, is also assumed to follow a Markov random field, and represented by the term s_i. A white noise error term u_i may also be included in the model for the IDDM rate φ_i.

Example 7.35 Attempted suicide and schizophrenia An example of such an epidemiological application involves rates of attempted suicide in 253 small areas of North East London over 1990–92. These are estimated via small area counts Y_i of admissions to hospital for deliberate self-harm (e.g. poisoning). One would expect a relationship between this outcome and the level of mental illness in communities. However, the latter is difficult to gauge accurately because complete population registers are not available. One is forced to rely on proxies, such as admissions to hospital for more severe forms of mental illness. Accordingly, let Z_i denote counts of schizophrenia admissions. Then if E_{zi} are expected schizophrenia admissions obtained from the usual standardisation procedures, we assume

$$Z_i \sim \text{Poi}(\theta_i E_{zi}).$$

Additionally $\log(\theta_i) \sim \text{N}\ (X_i, w)$, where X_i is taken to be the underlying measure of area mental health and to follow a spatially correlated prior with precision parameter τ_x. The relation between θ_i and X_i is subject to measurement error, with variance w. We take the precision of the latter model (the inverse of measurement error variance) to exceed the precision in the X_i themselves, so that $1/w = k\tau_x$, where $k > 1$. We assume $k = 10$ so that $1/w = 10\tau_x$, though a full analysis would compare a range of possible values for the multiplier k.

The model for the counts of attempted suicides then involves the latent psychiatric morbidity X. If E_{Yi} denote expected parasuicides then

$$Y_i \sim \text{Poi}(\varphi_i E_{Yi})$$
$$\log(\varphi_i) = \beta X_i + s_i + u_i$$

where s_i is a spatial process with precision τ_s and u_i are unstructured errors with precision τ_u. The results from the second half of a run of 10 000 iterations using *Program 7.35 Parasuicide* show a clear effect of latent morbidity on attempted suicides (Table 7.29).

Example 7.36 Repeated measures via different instruments Richardson (1996) provides an epidemiological example where information on the covariate measured with error is provided by readings on two instruments, A and B. Thus a discrete disease outcome Y_i is Bernoulli distributed with parameter π_i, where $\text{logit}(\pi_i) = \beta_0 + \beta_1 X_i + \beta_2 Z_i$, where X and Z are continuous. The first risk factor for the disease, namely X, is measured with error, and the second risk factor, Z, is known accurately. A prevalence model relates their mutual values, and is assumed to be bivariate normal in form, with mean $\mu = (\mu_X, \mu_Z)$ and

Table 7.29

	Mean	SD	2.5%	Median	97.5%
β	0.48	0.068	0.35	0.48	0.62
σ_u	0.32	0.024	0.28	0.32	0.37
σ_s	0.56	0.054	0.46	0.55	0.68
σ_x	1.30	0.064	1.19	1.30	1.44

dispersion matrix Σ. Throughout a sample of 1200 persons, a precise instrument (e.g. a questionnaire or clinical reading) is recorded, but has a bias in reproducing the X, such that

$$R_A \sim N(\phi + \psi X, \theta_A^{-1})$$

where ψ is the slope relating true exposure X and its proxy R_A. For a subsample of 200 persons, as well as recording R_A, another unbiased but less precise instrument is used. Thus

$$R_B \sim N(X, \theta_B^{-1})$$

where the assumption $\theta_A > \theta_B$ reflects the lesser precision of instrument B. Two readings with this instrument are made in the subsample. Thus information on X is provided by the prevalence model, relating the bivariate patterning of the two factors, by the use of two recording instruments, and by the replication on one of them.

Richardson generates data for this design with assumed parameters $\mu_X = 0.5, \mu_Z = -0.5$,

$$\Sigma = \begin{bmatrix} 1.02 & 0.56 \\ 0.55 & 0.96 \end{bmatrix}$$

and parameters $\phi = 0.8, \psi = 0.4, \theta_A = 0.9, \theta_B = 0.3, \beta_0 = -0.8, \beta_1 = 0.9$, and $\beta_2 = 1.2$.

We generate these data in Program 7.36 and then 're-estimate' the parameters. Because X and Z are adverse risk factors, our priors for β_1 and β_2 specify selection from positive values only, and usually the same applies to ψ when R_A is a direct proxy for X. Otherwise the priors are non informative. The second half of a run of 2000 iterations shows all the parameters to be well reproduced, except for the effect β_2 on $\Pr(Y_i = 1)$ of the accurately measured factor Z (Table 7.30). This parameter is higher than in the generating values. The intercept is also smaller absolutely than its true value. The effect on Y of Z is, however, estimated with greater precision (i.e. a smaller standard error) than that of X.

Table 7.30

Parameter	Mean	SD	Median
β_0	−0.59	0.16	−0.60
β_1	0.91	0.22	0.88
β_2	1.66	0.12	1.66
μ_1	0.48	0.07	0.48
μ_2	−0.53	0.03	−0.53
ϕ	0.79	0.04	0.80
ψ	0.41	0.06	0.41
θ_B	0.27	0.03	0.27
θ_A	0.77	0.05	0.77

7.12 SETS OF RELATIONSHIPS

Suppose that instead of a single equation model $y = \beta x + e$ we have more than one equation to estimate; examples might be investment functions for $i = 1, \ldots, M$ different firms, if time series data are available on each of the firms. Disturbances in the different equations for a given time point may reflect the impact of omitted factors having a similar impact on each firm's behaviour – so that there is contemporaneous correlation across errors in different equations at a given time point. This is to be distinguished from autocorrelation in the disturbances over time in a single equation.

Seemingly unrelated regression (SUR) methods extend to situations where a different regression relation is being estimated for M different samples (e.g. regressions are similar between firms but may differ in included predictors). Thus it may not be valid to assume the same regression applies across all M sets of equations but there may be contemporaneous correlation nevertheless. Seemingly unrelated regression issues also occur in the more general situation when there is more than one outcome variate $(y_{1i}, \ldots, y_{pi})$ for each individual i, and these outcomes are differentially related to nonidentical vectors of covariates $(x_{1i}, x_{2i}, \ldots, x_{pi})$; see Percy (1992). There may still be reasons to assume correlation between the outcomes for each individual, even though a different regression relationship applies to each outcome.

Consider first the univariate outcome situation but different regression relationships across the equations. There may be prior substantive information suggesting correlation between the relationships of different samples, for example if the model related to a set of firms facing similar economic conditions. These conditions would be expected to influence the sampling process by which the data are generated. Suppose we have sets of time series data for $t = 1, \ldots, T$ for $i = 1, \ldots, M$ firms on their gross investment levels y_{it}. We consider yearly variations in investment in relation to the firms' capital stock K_{it} at the start of each year, and the start of year stock market values V_{it} of the firms. The relationship for successive firms will take the form

$$y_{1t} = \alpha_1 + \beta_1 V_{1t} + \gamma_1 K_{1t} + e_{1t}$$
$$y_{2t} = \alpha_2 + \beta_2 V_{2t} + \gamma_2 K_{2t} + e_{2t}$$
$$\ldots$$
$$y_{Mt} = \alpha_M + \beta_M V_{Mt} + \gamma_1 K_{Mt} + e_{Mt}.$$

The error terms in each set of equations may be taken as uncorrelated with one another, with $e_{1t} \sim \mathrm{N}(0, \theta_1), e_{2t} \sim \mathrm{N}(0, \theta_2), \ldots, e_{Mt} \sim \mathrm{N}(0, \theta_M)$. So for $M = 2$ and $T = 10$ *(say)* we would have

$$y_i = \alpha 1_i + \beta V_i + \gamma K_i + e_i$$

where $y_i, 1_i, V_i, K_i$ and e_i are 10×1 vectors for $i = 1, 2$. Equivalently

$$y = X\delta + e$$

where y is 20×1, and X is a block diagonal 20×6 matrix,

$$X = \begin{bmatrix} X_1 & 0 \\ 0 & X_2 \end{bmatrix}$$

with 10×3 submatrices X_i containing the covariates for firm i, and δ is a 6×1 vector with $\delta' = (\alpha_1, \beta_1, \gamma_1, \alpha_2, \beta_2, \gamma_2)$. With the assumption of no correlations across the two sets of equations,

$$e = \begin{bmatrix} e_1 \\ e_2 \end{bmatrix} \sim \mathrm{N}\left(\begin{bmatrix} 0 \\ 0 \end{bmatrix}, \begin{bmatrix} \theta_1 I_{10} & 0_{10} \\ 0_{10} & \theta_2 I_{10} \end{bmatrix} \right) \tag{7.30}$$

where I_{10} and 0_{10} are 10×10 identity and null matrices respectively. However, separate estimation of each equation in this way ignores possible correlations of errors across equations. Allowing for them may improve the precision ('pool strength') of the estimates in each equation. Correlated errors would be specified by a model of the form

$$e = \begin{bmatrix} e_1 \\ e_2 \end{bmatrix} \sim \mathrm{N}\left(\begin{bmatrix} 0 \\ 0 \end{bmatrix}, \begin{bmatrix} \theta_{11} I_{10} & \theta_{12} I_{10} \\ \theta_{12} I_{10} & \theta_{22} I_{10} \end{bmatrix} \right). \tag{7.31}$$

Once such a correlated error structure is introduced further questions may be asked about the model in general; for example, are one or more of the regression coefficients equal across firms?

Example 7.37 Investment by firms Theil (1971) tabulates stock market values, capital stocks and investment levels by two major US firms for the period 1934–53, namely General Electric and Westinghouse. The data are in millions of dollars at constant prices. Least squares applied separately to each set of 20 data points, as in the error model (7.30), gives estimates (and standard errors)

$$\begin{aligned} y_{1t} &= -9.96 + 0.027 V_{1t} + 0.152 K_{1t} \\ &\quad\ (31.4)\quad (0.015) \qquad (0.026) \\ y_{2t} &= -0.51 + 0.053 V_{2t} + 0.092 K_{2t} \\ &\quad\ (8.0)\quad (0.016) \qquad (0.050). \end{aligned}$$

Allowing for correlation between the two sets of equations (Program 7.37) improves the precision of the estimates of the coefficients on V but diminishes the significance of the coefficient for K_2. The joint dispersion matrix of e_1 and e_2 is given by

$$\Phi = \Sigma \otimes I_{10} = \begin{bmatrix} 824 & 230 \\ 230 & 114 \end{bmatrix} \otimes I_{10}.$$

Introducing multivariate t rather than multivariate normal errors for e_{1t} and e_{2t} makes little difference to the results. The remaining parameters from a run of 10 000 with 500 burn-in, and using the above notation for the investment example are given in Table 7.31.

Table 7.31

Parameter	Mean	SD	2.5%	Median	97.5%
α_1	−16.4	22.5	−60.2	−16.9	28.0
β_1	0.034	0.012	0.010	0.034	0.060
γ_1	0.133	0.026	0.078	0.134	0.183
α_2	0.350	7.158	−13.660	0.409	13.730
β_2	0.056	0.015	0.028	0.056	0.086
γ_2	0.059	0.059	−0.058	0.059	0.176

7.13 SIMULTANEOUS EQUATIONS

If y depends on predictors $x_1, \ldots, x_k$, then the standard assumption is that the predictors are independent of the error term. One situation in which this assumption is violated is when there are measurement errors in the predictors. Another is when there is simultaneous dependence; as well as y depending on $x_1, \ldots, x_k$, the regressors may in turn depend on y and on one another. Each variable involved in this simultaneous or mutual dependence is known as an endogenous variable. Predictors which are determined outside the system are known as exogenous. The latter satisfy the usual assumption of regression, namely independence of the error terms.

Simultaneous equation models developed initially in econometrics, as in many economic relations the data are generated by feedback mechanisms involving several interrelated outcomes. Thus each variable is determined simultaneously in a system of equations, which in many circumstances is parallel to a real-world system. For example, in a simplified market for a product, the quantity demanded $q.d$ is determined by its price and in market equilubrium, the quantity demanded equals the quantity produced $q.s$. Suppose however there were an increase in demand so that $q.d > q.s$ at the existing price. This disequilubrium causes an increase in price which may both reduce demand and encourage greater production until equilubrium is restored. Suppose prices and quantities were observed over times $t = 1, \ldots, T$, then this system is represented by the two *structural equations*

$$q_{t.d} = a_1 + b_1 p_t + e_{t1}$$
$$q_{t.s} = a_2 + b_2 p_t + e_{t2}.$$

In equilubrium we have $q_{t.d} = q_{t.s} = q_t$. Because of the simultaneous determination of q_t and p_t the errors e_{t1} and e_{t2} are correlated. More broadly a demand function might include factors such as income Y_t, and a supply equation might include factor costs (such as wages) F_t. So the demand and supply functions are respectively

$$q_t = a_1 + b_1 p_t + c_1 Y_t + e_{t1}$$
$$q_t = a_2 + b_2 p_t + c_2 F_t + e_{t2}. \qquad (7.32)$$

We may regard Y_t and F_t as externally determined, or exogenous to the system being considered, and q_t and p_t as determined within, or endogenous to, the system. Note that as expressed above, both equations are *normalised* with respect to q (i.e. the coefficient of q_t is unity). We might also normalise the demand function with respect to price, so that

$$p_t = a_{1'} + b_{1'}q_t + c_{1'}Y_t + e_{t1'}.$$

Which of p and q is chosen as the normalising variable is largely a substantive question, in terms of which causal relationship (from p to q or from q to p) is regarded as paramount. Thus if quantity supplied is not responsive to price changes, the demand equation should be normalised with respect to price. Since price and quantity are determined by the interaction of demand and supply, it should make no difference to model estimation whether the equations are normalised with respect to p or to q.

The *reduced form equations* express each endogenous variable in terms of the exogenous variables alone (and in terms of all exogenous variables in the system). They play a central part in estimating simultaneous relationships. In the market example, solving equations (7.32) above leads to the reduced form equations, with redefined errors, as follows (and omitting time subscripts for simplicity)

$$q = (a_1b_2 - a_2b_1)/(b_2 - b_1) + c_1b_2Y/(b_2 - b_1) - c_2b_1F/(b_2 - b_1) + \eta_1$$
$$p = (a_1 - a_2)/(b_2 - b_1) + c_1Y/(b_2 - b_1) - c_2F/(b_2 - b_1) + \eta_2.$$

So the reduced form equations express the endogenous variables in terms of the exogenous variables and involve reduced form parameters,

$$\pi_1 = (a_1b_2 - a_2b_1)/(b_2 - b_1)$$
$$\pi_2 = c_1b_2/(b_2 - b_1)$$
$$\ldots$$
$$\pi_6 = c_2/(b_2 - b_1).$$

While the reduced form equations are always identifiable, the original structural parameters (a_1, b_1, c_1, a_2, b_2 and c_2 in this example) may not necessarily be.

The simplest rule for identifiability of the structural parameters is the order condition. In the above example there are $E = 2$ endogenous variables and two exogenous variables. The order condition relates to the number of variables of all types (endogenous or exogenous) missing from an equation as compared with $E - 1 = 1$. Thus in $q_t = a_1 + b_1p_t + c_1Y_t + e_{t1}, F_t$ is missing and in $q_t = a_2 + b_2p_t + c_2F_t + e_{t2}, Y_t$ is missing. Therefore both equations are just identified (or 'exactly' identified).

Under-identification occurs if there are less than $E - 1$ missing variables in an equation. For example, suppose we drop F_t completely from the model, so that the system is specified as

$$q_t = a_1 + b_1p_t + c_1Y_t + e_{t1}$$
$$q_t = a_2 + b_2p_t + e_{t2}.$$

In this case the demand function (the first equation) cannot be identified because there are no missing variables in its regression equation. The reduced form, now in terms of only one exogenous variable, is

$$q = (a_1 b_2 - a_2 b_1)/(b_2 - b_1) + c_1 b_2/(b_2 - b_1) Y + \eta_1$$
$$p = (a_1 - a_2)/(b_2 - b_1) + \{c_1/(b_2 - b_1)\} Y + \eta_2.$$

So there are only two reduced form coefficients in the demand function to identify three parameters, a_1, b_1 and c_1. Over-identification occurs if there are more parameters in the reduced form of an equation than in its original structural form. A more complete identification rule is based on the order condition (see e.g. Johnston and DiNardo, 1997).

In practice, we might use stochastic constraints on parameters, as expressed in priors on them, to ensure identifiability. These substitute for exact *a priori* constraints (Dreze and Richard, 1983), so that a Bayesian approach is potentially less constrained by identification conditions. Another advantage of a Bayesian approach is greater robustness in small sample size examples where there are potentially asymmetric posterior parameter densities (e.g. see the simulated analysis in Zellner, 1971). A recent review by Zellner (1998) confirms the advantages of Bayesian estimates of simultaneous equations in small sample data sets.

The potential to circumvent classical identification rules does not remove the usual necessity to specify a parsimonious model, with parameters well identified by the data. For example, where the observations are time series, it is generally assumed in simultaneous equation models that previous values of the endogenous variables can be counted as exogenous. In practice this may detract from stable identification.

More generally a simultaneous equation system such as (7.32) may be specified via the structural equations, for a sample of size N,

$$YB + X\Gamma = e$$

where Y is an $N \times E$ matrix of values on endogenous variables, and X is an $N \times k$ matrix of all the exogenous variables in the system. B and Γ are parameter vectors (likely to include identically zero cells) summarising the feedbacks in the system, and e is an $N \times E$ matrix of errors. Solving for Y gives the reduced form

$$Y = X\Pi + \eta$$

where $\Pi = -\Gamma B^{-1}$. Whereas Π contains Ek parameters, B and Γ may contain up to $E^2 + Ek$ parameters. Identifying restrictions must therefore be imposed. Note that normalisation, as discussed above, constitutes one form of restriction, and may of itself lead to identification.

7.13.1 Using instruments: Two- and three-stage least squares

To eliminate the correlation between regressors and errors in the structural model, instrumental variable methods may be used. The instruments $\hat{y}_j$ replace the original endogenous variables $y_1, \ldots, y_E$ where they appear on the right-hand side (regression terms) of a structural equation. Hopefully the correlation between $\hat{y}$ and the error is less than that between y and the error term. The two-stage least squares (2SLS) method is an instrumental variable method, and involves regressing each endogenous variable on all the exogenous variables and then substituting the resulting estimators $\hat{y}_1, \hat{y}_2, \ldots, \hat{y}_E$ in the regression terms. Two-stage last squares is asymptotically equivalent to what is known as limited information maximum likelihood, so called because it proceeds without considering information on (cor)relations across structural equations.

Suppose that in the supply–demand model above, we had estimated instruments $\hat{p}$ and $\hat{q}$ by regressing p and q on income Y and factor costs F. Then 2SLS is equivalent to a set of unrelated regressions for each outcome variable, so that the structural equations are redefined in terms of the instruments as follows:

$$p = f1(Y, \hat{q}) + u_1$$
$$q = f2(F, \hat{p}) + u_2$$

where u_1 and u_2 are independently distributed.

By contrast the three-stage least squares method for simultaneous equations allows correlation between the errors in these redefined structural equations – that is it takes account of the 'full information' conveyed by the set of equations, and in particular the likelihood of correlated errors. Thus u_1 and u_2 in the above example would be bivariate normal. Three-stage least squares applies the same principles as the SUR model to the set of structural equations, and has the same asymptotic properties as the method of full information maximum likelihood (FIML). Bayesian approaches to FIML have been proposed but are computationally demanding. Here we employ the 3SLS approximation based on combining the SUR model with 2SLS.

Because using instruments means dependent variables are used twice over it is necessary in BUGS to declare a given dependent variable twice, once for the 2SLS stage, once for the structural estimation stage. So differently named copies of each endogenous variable need to be included in the data file.

7.13.2 The consumption model

A simple form of interdependence occurs in the macroeconomic model for income and consumption. In this model, consumption C_t and income Y_t are jointly determined according to

$$C_t = \alpha + \beta Y_t + e_t$$
$$Y_t = C_t + I_t. \tag{7.33}$$

The first equation relates consumption to income through a model containing a structural parameter β, the marginal propensity to consume. This is assumed to lie between 0 and 1. The errors e_t are assumed normal with variance σ^2. The second equation is an identity, relating income to the total of consumption and investment. In this model, investment I_t is assumed exogenous. In practice the quantities C_t, I_t and Y_t will be deflated for price changes and expressed in per capita terms.

Example 7.38 The Haavelmo consumption model Haavelmo analysed US data on income, consumption and investment from 1929 to 1941 (i.e. $T = 13$ time points). He estimated the above model (7.33) by OLS, obtaining $\hat{\beta} = 0.73$. Chetty (1968) and Zellner (1971) reanalyse the model from a Bayesian perspective, adopting the prior $p(\alpha, \beta, \sigma) \propto 1/\sigma$.

Here we adopt the 2SLS approach (see Program 7.38) which in this simple model is equivalent to 3SLS. To form an instrument for Y_t we regress it on the single exogenous variable I_t in the system. Then we estimate the regression of C_t on $\hat{Y}_t$ to obtain α and β. We adopt vague priors on all three parameters, α, β and $1/\sigma^2$. To illustrate the potential of such models for prediction (i.e. forecasting) for future years, our data also includes I_t for 1942, but not consumption. We derive the predictive mean for c_{T+1}.

The estimated parameters (from the second half of a single run of 50 000 iterations) are as shown in Table 7.32, with the prediction of consumption to 1942 showing a wide variability.

Note that Zellner's approach to this problem, based on direct analytic results, obtains a more precise prediction. Extensions of this model are possible, including first order correlation in the error term e_t. This extension acts to reduces the estimated marginal propensity to consume.

Example 7.39 Pork market trends Merrill and Fox (1971) analyse trends in pork prices p_t and supply q_t in the US for 1922–41. The exogenous variables, as in (7.32), are income Y_t, in the demand function, and F_t, denoting general supply factors apart from price, in the supply function. All variables are transformed via logs, so a log-linear relation is being estimated. As above the structural equations are

$$p_t = a_1 + b_1 q_t + c_1 Y_t + e_{t1}$$
$$q_t = a_2 + b_2 p_t + c_2 F_t + e_{t2}.$$

Expressed in terms of instruments the system is

$$p_t = a_1 + b_1 \hat{q}_t + c_1 Y_t + u_{1t}$$
$$q_t = a_2 + b_2 \hat{p} + c_2 F_t + u_{2t}$$

where u_{1t} and u_{2t} are bivariate normal with 2×2 dispersion matrix Σ.

Table 7.32

Parameter	Mean	SD	2.5%	Median	97.5%
α	109	63	−15	109	234
C_{T+1}	581	38	502	581	655
β	0.68	0.13	0.43	0.68	0.94

We estimate the supply–demand model with the demand model normalised in terms of price (*Program 7.39 Pork Market*). Convergence here is assisted by using good starting values for the regression coefficients. Otherwise, a large number of iterations (10–20 thousand in the present example starting with null values for regression parameters) may be spent in reaching the region of the posterior.

In the present analysis, starting values are based on an earlier run, and posterior estimates based on iterations 5000–15 000 of a 15 000 iteration run (Table 7.33)

The signs of the structural coefficients are in the expected direction, according to economic theory. The effect of price on quantity supplied is only weakly positive, but in accord with expectation. The demand function shows prices fall as quantity supplied increases, but rising in line with income. The estimated dispersion matrix for the errors of the two equations shows a clearly negative covariance term $\Sigma_{12} = \Sigma_{21} = -0.0028$.

Example 7.40 Klein's model for a national economy Klein's model for economic fluctuations in the US in 1921–41 ($t = 1, \ldots, 21$) consists of three endogenous variables: consumption C_t, investment I_t, and demand for labour W_t (private sector wages total). Exogenous variables are government expenditure G_t, profits P_t, public sector wages W'_t, capital stock K_t, taxes T_t, and income net of taxes Y_t. For simplicity, the time subscript is omitted in the following three structural equations (with the subscript -1 then denoting $t-1$) and three identities:

$$C = \beta_1 P + \beta_2(W + W') + \beta_3 P_{-1} + \beta_4 + u_1$$
$$I = \beta_5 P + \beta_6 P_{-1} + \beta_7 K_{-1} + \beta_8 + u_2$$
$$W = \beta_9(Y + T - W') + \beta_{10}(Y + T - W')_{-1} + \beta_{11} t + \beta_{12} + u_3$$
$$Y + T = C + I + G$$
$$Y = W + W' + P$$
$$K = K_{-1} + I.$$

Table 7.33

Parameter	Mean	SD	2.5%	Median	97.5%
Σ_{11}	0.0130	0.0046	0.0067	0.0121	0.0245
Σ_{12}	−0.0028	0.0012	−0.0059	−0.0025	−0.0010
Σ_{22}	0.0013	0.0005	0.0007	0.0012	0.0025
a_1	−0.84	0.55	−1.66	−0.91	0.29
a_2	0.69	0.25	0.26	0.70	1.15
b_1	−0.76	0.16	−1.03	−0.78	−0.44
b_2	0.05	0.03	0.00	0.05	0.11
c_1	1.15	0.08	1.03	1.14	1.32
c_2	0.79	0.05	0.69	0.80	0.88

Since W is endogenous, so also is total wages $W + W'$; similarly writing $X = Y + T - W'$, X is endogenous because one of its constituents Y is endogenous. We then need instruments for the endogenous variables appearing on the right-hand side in the three structural equations, namely $X = Y + T - W'$, P and $W + W'$. We therefore regress them on the exogenous variable set consisting of P_{-1}, K_{-1}, X_{-1}, t, T and G. Allowing for error dependence across the equations in the 2SLS model involves a trivariate normal dispersion matrix, Σ.

The second half of a run of 40 000 iterations yields results as given in Table 7.34.

These are similar to those cited by Maddala (1981) from 2SLS, except that the coefficients on P_{-1} in the consumption equation and on X_{-1} and time in the private wage equation are less well identified. Maddala's 2SLS results are

$$C = \underset{(1.46)}{16.45} + \underset{(0.13)}{0.02}P + \underset{(0.04)}{0.81}(W + W') + \underset{(0.12)}{0.21}P_{-1}$$

$$I = \underset{(8.36)}{20.28} + \underset{(0.19)}{0.15}P + \underset{(0.18)}{0.62}P_{-1} - \underset{(0.04)}{0.16}K_{-1}$$

$$W = \underset{(1.89)}{0.06} + \underset{(0.06)}{0.44}X + \underset{(0.07)}{0.15}X_{-1} + \underset{(0.05)}{0.13}t.$$

Example 7.41 Simulated small sample simultaneous dependence Zellner (1971) presents a Monte Carlo simulation analysis of a two-equation simultaneous equation system. This and similar experiments are reviewed in Zellner (1998). The full analysis involved application of Bayesian exact estimation and classical sampling theory estimates to small and medium sized data sets.

Table 7.34

Parameter	Mean	SD	2.5%	97.5%
β_1	−0.14	0.28	−0.70	0.36
β_2	0.86	0.08	0.72	1.06
β_3	0.27	0.28	−0.19	1.02
β_4	16.26	2.38	11.13	21.21
β_5	0.19	0.23	−0.22	0.73
β_6	0.55	0.24	0.08	0.99
β_7	−0.15	0.04	−0.22	−0.06
β_8	19.47	8.77	0.77	32.22
β_9	0.63	0.11	0.44	0.81
β_{10}	−0.10	0.14	−0.36	0.14
β_{11}	0.13	0.13	−0.13	0.36
β_{12}	2.58	2.98	−3.33	7.90

Table 7.35

Parameter	Mean	SD	2.5%	97.5%
$X \sim N(0, 9)$				
γ	2.09	0.38	1.38	2.99
β	0.52	0.06	0.39	0.63
$X \sim N(0, 1)$				
γ	2.09	1.21	0.33	5.18
β	0.48	0.21	0.06	0.87

Parameter estimates using the two methods under repeated sampling from the model, which is for $t = 1, \ldots, T$, were made

$$Y_{1t} = \gamma Y_{2t} + u_{1t}$$
$$Y_{2t} = \beta X_t + u_{2t}.$$

The parameters β and γ are set to 0.5 and 2 respectively, while the errors (u_{1t}, u_{2t}) are assumed bivariate normal with mean vector $(0, 0)$ and dispersion matrix Σ. The exogenous variable X_t is assumed to be drawn under three possible alternatives, $N(0, 1)$, $N(0, 2)$ and $N(0, 9)$.

For small sample sizes ($T = 20$), especially with X_t drawn from a precise density, such as $N(0, 1)$, the sampling theory estimates are shown by Zellner (1971) to be prone to outlier estimates of the parameters (β, γ), while Bayesian posterior estimates are closer to the true values. The advantage of Bayes estimates persist for larger samples ($T = 100$).

As a small-scale illustration of some possibilities, we here take $T = 100$ and assume $\Sigma_{11} = 1, \Sigma_{12} = \Sigma_{21} = 1, \ \Sigma_{22} = 4$. For the covariate we assume two options for generating values, namely $X \sim N(0, 1)$ and $X \sim N(0, 9)$, though in the analysis X is treated as nonstochastic.

Estimates for the sample generated with $X_t \sim N(0, 1)$ show the inflation of the covariance matrix Σ estimates which results from assuming X is fixed. However, posterior means for β and γ are close to the values used to generate the sample (the estimates are based on the second half of a 10 000 run using Program 7.41). The coefficient on the endogenous variable is estimated less precisely than that on the exogenous variable. When the sample is generated with $X_t \sim N(0, 9)$ both coefficients are estimated precisely (Table 7.35).

NOTES

1. In particular, if the x_i are equally spaced at unit intervals then this smoother is

$$g(x_i) = \sum_{1}^{N} w_{ij} y_j$$

where the w_{ij} are approximately equal to $K\{(i-j)/h\}/h$, where $h = \lambda^{0.25}$, and where $K\{\}$ is a kernel function

$$K\{u\} = 0.5\exp(-|u|/2^{0.5})\sin(\pi/4 + |u|2^{0.5}).$$

2. Thus suppose y denotes a continous outcome, and x and z are covariates, the different 'copies' of the data are created when the cases are ordered according to x or z. Consider the example dataset with four cases:

y	x	z
3	1	7
3.5	5	5
4.3	4	1
4	6	3

Copy 1 of the data, denoted $\{y_1, x_1, z_1\}$, provides the relevant ordering for estimating the smooth on x. To incorporate the smooth on z in this estimation an index $\{iz\}$ of the ranking of z in copy 1 is required:

y_1	x_1	z_1	iz
3	1	7	4
4.3	4	1	1
3.5	5	5	3
4	6	3	2

Copy 1 of the data is based on an ascending ranking on x. The minimum value of x corresponds to the value of z with rank 4. Copy 2 of the data, denoted $\{y_2, x_2, z_2\}$ is based on ascending ranking on z and so provides the relevant ordering for the smooth on z. To incorporate the smooth on x, an index ix of the ranking of x in copy 2 is needed:

y_2	x_2	z_2	ix
4.3	4	1	2
4	6	3	4
3.5	5	5	3
3	1	7	1

Then with an intercept only in the systematic term, separate models for y_1 and y_2 are estimated with code such as the following:

```
y1 [i] ~ dnorm(g1 [i],P)
y2[i] ~ dnorm(g2[i],P)
g1[i] = alpha0 + f1[i,1] + f2[iz[i],1]
g2[i]= alpha0 + f1[ix[i],1] + f2[i,1]
```

The smooths f_1 and f_2 are based on x_1 and z_2 respectively, and so will involve different smoothing parameters.

CHAPTER 8

Multilevel Models, Multivariate Analysis and Longitudinal Models

8.1 INTRODUCTION

Multilevel models are a range of techniques developed to represent data sets with an intrinsically hierarchical nature. Their development has accompanied and been facilitated by statistical computing developments, including advances in full and empirical Bayes estimation for random effects models within nested data sets. Nested data structures have a long history in statistics in terms of univariate and multivariate analysis of variance techniques to assess the distribution of variability over different levels of the data. In multilevel analysis, the interest focuses more on adjusting the impact of covariates at all levels for the simultaneous operation of contextual and individual variability in the outcome. This may well involve analysis of variance random effects estimates of the impact of different hierarchical grouping indices, including modelling possible correlations between different types of group effect.

The simplest situation is for univariate outcome with a two-level data structure applied to cross-sectional data with individual subjects at the lower level (patients, pupils, employees) classified by a grouping variable or cluster at the higher level (hospitals, schools, firms, etc.). Thus two-level data y_{ij} are characterised by group $j = 1, \ldots, N$ and by individuals within groups $i = 1, \ldots, R_j$. Equivalently the data consist of observations Y_i for $R = \sum_j R_j$ cases and a grouping index G_i, taking on values between 1 and N.

The latter form of representing the data structure is useful if there are unequal numbers R_j in different groups, and the data file in BUGS would then typically include Y and G with dimension R, rather than the cross-indexed data. The first group variable may in turn be nested within a further categorisation of the groups j, denoted $k = 1, \ldots, M$, and so on to include possible further hierarchical nesting. Again the data for individuals may be represented as a single long string with appropriate indexing on all group variables, such as G_{1i} with N possible levels, G_{2i} with M possible levels, and so on. This structuring makes clear that crossed structures as well as nested structures can be modelled in broadly parallel ways. For example in a crossed structure for modelling school pupil attainment, the grouping indices might be area of residence and school, which are crossed unless schools have mutually exclusive catchment areas.

Whether formally nested or crossed, the group variables define a contextual setting which mediates the effect of individual characteristics on the outcome variable. The grouping variables may themselves have a direct, and substantively significant impact on the outcome, and define clustering of outcomes within which individual covariates become relevant. Such a structure is common in educational settings where performance varies across schools, even when variation in pupil ability is allowed for, and in health applications involving hospital league tables, where performance varies even when patient case-mix is adjusted for.

For example, individual pupils i are taught in classes j, which are in turn organised by school k. Pupil attainment or behaviour are likely to be influenced both by their own characteristics, and by the type of schools and classes they are in. The impact of pupil ability on attainment will be affected by class size and by the level of educational resourcing in the school. So when standardised for initial ability pupil outcomes tend to be correlated within classes and schools. Pupils tend to obtain results more similar to those of pupils in the same class than to those of pupils in different classes.

While in cross-sectional hierarchical data sets observations on individuals are clustered within organisational or other groupings, in longitudinal data settings the observations are additionally clustered in repeated measures on the subjects. When measurements are taken repeatedly on the same subjects, for example patients, or students, the measurement repetitions constitute the lowest level, and are nested within the subjects. Hence there are $t = 1, \ldots, T_i$ repeated observations on individuals $i = 1, \ldots, R$, which may in turn be nested in terms of clusters $j = 1, \ldots, N$. This raises modelling questions in terms of similarity of observations within clusters j or over time units t which we may model by assuming effects drawn from a common density.

As an illustration, we may have exam scores on the same pupils i at several time points t and by school j. In medical and growth curve settings we have repeated measures of patient state or physical measures, with subjects possibly further grouped by organisation, treatment or attribute. For repeated observations at occasions t on the ith subject the total possible effect of the sth regressor may vary over individuals:

$$(\beta_s + b_{is})x_{tis}.$$

Significant variation in the effect of a regressor over individuals may occur because there are unique elements in the growth path of subject i, or specific frailty influencing treatment outcome. Various broad approaches to longitudinal data are possible, with major differences in the nature of the assumed stochastic mechanism generating the data. For example in some repeated measures models all the measurements are treated as responses, rather than later measures being treated as conditional on earlier responses.

The most general notation for multilevel data therefore starts with a possible time level, with cross-sectional data a special case. For example, with Normal outcomes Y_{tij}, we might have time-varying explanatory predictors z_{tij}, and fixed predictors x_{ij}, in a model with regression coefficients fixed over both time and cluster

$$Y_{tij} = A + Bz_{tij} + Cx_{ij} + e_{tij} \tag{8.1}$$

with the analogous cross-sectional model simply being

$$Y_{ij} = A + Cx_{ij} + e_{ij.} \tag{8.2}$$

These are simple examples of the wide choice of modelling options, whether by fixed or random effects, to describe the joint or single impact of the stratifying variables, either directly on the level of the outcome or via regressors. In the above example, we might introduce cluster or time-specific intercepts, and cluster or time variability in the impact of both sets of covariates. With sufficient numbers of clusters or times, random effects models of time and cluster effects become more appropriate. This modelling flexibility extends to the general linear modelling framework with non-normal outcomes.

Suppose for example, we were modelling death counts O_{tij} by time t, small area i and region j, in terms of exposures E_{tij}, a constant small area deprivation measure x_{ij} and an updated (time varying) small area unemployment rate z_{tij}. Then we would take $O_{tij} \sim \text{Poi}(E_{tij}\mu_{tij})$ and our model might specify time-varying impacts of small area deprivation and unemployment, but region- and time-specific intercepts also:

$$\log(\mu_{tij}) = A_j + D_t + B_t z_{tij} + C_t x_{ij}.$$

We might additionally add region–time effects u_{1tj}, area-region effects u_{2ij}, or even observation level effects u_{3tij}.

8.1.1 Multivariate robustness and identification

The analysis of hierarchical data structures is naturally associated with multivariate forms of random variation, since contextual differences in the impact of several level one variables are likely to be correlated. For example, in the three-way categorisation of pupils above, if we had two or more class- level variables mediating the effects of individual ability, then these may be correlated and a multivariate prior appropriate to modelling their joint inpact.

Furthermore, while the most common applications of multilevel models are to univariate outcomes, in practice behaviour is frequently multivariate. Hence a multivariate outcome at subject level may be combined with contextual groupings at higher levels (Thum, 1997). A multivariate structure is also one modelling option for longitudinal or panel data analysis, especially when the outcome at the lowest level consists of T_i repeated measures for subject i, and $T_i = T$ for all subjects.

Multivariate errors and multilevel variance component models raise several issues with regard to model robustness and identifiability. There is evidence that fully Bayes methods may improve on empirical Bayes methods for multilevel models in that they take into account the uncertainty introduced in estimating the variances of higher level effects, and the influence this uncertainty has on estimates of fixed regression effects (Seltzer *et al.*, 1996). On the other hand, a fully Bayes method may show sensitivity to the prior on the cluster level covariance structure. Choice of prior may have interpretative implications; for example, the conventional multivariate normal assumption may lead to overshrinkage in terms of outlying schools or hospitals, when in fact one of the substantive goals of these multilevel applications is to identify extreme performance. This is especially so when the number of clusters is small.

The question of robustness to outlier units at higher levels has been considered in interlaboratory trials (Analytical Methods Committee, 1989) where $j = 1, \ldots, N$ laboratories each conduct T measurements on sets of n_j specimens. Estimates of the precision and overall mean of the analyte may be distorted by large variability between replicates within one or two ('outlier') laboratories. In this case a more robust alternative may be applied—via multivariate t densities, or via mixtures of the multivariate normal form.

Robustness issues also apply in cases where the outcome itself is multivariate, since as in univariate regression, outlying observations may unduly influence parameter estimates (Liu, 1996). In longitudinal data analysis choosing an unstructured covariance for the error term e_{it} (within subject variation) means estimation of $T(T-1)/2$ parameters, in the case where there are $T_i = T$ repetitions for each subject. This may well be associated with poor identification, and a more parsimonious and structured model may be preferable.

8.1.2 Ecological and atomistic fallacies

League table comparisons of educational or health performance provide numerous examples of individual behaviours or outcomes which are affected by contextual settings in a hierarchical data framework. For example, health outcomes at individual patient level are affected by that patient's characteristics or 'casemix' (age, severity of illness, etc.) but also vary by physician and according to the quality of care provided by the hospital or other facility where treatment occurs. Comparisons of performance between hospitals or physicians that do not allow for the casemix of patients they treat suffer from an 'ecological fallacy'. The ecological fallacy arises when results from an aggregate-level analysis are incorrectly assumed to apply at the individual level. However, comparisons of patients which do not allow for their contextual setting suffer from an 'atomistic fallacy'. The possibility of these fallacies has long been recognised in sociology and geography (especially the ecological fallacy) but the concepts have wide applicability. Recent reviews include Schwartz (1994) and Diez-Roux (1998).

Contextual effects may have a substantive role of their own and are not necessarily just an aggregated form of the individual effects (i.e. they are not just 'compositional'). Both contextual effects and individual-level variables may be relevant to the outcome, and ignoring one may bias the estimated effect of the other. In sociology, geography and epidemiology there has been extensive discussion of how far aggregated area correlations between variables or analysis at an area level can be used to make inferences about individuals. For example, Richardson *et al.* (1987) consider how far dose–response relationships obtained by geographical studies of epidemiological effects can be translated to individual level relationships, while Humphreys and Carr-Hill (1991) discuss whether area variations in health outcomes remain after controlling for individual characteristics. Liska (1990) draws attention to the wide range of outcome variables that may show variability between contextual settings, even though focus on a single dependent variable may show only a relatively small percent of variation due to variation between social units.

8.2 GENERAL LINEAR MULTILEVEL MODELS FOR HIERARCHICAL DATA SETS

With two or more level observations, the outcome variable or variables are likely only to be independent conditional on the clusters and possible higher level units. We consider the situation where cross-sectional outcomes Y_{ij} $(i = 1, \ldots, R,\ j = 1, \ldots, N)$ may be binomial or Poisson, and where the outcome is linked to the model for the mean (e.g. via log or logit links). To model dependence within clusters, N cluster-specific parameters or parameter vectors may be included in the linear predictor of the mean outcome. These are usually assumed to be random effects either singly drawn from independent densities, or jointly drawn from a common multivariate density. Most frequently, but not necessarily, they represent deviations at cluster level in the global regression effect β of a predictor. The average effect of the sth predictor for case i in cluster j is $\beta_s x_{ijs}$, but because the impact of this predictor is mediated by the cluster, the total impact is $(\beta_s + b_{js})x_{ijs}$ where b_{js} is a random effect specific to covariate s. If there are several covariates with cluster-specific effects, then let b_j denote a p variate density for such effects (which includes the possibility of cluster-specific intercepts).

The random effects structure of the two-level model also include an observation-specific random effect, equivalent to the error term e_{ij} in (8.2) for the normal model. This may be appropriate when discrete outcomes data are still overdispersed with cluster-specific effects alone (for example in the study of Poisson outcomes in a longitudinal study of epileptic patients by Gamerman, 1997). For both cluster and observation random effects, normal, Student t or non-parametric mixtures may be adopted. Nonparametric mixing via Dirichlet processes is also now being applied to such models (Hirano, 1998).

In the normal linear mixed model (Laird and Ware, 1982) the normal density for the cluster effect (scalar or vector) b_j is conjugate with the outcome and the posterior density of β and the variance–covariance structure of the b_j can be obtained analytically. In the general linear mixed model with possibly discrete outcomes MCMC techniques are advantageous in the presence of possible nonconjugacy. Among the first fully Bayesian applications of MCMC techniques to a discrete (binomial) outcome was that of Zeger and Karim (1991). Their model was an example of a general two-level framework in which a response y_{ij} is related to predictors x_{ijs} for observations $i = 1, \ldots, R_j$ within clusters $j = 1, \ldots, N$. Conditional on the cluster effect b_j, y_{ij} follows an exponential family density

$$f(y_{ij}\,|\,b_j) \propto \exp[\{y_{ij}\theta_{ij} - a(\theta_{ij}) + c(y_{ij})\}/\omega].$$

The conditional moments

$$\mu_{ij} = \mathrm{E}(y_{ij}\,|\,b_j) = a'(\theta_{ij})$$

and

$$V_{ij} = \mathrm{var}(y_{ij}\,|\,b_j) = a''(\theta_{ij})\omega$$

are related to covariates (possibly overlapping) x_{ij} and z_{ij}, whose impact is modelled via fixed and random effects respectively. Thus:

$$h(\mu_{ij}) = \beta x_{ij} + b_j z_{ij}$$
$$V_{ij} = g(\mu_{ij})\omega$$

where h and g define link and variance functions. The components of b_j usually correspond to particular components of the population parameter β, typically modelling cluster deviations with prior mean zero and covariance matrix φ. However their means may also be related to covariates w_j varying at cluster level.

Assuming only random cluster effects, the model specification is completed by conditional independence assumptions: namely for y_{ij} given b_j, β and φ, and for b_j given φ. The posterior density has the form

$$P(b_1, \ldots, b_N, \beta, \varphi) \propto \left\{\prod_{i=1}^{R_j}\prod_{j=1}^{N} L(y_{ij}\,|\,\beta, b_j, x_{ij})P(\beta)\right\}\left\{\prod_{j=1}^{N} P(b_j\,|\,\varphi)P(\varphi)\right\}.$$

The full conditionals are therefore

$$P(\beta\,|\,) \propto \prod_{i=1}^{R_j}\prod_{j=1}^{N} P(y_{ij}\,|\,\beta, b_j, x_{ij})P(\beta)$$

$$P(b_j|) \propto \prod_{i=1}^{R_j} P(y_{ij}|\beta, b_j, x_{ij})P(b_j|\beta)$$

and

$$P(\varphi|) \propto \prod_{j=1}^{N} P(b_j|\varphi)P(\varphi).$$

Example 8.1 Poisson model for small area deaths Congdon (1997) considers the incidence of heart disease mortality (9th International Classification of Disease Codes 410–414) in $i = 1, \ldots, 758$ small areas in the Greater London area over the three years 1990–92. These small areas (electoral wards) are grouped administratively into $j = 1, \ldots, 33$ boroughs (i.e. clusters). We have a single regressor z_{ij} at ward level, an index of socio-economic deprivation. This index is originally in standardised form with mean 0 but is transformed according to $x_{ij} = \log(10 + z_{ij})$. At borough level there is a single characteristic $w_j = 1$ for inner London boroughs and 0 for outer suburban boroughs. We assume cluster (i.e. borough) level variation in the intercepts and in the impacts of deprivation; this variation is linked to the category of borough (inner vs outer).

The observational model assumes death counts O_{ij} are Poisson with means $E_{ij}\mu_{ij}$ where E_{ij} are 'expectations' The expectation here is not a statistical expectation but the expected deaths based on the demographic technique of internal standardisation.

So the likelihood is

$$
\begin{aligned}
&O_{ij} \sim \text{Poi}(E_{ij}\mu_{ij}) \\
&\log(\mu_{ij}) = b_{1j} + b_{2j}(x_{ij} - \bar{x}) + \delta_{ij} \\
&\underset{\sim}{b} \sim \text{N}_2[(\beta_{1j}, \beta_{2j}), \varphi] \\
&\beta_{1j} = \Gamma_{10} + \Gamma_{11} w_j \\
&\beta_{2j} = \Gamma_{20} + \Gamma_{21} w_j.
\end{aligned}
$$

The terms δ_{ij} model residual variability (Poisson overdispersion) not explained by the regression part of the model.

The borough-level slopes b_{2j} representing the variable impact of small area deprivation within borough boundaries average 0.67 but the outer London average is given by 0.57 and the inner London average by $0.57 + 0.22 = 0.79$. There is clear evidence (see Table 8.1) that small area deprivation raises mortality, and that this differs by borough group, though the effect Γ_{21} has a 95% credible interval just straddling zero. There is also clear support for varying intercepts and slopes with the square roots of φ_{11} and φ_{22} estimated as 0.11 and 0.09. However, the correlation between slopes and intercepts (`corr1.2` in Program 8.1) does not appear as significant.

Including ward-level random variability δ_{ij} reduces the average Poisson GLM deviance (`tot.dev` in Program 8.1) to 808, with a 95% credible interval from 734 to 884. This is in line with the expected value of the GLM deviance for $N = 758$ areas if the Poisson model were appropriate. Without these

Table 8.1 Posterior estimates.

	Mean	SD	2.5%	97.5%
φ_{11}	0.112	0.021	0.076	0.161
φ_{22}	0.090	0.038	0.039	0.205
Correlation $(b1, b2)$	−0.165	0.545	−0.907	0.813
Deviance	810.6	39.9	733.9	883.8

observation-level effects the variability of the cluster-level effects b_{1j} and b_{2j} may be understated.

Example 8.2 Individual learning in organisational context: Classrooms and schools Bryk and Raudenbush (BR, 1992) consider a three-level model for changes in educational attainment (mathematics scores), as part of a study of individual learning variations within schools. This is in fact a panel data set with the lowest level consisting of repetitions over time – five occasions between the spring of the first grade and the spring of the third grade. There is substantive interest in the way scores fall off during the summer vacation. This is known as the 'summer drop' effect, the learning loss between spring and fall (modelled using a dummy index $SD = 1$ for the fall terms, and $SD = 0$ otherwise). However, major interest also lies in the variation over 72 schools in the attainment of 618 pupils, both in terms of learning rate and initial ability. Questions of robustness and sensitivity to priors about variance components are well illustrated in this example.

The academic achievement of pupil i in school j at time t is modelled as follows:

$$Y_{tij} = \pi_{0ij} + \pi_{1ij}T_t + \pi_{2ij}\text{SD}_t + e_{tij}$$

where $e_{tij} \sim \text{N}(0, \sigma^2)$, and σ^2 denotes overall observation level variation. T_t is a time index ($T_1 = 0$ at spring of 1st grade, ..., and $T_5 = 4$ at spring of grade 3). π_{0ij} denotes the pupil within school initial status, and π_{2ij} and π_{2ij} denote pupil within school learning rates and summer drop effects. At a minimum these vary at school level, and may also vary between pupils. Here the initial status and learning rate effects are assumed to vary over both pupils and schools, but the summer drop effect varies only over schools, such that

$$\pi_{0ij} = S_{0j} + r_{0ij}$$
$$\pi_{1ij} = S_{1j} + r_{1ij}$$
$$\pi_{2ij} = S_{2j}$$

with the pupil-level errors r_{0ij} and r_{1ij} assumed bivariate normal with covariance matrix $V.\pi_{[1:2,\,1:2]}$. At school level, the means on initial status and on learning rate covary with the summer drop effect. Thus

$$S_{0j} = v_0 + u_{0j}$$
$$S_{1j} = v_1 + u_{1j}$$
$$S_{2j} = v_2 + u_{2j}$$

with u_{0j}, u_{1j} and u_{2j} assumed to be trivariate normal with matrix $V.S_{[1:3,\,1:3]}$.

The full multivariate normal model at both levels, as implemented by BR, involved empirical Bayes estimation. This conditions on higher level variance estimates and so may understate uncertainty arising from this source. The school covariance estimates in a fully Bayes model with MVN priors at both levels are denoted `V.S[,]` in *Program 8.2 Organisational Change*. The posterior means of the diagonal terms of this covariance matrix are 1439 (variation in initial status school effects), 445 (variation in school learning rates) and 1307 (variation in summer drop effects). By contrast, BR estimate these variances as 173, 19 and 45. Under an MVN prior, the summer drop effect averages −35.4 and the learning rate averages 24.4 (compared with −27.8 and 28.5 obtained by BR). The posterior estimate of the covariance matrix at pupil level, denoted `V.pi[,]` in Program 8.2, is close to that of BR.

The picture is little changed if univariate normal priors are adopted for u_{0j}, u_{1j} and u_{2j}. The variances of S_{0j}, S_{1j} and S_{2j} are estimated as 1417, 474 and 1110. The averages v_1 and v_2 still show the summer drop rate of −34.5 offsetting the learning gain mean of 24.

The Bryk and Raudenbush analysis is replicated at school level in the original scale of the exam scores only if Student t priors (with 5 degrees of freedom) are adopted. This prior adapts to the outlier effect of certain schools and allows for the small numbers of pupils observed in some schools. Adopting this prior also changes the averages v_1 and v_2, making the negative effect of the summer drop effect virtually the same as the positive effect of the learning rate per academic year.

Alternatively if the interest is specifically in the correlations at school level then a scaled analysis in terms of standardised exam scores can be undertaken.

8.3 PANEL DATA MODELS AND GROWTH CURVES

Longitudinal studies investigate change over time in outcomes measured repeatedly for each individual subject. Examples include growth curve models for physical measures or biological processes, change in patient health outcomes in follow-up studies, and changes in educational or income attainment by time or age. In medical applications, longitudinal studies contribute to identifying risk factors involved in the aetiology of disease and the course of its development – for example, in the Framingham Heart Study (e.g. Allaire *et al.*, 1999) and the Six Cities Study of Air Pollution and Health (e.g. Chib and Greenberg, 1998). These and other longitudinal studies generally involve repeated measures on outcomes and covariates and the main aims are (a) to describe and model the evolution over time of the outcome per se and (b) to account for the impact of covariates on growth and change. Because of the repeated observations on each subject, the observations within subjects will generally be correlated. This within subject correlation should be accounted for in order that inferences can be correctly drawn about the impact of covariates.

8.3.1 Marginal and transition models

Alternative modelling approaches have developed for longitudinal data, encompassing both continuous and discrete outcomes. Marginal regression methods such as the generalised estimating equation approach (Liang and

Zeger, 1986) focus on the impact of covariates on the marginal expectation of each outcome, Y_{it}. For example, for an event count the marginal distribution of Y_{it} is Poisson and it is assumed that the Poisson mean μ_{it} is related to time-stationary or time-varying covariates:

$$\log(\mu_{it}) = \beta x_{it}.$$

To account for the time dependence within subjects under the marginal approach we may

(a) adopt a model for the within subject correlation matrix (see Liang and Zeger, 1986); or
(b) adopt a model for the joint distribution of the responses over the set of time points.

The latter approach is exemplified by models for repeated binary outcomes (Fitzmaurice and Lipsitz, 1995), which consider all possible combinations of outcomes over T periods (e.g. combinations $Y_{ir}Y_{is}$ for $r \neq s$, and $Y_{ir}Y_{is}Y_{it}$ for $r \neq s \neq t$). The goal of marginal methods is to model the mean (via the regression) separately from the time dependence within subjects.

An alternative strand of model development has been in conditional models in the sense that the regression function is no longer independent of the time correlation within subjects. There are two major ways to model such correlation in the conditional approach. One, the 'transition model', involves including previous outcome values in the regression model. For example, for a Poisson outcome, with lag one dependence:

$$\log(\mu_{it}) = \beta x_{it} + \delta Y_{i,t-1}.$$

This is analogous to a first order Markov assumption for transitions between states. Similar models can be stated for continuous outcomes, binary outcomes, and so on. A second order dependence would involve lags in terms for $t-1$ and $t-2$.

A model with lag one refers to unknown data for $t = 1$ and therefore the model cannot be estimated for that period in the same way as for $t \geq 2$. So one way to estimate these models is by conditioning on the initial periods. Alternatively, various prior assumptions for the latent data may be adopted so that the observed data for the initial periods is retained in the estimation. (Note that in this estimation context, conditionality refers to the treatment of the initial periods.)

8.3.2 Random effects models

The other way of conditioning is on latent random effects specific to each subject. The observations Y_{it} are conditionally independent given these latent effects. The conditional independence of observations is the first stage in a two-stage model for longitudinal data. The first stage specifies a probability distribution for the repetitions $t = 1, \ldots, T_i$ within subjects $i = 1, \ldots, R$. This distribution is the same (e.g. multivariate normal) for all subjects, but the parameters may differ between individual subjects.

At the second stage, subject parameters are assumed drawn from a common prior density, with mean subject effects possibly related to covariates w_i varying only at this level. Substantively, these effects may represent variations in growth patterns beyond those due to known attributes, or variations in patient susceptibility in a follow-up trial.

We may be interested in identifying outlier subjects on the basis of these random effects. Laird and Ware (LW, 1982) cite identification of sensitive subgroups or individual children following a pollution outbreak. In the study of LW, Y is a measure of forced exhalation before the pollution alert, during it, and on three successive weeks after the alert (i.e. $T_i = 5$ for the ith child).

In a typical growth curve model the vector Y_i of responses (of length $T_i = T$) is modelled in terms of covariates Z_i, frequently including terms z_{ti} (time, age, etc.) varying at the occasion level. Growth is modelled as a polynomial in these time terms. So if the effect of the latter varies between subjects

$$\underset{T\times 1}{Y_i} = \underset{T\times m}{Z_i}\ \underset{m\times 1}{b_i} + \underset{T\times 1}{e_i}$$

For example, if each subject is observed at times $t_{i1}, \ldots, t_{iT}$ and a linear growth curve (i.e. $m = 2$) is assumed, then

$$Z_i = \begin{bmatrix} 1 & t_{i1} \\ 1 & t_{i2} \\ \ldots & \ldots \\ 1 & t_{iT} \end{bmatrix}.$$

The error term is multivariate with mean zero and $T \times T$ covariance matrix Σ_i, and the vector b_i either consists of fixed effects or is also multivariate with mean α and $m \times m$ covariance matrix φ. If the $b_{1i} = \alpha_1$ and $b_{2i} = \alpha_2$ (i.e. fixed effects are assumed) then the covariance of the observations Y_i is also Σ_i. But if the b_i are random and uncorrelated with the e_i, then

$$\underset{T\times T}{Var(Y_i)} = \underset{T\times m}{Z_i}\ \underset{m\times m}{\varphi}\ \underset{m\times T}{Z_i'} + \underset{T\times T.}{\Sigma_i} \tag{8.3}$$

A small number of repeated observations may be better modelled via separate fixed (population) effects $\alpha_1, \ldots, \alpha_T$ (with $\alpha_1 = 0$ for identification). For example Fahrmeir and Tutz (1994) discuss the analysis of a dichotmous outcome for respiratory symptoms in Ohio schoolchildren at ages 7, 8, 9 and 10 in terms of a single covariate at child level (maternal smoking) and three fixed effects for different ages.

A more general specification, including but not limited to growth curves, is where some time-specific covariates X_{it} are modelled via fixed effects and the impact of other occasion-specific covariates is modelled via random effects. Thus,

$$\underset{T\times 1}{Y_i} = \underset{T\times m_1}{X_i}\ \underset{m_1\times 1}{\beta} + \underset{T\times m_2}{Z_i}\ \underset{m_2\times 1}{b_i} + \underset{T\times 1.}{e_i} \tag{8.4}$$

If the z_{it} are a subset of the x_{it} then some β coefficients will model the means of the b_i and the b_i will have mean zero. So if $e_i \sim \mathrm{N}_T(0, \Sigma_i)$ and $b_i \sim \mathrm{N}_{m_2}(0, \varphi)$, then $\mathrm{Var}(Y_i)$ has the form (8.3) above.

Alternatively the b_i may be related to covariates w_i varying only between subjects, these covariates possibly including higher level groupings of subjects (schools, treatments, etc.) or attributes of individuals (e.g. gender). Thus the structure, for a growth curve analysis, might be

$$\begin{aligned} y_{it} &\sim \mathrm{N}(\mu_{it}, \Sigma_i) \\ \mu_{it} &= \beta z_{it} + b_{i1} + b_{i2}\mathrm{Time}_{it} \\ b_{i[1:2]} &\sim \mathrm{N}(B_{i[1:2]}, \varphi) \\ B_{i1} &= \gamma_1 w_i \\ B_{i2} &= \gamma_2 w_i. \end{aligned}$$

Here β are population-wide regression effects and the b_i are subject-specific random effects.

8.3.3 Autocorrelation in errors

The other major element in two stage longitudinal models is modelling of the terms e_{it} denoting within subject variation over occasions. The unstructured form has $\Sigma_i = \Sigma$, where the latter contains $T(T-1)/2$ variance and covariance parameters. With a moderate or large number of occasions this model generates a high level of parameterisation. The greatest simplification is to assume the components of e_{it} are conditionally uncorrelated given the b_i and β of (8.4). Hence the model

$$\underset{T \times T}{\Sigma_i} = \sigma^2 \underset{T \times T}{I}$$

is known as conditional independence or uniform covariance. An intermediate alternative assumes temporal dependence, with first or higher order autocorrelation. For first order correlation,

$$e_{it} = \rho e_{i,t-1} + \eta_{it}$$

where η_{it} are uncorrelated with variance σ^2_η and $\sigma^2[=\sigma^2_e] = \sigma^2_\eta/(1-\rho^2)$. Then $\Sigma_i = \sigma^2 C$ where $C = \rho^{|s-t|}$ and s and t are different occasions.

An equicorrelated covariance structure occurs when the components of e_i are exchangeable over time, in the sense that

$$\mathrm{cov}(e_{is}, e_{it}) = \rho\sigma^2$$

for s not equal to t, and $\mathrm{var}(e_{it}) = \sigma^2$.

Temporal dependence in Poisson or binomial outcomes may be modelled in similar ways as for continuous outcomes, via correlated effects in the regression term, with an appropriate link to the natural parameter. For example, if O_{it} is a Poisson outcome, with expected events E_{it},

$$O_{it} \sim \text{Poi}(E_{it}\theta_{it})$$

then the model for the underlying rates θ_{it} can incorporate time dependence. Suppose we also have time-specific covariates $\underset{\sim}{x}_{it}$. Then we might have a first order autoregressive time dependence where for $t > 1$ the mean is modelled via the usual log-link

$$\log(\theta_{it}) = \gamma + \underset{\sim}{\beta}\underset{\sim}{x}_{it} + e_{it}, \quad t > 1$$

$$e_{it} = \rho e_{it-1} + \eta_{it}$$

with the u_{it} normal with mean zero and variance σ^2. The autoregressive parameter ρ may be assumed to be confined to the interval $[-1, 1]$ or allowed to take values absolutely exceeding 1.

For $t = 1$, the lagged error terms e_{i0} are unknowns and we can initialise the model in several ways. Thus in $e_{i1} = \rho e_{i0} + \eta_{i1}$, we might take $\lambda_i = \rho e_{i0}$ as either random or fixed effects, and assume η_{i1} are drawn from the same prior density as $n_{is}, s > 1$. Alternatively we could take $u_i = e_{i1}$ as a distinct random effect, distributed as (say) $\text{N}(0, \sigma_u^2)$. So

$$\log(\theta_{i1}) = \gamma + \beta x_{i1} + u_i$$

$$u_i \sim \text{N}(0, \sigma_u^2).$$

Example 8.3 Respiratory symptoms in Ohio school children An example of contrasting approaches to longitudinal data is an analysis of the Ohio data on wheeze status by child age (ages 7, 8, 9 and 10) and maternal smoking. Following Fitzmaurice (1998) we initially fit first and second order transition models, with age and maternal smoking as covariates. Thus the wheeze status at age t for child i is

$$Y_{it} \sim \text{Bernoulli}(\pi_{it}).$$

The model for times $t \geq 2$ is

$$\text{logit}(\mu_{it}) = \beta_0 + \beta_1\text{Age} + \beta_2\text{MatSmoke} + \gamma Y_{it-1}.$$

Fitzmaurice estimates this model by conditioning on the data for age 7. Here we retain these data by modelling the latent term γY_{i0} via an additional unknown β_3, so that

$$\text{logit}(\mu_{i1}) = \beta_0 + \beta_1\text{Age} + \beta_2\text{MatSmoke} + \beta_3.$$

We ascribe a vague prior to β_3 but in fact know that it is likely to be a positive lag effect times an unknown binary outcome, so could be specified more informatively.

We obtain the posterior estimates shown in Table 8.2.

The odds ratio associated with maternal smoking is estimated as $\exp(\beta_2) = 1.25$ with 95% interval from 0.97 to 1.63. By contrast, Fitzmaurice (1998), using a conditional estimation method, obtains a higher odds ratio for maternal smoking, namely $\exp(0.296) = 1.34$, and with a wider 95% interval,

Table 8.2 Posterior estimates.

	Mean	SD	2.5%	Median	97.5%
β_0	−2.461	0.111	−2.683	−2.460	−2.254
β_1	−0.247	0.100	−0.448	−0.244	−0.064
β_2	0.226	0.135	−0.034	0.224	0.486
γ	2.219	0.162	1.886	2.216	2.532
β_3	0.238	0.245	−0.250	0.245	0.719

namely from 0.99 to 1.83. Introducing a lag two effect means the model for times $t \geq 3$ is

$$\text{logit}\ (\pi_{it}) = \beta_0 + \beta_1 \text{Age} + \beta_2\ \text{MatSmoke} + \gamma_1 Y_{it-1} + \gamma_2 Y_{it-2}.$$

To avoid conditional estimation we add unknowns for $t = 1$ and $t = 2$, so that

$$\begin{aligned}
\text{logit}\ (\pi_{i2}) &= \beta_0 + \beta_1 \text{Age} + \beta_2\ \text{MatSmoke} + \gamma_1 Y_{i1} + \beta_3 \\
\text{logit}\ (\pi_{i1}) &= \beta_0 + \beta_1 \text{Age} + \beta_2\ \text{MatSmoke} + \beta_4.
\end{aligned}$$

The estimates for this model are given in Table 8.3.

It is apparent that the maternal smoking effect is broadly unchanged, but the age effect (which is collinear with the lag response) is enhanced.

Finally we adopt a random effects model

$$\text{logit}\ (\pi_{it}) = \beta_0 + \beta_1 \text{Age} + \beta_2\ \text{MatSmoke} + e_i, \quad t = 1, \ldots, T.$$

The variance of the e_i, assuming they are normal, is 4.92 with 95% interval from 3.5 to 6.8. The covariate effects given in Table 8.4 show an enhanced,

Table 8.3 Posterior estimates.

	Mean	SD	2.5%	Median	97.5%
β_0	−2.605	0.169	−2.940	−2.599	−2.282
β_1	−0.407	0.215	−0.812	−0.398	0.006
β_2	0.214	0.132	−0.044	0.215	0.473
γ_1	1.959	0.168	1.639	1.961	2.298
γ_2	1.157	0.220	0.730	1.156	1.594
β_3	0.036	0.352	−0.631	0.049	0.719
β_4	0.067	0.551	−0.973	0.087	1.106

Table 8.4 Posterior estimates, random effects model.

	Mean	SD	0.025	Median	0.975
β_0	−3.142	0.226	−3.608	−3.128	−2.729
β_1	−0.181	0.069	−0.319	−0.182	−0.052
β_2	0.394	0.270	−0.122	0.390	0.948

though less precise, maternal smoking effect when allowance is made for the intraclass correlation. However, as Fitzmaurice (1998) notes, covariate parameter estimates may be influenced by the assumed form of variability for the e_i.

Example 8.4 Haematocrit measures in hip replacement patients Hand and Crowder (HC, 1996, Chapter 7) present an analysis of four repeated measurements of haematocrit among 30 hip-replacement patients, one measurement before and three after a surgical intervention. They assume a normal outcome

$$y_{it} \quad (i = 1, \ldots, 30; t = 1, \ldots, 4)$$

with random individual-level intercepts b_i, and a set of fixed regression effects. Some observations are missing and treated as missing at random (the default BUGS option). The fixed regression effects relate to sex (1 for males, -1 for females), an effect for the first three periods (the fourth being a contrast), three sex–period interactions, an age effect, and an age–sex interaction. So the growth curve is modelled as a series of levels rather than via a smooth (e.g. linear) evolution. The model is

$$y_{it} = \mu + b_i + \alpha_1 x_{it} + \alpha_2 w_i + e_{it}$$

where x_{it} is of length 6, and w_i is of length 3. The assumed errors are $b_i \sim \mathrm{N}(0, \sigma_b^2)$, $e_{it} \sim \mathrm{N}(0, \sigma_e^2)$, and the within patient correlation ρ is then $\sigma_b^2/(\sigma_b^2 + \sigma_e^2)$. The covariance between y_{it} and y_{is} is $E[(b_i + e_{it})(b_i + e_{is})] - E[b_i + e_{it}]E[b_i + e_{is}] = \sigma_b^2$ for $s \neq t$ and $\text{cov}(y_{it}, y_{it}) = \sigma_b^2 + \sigma_e^2 = \sigma^2$. So the off-diagonal elements of the covariance matrix for patient i can be written $\rho\sigma^2$ (the 'equicorrelation model').

In BUGS the hierarchically centred form of parameterisation means the model is expressed as

```
for(i in 1:R) { for (t in 1:T){y[i,t] ~ dnorm(mu[i,t],tau.y);
                              mu[i,t] <- regression terms + b[i];}
                              b[i] ~ dnorm(0,tau.b);}
```

where `tau.y` and `tau.b` are precisions (see Program 8.4).

After 10 000 iterations with 5000 burn-in the means of the posterior densities of $\sigma_b^2 = 1/\tau_b$ and $\sigma_y^2 = 1/\tau_y$ are 14.61 and 3.84 respectively. These are slightly higher than those obtained by HC using maximum likelihood. The correlation coefficient has a posterior mean of 0.196 and a 95% credible interval (0.002, 0.46); the HC estimate is 0.23. The effect of being male has credible interval from -3.6 to 9.5 with mean 2.7, slightly higher than the HC estimate. The occasion effects are very close to those of HC, as are the occasion–sex interactions. In particular, there is a fairly clear male excess at the first reading but less clear sex differentiation subsequently. The age effect is clearly not important for this outcome.

Example 8.5 Heart rate and energy expenditure: A nonlinear model for repeated measures over physical tasks Crowder (1978) and Hand and Crowder (HC, 1996) consider data for seven subjects whose energy expenditure (calories per minute) and heart rate (beats per minute) are measured under four different physical tasks (lying, sitting, walking, skipping); this gives data E_{it} and H_{it} respectively ($t = 1, \ldots, 4; i = 1, \ldots, 7$), with H_{it} as the response. To scale energy, they took

$$x_{it} = E_{it}/w_i^{\gamma_1}$$

where w_i is the subject's weight in kilograms and the power γ_1 is here taken as a parameter. The value $\gamma_1 = 0.75$ is used as a reference value in metabolism studies.

The variation of heart rate with scaled energy is assumed by HC to be governed by linear regressions, with individual-specific intercepts and slopes. Specifically, the model is

$$y_{it} = a_i + b_i(x_{it} - \gamma_2) + u_{it}$$

where $y_{it} = H_{it}/10$, and a_i and b_i are bivariate normal with respective means A and B, and with 2×2 covariance matrix C. The parameter γ_2 is anticipated to be negative, since $E_{it} = x_{it} = 0$ represents a subject at rest but with the heart still active. Hand and Crowder report numerical difficulties in finding a stable solution, perhaps because of the small number of subjects and also because there is no effective covariation between slopes and intercepts. In Crowder (1978), who takes $Y_{it} = H_{it}$, the average intercept A is estimated to be around 50 beats per minute and the average slope to be around 180. If, as in Hand and Crowder (1996), $Y_{it} = H_{it}/10$, A and B should be accordingly scaled also.

For this small number of subjects, there are issues of robustness in parameter estimation and a risk of 'overfitting'. Both the above studies omitted the first subject as an outlier but an alternative is to adopt a prior density robust to outlying subjects, such as a t density for the random intercepts and slopes. Identifiability is also improved by adopting parameter constraints implied by the problem. However, under-parameterisation may also occur—for example, if intercepts are in fact randomly variable over subjects but are assumed equal.

A first analysis follows Hand and Crowder (1996) in allowing for covariation between intercepts and slopes, but retains the first case, and applies two restrictions, namely $\gamma_1 > 0$ and $\gamma_2 < 0$. In fact there seems little evidence for covariation in terms of the off-diagonal element of C which has median value effectively zero (see *Program 8.5 Heart Rate and Energy, First Analysis Program*). Only γ_1 and γ_2 are reasonably well identified; from the second half of a 50 000 iteration run, the power coefficient on weight is estimated as 0.48, and the origin shift γ_2 for the scaled energy as -1.2.

In a second analysis (*Program 8.5 Heart Rate and Energy, Second Analysis Program*) we take t densities with 5 degrees of freedom for a_i and b_i separately, with respective variances C_1 and C_2, and additionally constrain A and B to be positive. This improves identification to some degree, with the second half of a run of 50 000 iterations estimating the power coefficient at around 0.6, the origin shift at -0.5, and the average intercept and slope at 4.1 and 18.1 respectively (Table 8.5).

Table 8.5 Posterior estimates, heart rate and energy.

	Mean	2.5%	Median	97.5%
C_1	0.46	0.08	0.34	1.46
C_2	1.36	0.09	0.60	6.78
γ_1	0.61	0.05	0.65	1.05
γ_2	−0.47	−3.11	−0.17	−0.01
A	4.1	0.4	4.4	7.1
B	18.1	1.1	12.1	61.5

However, if additionally we constrain the intercepts to be fixed at a common value as in Crowder (1978), instability again emerges, with A not well identified.

Example 8.6 Red blood cell survival measured using radioactive tracing Brillinger and Preisler (BP, 1983) consider the modelling of red blood cell survival in sickle cell disease. Such haemolytic disorders lead to shortened survival of these cells. The task is to estimate the mean life span of red cells and model sampling fluctuations in the mean in order to assess effectiveness of interventions for haemolitic disorders. Estimates of red blood cell survival involve taking a blood sample, labelling the cells using a radioactive tracer, and then reinjecting them into the patient and counting surviving cells by drawing samples at successive times. The samples are taken on the same day as the tracing (day 0), on day 1, days 2, 3, 4, 7, 9, 11, 14 and so on, to a final reading on day 46. There are 18 sampling times in all, and an extra sample to obtain estimates of background radiation. The data are counts of gamma photons over 10-minute periods, with three replicated draws from a single blood sample from each patient put in a separate vial. Four readings on each vial are then made. The full data consist of counts, totalled over patients, Y_{ijk} for times $i = 1, \ldots, I$, where $I = 19$ (i.e at sample times between 0 and 46 days, plus the background readings), replicates by vial $j = 1, \ldots, 3$ at each time, and readings on each vial $k = 1, \ldots, 4$. (Only two replicates by vial were made for the background readings.)

Brillinger and Preisler model the totals over vial readings, namely $y_{ij} = Y_{ij\bullet}$, as Poisson variables with mean

$$\mu_{ij} = \pi + \rho_i u_{ij}.$$

Here the ρ_i are treated as *fixed* effects measuring the survival of red cells and the u_{ij} as randomly drawn from a single density, with mean 1, measuring variability in cell counts due to fluctuations in the volume of blood actually sampled (i.e. variability in pipetting). There are $18 \times 3 + 2 = 56$ data points for these data.

The u_{ij} are assumed Normal $u_{ij} \sim \mathrm{N}(1, \sigma^2)$, and the ρ_i taken as Normally distributed fixed effects with ρ_{19} assumed zero (see *Program 8.6 Red Blood Cell Survival*).

Table 8.6 Posterior estimates, red blood cell survival.

	Mean	SD	2.5%	Median	97.5%
π	3300	40.31	3222	3300	3380
ρ_1	17440	271.9	16920	17430	17980
ρ_2	17850	273.3	17330	17850	18400
ρ_3	16910	261.4	16400	16900	17440
ρ_4	17260	265.9	16740	17260	17790
ρ_5	15760	240.7	15290	15750	16240
ρ_6	14150	219.1	13720	14150	14590
ρ_7	14380	231.9	13930	14370	14840
ρ_8	13690	216.2	13270	13680	14140
ρ_9	12640	197.3	12270	12640	13050
ρ_{10}	13100	207	12690	13090	13520
ρ_{11}	11240	177.3	10900	11240	11600
ρ_{12}	11380	184.3	11030	11380	11760
ρ_{13}	10580	172.7	10250	10580	10940
ρ_{14}	8815	151.1	8520	8814	9117
ρ_{15}	8885	149.3	8596	8883	9188
ρ_{16}	8632	147	8340	8632	8923
ρ_{17}	7538	132.1	7282	7537	7800
ρ_{18}	6532	116.5	6302	6532	6761
σ	0.025	0.003	0.019	0.024	0.032

The posterior estimates approximate those obtained from the BP analysis. For this model, the estimates of ρ_i are close to those of BP but do not have the irregularity in standard errors $\mathrm{SE}(\rho_i)$ reported by them. The estimate here of σ is 0.025 (from a run of 50 000 iterations), compared with their estimate of 0.017 based on a maximum likelihood analysis involving numerical quadrature (Table 8.6).

Measuring fit by the simple Poisson deviance shows an average of 54.8, indicating an acceptable fit – this closeness of fit would be expected given the fullness of parameterisation.

8.4 DYNAMIC MODELS FOR LONGITUDINAL AND GROWTH CURVE DATA

In dynamic linear models for panel data, parameters describing slopes and means may evolve through random walk priors. For example, growth curve parameters may be taken as random effects drawn from a density with common evolution variance lead to a pooling of strength over time; often a specifically smooth or slow change is assumed, with this end in mind. We consider two types of application: to longitudinal discrete outcomes and robust modelling of growth curves. For example, suppose O_{it} is a Poisson outcome for subject i and time t, with expected events or periods at risk E_{it}, with underlying relative risks

θ_{it}, and time-specific covariates $\underset{\sim}{x}_{it}$ which have a time-varying effect on the outcome. The level of the series may also be changing. Then we might have a first order autoregressive time dependence

$$O_{it} \sim \text{Poi}(E_{it}\theta_{it}) \tag{8.5}$$

where for $t > 1$ the mean is modelled via the usual log-link

$$\log(\theta_{it}) = \gamma_t + \underset{\sim}{\beta}_t \underset{\sim}{x}_{it} + e_{it}, \quad t > 1 \tag{8.6}$$

$$e_{it} = \rho e_{it} - 1 + \eta_{it}.$$

The first period model is (as in 8.3.3 above)

$$\begin{aligned} \log(\theta_{i1}) &= \gamma_1 + \beta_1 x_{i1} + u_i \\ u_i &\sim \text{N}(0, \sigma_u^2). \end{aligned} \tag{8.7}$$

The time-varying slopes and intercepts we might model as fixed effects or via random walk priors, with variances either fixed or to be estimated. For example, suppose we have a single time- varying predictor x_{it}; then we might model its regression effect as

$$\beta_t \sim \text{N}(\beta_{t-1}, \sigma_\beta^2), \quad t > 1,$$

with the first period regression parameter β_1 assumed to be a fixed effect. For the binomial the development is the same except that a logit link is used.

An alternative approach uses the conjugate distributions for the Poisson and binomial. Thus if, as above, the outcome is Poisson with

$$O_{it} \sim \text{Poi}(E_{it}\theta_{it}) \tag{8.8}$$

then we model θ_{it} as a gamma variable. In the absence of regression effects, Harvey (1991) proposes

$$\theta_{it} \sim \text{G}(a_{it}, b_{it}) \tag{8.9}$$

and

$$\begin{aligned} a_{it} &= w a_{it-1} \\ b_{it} &= w b_{it-1}. \end{aligned} \tag{8.10}$$

The coefficient w has meaning both as an autocorrelation parameter, and if constrained to lie between 0 and 1 (as in Harvey), as a DLM discount factor. To allow for regression effects we take the parameterisation of the gamma as

$$\theta_{it} \sim \text{G}(b_{it}\mu_{it}, b_{it})$$

where

$$\log(\mu_{it}) = \gamma_t + \beta_t x_{it}$$

and μ_{it} is the mean of θ_{it}. To retain time dependence we take $b_{it} = w b_{it-1}$ and may model the initialising terms $\delta_i = b_{i0}$ as either random or fixed effects. In particular we may take δ_i to be log-normal.

Example 8.7 Suicide trends in London We examine changes in suicide mortality in 33 London boroughs in relation to time-varying impacts of deprivation (x_1) and community fragmentation or anomie (x_2) in four periods, 1979–82, 1983–86, 1987–90, and 1991–94. The counts for all ages suicide in these periods, O_{it} are compared with expected deaths E_{it}, based on England and Wales suicide mortality rates. Adopting firstly the log-linear model given by (8.5) to (8.7), we take a prior on ρ allowing values above 1 in absolute size, u_i as a separate random effect from e_{it}, $t > 1$, and the regression parameters β_{1t} and β_{2t} as fixed effects.

The autoregression parameter ρ is estimated as 0.62 (with 95% interval from 0.39 to 0.84) and shows the continuity of suicide differentials. Using *Program 8.7 Suicide Trends (Model A)*, we find a growing impact of deprivation on suicide and a declining impact of anomie (from iterations 1000–10 000 of a 10 000 run); see Table 8.7.

The DLM-conjugate model given by (8.8) to (8.10) with δ_i log-normal yields similar results on the changing impacts of covariates (see *Program 8.7 Suicide Trends, Model B*). The discount parameter w is estimated as 0.95 (Table 8.8).

Table 8.7 Posterior estimates, suicide trends.

Parameter	Mean	SD	2.5%	Median	97.5%
β_{11}	0.03	0.13	−0.22	0.03	0.29
β_{12}	0.03	0.11	−0.20	0.03	0.24
β_{13}	0.35	0.10	0.15	0.35	0.55
β_{14}	0.28	0.10	0.08	0.29	0.48
β_{21}	1.15	0.13	0.90	1.15	1.39
β_{22}	0.97	0.11	0.76	0.97	1.20
β_{23}	0.84	0.11	0.62	0.84	1.04
β_{24}	0.78	0.11	0.58	0.78	0.99
ρ	0.62	0.11	0.39	0.62	0.84

Table 8.8 Posterior estimates, suicide trends, DLM conjugate model.

Parameter	Mean	SD	2.5%	Median	97.5%
β_{11}	0.01	0.10	−0.19	0.01	0.22
β_{12}	−0.02	0.11	−0.22	−0.01	0.19
β_{13}	0.31	0.11	0.09	0.31	0.53
β_{14}	0.26	0.12	0.04	0.26	0.50
β_{21}	1.18	0.10	0.96	1.18	1.39
β_{22}	0.97	0.12	0.72	0.98	1.20
β_{23}	0.81	0.11	0.60	0.82	1.03
β_{24}	0.78	0.11	0.56	0.78	1.00
w	0.95	0.05	0.83	0.96	1.00

Table 8.9 Posterior estimates, suicide trends, state space regression parameters.

	Mean	SD	2.5%	97.5%
β_{11}	0.03	0.14	−0.24	0.30
β_{12}	0.04	0.12	−0.20	0.29
β_{13}	0.00	0.12	−0.23	0.24
β_{14}	0.07	0.12	−0.16	0.30
β_{15}	0.25	0.11	0.02	0.47
β_{16}	0.36	0.12	0.12	0.59
β_{17}	0.28	0.11	0.05	0.49
β_{18}	0.29	0.12	0.05	0.53
β_{21}	1.14	0.13	0.87	1.40
β_{22}	1.10	0.13	0.84	1.35
β_{23}	1.01	0.12	0.78	1.24
β_{24}	0.94	0.12	0.70	1.17
β_{25}	0.98	0.12	0.76	1.22
β_{26}	0.71	0.12	0.48	0.93
β_{27}	0.72	0.12	0.49	0.94
β_{28}	0.85	0.12	0.61	1.10
ρ	0.76	0.08	0.60	0.89

Finally we modify model A to allow state-space evolution in the regression parameters, assumed to vary over 2-year rather than 4-year periods (see model C), and with G(1, 0.1) priors on the evolution variances. This model enhances the autocorrelation parameter, though stationarity is still present (Table 8.9).

8.4.1 Growth curve models as DLMs

A frequently used model for growth processes (Hui and Berger, 1983) involves different levels and trend parameters for each individual i. Hui and Berger (1983) considered robustness to departures from linearity by introducing quadratic terms in time. A robust density alternative to the normal might also be considered in the event of possible outlying growth patterns. An approach considered by Gamerman and Smith (1996) is to include dynamic parameters to describe local random departures from the linear trend. The similarity between successive values of the population level and slope may be specified by a random walk pattern.

Thus for normal data Y_{it} with means λ_{it}, the baseline constant trend model at population level, with trend in relation to time rather than age, is as follows

$$\mu_t = E(\lambda_{it}) = \alpha + \gamma t.$$

This may be generalised to allow dynamic variability in both the level and the linear trend term at population level

$$\begin{aligned} \mu_t &= \mu_{t-1} + \gamma_t + u_{1t} \\ \gamma_t &= \gamma_{t-1} + u_{2t}. \end{aligned} \tag{8.11}$$

The average difference between successive γ_t is analogous to the slope γ in the constant trend model, since the latter describes average growth from t to $t+1$. The disturbances u_{1t} and u_{2t} may be taken as independently normal $\mathrm{N}(0, \sigma_j^2)$, $j = 1, 2$. Individual variability is modelled by a random term specific to both the individual i and time t, so that

$$\lambda_{it} = \mu_t + v_{it}.$$

We assume $v_{it} \sim \mathrm{N}(0, \sigma^2)$.
To allow for individual variability in growth to be incorporated, one option involves level and linear growth effects specific to individuals, as follows

$$\begin{aligned} Y_{it} &\sim \mathrm{N}(\lambda_{it}, \sigma^2) \\ \lambda_{it} &= \alpha_{it} + \gamma_{it} t. \end{aligned}$$

Evolution in level and growth is still dynamic at population level through the specification $\gamma_{it} = \gamma_t + \zeta_{1it}, \gamma_t = \gamma_{t-1} + \eta_{1t}$ and $\alpha_{it} = \alpha_t + \zeta_{2it}, \alpha_t = \alpha_{t-1} + \eta_{2t}$.

Another way to combine individual variability in growth paths with dynamic evolution in population parameters is through a mixture specification, with probability p on the population process and $(1-p)$ on the individual process. The mixture process applies to individual specific levels and trends:

$$\begin{aligned} \alpha_{it} &= (1-p)(\alpha_{i,t-1} + \gamma_{it}) + p\mu_t + v_{1it} \\ \gamma_{it} &= (1-p)\gamma_{i,t-1} + p\gamma_t + v_{2it} \end{aligned} \tag{8.12}$$

where μ_t and γ_t are as in (8.11) above. The mixture proportion p may be time dependent. A variation on this method is to allow the mixture to be in terms of distributions rather than means, so that

$$\begin{aligned} \alpha_{it} &\sim (1-p)\mathrm{N}(\alpha_{i,t-1} + \gamma_{it}, A_1) + \mathrm{N}(\mu_t, A_2) \\ \gamma_{it} &\sim (1-p)\mathrm{N}(\gamma_{i,t-1}, C_1) + p\mathrm{N}(\gamma_t, C_2). \end{aligned} \tag{8.13}$$

So the choice in the mixture is between an aggregate growth process (8.11) described by μ_t and γ_t and an individual-level growth process with level and trend parameters α_{it} and γ_{it}. The latter is obtained by setting $p = 0$ in equation (8.12) or (8.13).

This specification is a specialised one most suitable to moderately large samples and observed growth processes with steady evolution in means and perhaps small variation between individuals around the average growth path. It may well need simplification in specific examples to avoid being overparameterised; excess parameterisation may be evident, for example, in slow convergence.

Example 8.8 Potthof and Roy study An application is to the Potthof and Roy (1964) data on distance from pituitary to pteryomaxillary fissure (at ages 8, 10, 12, 14); see Table 8.10.

Table 8.10 Potthof and Roy growth data.

	Age			
Sex	8	10	12	14
M	21.0	20.0	21.5	23.0
M	21.0	21.5	24.0	25.5
M	20.5	24.0	24.5	26.0
M	23.5	24.5	25.0	26.5
M	21.5	23.0	22.5	23.5
M	20.0	21.0	21.0	22.5
M	21.5	22.5	23.0	25.0
M	23.0	23.0	23.5	24.0
M	20.0	21.0	22.0	21.5
M	16.5	19.0	19.0	19.5
M	24.5	25.0	28.0	28.0
F	26.0	25.0	29.0	31.0
F	21.5	22.5	24.0	27.5
F	23.0	22.5	24.0	27.5
F	25.5	27.5	26.5	27.0
F	20.0	23.5	22.5	26.0
F	24.5	25.5	27.0	28.5
F	23.0	20.5	31.0	26.0
F	27.5	28.0	31.0	31.5
F	23.0	23.0	23.5	25.0

For this case, neither C_1^{-1} nor C_2^{-1} in model (8.13) were well identified when assigned gamma priors, namely G(0.1, 0.1). Instead they were taken as fixed with $C_1 = C_2 = 0.001$, so favouring slow evolution of the trend parameters (see *Program 8.8 Fissure Data*). After discarding 50 000 iterations, the next 10 000 yielded values of γ, μ and the remaining hyperparameters as in Table 8.11 (with 95 % intervals for μ and 90 % intervals for the other parameters).

Table 8.11 Growth model, parameter summaries.

	Mean	SD	Credible interval	
μ_1	22.30	0.60	21.15	23.52
μ_2	23.21	0.53	22.14	24.23
μ_3	24.55	0.53	23.49	25.60
μ_4	25.74	0.60	24.59	26.92
γ_1	1.00	1.97	−2.33	4.17
γ_2	1.00	1.09	−0.75	2.63
γ_3	1.22	0.98	−0.34	2.75
γ_4	1.19	1.08	−0.51	2.86
$A_1^{0.5}$	2.12	0.81	0.50	3.04
$A_2^{0.5}$	1.08	1.18	0.25	2.99

These parameters provide evidence for individual variability in distances by age and show a clearer growth at older ages. The mixing probability p is estimated to have an average value of 0.67, so slightly favouring a population process.

8.5 MULTIVARIATE REGRESSION, FACTOR ANALYSIS AND STRUCTURAL EQUATION MODELS

Concepts of behaviour, health and beliefs are most frequently multivariate or multidimensional. However, analysis of univariate outcomes has been the predominant mode of statistical analysis. Separate univariate analyses of interrelated outcome variables may conceal informative substantive features, and by neglecting correlated outcomes may distort estimates of variance components and regression effects.

On the other hand, correlations between multivariate outcomes raise issues around sensitivity to the priors assumed to describe the covariation. Inferences about multivariate regression parameters and their posterior intervals may be sensitive to priors on variance components when sample sizes are small or events infrequent. For example, outcomes in spatial epidemiology are frequently correlated, and the impacts on them of risk factors are also correlated across outcomes. Thus inferences about relative health in different areas may benefit by pooling strength in terms of the underlying multivariate distributions.

Example 8.9 Mortality by Glasgow post-code As an example, Langford *et al.* (1999) describe univariate and multivariate models for cardiac and cancer deaths Y_{i1} and Y_{i2} in $i = 1, \ldots, 143$ Glasgow post code sectors. These areas vary widely in their populations and death event counts, so we expect extra-Poisson variation. We assume (see *Program 8.9 Glasgow Mortality*) an outcome model as follows

$$
\begin{aligned}
Y_{i1} &\sim \text{Poi}(E_{i1}\theta_{i1}) \quad \text{(cancer)} \\
Y_{i2} &\sim \text{Poi}(E_{i2}\theta_{i2}) \quad \text{(cardiac)} \\
\log(\theta_{i1}) &= a + v_{i1} \\
\log(\theta_{i2}) &= b + v_{i2}
\end{aligned}
$$

with E_{i1} and E_{i2} being expected deaths (indirect standardisation method). To allow for correlated risks, the cause-specific mortality terms v_{i1} and v_{i2} are initially assumed to be drawn from a multivariate normal density with mean 0 and covariance matrix Σ_N. As a robust alternative, a multivariate t assumption is made, with 5 degrees of freedom and covariance matrix Σ_T. The results (from iterations 5–25 thousand of a single chain) show smaller variances Σ_{jj}, $j = 1, 2$ under the t assumption. However, the central focus of this analysis is the posterior estimates of risk by area, and to this end the 95% credible intervals of the 143 SMRs

$$S_{ij} = 100 \exp(v_{ij}), \quad j = 1, 2$$

may be considered in relation to the average SMR of 100. These show 8 outliers for cancer and 22 for CHD under the t assumption (SMRs with 95% intervals entirely above 100 or entirely below 100). Under the normal assumption, there are 10 for cancer and 14 for CHD. Under the t assumption the correlation between the two risks is estimated as 0.66 with a 95% interval from 0.32 to 0.90.

Example 8.10 Tumour growth in mice Chatfield and Collins (1980) consider experimental data relating to the effect on implanted tumour growth in mice of (a) environmental temperature (either 4 °C, 20 °C, or 34 °C) and (b) sex. There are three outcome variables relating to weights in grammes, namely initial weight, final weight (net of tumour weight) and tumour weight. Three replicates (of all three variables) are made at each combination of temperature and sex. Thus the data are of the form x_{tsrj} where t denotes temperature, s denotes sex, $j = 1, \ldots, 3$ denotes the weight outcome variable, and r denotes replicate number. While random effects specifications for the effects of temperature and sex would be possible for a larger number of replications, it is not possible to reproduce the detail of a conventional multivariate analysis of variance with these data. The latter requires estimation of dispersion matrices over temperature, sex, and sex–temperature interactions, and would be sensitive to prior assumptions.

We consider a multivariate form for the outcomes

$$\underset{\sim}{X}_{tsr} \sim \mathrm{N}_3(\mu_{ts}, \Sigma)$$

where $\underset{\sim}{X}_{tsr}$ is the trivariate vector of outcomes observed at each replication, with Σ a 3×3 matrix (see *Program 8.10 Tumours in Mice*). The sex–temperature specific means

$$\mu_{t,s} = (\mu_{ts1}, \mu_{ts2}, \mu_{ts3})$$

for variables $j = 1, \ldots, 3$ are initially modelled via a regression on fixed effects of temperature;
sex;
and sex–temperature interactions
with low temperature and male as the reference categories for the first two sets of effects. Thus

$$\mu_{tsj} = v_j + \alpha_{tj} + \beta_{sj} + \gamma_{tsj}.$$

The results in Table 8.12, based on the second half of a run of 20 000 iterations, present the centred effects, $a_{tj} = \alpha_{tj} - \bar{\alpha}_j$ and $b_{sj} = \beta_{sj} - \bar{\beta}_j$.

Table 8.12 Mice tumours, parameter summary.

	Mean	SD	2.5%	Median	97.5%
$\Sigma_{1,1}$	1.734	0.812	0.771	1.546	3.764
$\Sigma_{1,2}$	0.611	0.651	−0.474	0.527	2.117
$\Sigma_{1,3}$	−0.017	0.029	−0.083	−0.015	0.030
$\Sigma_{2,1}$	0.611	0.651	−0.474	0.527	2.117
$\Sigma_{2,2}$	2.398	1.099	1.082	2.131	5.253
$\Sigma_{2,3}$	0.020	0.034	−0.041	0.017	0.097
$\Sigma_{3,1}$	−0.017	0.029	−0.083	−0.015	0.030
$\Sigma_{3,2}$	0.020	0.034	−0.041	0.017	0.097
$\Sigma_{3,3}$	0.005	0.002	0.002	0.004	0.010
Temperature effects					
$a_{1,1}$	−0.732	0.590	−1.964	−0.725	0.404
$a_{1,2}$	−0.536	0.737	−2.006	−0.515	0.875
$a_{1,3}$	−0.067	0.032	−0.131	−0.067	−0.004
$a_{2,1}$	0.875	0.598	−0.302	0.872	2.062
$a_{2,2}$	2.387	0.740	0.955	2.388	3.847
$a_{2,3}$	0.054	0.031	−0.008	0.054	0.118
$a_{3,1}$	−0.143	0.609	−1.344	−0.139	1.067
$a_{3,2}$	−1.851	0.719	−3.247	−1.852	−0.394
$a_{3,3}$	0.013	0.032	−0.050	0.013	0.075
Sex effects					
$b_{1,1}$	−0.329	0.512	−1.353	−0.318	0.681
$b_{1,2}$	−0.490	0.639	−1.776	−0.486	0.766
$b_{1,3}$	0.035	0.028	−0.019	0.036	0.091
$b_{2,1}$	0.329	0.512	−0.681	0.318	1.353
$b_{2,2}$	0.490	0.639	−0.766	0.486	1.776
$b_{2,3}$	−0.035	0.028	−0.091	−0.036	0.019
Interactions					
$\gamma_{2,2,1}$	−0.626	1.409	−3.433	−0.615	2.184
$\gamma_{2,2,2}$	−1.360	1.792	−5.030	−1.348	2.161
$\gamma_{2,2,3}$	−0.089	0.076	−0.242	−0.088	0.064
$\gamma_{3,2,1}$	−0.159	1.473	−3.149	−0.161	2.740
$\gamma_{3,2,2}$	0.685	1.810	−3.060	0.716	4.249
$\gamma_{3,2,3}$	−0.048	0.078	−0.202	−0.048	0.105
V_1	18.730	0.743	17.230	18.730	20.170
V_2	18.590	0.908	16.760	18.600	20.340
V_3	0.255	0.040	0.175	0.256	0.334

They show (in coefficients a_{13}, a_{22} and a_{32} respectively) that low temperatures are associated with low tumour weight, medium temperatures with the highest final weight, and high temperatures with low final weight. There is also some evidence that females have lower tumour weights (b_{23} has 95% interval from −0.09 to 0.02). Interaction effects appear nonsignificant, though female final weights under medium temperatures (coefficient γ_{222}) are lower. Excluding the interaction terms γ_{tsj} makes the sex effect on tumour weights much clearer.

Note that it would also be possible to analyse weight gain (final-initial weight) and tumour weight as a bivariate outcome to clarify these associations.

8.5.1 Multivariate regression for imputation of missing data

In practice, one application of multivariate regression methods often occurs when values on an outcome (univariate or multivariate) and some predictors may be missing, but there are one or more fully observed predictors. Then multivariate regressions of the outcome or outcomes y_i and of predictors subject to missingness $x_1, \ldots, x_m$ on the fully observed predictors $x_{m+1}, \ldots, x_o$ serves as a method for missing data imputation. This method involves making the set $(y, x_1, \ldots, x_m)$ jointly dependent on the set $(x_{m+1}, \ldots, x_o)$. The real regression of interest (e.g. y on some subset of $x_1, \ldots, x_o$) may then be carried out using the imputed values as necessary. For example, Liu (1996) describes methods for multiple augmentation using multivariate t regression and other robust regression methods in a situation where the missingness on both y and $x_1, \ldots, x_m$ may be arranged in a monotone pattern (if x_a has the greatest level of missingness, and x_b is subject to lesser degree of missing values, then all subjects with x_b missing have x_a missing as well, and no subject with x_b missing has x_a present).

Example 8.11 Creatinine clearance Liu (1996) discusses a clinical trial studying the relationship between a single outcome variable, endogenous creatinine clearance (CR), and three predictors: body weight (WT), age in years (AGE), and serum creatinine concentration in mg/decilitre (SC). Only age and creatinine clearance are fully observed for the $N = 34$ male subjects. A model recommended for this relationship (the regression of ultimate interest) is

$$\ln(\text{CR}) = \beta_0 + \beta_{Wt}\ln(\text{WT}) + \beta_{SC}\ln(\text{SC}) + \beta_{\text{AGE}}\ln(140 - \text{AGE}) + u_i.$$

To model the missingness, a multivariate normal regression is initially applied, relating (CR,WT,SC) to $X = \ln(140 - \text{AGE})$. This imputes the four missing values of SC and the two of WT on a missingness at random basis. This is called the imputation regression. Concurrently, the ultimate regression of CR on WT, SC and AGE is carried out with the imputed values substituted for missing values. In practice, the computation involves two copies, differently named, of the variables CR, WT and SC (see Program 8.11).

The resulting estimates for the regression of interest, in the form of posterior mean and (2.5 %,97.5 %) points are

$$\begin{aligned}\beta_{\text{Wt}} &= 1.04\ (0.56, 1.56)\\ \beta_{\text{SC}} &= -1.12\ (-1.36,\ -0.90)\\ \beta_{\text{AGE}} &= 0.44\ (-0.27, 0.96).\end{aligned}$$

To introduce a robust alternative, a contaminated mixture of multivariate normal regressions was applied in the imputation regression. This assumes a small average probability ($\pi = 0.05$) of outlying values with means on the three variables departing from the averages of the main population. All cases (except those with missing values) are potential outliers, with an indicator variable δ_i assigning case i as an outlier with probability π. Thus if outlier behaviour is confined to the level of the mean and not to slopes, we assume

$$\pi \sim \text{Beta}(5, 100)$$

$$\delta_i \sim \text{Bern}(\pi), i = 1, \ldots, N_o(\text{i.e. cases with nonmissing values on WT, SC})$$
$$\delta_i = 0, \qquad i = N_o + 1, \ldots, N(\text{cases with missingness on one or both of WT, SC})$$

and for $i = 1, \ldots, N$,

$$\begin{aligned}
y_{i,1:3} &\sim \text{MVN}(\mu_{i,1:3}, P^{-1}) \\
A_{i,1:3} &\sim \text{MVN}(0, Q^{-1}) \\
\mu_{ij} &= a_j + \delta_i A_{ij} + \beta_j \text{AGE} \qquad \text{for} \quad j = 1, 3 \\
P() &\sim \text{Wishart}(R(,), 3) \\
Q() &\sim \text{Wishart}(S(,), 3).
\end{aligned}$$

Case 27 is identified as an outlier with a probability (average δ_i over 10 000 iterations) of 0.956. The values of $A(27,j)$, $j = 1, \ldots, 3$ are estimated as 1.22 (on SC), –0.06 (on WT) and –2.03 (on CR).
The resulting ultimate regression coefficients are

$$\begin{aligned}
\beta_{\text{Wt}} &= 0.84\ (0.21, 1.35) \\
\beta_{\text{SC}} &= -1.11\ (-1.33, -0.89) \\
\beta_{\text{AGE}} &= 0.48\ (-0.01, 0.83).
\end{aligned}$$

Thus the age coefficient is estimated more precisely as compared with Liu (1996), and the WT coefficient is reduced.

8.6 FACTOR ANALYSIS AND STRUCTURAL EQUATION MODELS

In factor analysis we consider relationships involving a set of observed variables and another and smaller set of latent variables generated from the first set, and reproducing their essential structure. The motivations include

(1) data reduction, when the original number of variables is expected to show overlapping associations and is computationally extensive to analyse;
(2) detection of underlying constructs with substantive meaning which are not directly observable;
(3) identification of subgroups of correlated variables.

We regard each observed variable as dependent and predicted by smaller set of unobserved variables. The regression parameters relating observed variables to latent factors are known as factor loadings, and serve to identify the meaning of the factor and its statistical significance: typically, the major factors account for most of the original variance, and remaining minor factors will load significantly on just one or two of the original variables. A linear factor model for observed continuous variates X of dimension $N \times P$, relates the observed value X_{ij} for variable j and subject i to latent variables f of dimension $N \times M$.

The scale and origin of these latent variables is arbitrary and conventionally they are assumed to be standard normal, so that

$$f_{ij} \sim \mathrm{N}(0, 1), \quad j = 1, \ldots, M, i = 1, \ldots, N.$$

These latent variables are known as common factors since in combination they underlie ('explain') variability in all the observed indicators. Then the regression relation is defined in terms of factor loadings Λ of dimension $P \times M$ relating the observed variables to the M common factors. The communalities C_j of each observed variable are the proportions of their variation explained by the factors.

Also involved in the model are error terms u_{ij} unique to each case and variable, and so of dimension $N \times P$. The latter may be assumed independent of each other and of the common factors, and might then be assigned normal priors $\mathrm{N}(0, \varphi_j^2)$, $j = 1, \ldots, P$. The assumption that the residual error terms for each of the manifest variables are independent of each other is equivalent to assuming that, given the M common factors, the observed variables are independent of each other (i.e. are conditionally independent).

The factor analysis model may be replicated in BUGS by assuming independently normal random variation, with the systematic part of the model determined by the regression on the factors. Assume for instance that there are two latent factors. Then the full model for case i and observed variable j is

$$X_{\mathrm{ij}} \sim \mathrm{N}(\mu_{\mathrm{ij}}, \varphi_j^2)$$
$$\mu_{ij} = v_j + \lambda_{j1} f_{i1} + \lambda_{j2} f_{i2}$$

where v_j is the intercept for X_{ij}. Under this model the variation in each of the P original indicators explained by the common factors (the 'communalities' of the P indicators) is given by $C_j = \lambda_{j1}^2 + \lambda_{j2}^2$, $j = 1, \ldots, P$.

At a minimum, the factor loadings could be assigned similar priors to the regression parameters of usual regression analysis. However, if $M \geq 2$, no set of factor loadings obtained from this unconstrained model is unique since arbitrary rotation will produce solutions which also satisfy the model. A factor model with M factors requires $M(M-1)/2$ constraints on the parameters to ensure identifiability.

One approach to ensure identifiability is by adopting informative priors. For example, we might assume some loadings of a particular factor are negative and the others positive (giving a factor with a 'bipolar' form). Imposing prior structure on the factor loadings amounts to 'confirmatory' factor analysis where we postulate in advance certain model aspects which are to be tested for fit. These may include features such as the number of factors, direction or values of factor loadings or other parameters (Everitt, 1984). It might also involve fixing factor loadings (e.g. to 0 or 1), or constraints such as ordering their values, or equating their values. Formal constraints are also possible to uniquely identify factors, or to ensure orthogonal factors which are arranged in order of the original variability they explain. The constraint of orthogonal factors is not, however, an intrinsic feature of factor analysis.

The potential advantage of Bayesian methods to improve identifiability in factor analysis (and structural equation models) is mentioned, for example, by Press and Shigemasu (1989) and Scheines *et al.* (1999). With an informative prior the posterior can be used to make inferences for underidentified models. However, factor models are also subject to possible overparameterisation – selection of the correct number of factors in maximum likelihood factor analysis being one motive for the work of Akaike (1981) on the AIC criterion. Scheines *et al.* (1999) also find that informative priors for small samples may alleviate problems of multimodality.

Press and Shigemasu (1982) use the constraint that the variances (or covariance matrix) of the factor loadings

$$\Lambda_{(P \times M)} = \{\lambda_1, \ldots, \lambda_M\}$$

are proportional to the variances (dispersion matrix) of the sample data; typically we would assume the factor loadings have a higher precision (e.g. by a factor of ten) than the sample data. For example, suppose we assume the centred sample data are multivariate normal with precision matrix T, assumed to be Wishart with parameters (B, ν), then we would take Λ' to be multivariate normal with precision T_Λ where $T_\Lambda = 10T$.

If we also take the factor scores to be $N(0, 1)$ the essential coding in BUGS (for $M = 2$) is as follows:

```
{for (i in 1:N) {for (j in 1:M) {F[i,j] ~ dnrom(0,1)}
X[i,1:P] ~ dmnorm(mu[i,1:P] , T[1:P,1:P])
for ( k  in  1:P)  {mu[i,k] <- F[i,1]* Lam[1,k]+F[ i,2]* Lam[2,k]
for ( k in 1:M) {Lam[k,1:P] ~ dmnorm(Lam0[k,1:P],T.L[1:P,1:P])}
for (j in 1:P) {T.L[i,j] <- 10* T[i,j]}}
   T[1:P, 1:P] ~ dwish(B][,], nu) }
```

Simplifications on this (while retaining the variance proportionality assumption) would be to take the factor loadings as univariate normal, while the data was still assumed multivariate, or take both loadings and data to follow univariate priors.

The least complicated analysis is when a relatively small set of original variables are likely to be explained to a large extent by a single underlying factor. In this case identifying constraints are not usually needed. Suppose we form all the original variables in such a way that they are all positive (or all negative) indicators of an underlying construct (e.g. social status, nationalist belief, functional limitations on activities of daily living). Then the single factor underlying them is likely to be unipolar, and a prior constraint on the loadings may be imposed accordingly. Imposing arbitrary constraints such as orthogonality may (of itself) suggest secondary factors, which are likely to be bipolar; that is, with negative loadings on some of the original variables and positive loadings on others. In this situation rotation may be applied to clarify the structure.

Example 8.12 Open/closed book exam scores Factor analysis with $M = 1$ and $M = 2$ underlying factors is illustrated with the open/closed book examination score data of Mardia *et al.* (1979), and further analysed by Everitt and Dunn (1983). Here 88 students are tested on five exams, two involving no consultation with course books, the other three being 'open book' exams (the courses were 'mechanics' and 'vectors' which were closed, and algebra, analysis and statistics which were open). The maximum likelihood solution for $M = 1$ shows communalities ranging from 0.36 (closed mechanics) to 0.84 (open algebra). The analysis here, with $M = 1$, adopts priors for $1/\varphi_j^2$ relating to a standardised version of the original 88×5 data matrix (see Program 8.12). The communalities $C_1, \ldots, C_5$ (proportions of variation explained) are based on a single run of 10 000 iterations (excluding the first thousand as burn-in) and are higher than from the maximum likelihood analysis, ranging from 0.39 to 0.92. The loadings on the factor lead to a very similar interpretation to that from the ML analysis – namely a generic ability factor, loading highest on the open algebra exam. Estimates of communality and loadings (from the second half of a run of 10 000 iterations) are given in Table 8.13.

For the $M = 2$ analysis informative priors are adopted on the basis of the maximum likelihood solution. Thus the loadings on the first factor are constrained to be positive, in line with a unipolar factor representing overall ability. The loadings on the second factor are constrained to be of opposite sign for the closed exams as against the open exams – in line with the bipolar factor of the maximum likelihood analysis. The resulting solution shows a clear match to the ML solution in terms of the estimated loadings, but again shows higher communalities than the ML analysis (Table 8.14).

Table 8.13 Exam scores, parameter summary.

Parameter	Mean	SD	2.5%	Median	97.5%
C_1	0.39	0.14	0.17	0.38	0.70
C_2	0.49	0.15	0.24	0.47	0.82
C_3	0.92	0.18	0.62	0.90	1.34
C_4	0.64	0.16	0.37	0.63	1.01
C_5	0.57	0.16	0.30	0.56	0.92
λ_1	0.62	0.11	0.42	0.61	0.84
λ_2	0.69	0.11	0.49	0.69	0.91
λ_3	0.96	0.09	0.79	0.95	1.16
λ_4	0.80	0.10	0.61	0.79	1.00
λ_5	0.75	0.10	0.55	0.75	0.96

Table 8.14 Exam scores, two factor model parameter summary.

Parameter	Mean	SD	2.5%	Median	97.5%
C_1	0.66	0.27	0.27	0.61	1.25
C_2	0.68	0.23	0.32	0.65	1.20
C_3	0.90	0.18	0.59	0.89	1.30
C_4	0.70	0.19	0.39	0.68	1.10
C_5	0.61	0.17	0.33	0.59	0.99
λ_{11}	0.70	0.12	0.47	0.69	0.94
λ_{12}	0.33	0.24	0.01	0.27	0.82
λ_{21}	0.76	0.12	0.54	0.76	1.01
λ_{22}	0.24	0.18	0.01	0.20	0.67
λ_{31}	0.90	0.10	0.71	0.90	1.11
λ_{32}	−0.23	0.15	−0.56	−0.22	−0.01
λ_{41}	0.75	0.11	0.53	0.75	0.98
λ_{42}	−0.31	0.16	−0.65	−0.29	−0.04
λ_{51}	0.70	0.11	0.47	0.70	0.92
λ_{52}	−0.29	0.16	−0.63	−0.28	−0.04

Note, however, that compared with the single factor model, communalities actually decline for the open algebra variable. The interest in the pattern of loadings on the bipolar factor is in the contrast between the average loadings on the open book exams and those on the closed book exams. The loadings in themselves, around 0.2–0.4 in absolute size are not likely to be judged that significant. It may be noted that alternative priors on the factor loadings might have been adopted in this example.

Example 8.13 City crime rates Everitt (1984) presents crime rates (per 100 000 population, 1970) for 16 American cities and $P = 7$ different types of crime. These are respectively

(a) murder/manslaughter
(b) rape
(c) robbery
(d) assault
(e) burglary
(f) larceny
(g) autotheft.

We might expect to find a global 'crime factor' and accordingly find correlations between burglary and larceny, and between assault and rape both around 0.75. On the other hand, a more complex multifactorial structure is suggested by features such as near zero correlations between murder and larceny, and between murder and autotheft. Following Everitt (1984) we adopt a two-factor solution, and adopt a particular 'confirmatory' approach via informative priors (see Program 8.13, Model A). Thus, on the second factor we assume $\lambda_{12} = 0$ but otherwise loadings constrained to be positive using a truncated normal. The interest would then lie in the centred loadings for factor 2, namely

$$\kappa_{j2} = \lambda_{j2} - \bar{\lambda}_{.2}, \quad j = 1, \ldots, 7.$$

The estimated loadings in the first factor are found to be weighted towards violent crime as against nonviolent crimes such as burglary and autotheft. The second factor, considered in terms of the centred loadings κ_{j2}, shows a contrast between on the one hand, cities with high murder rates but low burglary and larceny rates, and on the other, cities with low murder and relatively high burglary and larceny (Table 8.15).

Alternatively we may impose a bipolar assumption on the loadings of the second factor and allow any one (and one only) of these loadings to be zero. Thus we could adopt a discrete prior with seven bins, each corresponding to a different model for factor 2. Each model is set up to reflect assumed bipolarity but has one loading of zero (see Program 8.13, Model B). Successive bins $j = 1, \ldots, 7$ set $\lambda_{j2} = 0$, and the other bins, when not set to zero, fall in two classes: one, the violent crimes with positive loadings, and the other, nonviolent crimes with negative loadings. Thus model 4 sets $\lambda_{42} = 0$ and has positive loadings on the remaining violent crimes, and negative loadings on the non-violent crimes.

The three most frequently selected models involve setting the loadings of autotheft, robbery or rape to zero, and by far the most frequent is the model with autotheft loading at zero. The resulting loadings, which average over the seven possible models, show a contrast between burglary and larceny on the one hand and murder and assault on the other (Table 8.16).

Table 8.15 City crime, two factor model, parameter summary.

	Mean	SD	2.5%	97.5%
κ_{11}	0.60	0.23	0.12	0.97
κ_{21}	0.67	0.20	0.15	0.98
κ_{31}	0.56	0.22	0.11	0.96
κ_{41}	0.81	0.17	0.34	0.99
κ_{51}	0.35	0.21	0.02	0.83
κ_{61}	0.25	0.21	0.01	0.83
κ_{71}	0.38	0.22	0.03	0.87
κ_{12}	−0.42	0.09	−0.60	−0.26
κ_{22}	0.02	0.19	−0.31	0.45
κ_{32}	−0.08	0.19	−0.39	0.36
κ_{42}	−0.09	0.19	−0.37	0.41
κ_{52}	0.27	0.20	−0.21	0.59
κ_{62}	0.35	0.21	−0.22	0.62
κ_{72}	−0.05	0.20	−0.37	0.40

Table 8.16 City crime, two factor model, weight selection, parameter summary.

	Mean	SD	2.5%	97.5%
κ_{11}	0.40	0.22	0.04	0.83
κ_{21}	0.73	0.17	0.35	0.98
κ_{31}	0.55	0.21	0.11	0.93
κ_{41}	0.78	0.16	0.40	0.99
κ_{51}	0.72	0.18	0.31	0.99
κ_{61}	0.73	0.19	0.30	0.99
κ_{71}	0.45	0.23	0.06	0.89
κ_{12}	−0.46	0.23	−0.86	0.03
κ_{22}	−0.21	0.18	−0.60	0.11
κ_{32}	−0.22	0.21	−0.71	0.14
κ_{42}	−0.45	0.17	−0.78	−0.09
κ_{52}	0.48	0.20	0.11	0.88
κ_{62}	0.66	0.20	0.21	1.01
κ_{72}	0.21	0.18	−0.07	0.65

Example 8.14 Job acceptability The approach of Press and Shigemasu (PS, 1989) is illustrated with data on job acceptability for 48 applicants; the applicants have been scored on $P = 15$ variables (e.g. appearance, lucidity, etc.) which are all framed in such a way that the first factor is likely to be unipolar but succeeding factors illustrate various facets of acceptability. PS propose a model based on four factors (based on an initial principal component analysis showing four components 'explaining' 82% of the variance). Also based on underlying theory they propose a multivariate normal prior for $\wedge'$ with mean prior loading matrix $\wedge_0'$, given in Table 8.17.

Table 8.17 $\wedge_0(15 \times 4)$.

	Factors			
Variables	1	2	3	4
1	0	0	0.7	0
2	0	0	0	0
3	0	0.7	0	0
4	0	0	0	0.7
5	0.7	0	0	0
6	0.7	0	0	0
7	0	0	0	0.7
8	0.7	0	0	0
9	0	0	0.7	0
10	0.7	0	0	0
11	0.7	0	0	0
12	0.7	0	0	0
13	0.7	0	0	0
14	0	0	0	0
15	0	0.7	0	0

Table 8.18 Job acceptability factor model, parameter summary.

	Mean	SD		Mean	SD
$\lambda_{1,1}$	0.17	0.09	$\lambda_{1,3}$	0.44	0.11
$\lambda_{2,1}$	0.17	0.11	$\lambda_{2,3}$	−0.29	0.13
$\lambda_{3,1}$	−0.01	0.08	$\lambda_{3,3}$	−0.14	0.12
$\lambda_{4,1}$	0.25	0.11	$\lambda_{4,3}$	0.15	0.11
$\lambda_{5,1}$	0.71	0.05	$\lambda_{5,3}$	−0.02	0.07
$\lambda_{6,1}$	0.85	0.10	$\lambda_{6,3}$	0.09	0.07
$\lambda_{7,1}$	0.13	0.04	$\lambda_{7,3}$	−0.01	0.07
$\lambda_{8,1}$	0.98	0.10	$\lambda_{8,3}$	0.05	0.08
$\lambda_{9,1}$	0.05	0.08	$\lambda_{9,3}$	0.71	0.11
$\lambda_{10,1}$	0.83	0.06	$\lambda_{10,3}$	0.14	0.09
$\lambda_{11,1}$	0.91	0.09	$\lambda_{11,3}$	−0.13	0.09
$\lambda_{12,1}$	0.74	0.06	$\lambda_{12,3}$	0.00	0.07
$\lambda_{13,1}$	0.83	0.08	$\lambda_{13,3}$	−0.13	0.07
$\lambda_{14,1}$	0.51	0.10	$\lambda_{14,3}$	0.38	0.09
$\lambda_{15,1}$	0.24	0.08	$\lambda_{15,3}$	0.21	0.08
$\lambda_{1,2}$	0.28	0.15	$\lambda_{1,4}$	−0.11	0.15
$\lambda_{2,2}$	0.28	0.14	$\lambda_{2,4}$	0.11	0.13
$\lambda_{3,2}$	0.31	0.15	$\lambda_{3,4}$	0.02	0.14
$\lambda_{4,2}$	0.00	0.12	$\lambda_{4,4}$	0.81	0.08
$\lambda_{5,2}$	−0.18	0.11	$\lambda_{5,4}$	0.03	0.09
$\lambda_{6,2}$	−0.01	0.11	$\lambda_{6,4}$	0.23	0.10
$\lambda_{7,2}$	−0.10	0.06	$\lambda_{7,4}$	0.86	0.07
$\lambda_{8,2}$	0.11	0.13	$\lambda_{8,4}$	−0.03	0.15
$\lambda_{9,2}$	0.62	0.13	$\lambda_{9,4}$	−0.08	0.08
$\lambda_{10,2}$	0.18	0.18	$\lambda_{10,4}$	−0.09	0.05
$\lambda_{11,2}$	−0.02	0.17	$\lambda_{11,4}$	−0.18	0.10
$\lambda_{12,2}$	0.26	0.09	$\lambda_{12,4}$	0.13	0.06
$\lambda_{13,2}$	0.27	0.15	$\lambda_{13,4}$	0.20	0.08
$\lambda_{14,2}$	−0.15	0.09	$\lambda_{14,4}$	0.24	0.12
$\lambda_{15,2}$	1.06	0.05	$\lambda_{15,4}$	−0.04	0.06

The Wishart parameters for the sample data were taken to be $B = 0.2I_{15}$ and $\nu = 33$. The degrees of freedom parameter is here *N-P*, though smaller values (between *P* and *N-P*) do not affect posterior inferences substantially. After a run of 3000 iterations (discarding the first 500) we obtain factor loadings as in Table 8.18. These are similar to those obtained via the nonsampling method of Press and Shigemasu.

8.6.1 Stochastic volatility by factor analysis

One use of factor analysis is in economic and financial applications, especially in multivariate time series where the outcomes are subject to changes in their variability, e.g. stock prices or exchange rates may be more volatile at some periods than others. Thus for a series $y_t \sim N(\mu_t, V_t)$ we consider models relating V_t to V_{t-1} or more usually $\log(V_t)$ to $\log(V_{t-1})$, as in the autoregressive model:

$$\log(V_t) = \varphi_0 + \varphi_1 \log(V_{t-1}) + \eta_t.$$

More generally, a number of authors have discussed models for multivariate time series of the form

$$Y_t = \beta f_t + \omega_t, t = 1, \ldots, N$$

where Y_t is $P \times 1$, $\beta = \{\beta_{ij}\}$ is $P \times M$, and f_t is $M \times 1$, with $f_t = \{f_{jt}\}$ a time-varying set of M factor variables, with $M < P$. The contemporaneous correlation among the P observed variables is thus modelled by a set of factors rather than by a multivariate normal dispersion methodology.

In turn each of the P components of the observational error term ω_t is subject to stochastic variability. Thus each $\omega_{it}, i = 1, \ldots P$ typically follows an evolution model of the DLM type:

$$\omega_{it} \sim \mathrm{N}(O, \exp(\alpha_{it}))$$

with $\alpha_{it} = \log(V_{it})$ an autoregressive series, such as

$$\alpha_{it} = \mu_i + \varphi_i(\alpha_{i,t-1} - \mu_i) + \eta_{it}.$$

Thus the log of the variance of the ω_{it} is related autoregressively to the log of the variance of $\omega_{i,\,t-1}$.
Similarly the factor scores are assumed to be stochastically variable over time, with

$$f_{jt} \sim \mathrm{N}(O, \exp(\varepsilon_{jt}))$$

with $\varepsilon_{jt} = \log(F_{jt})$ again an autoregressive series, but with a zero mean, reflecting the tying of factor score variation around the N(0, 1) baseline. Thus

$$\varepsilon_{jt} = \rho\varepsilon_{j,\,t-1} + u_{jt}.$$

For identifiability of the factor loadings we may take $\beta_{ii} = 1$ for $i = 1, \ldots, M$ and $\beta_{ij} = 0$ for $j > i$. Thus for a two-factor model with $P = 5$, we would have $\beta_{11} = 1$, β_{21} to β_{51} as free parameters, $\beta_{12} = 0$, $\beta_{22} = 1$, and β_{32} to β_{52} as free parameters. The entire approach has been designated the ISVN (independent stochastic volatility via normals) by Pitt and Shepherd (1999).

Example 8.15 US consumption and income We consider a stochastic volatility model for the US Consumption and Income Data of Chapter 7, between Winter 1947 and Fall 1980. So there are $P = 2$ variables, and we take $M = 1$ factors, with only one free factor loading parameter, β_{21}. We are interested in the changes in the variance $f_t = \exp(\varepsilon_t)$ of the single factor, and in comparing β_{21} with the fixed $\beta_{11} = 1$. In BUGS the parameterisation is in terms of precisions rather than variances, and the greatest volatility will be when the precision of the factor series is low, and the most stability will be when the precision is high.

Table 8.19 Parameter summary.

	Mean	SD	2.5%	97.5%
β_{21}	1.066	0.049	0.968	1.165
$\sigma.\alpha_1$	0.108	0.027	0.070	0.170
$\sigma.\alpha_2$	0.101	0.022	0.069	0.157
$\sigma.\varepsilon$	0.274	0.072	0.154	0.437

We follow Pitt and Shephard in assuming the autoregressions are stationary, so that ρ and φ lie between -1 and $+1$ (see *Program 8.15 US Consumption and Income*). Specifically we take $\phi_i = \phi$ and adopt the same Beta prior transformation as in their analysis,

$$\begin{aligned}\varphi^* &\sim \text{Beta}(18, 1)\\ \rho^* &\sim \text{Beta}(18, 1)\\ \varphi &= 2\varphi^* - 1\\ \rho &= 2\rho^* - 1.\end{aligned}$$

In the case of these two US macro-economic indices, the greatest stability occurs in the mid-1960s and the most instability in the final years of the series, 1979 and 1980.

The above coefficients, φ and ρ, are estimated close enough to 1 to suggest that a random walk prior be used instead. The factor loading β_{21} is estimated to be very close to one, emphasising the strong correlation between the consumption and income series. This loading and the standard deviations of the stochastic variance series are summarised as shown in Table 8.19 (from the second half of a run of 10 000 iterations).

8.6.2 Structural equation modelling (SEM)

Structural equation models offer wide versatility for modelling, and incorporate ideas from regression, path and factor analysis. While factor analysis allows the observed variables to be influenced by a smaller set of latent variables, structural equation models permit both predictors and response to be latent variables. In this way the summarisation of the original predictors by the latent predictors will also take in the anticipated causal relationship to the outcome. They allow the exploration of substantive and measurement relationships using survey, experimental and longitudinal data. The substantive questions are typically modelled via a structural equation model specifying a linear relation between one or more latent response variables and one or more latent explanatory variables.

For example if we have a single latent response denoted θ_i and three latent explanatory variables η_{1i}, η_{2i} and η_{3i}, then a simple linear model would just be

$$\theta_i \sim \mathrm{N}(\mu.\theta[i], \sigma^2_\theta)$$
$$\mu.\theta[i] = \beta_1\eta_{1i} + \beta_{2\eta 2i} + \beta_{3\eta 3i.}$$

Note that we assume all latent variables have mean zero, so a constant term is not needed.

The latent variables are modelled in terms of observed measures, $y_{1i}, y_{2i}, \ldots, y_{Pi}$ and $x_{1i}, x_{2i}, \ldots, x_{Qi}$ denoting indicators of the latent responses θ_i and of the latent predictors η_i respectively. The model relating observed to latent variables is the 'measurement model'. One goal of SEM is to routinely allow for measurement error in both the independent and response variables (Joreskog, 1973). To continue the above example, suppose we had two observed indicators of a single latent response and five indicators of the three latent predictors. Then assuming the observed data were continuous variables for which a normal probability model was appropriate, we might have

$$y_{1i} \sim \mathrm{N}(\mu_{1i}, \sigma^2_1)$$
$$\mu_{1i} = \gamma_{11} + \gamma_{12}\theta_i$$
$$y_{2i} \sim \mathrm{N}(\mu_{2i}, \sigma^2_2)$$
$$\mu_{2i} = \gamma_{21} + \gamma_{22}\theta_i.$$

If the observed indicators y_1 and y_2 were also centred we might omit the intercept parameters. For the first observed indicator x_{1i} of the latent predictors, the model might be

$$x_{1i} \sim \mathrm{N}(v_{1i}, \zeta^2{}_1)$$
$$v_{li} = \delta_{11} + \delta_{12}\eta_{li} + \delta_{13}\eta_{2i} + \delta_{14}\eta_{3i}),$$

with the models for the other observed x defined similarly. It may be noted that in both the structural and measurement models some coefficients may be defined as fixed, e.g. set to zero or one, both on substantive grounds and to improve identifiability.

A form of the model useful for longitudinal analysis is in terms of an autoregressive dependence in the structural model. Suppose we had a multivariate latent response θ_i $(i = 1, \ldots, N)$ of dimension M and a latent predictor η_i of dimension L. Also assume that η_i is constant in time, but that the response was time varying, viz. θ_{it}. So for times $t = 2, \ldots, T$ subsequent to the first, the structural model is

$$\theta_{it} = \beta_t\theta_{i,\,t-1} + \alpha_t\eta_i + \psi_{it}$$

while for the first period the model is

$$\theta_{i1} = \alpha_1\eta_i + \psi_{i1}.$$

Here the θ_{it} are $M \times 1$, the β_t are $M \times M$, the α_t are $M \times L$, and the ψ_{it} is an $M \times 1$ error vector, possibly intercorrelated at time t, but not correlated with errors at previous times. The measurement model relates P observed indicators of the response $(P > M)$ to the latent θ_{it} via a time-varying parameter set:

$$y_{it} = \mu_t + \gamma_t \theta_{it} + u_{it}$$

where y_{it} is $P \times 1$, μ_t is a time-varying intercept, γ_t is $P \times M$, and u_{it} is a time-specific error term. The measurement model for the Q observed indicators x_i of η_i is

$$x_i = v + \delta \eta_i + w_i$$

where v is a $Q \times 1$ fixed intercept, w_i is an error term, and δ is $Q \times L$.

Estimating a structural equation model, as for any model involving latent variables, involves specifying the scale of the latent variables. In factor analysis it is frequently assumed that the factors are in standard form, with mean zero and variance 1. In structural equation modelling an alternative strategy is often adopted, namely to assume the scale of each latent variable to be the same as one of the indicator variables. This is illustrated in the following example.

Example 8.16 Alienation over time Wheaton *et al.* (1977) discuss the stability of alienation θ, defined in terms of observed attitudes. These are related to an underlying measure of social status, η. Data on attitudinal variables were collected from two Illinois communities in 1967 and 1971: y_1, y_2 measuring anomie and powerlessness in 1967, and y_3, y_4 measuring the same concepts in 1971. The observed background indicators of social status were years of education (x_1) and Duncan's socio-economic index (x_2) (Duncan, 1961). The data used here are generated using the covariance matrix from Wheaton *et al.*, which is for centred versions of the variables y and x (see Program 8.16). The structural model is in terms of two latent indicators, anomie in 1967 (θ_1) and anomie in 1971 (θ_2), namely

$$\begin{aligned}
\theta_{i2} &= \alpha_2 \eta_i + \beta \theta_{i1} + \psi_{i2} \\
\theta_{i1} &= \alpha_1 \eta_i + \psi_{i1}.
\end{aligned}$$

The measurement model for the latent responses is

$$\begin{aligned}
y_{i1} &= \theta_{i1} + u_{i1} \\
y_{i2} &= \gamma_1 \theta_{il} + u_{i2} \\
y_{i3} &= \theta_{i2} + u_{i3} \\
y_{i4} &= \gamma_2 \theta_{i2} + u_{i4}
\end{aligned}$$

while the measurement model for the latent predictor (social status) is

$$\begin{aligned}
x_{i1} &= \eta_i + v_{i1} \\
x_{i2} &= \delta \eta_i + v_{i2}.
\end{aligned}$$

It is assumed that the scales for θ_1 and θ_2 are the same as those for y_1 and y_3, while the scale for η is the same as for x_1.

Table 8.20 Alienation model, parameter summary.

	Mean	SD	2.5%	97.5%
α_1	−0.67	0.06	−0.78	−0.56
α_2	−0.22	0.07	−0.35	−0.08
β	0.79	0.06	0.67	0.90
δ	6.78	0.35	6.10	7.46
γ_1	0.94	0.04	0.87	1.02
γ_2	0.88	0.03	0.82	0.94

The essential coding in BUGS is as follows

```
        for (i in 1:N) {# structural model
        theta1[i] ~ dnorm(mu.theta1[i],tau[1])
        mu.theta1[i] <- alpha[1]*eta[i]
        theta2[i] ~ dnorm(mu.theta2[i],tau[3])
        mu.theta2[i] <- beta*theta1[i]+alpha[2]*eta[i]
        eta[i] ~ dnorm(0,tau[5])
# measurement models for indicators of theta1 and theta2
    y1[i] ~ dnorm(theta1[i],tau[1])
    y2[i] ~ dnorm(mu2[i],tau[2])
    mu2[i] <- gamma[1]*theta1[i]
    y3[i] ~ dnorm(theta2[i],tau[3])
    y4[i] ~ dnorm(mu4[i],tau[4])
    mu4[i] <-gamma[2]* theta2[i]
# measurement models for indicators of eta
    x1[i] ~ dnorm(eta[i],tau[5])
    x2[i] ~ dnorm(nu2[i],tau[6])
    nu2[i] <- delta*eta[i]}
```

The coefficients obtained from a single run of 5000 iterations (discarding the first 1000) match those obtained by the LISREL program of Joreskog and Sorbom (1981) – see also Scheines *et al.* (1999). The continuity coefficient β shows the stability of alienation over time, while higher socio-economic status has a negative influence (via α_1 and α_2), though diminishing, on alienation (Table 8.20).

The large size of δ in the measurement equation for the indicators of the predictor reflects the discrepant scale of x_2 compared with the remaining observed indicators.

Example 8.17 Tests of thinking style Child (1970) presented results X_{ij} from $j = 1, \ldots 8$, intelligence tests for $i = 1, \ldots, 306$ children. Using the correlation matrix reproduced by Everitt (1984, p. 45), 306 'observations' are generated via the Sample Generate Model in Program 8.17. The first two of the eight tests are intended to measure convergent or conventional intelligence, while the other six are designed to measure divergent thinking. Tests 3–6 stress verbal ability, while tests 7 and 8 stress nonverbal responses. Three latent variables η_{ik} are posited to underlie the eight manifest variables. All are drawn from standard normal priors but the first two are assumed correlated. The first underlies variations in responses 3–6, the second variability in tests 7 and 8, and their correlation ρ derives from their both representing divergent thinking. The third latent variable explains variability in the first two tests and is assumed independent of the other two.

There are notionally 17 parameters to estimate: the correlation parameter ρ in the bivariate normal for $\eta 1$ and $\eta 2$, the eight unique variances σ_j^2 of the X_j, and the factor loadings $\lambda_1, \ldots, \lambda_8$ relating $X_1, \ldots, X_8$ to $\eta_1, \ldots, \eta_3$ in the following

$$
\begin{aligned}
X_{ij} &\sim \mathrm{N}(\mu_{ij}, \sigma_j^2), \quad j = 1, \ldots, 8 \\
\mu_{i1} &= \lambda_1 \eta_{i3} \\
\mu_{i2} &= \lambda_2 \eta_{i3} \\
\\
\mu_{i3} &= \lambda_3 \eta_{i1} \\
\mu_{i4} &= \lambda_4 \eta_{i1} \\
\mu_{i5} &= \lambda_5 \eta_{i1} \\
\\
\mu_{i6} &= \lambda_6 \eta_{i1} \\
\mu_{i7} &= \lambda_7 \eta_{i2} \\
\mu_{i8} &= \lambda_8 \eta_{i2.}
\end{aligned}
$$

In practice, λ_1 and λ_2 are not separately identifiable without an additional constraint; similar remarks apply to σ_1^2 and σ_2^2 (see Everitt, 1984). One possible constraint is to assume $\lambda_1 = \lambda_2$ and $\sigma_1^2 = \sigma_2^2$. There are then 15 free parameters. To select an appropriate standard bivariate normal density for η_{i1} and η_{i2} a discrete prior with six possible correlation matrices is used (namely those with $\rho = 0.1$, $\rho = 0.2$, $\ldots \rho = 0.6$). For interpretative purposes, the coefficients $\lambda_1 (= \lambda_2), \lambda_3, \ldots, \lambda_8$ are assumed positive (Table 8.21).

The resulting parameter estimates are from a run of 10 000 iterations and are close to those cited by Everitt (1984), except that the correlation between the first two factors is slightly lower at around 0.35.

8.6.3 Different group means and covariances on latent variables

If there are a set of indicators X_{ikg} observed for indicators i, subjects k and groups g, we may often believe a *single* latent factor operates across the groups to explain variability in the observed indicators. However, the group means on this latent variable are likely to differ. Since the scale of the latent factor is arbitrary, we need to impose constraints on the mean factor scores in the groups in order to identify any differences in group means.

For example, suppose scores on a single factor in group g are denoted z_{kg}. We set the mean latent factor score to zero in group 1, and allow the other

Table 8.21 Thinking styles, parameter summary.

Parameter	Mean	SD	2.5%	Median	97.5%
$\lambda_1(=\lambda_2)$	0.65	0.04	0.57	0.65	0.73
λ_3	0.76	0.06	0.65	0.76	0.87
λ_4	0.71	0.06	0.59	0.71	0.83
λ_5	0.69	0.06	0.57	0.69	0.81
λ_6	0.45	0.06	0.33	0.45	0.57
λ_7	0.84	0.07	0.66	0.85	0.96
λ_8	0.51	0.07	0.39	0.51	0.67
ρ	0.35	0.07	0.20	0.30	0.50
σ_1^2	0.45	0.04	0.38	0.45	0.52
σ_3^2	0.41	0.06	0.30	0.41	0.53
σ_4^2	0.54	0.06	0.42	0.54	0.67
σ_5^2	0.53	0.06	0.42	0.53	0.66
σ_6^2	0.71	0.06	0.59	0.71	0.85
σ_7^2	0.15	0.10	0.03	0.12	0.42
σ_8^2	0.66	0.07	0.51	0.66	0.79

group means to be parameters. The variances of the latent factor scores in each group are also parameters. Thus

$$z_{kg} \sim \mathrm{N}(\eta_g, \delta_g) \tag{8.14}$$

where $\eta_1 = 0$. For $G = 2$ groups we thus have a single difference parameter to estimate.

We might posit the following structure, with factor loadings λ_i for variables $i = 1, \ldots, I$, for groups $g = 1, \ldots, G$, and subjects $k = 1, \ldots, n_g$ within groups

$$\begin{aligned} X_{ikg} &\sim \mathrm{N}(\mu_{ikg},\ \tau_{ig}) \\ \mu_{ikg} &= v_i + \lambda_i z_{kg}. \end{aligned} \tag{8.15}$$

The first factor loading λ_1 is set to 1 for identifiability, and equal loadings $\lambda_{ig} = \lambda_i$ are assumed across groups for variables $i > 1$. The goodness of fit of structural equation models may be assessed via comparing the covariance matrix of the actual data with the covariance matrix resulting from the model (see Everitt, 1984; Scheines *et al.*, 1999). Alternatively we can apply deviance-based tests to compare actual observations X_{ikg} and means μ_{ikg}. The hypothesis of equal loadings across groups may be rejected on this basis, and group-specific loadings λ_{ig} applied.

Such comparisons across groups may be extended to situations where more than one factor is postulated. Assume the data is centred (though not necessarily standardised). Then suppose there are $M = 2$ factors, one underlying variations in a subset of the original variables $i = 1, \ldots, I_1$ and the other explaining variation in variables $i = I_1 + 1, \ldots, I$. We might posit the following structure: for groups $g = 1, \ldots, G$, and subjects $k = 1, \ldots, n_g$ within groups

$$X_{ikg} \sim \mathrm{N}(\mu_{ikg}, \tau_{ig}), \quad i = 1, \ldots, I$$
$$\mu_{ikg} = \lambda_{ig1} z_{k1g}, \quad i = 1, \ldots, I_1$$
$$\mu_{ikg} = \lambda_{ig2} z_{k2g}, \quad i = I_1 + 1, \ldots, I. \qquad (8.16)$$

The M factor scores z_{kmg}, $m = 1, \ldots, M$ will typically be assumed multivariate standard normal – so bivariate normality is appropriate in the above case. If the covariance matrices S_1, S_2,...,S_G of the original data are unequal, then we may postulate a structural equation model to explain part of this covariance heterogeneity.

Example 8.18 Subjective latent class scores A study by Kluegl *et al.* (1977) sampled 800 adults, 432 whites and 368 blacks, on four indices of subjective class: an occupation index, an income index, life style and influence. We use the group-specific means and standard deviations on these indices and the correlations between them to generate a comparable sample (see *Program 8.18 Subjective Class*).

Suppose a single latent subjective class variable underlies variability in the four observed indicators, though with possible different latent indicator variances by group and different means. As stressed above, we can only identify contrasts in means on the latent variable, here $\kappa = \eta_2 - \eta_1$.

We apply the above model (8.15) for $G = 2$, $n_1 = 432$, $n_2 = 368$ and $I = 4$, and obtain similar answers to those cited by Everitt (1984) using the LISREL

Table 8.22 Subjective class, parameter summary.

Parameter	Mean	SD	2.5%	Median	97.5%
κ	0.25	0.04	0.18	0.25	0.33
λ_2	1.41	0.09	1.25	1.41	1.59
λ_3	1.40	0.09	1.24	1.40	1.59
λ_4	1.19	0.08	1.04	1.19	1.37
Observation variances					
$\tau(1,1)$	0.30	0.02	0.26	0.30	0.35
$\tau(1,2)$	0.34	0.03	0.28	0.34	0.40
$\tau(2,1)$	0.20	0.02	0.17	0.20	0.24
$\tau(2,2)$	0.26	0.03	0.21	0.26	0.32
$\tau(3,1)$	0.13	0.02	0.10	0.13	0.16
$\tau(3,2)$	0.18	0.03	0.13	0.18	0.23
$\tau(4,1)$	0.25	0.02	0.22	0.25	0.30
$\tau(4,2)$	0.46	0.04	0.39	0.46	0.54
σ^2_{z1}	0.11	0.01	0.09	0.11	0.14
σ^2_{z2}	0.24	0.03	0.18	0.24	0.31
ν_1	1.28	0.04	1.21	1.29	1.35
ν_2	1.17	0.05	1.08	1.17	1.26
ν_3	1.20	0.04	1.11	1.20	1.28
ν_4	1.30	0.04	1.22	1.31	1.38

program. The group mean difference on the latent variable between whites and blacks is denoted `kappa[]` in Program 8.18. Its posterior mean is 0.25, with 95 % interval from 0.18 to 0.33. The variances, as in (8.14), of the latent variable $\delta_j = \sigma^2_{Zj}$ for groups $j = 1, 2$, show greater variability among black subjects. The identifiable factor loadings $\lambda_2, \ldots, \lambda_4$ are all in line with a unipolar class identity factor (Table 8.22).

Example 8.19 Cognitive tests Yule *et al.* (1969) present results on scores for 10 cognitive tests for 150 children. By means of a separate test of reading ability, the children were divided into two groups, poor readers and good readers (containing respectively $n_1 = 80$ and $n_2 = 70$ pupils). The 10×10 covariance matrices of these groups on the cognitive tests were used to generate comparable data.

The tests can be divided into verbal tests (1–5) and performance tests (6–10) and so two factors are posited to underlie variability in the $I = 10$ original indicators. We adopt a model of type (8.16) with factor loadings equal across the two groups of children, so that

$$\lambda_{i11} = \lambda_{i21} = \lambda_{i1}$$
$$\lambda_{i12} = \lambda_{i22} = \lambda_{i2}$$

Note that $\lambda_{i1} = 0$ for variables 6–10, and $\lambda_{i2} = 0$ for variables 1–5. We assume the group-specific factor scores z_{k1g} and z_{k2g}, $k = 1, \ldots, n_g$ are derived from group-specific standard bivariate normal priors.

We use a discrete prior with 10 bins to model various options for the correlation parameters ρ_g $(g = 1, 2)$ in these bivariate normal densities (only positive correlations $0.1, 0.2, 0.3, \ldots, 0.9, 0.99$ are considered). See Table 8.23.

The results are close to those obtained via the LISREL program with ρ_1 estimated as 0.56 and ρ_2 at 0.86. Everitt (1984) obtains values 0.49 and 0.78.

Table 8.23 Cognitive tests, parameter summary.

Parameter	Mean	SD	2.5%	Median	97.5%
λ_{11}	2.22	0.21	1.82	2.22	2.65
λ_{21}	2.21	0.24	1.75	2.20	2.71
λ_{31}	1.73	0.19	1.36	1.73	2.12
λ_{41}	1.14	0.19	0.77	1.14	1.53
λ_{51}	1.48	0.19	1.10	1.48	1.87
λ_{62}	0.77	0.22	0.33	0.77	1.19
λ_{72}	1.54	0.20	1.16	1.53	1.94
λ_{82}	1.74	0.25	1.25	1.74	2.24
λ_{92}	1.16	0.27	0.63	1.15	1.71
$\lambda_{10,2}$	1.92	0.20	1.54	1.91	2.31
ρ_1	0.56	0.13	0.30	0.60	0.80
ρ_2	0.86	0.06	0.70	0.90	0.90

8.7 MULTIVARIATE HIERARCHICAL MODELS

Multivariate outcomes often occur in combination with nested multilevel data structures, in longitudinal data sets with replications on each subject, or in data sets with crossed strata. For example, pupil performance may be measured by multiple outcomes, with pupils themselves classified by class, school or education agency. In follow-up studies, repeated observations are typically of several interrelated outcomes.

Suppose we consider the longitudinal case and have $i = 1, \ldots, n_j$ repeated observations y_{ij} for cases $j = 1, \ldots, N$ on P related outcome variables. For continuous outcomes, and assuming a normal observation model, we may allow for covariation among the outcomes. Additionally we may consider the impact of (a) the time variable itself at measurements $1, 2, \ldots, n_j$; (b) covariates $x_{ij1}, x_{ij2}, \ldots$, other than time, which vary over replications $1, \ldots, n_j$; and (c) fixed covariates w_j for cases j. So

$$y_{ijv} \sim N_P(\mu_{ijv}, \Theta)$$

where $v = 1, \ldots, P$. If the number of repetitions is the same ($n_j = n$) for all individuals and repetitions are equally spaced, then the covariance matrix Θ might allow for changing variation over time and be of dimension $P \times P$ for each time period. If there were a large number of repetitions for each individual we might envisage a unique covariance matrix between the outcomes for each subject. A simpler initial structure than either of these cases assumes Θ fixed over times and individuals and of dimension $P \times P$, with nonzero off-diagonal elements to allow for correlations among the outcomes.

In a longitudinal setting, the means μ_{ijv} combine fixed and random regression effects on time t_{ij} itself and on covariates specific to times and cases $\{i, j\}$ or to cases j only. Suppose we have a single covariate x_{ij} with a regression effect fixed over all cases though varying according to outcome; a single covariate z_{ij} with regression effect randomly varying over cases j as well as over the outcomes, a random linear growth effect in time t_{ij}, and a random intercept. Then

$$\begin{aligned}
\mu_{ij1} &= \alpha_1 x_{ij} + \beta_{1j1} + \beta_{2j1} t_{ij} + \beta_{3j1} z_{ij} \\
\mu_{ij2} &= \alpha_2 x_{ij} + \beta_{1j2} + \beta_{2j2} t_{ij} + \beta_{3j2} z_{ij} \\
&\cdots \\
\mu_{ijP} &= \alpha_P x_{ij} + \beta_{1jP} + \beta_{2jP} t_{ij} + \beta_{3jP} z_{ij}.
\end{aligned} \tag{8.17}$$

Thus for only one outcome, we have for replications $i = 1, \ldots, n_j$ and individuals $j = 1, \ldots, N$, a mean defined as

$$\mu_{ij} = \alpha x_{ij} + \beta_{1j} + \beta_{2j} t_{ij} + \beta_{3j} z_{ij}.$$

More complex time effects are possible (e.g. polynomial curves or splines), not necessarily all randomly varying over individual subjects. In other multilevel settings, without a time dimension, the index i might denote individual cases nested within organisational or other clusters j. The random effects are then specific to the clusters.

In general, suppose we have Q covariates, the impact of which varies randomly over individuals, including possibly random intercepts and/or growth curves. In the second stage of the multivariate hierarchical model, the random effects

$$\beta_{cjv} \quad (c = 1, \ldots, Q; j = 1, \ldots, N; \; v = 1, \ldots, P)$$

are assumed to be drawn from a common hyperdensity, such as the multivariate t or normal densities. The means of the β_{cjv} may be related to fixed individual level covariates w_j via coefficients γ_{cv} specific to covariate $c = 1, \ldots, Q$ and outcome $v = 1, \ldots, P$. The covariation among the β_{cjv} is typically described by a $QP \times QP$ variance–covariance matrix Ψ, with elements $\Psi_{cv,c'v'}$. The latter might be simplified, for example by assuming

$$\Psi = \text{diag}\left(\sigma^2_{\psi 11}, \sigma^2_{\psi 12}, \ldots, \sigma^2_{\psi QP}\right).$$

Example 8.20 Repeated measurements of paired outcomes A simple example of multivariate analysis of repeated measures is provided by the analysis by Bland and Altman (1995) of readings on $P = 2$ biochemical outcomes, intramural pH and Paco2. There are 47 readings relating to eight subjects and we aim to model the correlation between the two sets of readings. If we are interested in whether patients with high pH also have high Paco2, then a correlation between the eight subject means is relevant. If the interest is focused on whether an increase in pH within an individual patient is associated with a change in Paco2, we would consider observations without regard to the grouping into subjects.

Bland and Altman apply regression analysis to the question of covariation within subjects – this involves removing variation between subjects and expressing that part of the remaining variation due to correlation between the two variables. To calculate association between means at the patient level, they used weighted correlation (with weights based on the number of replications n_j for subject j).
To address these questions we assume a bivariate observation model

$$y_{jiv} \sim \text{N}_2\left(\mu_{ijv}, \Theta\right)$$

where Θ is 2×2, with $v = 1$ for pH, and $v = 2$ for Paco2 (see Program 8.20). The subject index j runs from 1 to 8 and the number of repetitions n_j varies from 3 to 9. The correlation at observation level is measured by $\Theta_{12}/(\Theta_{11}\Theta_{22})^{0.5}$. The model for the means β_{jv} at subject level is, since $Q = 1$.

$$\mu_{ijv} \sim \text{N}\left(\beta_{jv}, \Psi\right)$$

where Ψ is also 2×2. The correlation at subject level is measured by $\psi_{12}/(\psi_{11}\psi_{22})^{0.5}$.

Both Ψ^{-1} and Θ^{-1} are assigned Wishart priors with 2 degrees of freedom, with diagonal elements in the prior precision matrix set to 0.1, and off-diagonals set to 0.005. Following a run of 10 000 iterations, the dispersion matrix Ψ has posterior mean

$$\begin{bmatrix} 0.116 & 0.038 \\ 0.038 & 0.535 \end{bmatrix}$$

and the average of the correlation at subject level is 0.156. (This compares with 0.08 via the weighted correlation method.) However, neither the covariance matrix elements or the correlation are precisely identified, with the correlation having 95% credible interval (−0.56, 0.77). At observation level, there is an inverse correlation between outcomes averaging −0.46 (with 95% interval from −0.68 to −0.19). This compares with the correlation coefficient within subjects of −0.51 identified by Bland and Altman.

Example 8.21 Cataract outcomes Bland (1995) presents data on a pre- and post-surgery measures of visual function among 17 cataract patients (i.e. $n_j = 2$ all j). The first measure is a 'visual acuity (VA) score', in which the second number in the score (after the slash) is the minimum size of letter which can be read from a distance of six metres; thus 6/12 means a letter of size 12 can be read at six metres, and higher numbers after the slash denote worse vision. The second measure is a contrast sensitivity (CS) test, in which high numbers denote better vision. Bland poses the questions (a) how to compare the two measures pre- and post-treatment (to assess treatment effectiveness) and (b) how to assess the interrelation between the two measures after treatment.

The observation model cannot be simply represented as bivariate normal, as in the preceding example, unless the original VA scores are assumed to approximate a continuous scale. Instead a cumulative logit model for the VA score is adopted, with the original seven scale categories, namely (6/6, 6/9, 6/12, 6/18, 6/24, 6/36, 6/60) combined into three to improve identifiability of the latent cut-points. These are (1) 6/6 and 6/9; (2) 6/12 and 6/18; and (3) higher scores. This device might not be needed for a larger sample of patients. The observation model for the CS scores is a univariate normal model with a conventional measurement error, but with an additional error term correlated with an error term in the means for the VA cumulative probabilities

The association between the two scales is measured in terms of these random effects which are comparable with the terms β_{1j1} and β_{1j2} ($j = 1, \ldots, 17$) in the model (8.17) above. Treatment gain between the two measures involves comparing the terms β_{21} and β_{22} in the above framework; for only two points there is no real question of modelling patient variability in treatment gain. Using Program 8.21, the posterior estimates of the cut-points, the correlation between β_{1j1} and β_{1j2}, and the treatment effects are given in Table 8.24 (from the second half of a run of 30 000 iterations).

Table 8.24 Cataract outcomes, parameter summary.

	Mean	SD	2.5%	Median	97.5%
Change in VS	−2.640	1.107	−5.085	−2.531	−0.793
Change in CS	0.336	0.061	0.216	0.337	0.455
κ_1	−2.458	1.190	−5.328	−2.300	−0.596
κ_2	1.979	1.102	0.161	1.861	4.516
ρ	−0.721	0.174	−0.937	−0.762	−0.276

They show a clear treatment gain under both measures. A negative association (parameter ρ) between the measures is also apparent – this has no clinical significance and is simply because one is a positive measure of good function and the other a negative measure. Bland obtains a Spearman's rho of −0.49 for this association, though using only post-surgery measures. The cut-points κ_j of the VA scale show a clear demarcation between successive categories.

Example 8.22 Written and coursework exams Goldstein *et al.* (1998) analyse bivariate data on written exam and continuous assessment ('coursework') scores in science. The data are for $N = 1905$ students in $M = 73$ schools, and the exams took place as part of the 1989 UK General Certificate of Education tests. They first consider a model with school-specific intercepts, and with gender as an effect fixed over all students. Let $G_i = 1$ for males and $G_i = 2$ for females, and S_i denote the students school. With y_{i1} denoting scores on the written exam and y_{i2} scores on coursework, and assuming multivariate normality in the school effects, the model is

$$\underset{\sim}{y}_{i,1:2} \sim \mathrm{N}_2(\underset{\sim}{\mu}, P)$$

where P is a 2×2 dispersion matrix for the observations, and

$$\begin{aligned} \mu_{i1} &= \beta_{Si,1} + \gamma_{1,Gi} \\ \mu_{i2} &= \beta_{Si,2} + \gamma_{2,Gi} \end{aligned} \tag{8.18}$$

for students $i = 1, \ldots, N$, with $\gamma_{1,j}$ and $\gamma_{2,j}$ denoting fixed effects, and β_{j1} and β_{j2} random effects.

Thus for the $j = 1, \ldots, M$ schools the model assumes varying performance over all students as follows:

$$\underset{\sim}{\beta}_j \sim \mathrm{N}_2(\underset{\sim}{\nu}, Q)$$

where Q is a 2×2 dispersion matrix for the school effects. The remaining fixed effects are the overall means ν_1 and ν_2 for the two exams, and the gender effects. We set $\gamma_{11} = \gamma_{21} = 0$ so the gender effects describe the female vs male difference in exam scores.

A more complex model allows for schools to have differential effects on the performance of each gender. Thus we make the female–male difference a random effect at school level, as follows

$$\begin{aligned}\mu_{i1} &= \beta_{Si,1} + \delta(G_i = 2)\beta_{Si,3} \\ \mu_{i2} &= \beta_{Si,2} + \delta(G_i = 2)\beta_{Si,4}\end{aligned} \tag{8.19}$$

for $i = 1, \ldots, N$, where $\delta(L) = 1$ if L is true. For the $j = 1, \ldots, M$ schools the model now involves four effects

$$\underset{\sim}{\beta}_j \sim \mathrm{N}_4(\underset{\sim}{\nu}, Q).$$

The results obtained with school intercepts only (as in (8.18)) are close to those of Goldstein *et al.* (1998). However, the multivariate normal model with both school and school–gender effects, as in (8.19), has smaller diagonal elements in the dispersion matrix Q as compared with Goldstein *et al.* Instead suppose we adopt a multivariate t with 5 degrees of freedom,

$$\underset{\sim}{\beta}_j \sim \mathrm{T}_4(\underset{\sim}{\nu}, Q, 5).$$

This modification leads to a replication of the results obtained by Goldstein *et al.* with one major exception; namely, there is evidence for variation among schools in the gender difference for both the written and coursework components, whereas Goldstein *et al.* found only coursework results to show school-level variation.

8.8 SMALL AREA AND SURVEY DOMAIN ESTIMATION

Frequently survey data are the only source of information regarding behavioural or health indicators for small geographical regions or subpopulations, such as populations in a specific age, sex and ethnic group. These subpopulations are generically called 'small areas' or 'domains'. The usual direct survey estimates, using survey data specifically relating only to those domains, are likely to have low precision, or may not even be possible to estimate, especially when the sample in the domain is small. Small area estimation describes a set of procedures, including fully Bayesian hierarchical methods, to combine information or pool strength over similar small areas by incorporating ancillary information. The latter would typically be from non-survey sources, and might be Census, administrative or vital statistics data. An increasing area of application is in terms of socio-demographic profiling where the 'areas' may consist of sets of geographic areas defined by a combination of Census and other characteristics (e.g. income, housing zone, etc.). In the UK there are well-established area clusters based on multivariate classification of Census data.

Small area models including random area-specific effects overcome some of the limitations of synthetic and related estimators of small domain characteristics. For example, suppose the survey outcome is continuous (e.g. income) or is a count of subjects exhibiting a certain characteristic (e.g. the survey totals of smokers and nonsmokers). We wish to use the survey to provide estimates for each small area i of means of continuous variables or proportions for binary variables. For example, Malec *et al.* (1997) consider small area estimation of the probability of visiting a doctor in the last year, using data from the United States National Health Interview Survey. Like them, we seek to use hierarchical random effects models to combine information across geographical areas and/ or socio-demographic groups to produce estimates of domain proportions or means and measures of their precision.

Synthetic estimators rely on survey information available for large domains g (e.g. for large geographic regions or for highly aggregated groups, such as all males). Suppose the survey provides totals y_g, based on sample sizes n_g, large enough to make reliable inferences about these domain population totals Y_g, means $\bar{Y}_g$, or population proportions P_g, when the outcome is binary. The small areas i cut across these larger domains such that $Y_g = \sum_{ig} Y_{ig}$, where Y_{ig} is the population total for cell (i, g). Suppose ancillary data about these cells, namely X_{ig}, were available from nonsurvey sources. Then a synthetic estimator of the population total for small area i, Y_i, would be

$$Y_i^S = \sum_g (X_{ig}/X_{.g}) Y_g^D$$

where $X_{.g} = \sum_i X_{ig}$, and where Y_g^D is a direct estimate for group g based on the survey information y_g.

Often the ancillary information consists of population counts $X_{ig} = N_{ig}$ for cells (i, g). In these circumstances the design bias of Y_i^S only disappears under the assumption that the small area means $\bar{Y}_{ig}$ are equal to the overall domain mean $\bar{Y}_g$ (Ghosh and Rao, 1994). Another possibility is weighted composite estimators, combining direct estimators such as

$$\hat{Y}_{ig} = y_{ig}/n_{ig}$$

having weight $f_{ig} = n_{ig}/N_{ig}$, with synthetic estimates having weight $1 - f_{ig}$. Hence the composite estimator tends towards the synthetic estimator when sampling fractions n_{ig}/N_{ig} are small, whatever the ratio of the between small area to within area variability.

8.8.1 Modelling small area rates

The modelling approach typically assumes a regression form with a small area random effect and using auxiliary information. The most general formulation would be for an outcome y_{igj}, defined by elements within populations $j = 1, \ldots, N_{ig}$, and by domains g and survey areas i, with the mean effect modelled by domain, area and element terms. Regression modelling potentially involves both element-specific auxiliary data (for individual survey respondents) combined with area- and domain-specific data from nonsurvey sources. In practice models are more specific in nature, and defined by the application.

Thus Malec *et al.* (1997) illustrate the potential modelling approaches for a binary outcome defining a health behaviour, when population totals N_{ig} are available for small areas i, $i = 1, \ldots, L$ (about 3000 US counties) crossed by domain g, namely 72 age, sex and race categories. We then require an estimate of Y_{ig} the number exhibiting the health behaviour in the population. Suppose that in the survey the count from cell (i, g) is n_{ig} and of these y_{ig} have the behaviour concerned. Then the binomial probability p_{ig} is related to covariates which are based on the domain definition, and which are the same for all individuals in that domain regardless of county of residence. Variation in these covariates' effects over areas may then be modelled using hierarchical random effects.

The covariates relating to domains could be categorical factors or more specialised functions (e.g. spline curves for age effects, if age is a component of the domain definition). Suppose we had an age factor in the domain definition, with five categories (denoted $X_1, \ldots, X_5$), for ages 15–29, 30–44, 45–59, 60–74 and 75+ (with 0–14 as the reference category). The remaining domain indicators are binary, X_6 for sex (1 =males) and X_7 for white/black ethnicity (1 =white). Then a domain defined by black females, aged over 75, would have a value of zero on all indicators except X_5. The relation between the outcome and these covariates, described by coefficients $\beta_1, \beta_2, \ldots, \beta_7$ may be differentiated between counties. So also may be the level of the outcome itself, as reflected in the intercept β_0. So a model might be

$$y_{ig} \sim \text{Bin}(p_{ig}, n_{ig})$$
$$\text{logit}(p_{ig}) = \beta_{0i} + \beta_{1i}X_{1g} + \beta_{2i}X_{2g} + \beta_{3i}X_{3g} + \ldots + \beta_{7i}X_{7g}.$$

Variability in the coefficients $\beta_{ji}, j = 0, \ldots, 7, i = 1, \ldots, L$ could be related to Q area-level covariates Z_i, such as county per capita income in the example of Malec *et al.* For example

$$\beta_{ji} \sim \text{N}(\gamma_j Z_i, \Delta)$$

for $j = 0, 1, \ldots, 7$ and $i = 1, \ldots, L$, and where Δ is 8×8, and γ is $8 \times Q$. Prediction of the population total Y_{ig} in area i and domain g would then be based on summing the observed number y_{ig} and the predicted Y^*_{ig} based on applying the predicted rate p_{ig} to a population of size $N_{ig} - n_{ig}$.

An example with a continuous outcome and without a domain dimension is provided by the nested error regression model of Ghosh and Rao (1994). Thus let y_{ij} denote the survey outcome for element j in area i, for n_i survey responses in $i = 1, \ldots, L$ areas. Suppose element-specific covariates $x_{ij} = (x_{ij1}, x_{ij2}, \ldots, x_{ijQ})$ are available. Then for a single covariate ($Q = 1$) the model could be

$$y_{ij} \sim \text{N}(\mu_{ij}, \zeta_{ij})$$
$$\mu_{ij} = \beta_{0i} + \beta_{1i}x_{ij}.$$

To allow for design effects, the variances ζ_{ij} could be modelled in terms of known constants k_{ij} defined by x_{ij} such that

$$\zeta_{ij} = k_{ij}\text{Var}(\varepsilon_{ij})$$

with $\text{Var}(\varepsilon_{ij})$ being constant.

Stroud (1987) also considers continuous survey responses y_{ij} for $i = 1, \ldots, L$ and $j = 1, \ldots, n_i$ but with auxiliary covariates x_i defined only at area level. Thus for a single area covariate

$$\begin{aligned} y_{ij} &\sim \mathrm{N}(\mu_i, \sigma^2) \\ \mu_i &\sim \mathrm{N}(v_i, \tau^2) \\ v_i &= \alpha + \beta x_i. \end{aligned} \tag{8.20}$$

Let

$$\psi_i^{-1} = \tau^2/\sigma^2 + 1/n_i$$

so that $0 < \psi_i < n_i$ with the upper limit approached when $\sigma^2 \gg \tau^2$.

Then the preceding model implies that the conditional expectation of μ_i is a weighted average of the observed $\bar{y}_i$ with weight $1 - \psi_i/n_i$ and of the predictions $\widetilde{y}_i$ with weight ψ_i/n_i. The term $\widetilde{y}_i$ is a weighted combination of all the area means with weights defined by the auxiliary regression (8.20). For the case $n_i = n$ with equal sample sizes in all areas the pooling model is defined by priors on the regression parameters, on $1/\sigma^2$, and on

Table 8.25 Home trips data set.

Municipality	Student estimate	Municipality mean	Inverse distance squared
Belleville	3	12.33	0.1125
	5	12.33	
	29	12.33	
Calgary	1	1.33	0.0171
	1	1.33	
	2	1.33	
Montreal	7	4	0.0579
	1	4	
	4	4	
Oakville	3	4.33	0.0587
	2	4.33	
	8	4.33	
Ottawa	6	4.67	0.0774
	6	4.67	
	2	4.67	
Sault Ste Marie	2	3.67	0.0337
	2	3.67	
	7	3.67	
Toronto	8	4	0.0624
	2	4	
	2	4	
Vancouver	1	1	0.0149
	1	1	
	1	1	

$$\psi^{-1} = \tau^2/\sigma^2 + 1/n$$

specifically via a beta prior on ψ/n.

Example 8.23 Home trips We illustrate the above model with a sample of expected trips home during an academic year among 24 students of Queens University (Kingston, Ontario) with equal sample sizes $n = 3$ in $L = 8$ municipalities. The survey question was 'how many trips do you estimate you will have taken by the end of the academic year'. The auxiliary variable for each municipality is x_i defined as the inverse of the distance squared from Kingston to the students home – the two furthest municipalities considered were Calgary and Vancouver (Table 8.25).

We adopt a beta B(0.5, 0.5) prior for ψ/n and vague priors for the regression parameters and $1/\sigma^2$. Our posterior estimate of ψ from a run of 10 000 iterations is 2.35 (SD = 0.52) compared with 2.17 obtained by Stroud. Because of the small sample sizes in each home town the posterior means μ_i are not precisely estimated but illustrate the prior assumption that they are not exactly equal to $v_i = \alpha + \beta x_i$ but fluctuate around it (Table 8.26).

Example 8.24 Wage and salary survey of firms Ghosh and Rao (GR, 1994) consider the problem of estimating area average wages and salaries $\bar{Y}_i$ of $N = 114$ firms in $L = 16$ areas. A random survey has been taken of 38 firms providing both wage and salary bills y_{ij} $(i = 1, \ldots, L,\ j = 1, \ldots, n_i)$ and an auxiliary variable x_{ij}, gross business income. The samples sizes in each area n_i vary from 0 to 10, with three areas having no firms sampled (thus $n_1 = n_4 = n_{13} = 0$). The following regression model is postulated relating the y_{ij} to the x_{ij}:

$$y_{ij} = x_{ij}^T\beta + v_i + e_{ij} \tag{8.21}$$

where $e_{ij} = k_{ij}\eta_{ij}$ is a function of (a) known constants k_{ij} and (b) an error term η_{ij}, normal with mean 0 and variance σ^2_η. This form is adopted in the face of

Table 8.26 Home trips, parameter summary.

	Average $\bar{y}_i$	Fitted regression	$E(\mu_i\|y)$	$V^{0.5}(\mu_i\|y)$	2.5%	97.5%
Belleville	12.33	10.14	10.55	2.477	5.625	15.48
Calgary	1.33	0.75	1.01	2	−2.992	5.039
Montreal	4	4.77	4.662	1.676	1.176	7.902
Oakville	4.33	4.85	4.776	1.693	1.242	8.039
Ottawa	4.67	6.69	6.252	1.857	2.387	9.712
Sault Ste Marie	3.67	2.39	2.751	1.794	−0.7776	6.435
Toronto	4	5.21	4.98	1.718	1.361	8.278
Vancouver	1	0.54	0.7908	2.061	−3.318	4.928

likely heteroscedasticity in the outcome. Specifically $k_{ij} = \sqrt{x_{ij}}$. The random effect v_i with mean 0 and variance σ_v^2 models correlations within areas. The regression effect (of the auxiliary variable gross business income) is assumed fixed over areas.

If $f_i = n_i/N_i$ is the sample fraction in area i, then the estimate of the mean wage/salary for area i is then

$$\widetilde{Y}_i = f_i\bar{y}_i + (1 - f_i)\bar{y}_i^*$$

where $\bar{y}_i^*$ is the predicted sample mean, based on the nested errors regression (8.21), for the nonsampled population. The latter is based on predictive sampling from the regression model mean, based on the sampled population of $n = \sum_i n_i = 38$ firms. The regression mean for predictive sampling uses the known auxiliary variable $\bar{x}_i$ in each area.

The 'sample data' of GR were in fact randomly drawn from a model estimated from a population of all firms, namely

$$y_{ij} = -2.47 + 0.20x_{ij} + v_i + e_{ij}$$

where $e_{ij} \sim N(0, 0.47x_{ij})$ and $v_i \sim N(0, 22.14)$ for $L = 16$ areas.

The predictive analysis may show some sensitivity to the choice of prior for the variance terms – see Program 8.24. Here G(0.001, 0.001) priors were used for the precisions $1/\sigma_\eta^2$ and $1/\sigma_v^2$. The posterior means of the variance terms, from the second half of a run of 10 000 iterations, are 6.1 and 0.7, and these estimates reflect the lesser intra-stratum clustering in the sample than population data, though the 95% interval for σ_v^2 is from 0.001 to 36 and so includes the true value. The posterior means of the regression coefficients used to generate the data in (8.21) are −3.1 and 0.186. The 95% credible intervals of these parameters include the

Table 8.27 Survey of firms, parameter summary.

Area	N_i	$\bar{X}_i$	$\bar{Y}_i$	$\widetilde{Y}_i$
1	1	137.7	24.22	22.19
2	6	100.84	20.43	13.80
3	4	47.72	5.48	5.04
4	1	45.64	6.55	5.61
5	8	108.53	20.55	15.21
6	6	65.68	14.85	11.50
7	6	116.34	21.46	19.95
8	6	92.74	13.4	12.46
9	27	97.58	15.56	14.56
10	5	76.04	5.88	7.55
11	12	90.15	15.2	14.28
12	7	86.24	13.4	11.23
13	4	164.28	26.06	27.00
14	6	164.7	22.44	15.69
15	13	83.86	9.4	9.43
16	2	134.49	29.49	37.71

true values, −2.47 and 0.20. The goodness of fit comparing $\bar{Y}_i$ and $\widetilde{Y}_i$ is monitored by average relative errors and average squared errors; their average (minimum) values are respectively 20.3 (11.5) and 17.7 (9.3).

Total numbers of firms by area, known auxiliary (gross business income), and actual and predicted outcome (average wages and salaries), are as in Table 8.27.

CHAPTER 9

Life Table and Survival Analysis

9.1 INTRODUCTION

Social and health scientists frequently analyse survival time and event history data in terms of an underlying stochastic process which is postulated to generate them. We may consider time as a single 'terminating' event such as death, or waiting times between repeated events such as changes of residence or job.

Thus event history analysis generally refers to a set of recurrent events, like spells of residence in different homes or times between nonfatal health events, with more than one spell possible on each individual (Tuma *et al.*, 1979). By contrast, survival analysis generally refers to the duration till a single event (Chiang, 1968). The events are defined by temporary or permanent changes of state, and may include transitions to several alternative states. Such data can be studied either in terms of the distribution of waiting times or in terms of the rate of change between states in given time intervals.

Among the questions that occur in survival analysis are (a) the impact of covariates on the length of survival or the rate of changing states; (b) the shape of the hazard rate (the rate of transition between states), for example, whether it increases or decreases monotonically with time spent in the current state; and (c) whether to model any such time dependence parametrically or nonparametrically. Social scientists may be interested in dynamic event history models analysing the rate at which marriage formation and dissolution occur (Vuchinich *et al.*, 1991), in the factors influencing career or job histories (Kandel and Yamaguchi, 1987; Hayward *et al.*, 1989), or in the relation of migration to time in the current state of residence and age. The question of time dependence is often of secondary interest; while parametric models for time effects were predominant till the 1970s, the Cox regression model and extensions to counting processes, both of which focus on covariate or treatment effects, have become increasingly popular.

In medical and reliability applications, the event under consideration is often 'absorbing', in the sense that repetition of the event is either not possible, or excluded from consideration; for example time to death, or disease recurrence or component failure. In medical applications, the observation on a case might start with diagnosis, onset of disease, or commencement of a clinical trial or treatment regime. The focus is then on time until a single unrepeatable event, such as death or relapse. In reliability applications also, the single possible event may be the failure of a component and the distribution of interest is of the component lifetime.

Often survival times are recorded only for grouped time units (e.g. in days or months) even though the timing of the event is theoretically available to much greater accuracy (Fahrmeir and Knorr-Held, 1997; Lewis and Raftery, 1999). In certain applications a grouped time unit may be the natural one (e.g. Sheike and Jensen, 1997). If survival times are grouped into discrete intervals, then the natural framework for analysis is provided by life tables defined on each of the discrete intervals (or possibly groupings of the original intervals). These may be used to compare the survival experiences of two or more samples in terms of expected lifetimes or proportions surviving to certain times. So within the scope of survival analysis are actuarial life tables when survival time is replaced by age, and large human populations are compared, for example in terms of life expectancies at different ages.

Whether the framework of analysis is in discrete or continuous time, additional issues occur concerning first, the possibility of multiple types of exit, or choices of new state, leading to multiple decrement or competing risk models; and second, the impact that unobserved differences between subjects may have on survival chances or duration times, and how allowing for them may change the estimates of the survival curve and of the impacts of observed covariates. This raises issues analogous to those of Chapters 5 and 6 in terms of suitable random effects or mixture models for variability in proneness or frailty but in the context of event times. Moreover, there is considerable debate regarding sensitivity of inferences to the method (for example, whether parametric or nonparametric) adopted for modelling unobserved heterogeneity.

Below we initially consider the basic densities central to modelling waiting times, how they may be modelled in simplified parametric terms, and how such modelling may be carried out in BUGS. Bayesian sampling methods for survival analysis may have advantages in facilitating the use of prior information about covariate effects, often the main focus of interest as compared with time effects. In parametric survival modelling, the analysis of censored waiting times is simplified in Bayesian models by considering them as extra unknowns. The full conditionals for the remaining parameters can then be updated as if all the missing t_i were in fact observed (Kuo and Smith, 1992). The Bayesian approach is a variant on EM and data augmentation methods for estimating the unobserved failure times (see Cox and Oakes, 1984, on the use of the EM method).

A number of papers discuss Bayesian estimation of Weibull parametric survival models (e.g. Dellaportas and Smith, 1993), with Abrams *et al.* (1996) considering priors for treatment effect parameters in clinical trials. They suggest, for example, a meta-analysis of previous findings on the treatment effect to establish a prior for the current survival study. Kalbfleisch (1978) considers the Cox semi-parametric regression method where the baseline hazard consists of a number of disjoint time intervals, the hazard being constant within each interval. Prior information about the baseline hazard then follows a gamma process. Ducrocq and Casella (1996) consider the addition of frailty to both the Weibull hazard and the Cox model from a Bayesian perspective – though adopting the Laplace approximation rather than full Gibbs sampling.

Clayton (1991) considers frailty effects in counting process models which are a generalisation of the Cox model. These models involve a process $N(t)$ counting events till time t and at risk process $Y(t)$. The latter is incorporated in a multiplicative intensity

$$L(t) = Y(t)\lambda_0(t)\exp(\beta x)$$

where $\lambda_0(t)$ is the baseline intensity. The corresponding likelihood is analogous to the Cox likelihood. An additional Bayesian perspective in survival modelling is provided by state-space models, with Fahrmeier and Knorr-Held applying dynamic linear model parameterisations to a discrete time survival model.

9.2 SURVIVAL ANALYSIS IN CONTINUOUS TIME: PARAMETRIC MODELLING OF DURATION AND COVARIATES

Let T denote a random variable in continuous time representing a survival time or length of stay. Let the survival time for individuals in a sample follow a density $f(t|\theta)$ where θ denotes parameters defining how the event rate changes with time. For simplicity we initially omit such parameters and consider the distribution function, or proportion of the population changing state by time t. Thus denote the distribution function of T by

$$F(t) = P(T < t) = \int_0^t f(u)du \tag{9.1}$$

which is simply the probability that the lifetime does not exceed a value t. The complement of this function is therefore the probability that the lifetime is t at a minimum,

$$S(t) = 1 - F(t) = P(T \geq t). \tag{9.2}$$

This is the fraction of the population still not having died or changed state by time t, variously known, according to the application, as the stayer proportion, or the survival rate. The hazard function, which is analogous to the death rate in discrete time, is the probability of an event (e.g. death, component failure) at time t. For this to be possible (for the subject to be at risk) the lifetime must be at least t, i.e. $T \geq t$. So imagine the chance of an event in a short interval $(t, t + \Delta t)$, given that $T \geq t$, namely

$$P(t < T < t + \Delta t | T \geq t).$$

The hazard function is the ratio of this probability to the length of the interval Δt, the ratio being taken to a limit as the interval length tends to zero. By the usual probability rules, the numerator of this ratio can be written

$$P(t < T < t + \Delta t)/P(T \geq t).$$

In terms of the distribution and survival functions, this becomes

$$\frac{F(t+\Delta t) - F(t)}{S(t)}.$$

So the hazard function is

$$\begin{aligned} h(t) &= \lim_{\Delta t \to 0} \frac{F(t+\Delta t) - F(t)}{\Delta t}[1/S(t)] \\ &= f(t)/S(t) \end{aligned}$$

since the limit term is just the derivative of $F(t)$. Equivalently the limit term is the derivative of $1 - S(t)$, so

$$\begin{aligned} h(t) &= -S'(t)/S(t) \\ &= -d/dt\{\log S(t)\}. \end{aligned}$$

It follows that

$$S(t) = \exp[-H(t)]$$

where $H(t) = \int_0^t h(u)du$ is the integrated or cumulative hazard rate.

Observed survival times and cumulative densities will typically be jagged with respect to time, and nonparametric methods for plotting and analysing survival reflect this. However, parametric lifetime models are also often applied for reasons of model parsimony, to smooth the observed survival curve, and to test whether certain basic features of time dependence are supported by the data. The first and most obvious is whether the exit or hazard rate is in fact a clear function of time. In the exponential model, the leaving rate is constant, defining a stationary process, with a hazard

$$h(t) = \lambda,$$

a survival function

$$S(t) = \exp(-\lambda t)$$

and a density

$$f(t) = \lambda \exp(-\lambda t).$$

Commonly used parametric forms for lifetime distributions which exhibit time dependence include:

(1) the Weibull, with a hazard rate monotonic in time and governed by a single time exponent parameter $\alpha > 0$,

$$h(t) = \lambda \alpha t^{\alpha-1},$$

and with a survival function

$$S(t) = \exp(-\lambda t^{\alpha}).$$

Here values of α exceeding 1 correspond to positive duration dependence and values between 0 and 1 to negative duration dependence (sometimes called 'cumulative inertia' in sociological applications).

(2) the gamma $G(\gamma, \lambda)$, with a hazard rate governed by the gamma shape parameter $\gamma > 0$, and density function

$$f(t) = \frac{\lambda}{\Gamma(\gamma)}(\lambda t)^{\gamma-1}\exp(-\lambda t).$$

Here values of γ between 0 and 1 mean the hazard decreases monotonically with time, and for $\gamma > 1$ the hazard increases from 0 to λ as time increases from 0 to ∞.

9.2.1 Modelling covariate impacts and time dependence in the hazard rate

Usually the hazard will depend both on time t itself and on covariates x, which may be fixed throughout the observation period or time varying. The question is then whether interactions exist between time and covariate effects and if so, how to model them. A simplifying assumption, analytically applicable to most survival densities (though still an empirical assumption), is known as proportional hazards, under which the mean function of covariates, $B(x) = \exp(x\beta)$ is independent of the time function. Thus

$$h(t) = \lambda_0(t)B(x) \tag{9.3}$$

where $\lambda_0(t)$ is the baseline hazard rate as a function of time only and where the exponential is used to ensure the overall impact of covariates is positive. Under proportional hazards, the hazard ratio comparing two individuals will be constant over time, providing their relevant descriptives x do not change. The exponential, Weibull and gamma models are all compatible with proportional hazards forms, whereas the logistic hazard

$$h(t) = 1/\sigma\left[e^D/(1 + e^D)\right]$$

where $D = (t - \beta x)/\sigma$, is an example of a nonproportional model.

For the exponential distribution of lifetimes, the hazard may be written

$$h(t) = 1/\mu$$

where $\mu = 1/\lambda$ is the average survival time. So $H(t)$ is simply t/μ. If 'death' or 'failure' is replaced by an event which can occur repeatedly, then such events occur in a Poisson process with constant rate λ when inter-event times are exponential with mean μ. So the probability of m events in time t is Poisson with mean λt. Sometimes it may be possible to subdivide the set of durations such that within subsets the frequency of events does not vary, i.e. the exponential process is restored when shifts in the series are allowed for.

The impact of time itself can therefore be illustrated by comparison with the 'timeless' exponential hazard. Thus if $\alpha = 1$ in the Weibull model

$$h(t) = \lambda \alpha t^{\alpha-1}$$

then the Weibull and exponential hazards are the same, and time itself has no effect. In practice, a posterior density for α which straddled the value 1 might be taken to indicate an exponential model would be suitable.

9.2.2 Censored time observations

A distinguishing feature of survival and event history analysis is censoring: an individual's lifetime is only partially observed and not necessarily followed through to its completion. This would be the case in a clinical trial if some individuals withdrew from observation or were (say) still alive at the end of the trial. In some applications (e.g. models for marital status or job change) it is possible that a move never occurs and censoring is also present then.

Censoring leads to data truncated from below when censoring is to the right: the observed time is less than the actual (unobserved) complete survival time. Less frequently survival data may be truncated from above (left censoring), when the observed time is greater than the actual time when the state commenced. For example, population census data may record long- term illness status by current age but not by age when illness commenced.

For right-censored data, the likelihood consists only of $S(t_i)$ for censored cases and of $f(t_i) = h(t_i)S(t_i)$ for observed failures. If we include the censoring indicator $\delta_i = 1$ for observed failures and $\delta_i = 0$ for censored cases, the likelihood is a product over cases of the terms

$$h(t_i)^{\delta_i} S(t_i).$$

9.2.3 Survival Analysis in BUGS

In BUGS direct parametric analysis of survival data is possible for the exponential, Weibull, gamma and log-normal densities. In these cases right censoring is handled by creating a vector of censored times (`centime[]`) which are coded 0 for individuals whose survival times are actually observed, and another vector (`survtime[]`) of actual survival times, coded as unobserved or NA for individuals who are censored. Thus the censored survival times are regarded as missing data. In BUGS, both `centime[]` and `survtime[]` are included in the data file, and estimates will be made of the completed survival times for censored cases. It is also possible but not usually necessary to create starting values for the censored or NA values of `survtime[]` in the inits file.

So if a Weibull distribution with time parameter `r` and mean `lambda[]` were assumed, then right censoring would be denoted

```
survtime [i] ~ dweib (r, lambda[i]) I(centime[i],)
```

For left censoring we would have

```
survtime [i] ~ dweib (r, lambda[i]) I(, obstime [i])
```

The simplifying assumption is usually made that censoring is non informative, i.e. that censoring of lifetimes is not associated with the length of survival time which would have been observed on completion. This is analogous to the assumption of data missing at random (Rubin, 1987).

For distributions not included in BUGS, such as the log-logistic, survival analysis involves defining case indices C_i for all sample members and then modelling C_i as Bernoulli with probabilities derived from the relevant likelihood, using exit times and censoring indicators as described above.

Example 9.1 Remission from leukaemia An example of a choice between an exponential and Weibull distributions is provided by remission time data analysed by Gehan (1965), Aitkin *et al.* (1989), McCullagh and Nelder (1989) and others. These data consist of remission times for two samples of 21 leukaemia patients, the first sample treated with an experimental drug and the second assigned a placebo; these are assigned vectors `treat[i]=2` and `treat[i]=1` respectively in Program 9.1. There is therefore a single dummy treatment covariate x, to be evaluated under exponential and Weibull models, with respective hazards

$$h(t) = \lambda \exp(\beta x)$$
$$h(t) = \lambda \alpha t^{\alpha-1} \exp(\beta x).$$

Comparison of these models shows a stronger treatment effect under the Weibull option. The hazard ratio for the untreated group is then, using *Program 9.1 Remission Times*, estimated as exp $(1.76) = 5.8$ as compared with the treated group. However, comparing the usual deviance statistics (minus twice the log-likelihood) suggests only a small gain in goodness of fit in departing from the exponential; the average deviance falls by only about 3 at the cost of an extra parameter.

Adopting a discrete prior on the Weibull parameter α is another way to assess the value of departing from the exponential. For example suppose we assume 20 possible values of α from 1 to 1.95, with equal prior probability of 0.05. A run of 15 000 iterations, after a burn-in of 2500, gives 1.3 % taking the value 1. This implies a Bayes factor against the default exponential value of $\alpha = 1$ of about 75.

Example 9.2 Survival after nephrectomy Collett (1993) presents survival data from time of diagnosis on patients with kidney tumour, of whom 36 controls received chemotherapy and immunotherapy, but 29 treatment patients additionally underwent nephrectomy (kidney removal). Of interest are the effects on survival of patient age group (ages < 60, 60–70 and over 70) and the treatment by nephrectomy. Collett fits age, treatment and age by treatment interactions. However, given the small number of control patients, interaction effects may be poorly identified (there is only one control patient aged over 70). Collett assumes a parametric Weibull model with a homogenous survival time effect over all patients, namely

$$h(t) = e^{\beta x} \alpha t^{\alpha-1},$$

and obtains the value $\alpha = 1.55$. This means that patients at increasing times after diagnosis are more likely to die.

Here minimally informative priors are adopted for the overall constant c, and for the free age and treatment effects A_i and $B_j (i = 2, 3; j = 2)$, where $B_2 = 1$ for treated cases (see *Program 9.2 Nephrectomy*). A gamma G(1,0.001) prior is adopted for the Weibull parameter. After a burn-in of 5000 iterations, a further 5000 iterations yields the posterior mean for α of 1.46 with a 95% interval from 1.07 to 1.89.

Of the age effects, only that of age over 70 is apparently positive, with the parameter A_3 averaging 1.58; so older age increases the chance of mortality by around 5 times compared with younger patients. The treatment effect (B_2) is estimated to average −1.66 with a 95% interval −2.72 to −0.59; so subject to the small sample size in the comparison, nephrectomy reduces the chance of early mortality. The deviance of this model is estimated at 198.7 (with a 95% interval from 190.7 to 210.3).

Example 9.3 Aluminium coupon lifetimes: A reliability application An illustration of gamma survival analysis is provided by the lifetimes of 101 strips of aluminium coupon (Birnbaum and Saunders (BS), 1958). The strips were subjected to periodic loading till all failed. The shape parameter of the gamma density $G(\gamma, \lambda)$ is estimated here as $\gamma = 9.69$ with 95% interval 7.5 to 12.5 (see *Program 9.3 Aluminium Coupon*). The parameter λ has a small positive value and we adopt a prior $b \sim N(-10, 1/V)$ where

$$\log(\lambda) = b.$$

Taking $V = 0.0001$ gives an estimated b of −5. BS obtain a larger value of γ (of 11.8) and a smaller value of $\lambda = e^b$. Comparison of observed and fitted cumulative distribution functions shows a reasonably close fit. This may be verified by prediction of new survival times *t.new* from the posterior lifetime density and cumulating them for intervals of $t = 50$ from 350 to 2800.

Example 9.4 Vietnam veterans lung cancer survival Aitkin *et al.* (1989) and Venables and Ripley (1999) both analyse survival time data on 137 lung cancer patients from a Veterans' Administration Lung Cancer Trial. The available covariates are treatment (standard or test), cell type (1 squamous, 2 small, 3 adeno, 4 large), the Karnofsky score of health state (higher for less ill patients), MFD (months from diagnosis) and PT (prior therapy, 1 = No, 2 = Yes).

Aitkin *et al.* apply a Weibull hazard because of an apparent positive relationship between log hazard and log time under a piecewise exponential model. They estimate the Weibull parameter as $\alpha = 1.08$, suggesting that an exponential distribution for survival times is in fact suitable. Their final model includes the Karnofsky score, cell type, prior therapy and an interaction between the Karnofsky score and prior therapy. This model is estimated with minimally informative priors on the covariate effects and a gamma prior $G(1, 0.001)$ on the Weibull parameter (see *Program 9.4 Cancer Veterans*).

The Weibull parameter is estimated to average 1.12 with posterior standard deviation 0.08 and 95% interval from 0.97 to 1.28. So an exponential might on this evidence be considered appropriate. The posterior summaries of the other parameters generally replicate those of Aitkin *et al.*, though the prior therapy effect averages 2.04, exceeding that of 1.76 in Aitkin *et al.* See Table 9.1.

Example 9.5 Coalmining disaster intervals Lindsey (1992) presents data on intervals in days $\{t_i, i = 1, \ldots, 190\}$ between UK coalmining disasters between 1851 and 1962. These are therefore fully recorded without censoring. We model them as due to a gamma density, adopting the form

$$f(t \mid \nu, \mu) \propto t^{\nu-1} \exp(-t/\mu)$$

where μ is denoted the scale parameter, and ν as the shape parameter. A posterior density of ν spanning the value $\nu = 1$ suggests that an exponential distribution is appropriate. If we fit the same model over the entire span of years we obtain a value ν different from 1, with posterior credible interval (0.61,0.86).

Suppose, however, we model the disaster rate $1/\mu_i$ as a function of covariates, specifically a shift variable S_i defined as 1 after the 125th disaster, and $S_i = 0$ for $i < 125$. Then the disaster rate is lower in the second period with estimated model (from the second half of 10 000 iterations using model B in *Program 9.5 Coalmine Disasters*) being

$$\begin{aligned} 1/\mu_i = &\ 0.0082 \quad - \quad 0.0058 S_i \\ &(0.00097) \quad (0.00089) \end{aligned}$$

with the 95% interval for ν now (0.79,1.10).

The exponential nature of the renewal process is confirmed if lags in $1/t_{i-1}, 1/t_{i-2}$, etc. are added to the model for $1/\mu_i$. A single lag in $1/t_{i-1}$ and a model (model C in Program 9.5) defined for intervals $i = 2, 190$ yields a nonsignificant lag effect, and increases the negative coefficient on S_i as follows

$$\begin{aligned} 1/\mu_i = &\ 0.0087 \quad - \quad 0.0062 S_i - 0.00161/t_{i-1}. \\ &(0.0012) \quad (0.0011) \quad (0.0019) \end{aligned}$$

Table 9.1 Posterior means, Vietnam veterans survival model (iterations 500–5000 of single run).

Parameter	Mean	SD	2.5%	97.5%
Karnofsky	−0.25	0.06	−0.36	−0.14
Prior therapy (PT)	2.04	0.64	0.78	3.35
Cell type 2	0.72	0.24	0.24	1.18
Cell type 3	1.17	0.29	0.59	1.73
Cell type 4	0.30	0.27	−0.23	0.82
Karnofsky-PT interaction	−0.34	0.11	−0.57	−0.13
Weibull time	1.12	0.08	0.97	1.28

9.3 NON-MONOTONIC HAZARDS

Many types of survival analysis refer only to events with monotonic trends over time, with the hazard rate either increasing at higher values of t or decreasing. Thus the Weibull model, and related models used in life table modelling (e.g. the Gompertz), assume that the probability of an event is a monotonic function of time. However, these models are inappropriate if the probability of an event at first increases but after reaching a maximum value tails off again, possibly to zero. For example, job stays may approximate this pattern; people in jobs for a very short time show a low rate of mobility, but their rate of job change increases initially as they accumulate transferable skills; however, people who stay in jobs a long time are increasingly unlikely to change.

Among parametric models which can be applied to such behaviour are the log-logistic model and the sickle model (Bennett, 1983; Diekmann and Mitter, 1983; Li, 1999) The former model has a hazard and survivor rates

$$h(t) = \alpha(t/\theta)^{\alpha-1}/[\theta + t^{\alpha}\theta^{1-\alpha}]$$
$$S(t) = 1/[1 + t^{\alpha}\theta^{-\alpha}]$$

where α and θ are both positive. The sickle model has rates

$$h(t) = cte^{-t/\lambda}$$
$$S(t) = 1 - \exp[-\lambda c\{\lambda - (t + \lambda)\exp(-t/\lambda)\}]$$

with both c and λ positive. These models contain counteracting influences: one part of the hazard function rises with time, the other diminishes. An example of a Bayesian application of the log-logistic model is provided by the analysis of Li (1999) on exits by firms from bankruptcy, and involving a Laplace approximation approach.

Example 9.6 Austrian occupational mobility Diekmann and Mitter (DM, 1983) present data from an Austrian occupational mobility study on times of moves in years (+ part years) for $N = 2660$ respondents. Thus 53 persons moved in their first year with average stay (assumed) of 6 months, 252 moved in year 1 with average stay 18 months and so on. For 118 individuals we know they moved at length of stay 17.5 or more years, and 422 individuals were censored (they had not changed jobs, and their stays exceeded 17.5 years). We take a 10% sample ($N = 266$) of these individuals, with 54 persons either censored at 17.5 years or known to have moved at more than 17.5 years of stay, but with time of move not exactly specified. We treat all 54 as censored at 17.5 years.

DM point to the nonmonotonic job change intensity among these individuals and use both sickle and log-logistic model to describe it. Since the log-logistic is not available to sample from in BUGS, we use the device of assigning dummy indices of 1 to all 266 cases, and then modelling these indices as Bernoulli with likelihood probabilities π_i determined by the log-logistic model. Thus for the 'mover' cases 1–212 the π_i are equal to $h(t_i)S(t_i)$ and for 'stayer' individuals 213–266 the π_i are given by $S(t_i)$. This will involve Metropolis sampling to find the (approximate) mode of the nonstandard likelihood, or more accurately sample the density around it.

We obtain, using *Program 9.6 Occupational Mobility* results virtually the same as in DM with $\alpha = 1.48$ (95% interval 1.32 to 1.65) and with $\theta = 7.4$ (with a single run of 10 000 iterations and 500 burn-in). The first parameter governs the ascent, the second the descent from the maximum. The fitted hazard for years of stay 0.5 to 50.5 shows the nonmonotonic pattern of movement. The total of job moves $H(\infty) = \int_0^\infty h(u)du$ is approximated by the trapezoid integration rule up to stay lengths of 50.5 years and averages 2.8.

9.4 ANALYSING COMPETING RISKS IN CONTINUOUS TIME

The modelling of survival or event times can be extended to processes with several possible causes of exit or death. In human mortality applications we may be interested in competing causes of death (e.g. cardiovascular diseases, cancers and other causes) (Lai and Hardy, 1999). More generally in event histories we may frequently be interested in rates of movement of different types or change to several different destinations (Hachen, 1988). In applications to human behaviour (e.g. migration, job mobility) the destinations may be alternative distance bands, occupation groups and so on. In the latter case it may be possible to effectively never move (i.e. be a permanent 'stayer').

Let J_i be the cause of exit or type of move, with $1, \ldots, C$ possible states, where causes or types are mutually exclusive and exhaustive. Then the survival process governing transitions to states $j = 1, \ldots, C$ for individual i is specified by a destination-specific hazard

$$h_j(t_i)dt = \Pr(t_i \leq T < t_i + dt, J_i = j | T \geq t_i) \tag{9.4}$$

with the total hazard, assuming independence between destinations, given by

$$h(t_i) = \sum_{j=1}^{C} h_j(t_i).$$

This simply expresses the fact that exit or change, if it occurs, is to one of the C causes or destinations. The survivor or stayer function is then

$$\begin{aligned} S(t_i) &= \exp[-\int_0^{t_i} h(u)du] \\ &= \exp[-\int_0^{t_i} h_1(u)du - \int_0^{t_i} h_2(u)du - \ldots - \int_0^{t_i} h_c(u)du]. \end{aligned} \tag{9.5}$$

The density $f_j(t_i)$ governing waiting times till the jth types of exit or destination is therefore

$$f_j(t_i)dt = \Pr(t_i \leq T < t_i + dt, J_i = j \,|\, T \geq t_i)\Pr(T \geq t_i)$$
$$= h_j(t_i)S(t_i)$$

The likelihood is taken over both individuals and all possible causes with censoring indicators $\delta_{ij} = 1$ if individual i exits for cause j, and $\delta_{ij} = 0$ otherwise. For an individual with $\delta_{ij} = 1$ survival times on other possible causes than j are regarded as censored. A 'stayer' is censored on all possible causes. This leads on to a latent failure time interpretation where the observed waiting time is the minimum of C possible latent failure times.

If covariates are available their effects may be differentiated according to type of move or exit. A general destination dependence model (Davies, 1983) then also specifies the parameters of the hazard to be destination specific. For example, if we have a Weibull hazard, then we would parameterise it as

$$h_j(t_i) = \alpha_j t_i^{\alpha j - 1}\exp(\beta_j x_i).$$

In behavioural applications, the competing risk model is most sensible when the decision to leave the current state and the choice of the future state are interdependent – for example, in voluntary job exits (Hachen, 1988).

Example 9.7 Competing risks in occupational mobility Seeber (1984) reports on a competing risks analysis of occupational history data obtained in September 1972 from 80 000 Austrian residents. A subsample consisted of 322 males who started their career between 1950 and 1960 as an employee with medium skill activities, and Seeber's focus was on the first change of occupation, and on its direction: whether upwards or not in status terms. A single covariate was used to predict advancement, namely a binary variable x coded 1 for a person with university or college degree, and 0 otherwise. The hazard rate model Seeber estimated was of the form

$$h_j(t_i) = \alpha_j t_i^{\alpha_j - 1}\exp(b_{1j} + b_{2j}x_i), \quad j = 1, 2$$

where t denotes waiting time in years before a move, $j = 1$ for advancement and $j = 2$ for a change of job with no advancement. Of the 332 people in the original sample, 134 experienced upward moves, 75 other sorts of move and 123 were censored.

Using the data provided by Seeber on the relationship between mobility and education, and the estimated parameters in the cause-specific hazard functions, parallel data were generated via a single run. The parameter estimates from the original study were (with standard errors)

$$\alpha_1 = 1.29 \ (0.07),$$
$$\alpha_2 = 0.99 \ (0.05),$$
$$b_{11} = -4.197 \ (0.21),$$
$$b_{21} = 0.617 \ (0.18),$$
$$b_{21} = -3.561(0.19),$$
$$b_{22} = -0.336(0.24).$$

Thus upward mobility is positively related to tertiary education, and shows a stronger duration effect above the default value of 1 (this default being equivalent to an exponential distribution of occupational waiting times).

The generation of parallel sample 'data' assumes a uniformly random entry point during 1950 to 1960 and uses the education data, as in the covariate x, as supplied by Seeber. To generate waiting times for upward and other moves requires sampling from Weibull densities with appropriate Weibull parameters α_1 and α_2 and regression parameters. The type of move (`J[]` in the code below) depends on which is the minimum of t_{i1} (time to advancement) and t_{i2} (time to other move). If both these times exceed the maximum possible (`Tmax[]`) given the date of the survey and the entry year, then the person's history is censored. The coding is:

```
{for (i in 1:332) {entyr[i] ~ dunif (1950,1960)
                   Tmax[i] < - 1972-entyr[i]
                   mu[i,1] < - exp(-4.197+0.6165*z[i])
                   mu[i,2] < - exp(-3.561-0.3355*z[i])
                   t[i,1] ~ dweib(1.2936,mu[i,1])
                   t[i,2] ~ dweib(0.9901,mu[i,2])
                   T[i] < - min(t[i,1],t[i,2])
# select destination according to min surv time
        J[i] < -2*step(t[i,1]-t[i,2])+step(t[i,2]- t[i,1])
        cen[i] < - step(T[i]-Tmax[i])
# lower limit for Weibull sampling (for stayers or non-chosen
destination)
        D[i] < - equals(cen[i],1)*Tmax[i]+equals(cen[i],0)*T[i]}}
```

In the parallel data generated as above, 102 persons were censored (i.e. were 'stayers'), 140 had upward moves and 89 had other types of move.

To then estimate in BUGS the parameters appropriate to these parallel data we need to create censored and observed waiting times for each of the competing risks. Persons making an upward move are censored from the viewpoint of other types of move. So we define `t.adv[]`, `t.adv.cen[]`,`t.noadv[]`, `t.noadv.cen[]`. For the 102 censored persons both `t.adv.cen` and `t.noadv.cen` are set to the maximum possible time between year of entry and 1972. For the 140 upward movers, `t.adv.cen` is set to zero, `t.noadv.cen` is set to their observed waiting time before the upward move, `t.adv` is set also to the observed waiting time, and `t.noadv` is missing.

Our estimates for the parallel data (from iterations 501–5000 of a single run) are made using *Program 9.7 Occupational Advancement*. They are given in Table 9.2.

They show a clear education effect but less contrast in the waiting time parameters $\alpha_j, j = 1, 2$ – this may be largely a chance outcome from the single sample selected.

Example 9.8 Types of death in the Stanford Heart Transplant Study The competing risk model may be illustrated with the Stanford Heart Transplant

Table 9.2 Posterior estimates: Occupational advancement survival analysis.

Parameter	Mean	SD	2.5%	Median	97.5%
b_{11}	−4.26	0.28	−4.81	−4.27	−3.69
b_{12}	0.50	0.18	0.15	0.50	0.85
b_{21}	−4.02	0.34	−4.76	−4.00	−3.44
b_{22}	−0.53	0.23	−0.96	−0.52	−0.10
α_1	1.30	0.10	1.11	1.31	1.50
α_2	1.23	0.12	1.02	1.22	1.48

data relating to 65 patients (Crowley and Hu, 1977). Survival times after transplant have previously been analysed in terms of prior open-heart surgery (SURG=1 or 0), age at transplant (AGE) and a mismatch score (MM) relating to the concordance between recipient and donor tissue type. Following Aitkin *et al.* (1989) we distinguish two types of death: those associated with rejection of the new heart and other deaths. There are 41 deaths and 24 censored survivals. As noted by Aitkin *et al.*, the hazard function of the rejection deaths shows a nonmonotonic pattern, with a rapid rise between 10 and 65 days, but a constant hazard from around 130 days. The nonrejection deaths by contrast have a steadily declining hazard with the log of time after transplant, suggesting a Weibull density.

Suppose we initially adopt a competing risks model as above (see model A in Program 9.8):

$$h_j(t_i) = \alpha_j t_i^{\alpha j-1} \exp(\beta_j x_i), \quad j = 1, 2.$$

This is an obvious simplification for the rejection deaths, but shows the effects of age and mismatch score on such deaths to be significantly positive, and that of previous surgery to be negative (results are from the second half of a run of 10 000 iterations).

The decline in deaths with time among the remaining deaths is apparent in a Weibull parameter of 0.36, though none of the covariate effects are well identified with the vague priors adopted. There may be grounds for example in positing a positive effect of mismatch on reject deaths, but a negative effect on nonreject deaths.

As a step to greater realism we introduce a 'crisis' covariate, modelling the period at which the reject group appear at greatest risk of dying early (model B in Program 9.8). We include this as a covariate only for the reject deaths. We allow the start of the crisis to be uniform between 0 and 50 and the end of the crisis to be up to 20 days later. The fact that the form of the covariate is being simultaneously identified and a regression parameter is then applied to it will reduce its regression 'significance'; we could alternatively define a dummy variable with preset lower and upper limits. Here the lower limit is estimated reasonably precisely as 45 days and the upper limit as 65–70 days. The Weibull parameter for the rejection group now exceeds 1, with a posterior mean around 1.6.

Example 9.9 Use of second homes for retirement Davies (1983) applies the destination-specific Weibull model to tenure changes by second-home owners in Wales. Of 493 initial second-home owners, $n_1 = 26$ retired to these homes within a 16-year follow-up period (i.e. exited from second home status), and $n_2 = 124$ relinquished them by selling or other means. The remaining 343 were still owned as second homes. There are no covariates. Davies compares a destination dependent model (model 1)

$$h_j(t_i|x_i) = \alpha_j t_i^{\alpha_j - 1} \lambda_j, \quad j = 1, 2$$

($j = 1$ for retirees, $j = 2$ for sellers) with a destination independent model (model 2)

$$h_j(t_i|x_i) = \alpha t_i^{\alpha - 1} \lambda_j, j = 1, 2$$

and finds little loss of fit in moving to the restricted model with destination independence.

If the independence model applies and the duration of second-home tenure is independent of outcome, then the retirement proportion can be modelled as a binomial

$$n_1 \sim \text{Bin}(p_1, n_1 + n_2).$$

If, however, the duration effects are unequal, with $\alpha_1 \neq \alpha_2$, such that (for example) second homes intended for retirement were retained longer, then the effect of censoring would be that observed retirements underestimate the retirement proportion. In fact the data are rather sparse to reach firm conclusions on this, and seem to show nonmonotonic dependencies in the hazard rates for both outcomes. The assumed priors may influence conclusions regarding the hypothesis of independence.

However, we follow Davies in adopting a framework allowing for a model choice, and assuming monotonic time effects. Thus take the first group (retirees) to have time dependence

$$\alpha_1 \sim E(1)$$

and

$$\alpha_2 = \alpha_1 + \delta,$$

where δ is assigned a discrete prior with 17 possible values, ranging from -0.4 to 0.4. The ninth value is zero, corresponding to strict destination independence. The resulting estimates of the duration effects broadly support the dependence hypothesis in the Weibull parameters. The second half of a run of 5000 iterations using *Program 9.9 Tenure Changes* leads to estimates of α_j and λ_j in Table 9.3. The estimates with this prior show a stronger positive duration effect on the changes to retirement homes.

An alternative approach uses a $N(0, 1)$ prior on $\log(\alpha_1)$ and takes

$$\log(\alpha_2) = \log(\alpha_1) + \delta G_1 + (1 - \delta) G_2$$

Table 9.3 Exits from second homes.

	Mean	SD	2.5%	97.5%
Discrete Prior				
δ	−0.247	0.054	−0.350	−0.200
λ_1	0.0022	0.0007	0.0012	0.0041
λ_2	0.0176	0.0038	0.0118	0.0266
α_1	1.66	0.10	1.44	1.82
α_2	1.42	0.09	1.21	1.58
Beta Prior				
λ_1	0.0039	0.0017	0.0015	0.0075
λ_2	0.0150	0.0038	0.0088	0.0240
α_1	1.40	0.19	1.11	1.77
α_2	1.49	0.11	1.28	1.71

where δ has a beta prior, G_1 is positive and G_2 is negative. This approach is less conclusive. Table 9.3 shows the results from iterations 5000–20 000 of a single run with this prior (note that Metropolis sampling is appropriate to this model and BUGS only samples from iteration 4000).

9.5 VARIATIONS IN PRONENESS: MODELS FOR FRAILTY

Comparison of event histories and survival times between members of a population may well suggest heterogeneity among them in their underlying risk. The latter source of variability is variously known as proneness, susceptibility, or frailty. Thus in medical studies with death or relapse as an end-point, some patients will survive or stay healthy relatively long despite adverse observable risk factors, whereas some will survive shorter than expected. This heterogeneity may distort the estimation of the hazard or exit rate so that it does not represent the hazard of any particular individual or subgroup in the population.

Suppose we envisage a relatively few subpopulations defined by different levels of frailty (risk of undergoing the event or events concerned). For survival data (with an absorbing exit) the more frail will tend to undergo the event earlier, so that with increasing time the overall hazard rate will descend to that of the subgroup with the lowest frailty. So the observed hazard rate may decline even though the hazard rates for the subpopulations are constant. Yashin *et al.* (1985) give the simplest possible example of a population with only two subgroups, one with hazard rate $h_1(t)$ and the other with rate $h_2(t)$. The survivorship rates for the two groups will then be

$$S_i(t) = \exp[-\int_0^t h_i(u)du], i = 1, 2.$$

If $p_1(0)$ and $p_2(0)$ denote the initial proportions in the population with $p_1(0) + p_2(0) = 1$, then the proportion of the surviving cohort at time t that comes from the first subgroup is

$$p_1(t) = p_1(0)S_1(t)/[p_1(0)S_1(t) + p_2(0)S_2(t)].$$

The hazard rate for the entire cohort at time t will then be

$$h_e(t) = p_1(t)h_1(t) + p_2(t)h_2(t).$$

If the first subgroup is the more robust then it will come to dominate the population hazard rate. For example, the recidivism rate after release from prison may decline as time from release increases. This may be because of subgroups with varying chances of re-offending, but each with a constant hazard rate.

A simple model for such subgroups is the mover–stayer model: one subpopulation has high risk of divorce, migration, job change, or disease and another has zero or low risk. If the hazard rate for the susceptible groups is increasing then the observed hazard for the entire population may at first rise and then fall. More generally, we can model unobserved variations in frailty by using discrete mixtures or by assuming a continuous density for frailty.

One impact of neglected heterogeneity is that covariate effects may be estimated too precisely. For example in mortality analysis if risks are not homogenous over individuals, then the effect of age on individual mortality is greater than that on population mortality.

Heckman and Singer (1984) present evidence that using different parametric mixture options (normal, gamma, etc.) in the face of unobserved heterogeneity in survival analysis leads to different inferences about economic parameters. They argue that the standard assumption of random effects for unobserved heterogeneity is impeded by the absence of theoretical knowledge regarding an appropriate functional form and that random effects models overparameterize econometric duration models. They instead use artificially generated data to demonstrate that a nonparametric mixture approach reliably identifies the structural model parameters (the duration effect and the effects of economic covariates).

They consider samples of observations generated with a Weibull hazard and single covariate x_i, of the form

$$h(t_i) = \alpha t_i^{\alpha-1}\exp(\beta x_i + \theta_i) \tag{9.6}$$

with the unobserved heterogeneity θ_i generated as a multinomial mixture or via a symmetric density (e.g. the normal), or with $\phi_i = \exp(\theta_i)$ generated via a gamma density, and with the durations t_i generated without truncation or censoring. The heterogeneity may be at cluster level, θ_{G_i} with G_i denoting the cluster for subject i. With some analogies to the DPP prior methodology, they then use a nonparametric strategy with a large number (e.g. 10) of potential classes to regenerate the structural parameters (α, β). They show that such a procedure robustly estimates main objects of theoretical interest, namely the structural economic parameters and the cumulative density of the durations, whatever assumptions were made about the underlying density of θ_i. However, the mixing distribution itself is not well estimated.

Example 9.10 Survival after nephrectomy (continued) Examination of the survival data following nephrectomy suggests considerable heterogeneity in survival chances. For example, among the four highest times, namely those over 100 months, two are censored (i.e. the individuals were still alive).

A multiplicative random effects form for this heterogeneity

$$h(t \mid \phi, x) = \phi e^{x\beta} h(t)$$

is adopted with $\phi \sim G(\omega, \omega)$ and $\omega \sim E(1)$. Thus the relative frailties average 1 and have variance 1 also.

With null starting values in Program 9.10, convergence is obtained after some 15 000 iterations. The second half of a run of 30 000 iterations shows the mixture parameter, monitored in the form $\log(\omega)$, estimated to average 0.82 (with a 95% interval from 0.22 to 2.5). The Weibull time parameter and the treatment effect appear less strongly identified as compared with the homogenous model of Example 9.2. The former still shows mortality increasing with time from diagnosis, and the 95% interval now runs from 1.55 to 4.0.

The treatment parameter is negative at −2.35 (with a 95% interval from −4.4 to −0.6), and increased in absolute size compared with the analysis above without a frailty effect.

Median survival times for age and treatment combinations may be monitored directly or using the analytic result for the median

$$T_{ij}(50) = [\log 2/(\lambda \exp\{A_i + B_j\})]^{1/\alpha}$$

where $\lambda = \exp(c)$. The longer survival times of nephrectomy patients are apparent (see Table 9.4).

Example 9.11 Veterans lung cancer survival (continued) To allow for heterogeneity in survival, a discrete mixture is an alternative approach and allows

Table 9.4 Nephrectomy, posterior summaries including median survival times by patient category (frailty model).

			Credible Interval	
	Mean	SD	2.5%	97.5%
Treatment effect	−2.34	0.94	−4.44	−0.58
Weibull parameter	2.50	0.61	1.55	4.02
Gamma mixture parameter	0.85	0.60	0.22	2.46
Median survival times				
Ages < 60 and no nephrectomy	12.57	4.67	5.77	23.77
Ages < 60 and nephrectomy	31.93	9.19	15.53	51.63
Ages 60–70 and no nephrectomy	10.45	4.01	4.64	19.91
Ages 60–70 and nephrectomy	26.74	9.00	13.16	48.17
Ages > 70 and no nephrectomy	4.09	2.05	1.29	9.11
Ages > 70 and nephrectomy	10.18	4.07	4.24	20.02

flexibility in terms of extension to include mixing on the time and regression parameters, as well as just the level. Here only the intercept (i.e. the overall level of frailty) is allowed to vary between groups, and a two-group mixture is adopted (see model A in Program 9.11). A Dirichlet prior on the mixing proportions P_1 and P_2 is used with equal prior weights of 1 on each group.

The second half of a run of 10 000 iterations leads to estimates of $P_1 = 0.25$, $P_2 = 0.75$ with $c_1 = -6.36$ and $c_2 = -4.26$. So a smaller low-mortality group is distinguished. The Weibull parameter becomes more clearly above one, with an average of 1.44 and 95% interval from 1.14 to 1.83. Among the covariate effects, the impact of the Karnofsky score in particular is enhanced (see Table 9.5).

Model B in Program 9.11 illustrates the extension to mixing on all parameters. With this relatively small data set this extension is not well identified.

Example 9.12 Stanford Heart Transplant Study Turnbull (1974) and Tanner (1996) analyse data from the Stanford Heart Transplant Study for survival in days after transplant for 45 patients. The available data included survival status at February 1980, age in years at the time of the transplant and a mismatch score comparing the donor and recipient tissues with regard to HLA antigens.

Table 9.5 Veterans cancer data, posterior summaries of parameters.

	Mean	SD	2.5%	Median	97.5%
Single-group model					
Constant	−4.257	0.545	−5.347	−4.257	−3.133
Karnofsky score	−0.255	0.058	−0.368	−0.256	−0.136
Prior therapy (PT)	1.945	0.652	0.654	1.959	3.206
Small cell type	0.715	0.247	0.229	0.720	1.201
Adeno cell type	1.163	0.291	0.578	1.166	1.726
Large cell type	0.298	0.265	−0.238	0.303	0.811
PT × Karnofsky	−0.320	0.110	−0.534	−0.322	−0.103
r (Weibull parameter)	1.110	0.074	0.961	1.111	1.254
Two-group model					
Probability (grp 1)	0.262	0.144	0.05572	0.2447	0.6399
Probability (grp 2)	0.737	0.144	0.3601	0.7553	0.9443
Constant (grp 1)	−6.35	0.902	−8.076	−6.376	−4.526
Constant (grp 2)	−4.25	1.083	−5.699	−4.299	−2.94
Karnofsky score	−0.430	0.116	−0.6473	−0.4295	−0.2216
Prior therapy (PT)	2.326	0.901	0.6359	2.301	4.173
Small cell type	0.813	0.365	0.1428	0.7914	1.60
Adeno cell type	1.041	0.407	0.2212	1.046	1.844
Large cell type	0.096	0.401	−0.7514	0.1188	0.8414
PT × Karnofsky	−0.366	0.147	−0.6477	−0.3608	−0.1013
r (Weibull parameter)	1.430	0.184	1.078	1.437	1.771

Turnbull *et al.* propose that each patient has an exponential lifetime (constant hazard) on starting treatment, but with the rate varying between patients according to a gamma density.

A homogenous exponential model (model A in Program 9.12) shows that both age and mismatch score are associated with higher and earlier risk of death. Since three mismatch scores are missing it is assumed that the mismatch scores are drawn from a common density; this is a 'missing at random' approach to providing estimates or imputations of the missing values. Allowing for gamma heterogeneity with mean 1 and variance d leads to a clear gain in fit (measured by the average deviance).

This generalisation is implemented in model B in Program 9.12. It is also apparent that the impact of the mismatch score is much attenuated when gamma frailty is introduced. The variance d is estimated at 1.54, with 95% interval from 0.61 to 2.73.

Example 9.13 Non-parametric frailty to estimate structural parameters As noted above, Heckman and Singer (HS, 1984) propose a discrete mixture (nonparametric) strategy with a large number (e.g. $C = 10$) of potential classes to represent frailty and so to regenerate the structural parameters (duration and regression coefficients) regardless of the actual frailty mechanism. This involves postulating different intercepts for each latent frailty class, but not usually differences on other parameters – so there is still a common duration parameter α, and common effects of covariates, but the intercepts are differentiated by latent class, $\beta_{0j}, j = 1, \ldots, C$.

They argue that such a procedure robustly estimates the main objects of theoretical interest, namely the structural economic parameters and the cumulative density of the durations, whatever assumptions were made about the underlying frailty in generating the data. However, the mixing distribution itself is not well estimated and indeed it is not the intention of the procedure to reproduce it.

Here we illustrate this approach by generating 500 observations from a Weibull hazard

$$h(t|x, \alpha, \beta) = \alpha t^{\alpha-1}\exp(\gamma + \beta x + \theta)$$

with $\alpha = 2, \gamma = 0, \beta = 0.5$ and with a single covariate x, whose values are drawn from a standard normal. The unobserved variations θ are generated under two possible mechanisms: first, a multinomial mixture, with two classes and weights $P_1 = 0.1$, $P_2 = 0.9$ such that

$$S_i \sim \text{categorical}(P_{1:2}), \quad \theta_1 = 0.25, \theta_2 = 0.75.$$

The second assumption is that $\theta \sim \text{G}(1.5, 1)$ as in HS (1984, Table II). The programs used to generate the observations are listed in Program 9.13. For example, the code for the multinomial sampling model is as follows

```
model {for (i in 1:500) { s[i] ~ dcat(p[1:2])
                          x[i] ~ dnorm(0,1)
                          t[i] ~ dweib(2,mu[i])
mu[i] <- exp(0.5*x[i] + theta[s[i]])}
p[1] <- 0.1; p[2] < - 0.9
theta[1] <- 0.25; theta[2] <- 0.75}
```

Presented only with the data thus sampled, we aim to reproduce the structural parameters (the Weibull parameter and the covariate effect) using a mixture approach on θ with $C = 5$ potential classes, assigned equal prior Dirichlet weights of one. The observed data from the above assumptions are durations t and covariate values x. We predict new durations *t.new* using the estimated (α, β, θ) at each iteration and then compare the resulting forecast of the cumulative density of durations with the observed cumulative density. For the data generated under the multinomial option for the unobserved effects, a single run of 10 000 iterations (with 500 burn-in) gives estimates, with 95% credible intervals

$$\alpha = 2.29\ (1.87, 2.82)$$
$$\beta = 0.54\ (0.39, 0.71).$$

The duration parameter is slightly overestimated. The last 500 iterations are used to sample from the predicted cumulative density of the new durations *t.new*, in time intervals up to 0.1, 0.2, 0.3, ...,2.5. These show a close match to the observed cumulative density of durations, taking breaks 0.1, 0.2,. . .up to 2, as in Table 9.6.

For data generated under the gamma heterogeneity option the estimate of the duration slope is close to the true value of 2, but the regression parameter is underestimated, at just under 0.4. Here the range of observed durations t is up to 5 and a close match is again obtained.

9.6 COUNTING PROCESS MODELS

In survival analysis for recurrent events or events of different types an alternative framework is provided by counting process models, first used by Aalen (1976) and further discussed in Fleming and Harrington (1991) and Andersen *et al.* (1993). These extend the nonparametric hazard approach of the Cox regression model, but reformulate the observed data in such a way as to enhance the link of survival modelling to the broader class of general linear models and to facilitate the formation of model residuals.

Consider a time W until an event of a certain type occurs, and a time Z to another event or end of follow-up (including censoring and withdrawals). Then the observations for a case consist of a duration or survival time $T = \min(W, Z)$, and an event type indicator, with $E = 1$ if $T = W$ and $E = 0$ if $T = Z$. The functions of concern in the counting process approach are the counting process $N(t)$ itself

Table 9.6 Predicted and observed cumulative density of durations.

Duration	Observed	Predicted	SD	2.5%	Median	97.5%
0.1	0.038	0.034	0.010	0.016	0.034	0.054
0.2	0.106	0.111	0.016	0.076	0.112	0.140
0.3	0.192	0.208	0.022	0.162	0.210	0.246
0.4	0.315	0.323	0.026	0.270	0.322	0.372
0.5	0.423	0.443	0.026	0.388	0.444	0.502
0.6	0.543	0.555	0.024	0.500	0.556	0.602
0.7	0.638	0.655	0.023	0.608	0.656	0.702
0.8	0.725	0.736	0.022	0.688	0.738	0.772
0.9	0.818	0.802	0.020	0.756	0.806	0.838
1	0.857	0.854	0.018	0.814	0.856	0.886
1.1	0.894	0.893	0.015	0.866	0.896	0.920
1.2	0.923	0.922	0.014	0.894	0.924	0.946
1.3	0.957	0.943	0.012	0.916	0.944	0.968
1.4	0.967	0.959	0.010	0.936	0.960	0.980
1.5	0.975	0.971	0.008	0.952	0.972	0.984
1.6	0.982	0.978	0.007	0.962	0.978	0.990
1.7	0.99	0.984	0.006	0.968	0.984	0.994
1.8	0.991	0.989	0.006	0.976	0.988	0.998
1.9	0.991	0.992	0.005	0.982	0.992	1.000
2	0.995	0.994	0.004	0.984	0.994	1.000

$$N(t) = I(T \leq t, E = 1)$$

and the at-risk function

$$Y(t) = I(T < t)$$

where $I(U)$ is the indicator function with value 1 if U is true and 0 otherwise. These two functions simple re-express the information contained in the survival times T_i and event (or censoring) indicator E_i. So the observed event history for subject i is $N_i(t)$, denoting the number of events which have occurred up to continuous time t. If only a single event (e.g. mortality) can be observed, then $N_i(t)$ has value 0 until the event is observed and value 1 thereafter. Let $dN_i(t)$ be the increase in $N_i(t)$ over a very small interval $(t, t + dt)$, such that $dN_i(t)$ is (at most) 1 when an event occurs and zero otherwise.

The expectation of the increment in $N(t)$ is given by the intensity function

$$L(t)dt = Y(t)h(t)dt$$

where $h(t)$ is the usual hazard rate defined by

$$h(t)dt = \Pr(t \leq T \leq t + dt, E = 1 | T \geq t).$$

In the counting process approach, it is the intensity function that is modelled as a function of possibly time-specific covariates, rather than the conditional hazard as in traditional survival analysis. The intensity process is analogous to an expected number of binomial events at time t, with $Y(t)$ the number at risk and $h(t)$ the probability of an event. The predicted total of events to time t is obtained by integrating the intensity process, giving a cumulative intensity process A (or compensator process):

$$A(t) = \int_0^t L(u)du.$$

This is used in defining residuals $M(t) = N(t) - A(t)$ between actual and predicted cumulative events.

If there are covariates then the proportional hazards assumption gives the intensity model

$$L(t) = Y(t)\lambda_0(t)\exp(\beta x)$$

where $h(t) = \lambda_0(t)\exp(\beta x_i)$ is the usual Cox proportional hazard model, with baseline hazard $\lambda_0(t)$ modelled nonparametrically. Denoting the integrated hazard by

$$\Lambda_0(t) = \int_0^t \lambda_0(u)du$$

the intensity may be written

$$I_i(t) = Y_i(t)d\Lambda_0(t)\exp(\beta x_i).$$

Hence the intensity function may be parameterised in terms of the integrated hazard Λ_0 and the regression parameter β.

Under a Bayesian approach with β and Λ_0 *a priori* independent, the joint posterior, with data $D = (N_i(t), Y_i(t), x_i)$ is

$$P(\beta, \Lambda_0 \,|\, D) \propto P(D \,|\, \beta, \Lambda_0)P(\beta)P(\Lambda_0).$$

Since the conjugate prior for the Poisson mean is the gamma, it is convenient to adopt a prior for $d\Lambda_0$ as follows

$$d\Lambda_0 \sim G(cH(t), c)$$

where $H(t)$ can be thought of as a guess at the unknown hazard rate per unit time and $c > 0$ is higher for stronger belief in this guess. The mean hazard under this prior is $cH(t)/c = H(t)$ and its variance is $H(t)/c$ which increases as c becomes smaller. Since the prior actually relates to $d\Lambda_0$ rather than Λ_0 itself it is often termed an independent increments prior.

Conditional on β, the posterior for Λ_0 is again of independent increments form on $d\Lambda_0$ rather than Λ_0 itself, namely

$$d\Lambda_0(t) \sim \mathrm{G}(cH(t) + \sum_i dNi(t), c + \sum_i Y_i(t)\exp(\beta x_i)).$$

This model may be adapted to allow for unobserved covariates or other sources of heterogeneity ('frailty'). This frailty effect may be at the level of observations or for some form of grouping variable. For example, the observations i might in fact denote repetitions for a smaller number of individuals, e.g. if $i = 1, 2, 3$ for three repeated events for individual $1, i = 4, 5$ for two repeated events for individual 2, and so on. Alternatively the grouping variable might be institutional (patient survival times grouped by hospital), or be defined by study design with observations paired or otherwise matched, and the matched group being the level at which frailty is modelled.

One feature of the counting process model is the relative ease with which nonproportionality can be assessed by defining suitable time-varying regressors in hazard models of the form

$$hi(t) = \lambda_0(t)\exp\{\beta x_i + \gamma w_i(t)\}.$$

For example, $w_i(t) = x_i g(t)$ could be the product of one or more covariates with a dummy time index $g(t)$ set to 1 up to time t^* and zero thereafter. This is equivalent to proportional hazards if $\gamma = 0$. Cox (1972) proposed a function $g(t) = \ln(t)$, the power of which was investigated by Quantin *et al.* (1996). The main alternative method for assessing proportional hazards involves residuals specific for each case and each covariate.

The counting process approach offers a way to generalise to a regression setting the nonparametric survival estimators such as the Kaplan–Meier. Suppose uncensored waiting or survival times for n cases are arranged in ascending order $t_1, \ldots, t_n$. Then the Kaplan–Meier estimate of the proportion surviving beyond time t_k is estimated as

$$p_k = \prod_{i=1,\ldots,k} (r_i - d_i)/r_i$$

where r_i is the number of subjects still alive just before t_i, and d_i is the number actually dying at t_i. However, when there are censored observations present it is necessary to calculate an effective sample size

$$n'_i = (r_i - d_i)/p_i$$

each time a death occurs (Peto *et al.*, 1977).

Consider the data on 25 colorectal cancer patients treated by γ-linolenic acid and presented by Machin and Gardner (1988). The survival times in months, with censored cases starred, are as follows $\{1^*, 5^*, 6, 6, 9^*, 10, 10, 10^*, 12, 12, 12, 12, 12^*, 13^*, 15^*, 16^*, 20^*, 24, 24^*, 27^*, 32, 34^*, 36^*, 36^*, 44^*)$. So the numbers at risk r_i just before the five distinct death times $\{6, 10, 12, 24, 32\}$ are $r = \{23, 20, 17, 8, 5\}$ and the survivors after these times are $s = \{21, 18, 13, 7, 4\}$. We can reproduce theKaplan–Meier estimate using beta sampling based on the effective sample sizes, in this case via a BUGS coding such as

```
        d[1] <- r[1]-s[1]
        for (i in 2:5) {d[i] <- E[i]-s[i]
# Effective Sample Size
                      E[i]- r[i]/p[i-1]}
        for (i in 1:5) {p[i] ~ dbeta(s[i],d[i])}
```

We can extend this procedure to several strata (e.g. treatment groups). However, a counting process approach generalises more readily to include other explanatory variables, including metrical or time-varying ones.

Example 9.14 Cancer drug trial follow-up We apply the effective sample size method to comparing survival experience between the treatment group just described and a control group of 23 patients. The object is to compare survival at two years (see Program 9.14 Model A) between the two groups. The fourth distinct survival time among the treatment patients was at 2 years, as was the fifth distinct survival time among the controls. Differential survival at two years is estimated as

$$\text{p.treatment} - \text{p.control} = 0.037$$

with a standard deviation of 0.26, showing no advantage to the drug treatment.

By contrast, applying the counting process approach (Program 9.14 Model B) yields a slightly greater value of the survival difference in favour of treatment, namely $p.treatment - p.control = 0.085$ with a standard deviation of 0.14. The overall density of the treatment effect β points towards a survival advantage, but the values of this parameter straddle zero and offer no conclusive evidence.

Example 9.15 Clinical trial of liver disease drug Fleming and Harrington (FH, 1991) present a counting process analysis of clinical trial data concerning a drug treatment for primary biliary cirrhosis (abbreviated PBC), a liver disease. A total of 312 patients were randomised between treatments, and the initial interest is in the impact of the drug treatment in improving survival chances. Following FH, we analyse survival chances for the drug and placebo groups with a single treatment covariate, with $x = 1$ for the drug groups and $x = 0$ otherwise in the model

$$L(t) = Y(t)\ \lambda_0(t)\ \exp(\beta x).$$

Survival times are in days, with the number of distinct and uncensored survival times being 122 (see Program 9.15 Model A). We find, as do FH, no clear treatment effect with a 95% credible interval for β straddling zero.

It should be noted that a test of non-proportionality based on adding the time-specific covariate $w_i(t) = x_i \ln(t)$ provides more support for a significant treatment effect. Iterations 500–3000 of a single run yield a treatment parameter averaging $\beta = -0.89$ with a 95% credible interval $(-2.3, 0.5)$. The interaction parameter itself is biased towards positive values though not precisely estimated:

FH also undertake a multivariable analysis of survival using biochemical and prognostic indicators, and excluding the treatment variable. Here we reduce the time scale to quarter years to facilitate the analysis (reducing the number of distinct uncensored survival times to 42). The variables used by FH are patient age in days (increasing age being expected to have a positive influence on mortality), log(albumin), log(bilirubin), presence of oedema and log(prothrombin time). The oedema variable is coded as follows:

0 = no oedema and no diuretic therapy for oedema;

0.5 = oedema present without diuretics, or oedema resolved by diuretics;

1 = oedema despite diuretic therapy.

Unlike FH we take the log of the age in days variable. We consider the predicted survival experience of two patients, both aged 52 (taken as 18 993 days), but differing in their prognostic indices (see Program 9.15 Model B). The high-risk patient survival is evaluated up to the end of quarter 4, and the low-risk patient up to the end of quarter 20 (i.e. 5 years after the start of the trial).

The covariate effects in the multivariate analysis generally reproduce those of FH, despite the change in time interval – the exception is the lowered effect of log(prothrombin time) which FH estimate as 3.02 with a standard error of 1.02. The survival rate of the low-risk patient is estimated as 0.966 compared with 0.97 by FH, but the high-risk patient is estimated to have a 0.46 survival probability at one year, higher than the FH estimate of 0.39. The estimates in Table 9.7 are based on a 3000 iteration run, excluding the first 500.

Example 9.16 Rat survival Clayton (1991) considers survival times in weeks of 150 rats, with one rat exposed and two controls in each of 50 litters. The exposure is to a putative carcinogen. Here we introduce frailty effects modelled at litter level. Maximum follow-up is two years ($T = 104$) and 40 rats died. The estimate of the overall hazard rate (*Program 9.16 Rat Survival*) is based on the number of deaths in relation to the total number of rats and their average exposure in weeks.

As in Clayton (1991) subsampling at each tenth iteration is carried out, and 10 000 iterations were made after a burn-in of 500. A relatively informative prior for the precision of the litter random effects is adopted for numerical stability, namely $\tau \sim \mathrm{G}(0.1, 0.1)$, but a vague prior is taken on the effect β of the exposure index ($x = 1$ for exposed, $x = 0$ otherwise). Thus the hazard function is

$$\lambda(t_i \mid x_i) = \lambda_0(t_i) \exp(\beta x_i + \varepsilon[L_i])$$

Table 9.7 PBC survival analysis, posterior summary.

	Mean	SD	2.5%	97.5%
Log(age)	0.454	0.206	−0.053	0.751
Log(albumin)	−3.292	0.653	−4.393	−1.814
Log(bilirubin)	0.825	0.103	0.625	1.026
Oedema	0.781	0.283	0.215	1.344
Log(prothrombin time)	2.172	0.770	1.060	3.735
One-year survival rate, high-risk patient	0.458	0.096	0.265	0.641
Five-year survival rate, low-risk patient	0.966	0.009	0.946	0.982

where $j = L_i$ is the litter of the ith rat, and the litter effects are

$$\varepsilon_j \sim \mathrm{N}(0, \sigma^2).$$

for $j = 1, \ldots, 50$. We find a clear exposure effect, with β averaging 0.9 (95% interval from 0.28 to 1.57). There is also significant variability in survival chances between litters, with a median σ^2 of 0.52. Both effects combine to produce different survival curves between exposed and controls.

9.7 DISCRETE TIME SURVIVAL MODELS AND LIFE TABLES IN DISCRETE TIME

9.7.1 Discrete time event history models

Allison (1984) reviewed discrete time models for event histories and argues that 'under some circumstances discrete-time models and methods may be quite appropriate or when less appropriate, highly useful'. Even when events occur in continuous time, many event histories actually only record the nearest month or year (e.g. marital or job histories). In medical trials, follow-up may occur at relatively widely spaced intervals. Sometimes durations may be grouped by definition – for example the number of menstrual cycles (in terms of months) to conception after marriage.

However, event history approaches to data recorded in discrete time may still assume an underlying continuous time model. Suppose this continuous distribution with density $f(t)$, survivorship $S(t)$ and hazard $h(t)$ is grouped into intervals $B_j = (b_{j-1}, b_j), j = 1, \ldots, m$ with $b_0 = 0$, $b_r = \infty$ and with durations or failure times between b_{j-1} and b_j recorded as a single value t_j. If $x[t_j]$ denotes interval-specific covariates then the regression effect is fixed within intervals but may vary between intervals. If the underlying continuous model follows the proportional hazard form, then the likelihood of an event after duration t_m is

$$\left[1 - \alpha_m^{\exp(\beta \mathrm{x}[\mathrm{t_m}])}\right] \prod_{j=1}^{m-1} \alpha_j^{\exp(\beta x[t_j])}. \tag{9.7}$$

The α_j are generally estimated as fixed effects and are equivalent to conditional survival probabilities in interval B_j for a standard individual (with $x[t_j] = 0$). The full set of parameters then consists of $\{\alpha_1, \ldots, \alpha_{r-1}\}$ and β. Time dependence in β can be introduced by making regression coefficients specific to intervals or subgroups of intervals. Since the parameters α_j are between 0 and 1, they can be reparameterized to remove range restrictions by a complementary log-log transform:

$$\gamma_j = \log(-\log\alpha_j).$$

The typical likelihood contribution is then

$$\delta_m \log[1 - \exp(-\exp\{\gamma_m + \beta x[t_m]\})] - \sum_{j=1}^{m} \exp\{\gamma_m + \beta x[t_m]\}$$

where the censoring indicator $\delta_m = 1$ if an event occurs in the mth interval and 0 otherwise. An individual never experiencing the event will be coded $\delta_j = 0, j = 1, \ldots, m_i$, while an individual undergoing a first event at time m_i will be coded

$$\delta_j = 0 \quad \text{for } j = 1, \ldots, m_{i-1}, \text{and } \delta_{m_i} = 1.$$

Taken over all individuals $i = 1, \ldots, I$, observed for $j = 1, \ldots, m_i$ intervals, this framework is equivalent to Bernoulli sampling over individuals and intervals with probabilities π_{ij} modelled via a complementary log-log link, and with the censoring indicator forming the response. Thus

$$\begin{aligned} &\delta_{ij} \sim \text{Bernoulli}(\pi_{ij}), \qquad\qquad i = 1, \ldots, I, j = 1, \ldots, m_i \\ &\log\{-\log(1 - \pi_{ij})\} = \gamma_j + \beta x[t_{ij}]. \end{aligned}$$

Example 9.17 Leukaemia survival Feigl and Zelen (1965) present data on survival times in weeks among acute myelogeneous leukaemia patients. The covariates are white blood cell count in thousands (WBC) and AG-factor (1 =positive, 2 =negative). All individuals are followed to death, so there is no final censoring. We define the censoring indicators in each week, and the Bernoulli outcomes, with the following code

```
for (i in 1:n) { delta[i,m[i]] <- 1
                 for (j in 1:m[i]-1) delta] [i,j] <- 0}
                 for (j in 1:m[i]) {delta[i,j] ~ dbern ([pi[i,j])}}
```

The complementary log-log transform of pi[i,j] is then modelled in terms of a factor for number of weeks, the two covariates (with LWBC being log of blood cell count), and their interaction

```
for (i in 1:n) {for (j in 1:m[i]) {cloglog (pi[i,j])<- b1[AG[i]] + b2*
        LWBC[i] + b3*LWBC[i]* equals(AG[i],2) + a[j]}}
```

The results show a significant effect of both the main effects and interaction term, see Table 9.8.

Table 9.8 Leukaemia survival.

	Mean	SD	2.5%	97.5%
AG negative	2.98	1.05	1.01	5.12
LWBC	0.77	0.24	0.33	1.27
Interaction	−0.62	0.30	−1.22	−0.06

9.8 LIFE TABLES: DEMOGRAPHIC AND MEDICAL APPLICATIONS

In its most basic form, the life table shows the number of survivors from an initial batch of new entrants at successive discrete intervals. This framework fits a wide range of applications: from new engineering components followed up weekly, to human births followed up at successive years of age. The following analysis focuses on human survival in clinical and epidemiological contexts but extends with relevant modifications to reliability applications.

9.8.1 Clinical and epidemiological applications

Epidemiological and clinical settings where life table methods are applied often involve comparisons of length of survival, time to disease onset, or time to remission between a treatment or control group, or between several groups receiving different treatments. These comparative settings include randomised trials with prolonged follow-up after treatment, and summarisation of the mortality risk of groups subject to differential risk exposures (e.g. life tables for smokers as against nonsmokers, or for different occupational groups).

When the period of follow-up observation ends, there will usually be some patients still alive for whom mortality or remission data is incomplete. First, some patients may still be alive (or well); second, some may have died from other causes other than that being studied; and third, some patients will be lost to follow-up for various reasons. Suppose N_0 patients are initially admitted to the study and survival is the outcome of interest. Then unless they withdraw or are lost to follow-up, patients are followed either till death or the end of the study, whichever occurs first.

For illustration, suppose we subdivide follow-up into yearly intervals, and let $N_x = P_x N_0$ be the number of patients still alive at the start of the xth year after the start of the study, i.e. N_x patients are alive at the start of the year-long interval $(x, x+1)$. Let P_x denote the probability of surviving to the start of the xth year (this may sometimes be denoted ${}_xP_0$). If q_x denotes the probability of death in the xth year, and p_x is the probability of surviving a single year, with $p_x + q_x = 1$, then the cumulative probability of surviving x years is

$$P_x = \prod_{z=0}^{x} p_z.$$

Decreases in N_x occur through death or withdrawal. Following Chiang (1968, Chapter 12) suppose the $N_x = w_x + m_x$ individuals beginning the yearly interval $(x, x+1)$ comprise first, m_x observed for the entire interval $(x, x+1)$, and second w_x individuals who withdraw in $(x, x+1)$. Of the w_x patients, d'_x will die before the end of the interval, and s'_x will survive to its end; and of the m_x patients, s_x will survive the interval and d_x die during it. The sum $D_x = d_x + d'_x$ is the total of deaths in the interval.

If there is a constant risk of death λ_x in interval $(x, x+1)$ then the probability of surviving to its end is

$$p_x = \exp(-\lambda_x)$$

and for a subinterval $(x, x+t)$, $0 < t \leq 1$ the probability of surviving is

$$p_x^t = \exp(-\text{t}\lambda_\text{x}).$$

If withdrawals occur at random in the interval, then the latter survival probability is approximately $r_x = p_x^{0.5}$ (Chiang, 1968, p. 272). So the binomial likelihood governing the $m_x = s_x + d_x$ patients observed to the end of the interval, or known to have died during it, is proportional to

$$p_x^{s_x}(1 - p_x)^{d_x}.$$

The likelihood for the $w_x = s'_x + d'_x$ withdrawals is correspondingly

$$r_x^{s'_x}\ (1 - r_x)^{d'_x}.$$

Suppose there are several competing risks $d = 1, \ldots, C$, such that the total risk of death in $(x, x+1)$ can be decomposed into constant intensities during the interval for each risk:

$$\lambda_x = \lambda_{x1} + \ldots + \lambda_{xC}.$$

The crude probability that a patient alive at the start of interval $(x, x+1)$ will die at time t during the interval $(0 < t \leq 1)$ is then

$$Q_{xd}^t = \{\lambda_{xd}/\lambda_x\}[1 - p_x^t].$$

Combining the last two relations gives

$$Q_{x1}^t + Q_{x2}^t + \ldots + Q_{xC}^t + p_x^t = 1. \tag{9.8}$$

For those observed till the end of the interval the likelihood is defined by p_x^t and Q_{xd}^t for a complete interval of length $t = 1$. So for observations $\{d_{x1}, \ldots, d_{xC}, s_x\}$, we have multinomial probabilities $\{Q_{x1}, Q_{x2}, \ldots Q_{xC}, p_x\}$ with index m_x. For the w_x withdrawals, we assume withdrawal on average at $t = 0.5$, and the relevant cause-specific crude probabilities are, from (9.8), given by

$$V_{xd} = Q_{xd}[1 + r_x]^{-1} \quad \text{for } d = 1, \ldots, C$$

with survival probability r_x. The likelihood for withdrawals is then multinomial with index n_x and probabilities

$$\{V_{x_1}, V_{x_2}, \ldots V_{x_c}, r_x\}.$$

The usual follow-up design involves a cohort followed through time, without additional recruitment to the study, and we would not observe the components d'_x and s'_x of N_x. Chiang (1968) considers a design in which recruitment is possible at all times, but the observations at interval $(x, x+1)$ includes 'withdrawals' in the sense that they were recruited too late to the study in order to be observed till the interval's completion. Thus he considers 20 858 breast cancer patients recruited to a California Tumor Registry, with the study finishing on 31 December 1963, and recruitment ending on 31 December 1962. Recruitment started some 20 years earlier, on 1 January 1942. So patients admitted between 1 January 1962 and 31 January 1962 are considered as 'withdrawals' with regard to the interval $(x, x+1) = (1, 2)$ between 1 and 2 years since admission, even though in fact their mortality and survival experience were completely recorded till 31 December 1963.

Example 9.18 Framingham study The Framingham Cardiac Risk Study provides data on survival among men and women with different levels of exposure to a number of risk factors. Here we focus on survival among men aged 55–59 in terms of their systolic blood pressure at the initial exam and their survival chances in a follow-up period of 18 years.

Suppose initial numbers at risk N_x, and deaths d_x are recorded for five intervals: 0–4 years, 4–8 years, 8–12 years, 12–16 years, and 16–18 years. Also recorded at each interval are withdrawals w_x (i.e. lost to follow-up among the initial group of size N_x). Then we need to adjust the N_x for withdrawals, and the usual procedure assumes such withdrawals are exposed for on average half the period being considered. We may convert this assumption into a stochastic mechanism by assuming the average size of withdraw group exposed to risk of the event is

$$e_x \sim \text{Bin}(0.5, w_x).$$

Then deaths d_x are binomial in relation to a population at risk of size $N_x - e_x$:

$$D_x \sim \text{Bin}(q_x, N_x - e_x)$$

where x is age at the beginning of each interval. In this application an additional aspect of notation would be to signify the length of the interval that survival or death related to. For example, in the first interval of 4 years we could write the death probability as ${}_4q_0$, and in the last interval as ${}_{18}q_{16}$. The cumulative probability of surviving throughout the follow-up is obtained as

$${}_4P_0 \cdot {}_8P_4 \cdot {}_{12}P_8 \cdot {}_{16}P_{12} \cdot {}_{18}P_{16}$$

and the probability of surviving to the end of the fourth interval by

$${}_4P_0 \cdot {}_8P_4 \cdot {}_{12}P_8 \cdot {}_{16}P_{12}.$$

In the current example the survival probabilities at year 16 show a distinct advantage for those with low SBP (group 1 in *Program 9.18 Framingham Heart Study)*, as shown in Table 9.9.

The relative risks across the two groups can be summarised by a Mantel-Haensel procedure. Let c_x denote deaths in interval x in the high-risk SBP group (group 2), and d_x deaths in the low-risk group. Survivors are denoted s_x and t_x. The total risk population, including adjustment for withdrawals, is $R_x = c_x + d_x + s_x + t_x$. The ratios $(c_x t_x / R_x)$ and $(s_x d_x / R_x)$ are calculated for each follow-up interval, with coding as follows:

```
R[x] <- Obsadj[i,1] + Obsadj [x,2]
c[x] <- death[x,2]
s[x] <- Obsadj [x,2]-death[x,2]
d[x] <- death[x,1]
t[x] <- Obsadj[x,1]-death[x,1]
OR.numerator[x] <-c[x]*t[x]/R[x]
OR.denominator[x] <-s[x]*d[x]/R[x]}
```

The overall statistic is then

$$\sum_x (c_x t_x / R_x) / \sum_x (s_x d_x / R_x).$$

Table 9.9 shows the clear excess risk for high SBP shown by this combined odds ratio statistic.

Example 9.19 Breast Cancer Survival: Observations from a tumor registry We consider the dataset of 20 858 breast cancer patients recruited to a California Tumor Registry, and with the design outlined above. The observed deaths by age x and risk $j = 1, \ldots, C$ are, first, $\{d_{x1}, \ldots, d_{xC}, s_x\}$ for those fully observed through intervals $(x, x+1)$, and second, the corresponding data $\{d'_{x1}, \ldots, d'_{xC}, s'_x\}$ for 'withdrawals'. We consider the deaths from breast cancer itself $(j = 1)$, from other cancers $(j = 2)$, from causes other than cancer $(j = 3)$, and 'lost cases', considered as a fourth mortality decrement $(j = 4)$.

This is interesting from a modelling perspective as the probabilities for the two groups of patients are interrelated, with the multinomial likelihood for the data $\{d'_{x1}, \ldots, d'_{xC}, s'_x\}$ being defined by probabilities

$$Q_{x1}(1 + p_x^{0.5})^{-1}, Q_{x2}(1 + p_x^{0.5})^{-1}, \ldots, Q_{xC}(1 + p_x^{0.5})^{-1}, p_x^{0.5}$$

and the likelihood for $\{d_{x1}, \ldots, d_{xC}, s_x\}$ defined by probabilities

$$Q_{x1}, Q_{x2}, \ldots, Q_{xC}, p_x.$$

We aim to estimate net probabilities of death from cause d acting alone, and the net probability of death from any cause when a particular cause d is eliminated. These are respectively

$$q_{xd} = 1 - p_x^{D_{xd}/D_x}$$

and

Table 9.9 Test statistics; Framingham study.

	Mean	SD	2.5%	Median	97.5%
Mantel–Haensel odds	3.08	0.12	2.88	3.07	3.35
Survival rate till 16 years (low SBP)	0.726	0.003	0.721	0.727	0.731
Survival rate till 16 years (high SBP)	0.437	0.015	0.40	0.439	0.458

$$q_{x \cdot d} = 1 - p_x^{(D_x - D_{xd})/D_x}.$$

Thus consider the probabilities of surviving till the xth anniversary P_{x1} when the risk is breast cancer alone, namely

$$\prod_{z=0}^{x} (1 - q_{z1}).$$

and the probabilities $P_{x.4}$ of surviving till the xth anniversary when all risks operate except $d = 4$ (i.e. being lost is eliminated as a decrement), namely

$$\prod_{z=0}^{x} (1 - q_z.4).$$

The last 1000 iterations from a single run of 5000 gives estimates of P_{x1} and $P_{x.4}$ (`P1[]` and `p.4[]` in Program 9.18) as in Table 9.10.

By comparison, Chiang (1968) obtains 0.486 for the probability $P_{10.1}$ of surviving till the tenth anniversary when breast cancer is the sole risk (see also Kahn and Sempos, 1989, p. 181). The standard deviations in Table 9.10 are in fact smaller than those of Chiang which were obtained by asymptotic approximations.

Table 9.10 Survival probabilities (all causes and breast cancer alone).

	Mean	SD		Mean	SD
$P_{1.4}$	1.0000		$P_{1.1}$	0.8826	0.0018
$P_{2.4}$	0.8505	0.0022	$P_{2.1}$	0.7887	0.0025
$P_{3.4}$	0.7410	0.0029	$P_{3.1}$	0.7177	0.0027
$P_{4.4}$	0.6574	0.0032	$P_{4.1}$	0.6591	0.0029
$P_{5.4}$	0.5871	0.0033	$P_{5.1}$	0.6176	0.0030
$P_{6.4}$	0.5354	0.0034	$P_{6.1}$	0.5800	0.0030
$P_{7.4}$	0.4870	0.0034	$P_{7.1}$	0.5525	0.0031
$P_{8.4}$	0.4520	0.0035	$P_{8.1}$	0.5282	0.0032
$P_{9.4}$	0.4185	0.0036	$P_{9.1}$	0.5039	0.0033
$P_{10.4}$	0.3874	0.0036	$P_{10.1}$	0.4862	0.0033
$P_{11.4}$	0.3613	0.0038	$P_{11.1}$	0.4699	0.0034
$P_{12.4}$	0.3400	0.0038	$P_{12.1}$	0.4564	0.0034
$P_{13.4}$	0.3185	0.0038	$P_{13.1}$	0.4417	0.0034
$P_{14.4}$	0.2973	0.0037	$P_{14.1}$	0.4327	0.0034
$P_{15.4}$	0.2790	0.0038	$P_{15.1}$	0.4226	0.0035
$P_{16.4}$	0.2610	0.0040	$P_{16.1}$	0.4129	0.0036
$P_{17.4}$	0.2453	0.0040	$P_{17.1}$	0.3985	0.0038
$P_{18.4}$	0.2294	0.0041	$P_{18.1}$	0.3871	0.0039
$P_{19.4}$	0.2159	0.0042	$P_{19.1}$	0.3854	0.0039
$P_{20.4}$	0.2067	0.0043	$P_{20.1}$	0.3829	0.0039
$P_{21.4}$	0.1897	0.0049	$P_{21.1}$	0.3730	0.0043
$P_{22.4}$	0.1731	0.0056	$P_{22.1}$	0.3730	0.0043

9.9 POPULATION LIFE TABLES

In the case of human mortality the information on survivors (at ages 1, 5, 15, etc.) can in principle be obtained from direct observation. Thus members of a single birth cohort (e.g. all people born in 1900) could be followed through their lives till they have all died. Such cohort life tables are, however, impractical to compile and unsuitable for representing current expectations of mortality. The latter are best described by a period or current life table, involving the device of tracing a hypothetical cohort of infants through their lifetime, as if current mortality rates used in the table applied to the cohort throughout their lives.

Despite this change of emphasis, cohort principles still guide the construction of period life table death rates. If a cohort of people is followed through time, then the appropriate denominator population to describe deaths among the cohort is the population at the start of a period. Consider a cohort of 600 people born on 1 January 1930, of whom $l_{65} = 500$ were still alive at 1 January 1995. If $d_x = 10$ die during 1995, the noncentral or life table death rate, denoted q_x for age $x = 65$, would be 20 per 1000. This death rate relates to the probability of death during a yearly period of those alive at the start of the period, with binomial sampling $d_x \sim \text{Bin}(q_x, l_x)$.

The central mortality rate by contrast refers to deaths divided by the denominator population at mid-year which here (in the absence of further information) could be estimated as $P_x = 495$, assuming deaths occur approximately uniformly through the year. The central death rate for the year 1995, m_x for $x = 65$, would then be 10/495, or 20.2 per thousand.

More generally, suppose the observed data are deaths D_x by single year of age x and mid-period populations $P_x (x = 0, 1, 2, \ldots)$. Then the likelihood for the central death rates is

$$D_x \sim \text{Bin}(m_x, P_x)$$

and the noncentral death rate (life table probability of death) is approximated as

$$q_x = D_x/(P_x + 0.5D_x) = m_x/(1 + 0.5m_x).$$

In the typical case, a life table is constructed using the observed data $\{D_x, P_x\}$ on a basis of a hypothetical birth cohort usually of size $l_0 = 100,000$. At ages x above zero, l_x is the number remaining alive to exact age x from the original l_0. Given that q_x is the probability of dying between exact ages x and $x + 1$, the number from the original cohort surviving to age 1 is expected to be

$$l_1 = l_0(1 - q_0) = l_0 p_0$$

and deaths between ages 0 and 1 are then

$$l_1 - l_0 = l_0 q_0.$$

This relation holds at later ages, namely

$$l_{x+1} = 1_x(1 - q_x) = 1_x p_x \tag{9.9}$$

where $p_x = 1 - q_x$ is the probability of surviving from age x to $x + 1$.

While for most age groups the assumption is adequate that deaths are uniformly spread through the year, for very young ages it is necessary to reflect the skewed concentration of death in the first few weeks. Thus let a_x be the average proportion of a year lived by those aged x at the year's start and dying during the year. If deaths at age x are uniformly distributed throughout a period of observation, then values of a_x can be taken as 0.5. However, a_0 is typically set to 0.1 to reflect the concentration of infant deaths in the first few weeks after birth.

Either for simplicity or data availability, the life table calculations may be done not on the basis of deaths by single year of age x but using deaths and populations in five- or ten-year age groups ('abridged' life tables). If we work with five-year age groups, then instead of q_x we would have obtained the probability that a person in age group x dies within 5 years, written ${}_5q_x$. In general suppose the interval is of length n_x with an average proportion a_x of the interval lived by those dying within it. The conversion from central to life table death rates now becomes

$$q_x = n_x m_x / [1 + (1 - a_x) n_x m_x].$$

Frequently we need to know the relative proportions of people in different age groups and the average future lifetimes of people at different ages. This involves a new function, L_x, the number of persons alive between exact ages x and $x + 1$ in the stationary population. The values of L_x can be calculated by taking the average of adjacent values of l_x when deaths are approximately uniformly distributed over a year (or more generally over the length of the age group in the lifetable). So for a one-year age group life table

$$L_x = \frac{1}{2}(l_x + l_{x+1}).$$

For an abridged lifetable using n-year age group data (e.g. $n = 5$ for five-year age bands)

$${}_nL_x = \frac{n}{2}(l_x + l_{x+n}).$$

From the persons aged x function L_x can be calculated the expected number of person-years lived until death by the l_x persons now aged x, namely

$$T_x = L_x + L_{x+1} + L_{x+2} + \ldots.$$

If l_x people expect to live a total of T_x years of life, it follows that their average life span (or 'life expectancy') is T_x/l_x. It should be remembered that in a current life table this expectation is based on currently prevailing mortality rates. Cause-specific expectancies (relating to death from a particular cause) may be radically changed by medical advances, and total life expectancies (from all causes of death) improve due to medical advances and raised standards of living.

Example 9.20 Life table for small area human populations Mortality variation at small area level is often used as a proxy for assessing health resourcing. However, small populations at risk, uncertainty in population estimates outside census years, and the small number of deaths, all increase the uncertainty of estimated life table functions.

Suppose we take the mortality experience of an electoral ward (administratively defined small area) in London[1] over a three-calendar-year period 1995 to 1997. So deaths are accumulated over three years and related to an average population at risk, usually the mid-period population (here that for 1996). A total of 132 deaths are observed over the three years in a population of just under 5000. Between ages 5 and 14 no deaths occurred in this small area during this period.

To allow for such a sampling variability, we may adopt a binomial or Poisson model for the observations, with age-specific random effects to pool strength over the age categories of the life table (here 0, 1–4, 5–9, ..., 80–84 and 85+). The Poisson model for observed death totals m_x at age group x among population at risk P_x takes the form

$$m_x \sim \text{Poi}(\gamma_x)$$
$$\gamma_x = \lambda_x P_x.$$

If a log link for the Poisson mean is used then

$$\log\gamma_x = \log P_x + \log\lambda_x.$$

If the deaths relate to more than one calendar year but we require annual death rates, then our exposed populations would be approximately AP_x where A is the number of years over which deaths are cumulated, and P_x is the population at the middle of the period. So

$$\log\gamma_x = \log(AP_x) + \log\lambda_x.$$

A fixed effects Poisson (with no pooling of strength over age groups) would typically involve a gamma prior such as

$$\lambda_x \sim \text{G}(c, d)$$

where values such as $c = d = 0.001$ would be minimally informative.

However, a random effects model would refer the age-specific death rates to an overall hyperdensity. For example we might specify

$$\lambda_x \sim \text{G}(a, b)$$
$$a \sim \text{E}(1)$$
$$b \sim \text{G}(0.1, 0.1)$$

as in the mixed Poisson model of George *et al.* (1993).

[1] Abbey ward in the London Borough of Barking.

The code for *Program 9.20 Small Area Life Table* includes both fixed and random effects options, and Poisson and binomial sampling models. The data are in Table 9.11 and the estimated life table survivors l_x, as in (9.9), under a fixed effects Poisson in Table 9.12. Life expectancies under the four-model (and prior) choices are given in Table 9.13. It is notable that random effects approaches lead to slightly lower life expectancies at birth and lower ages (as might be expected given the instability of fixed effects methods for small area populations, and small even totals at lower ages). At older ages, random effects priors slightly increase expectancies.

9.10 COMPETING RISKS POPULATION LIFE TABLES

In human mortality contexts, competing risks analysis is often applied to assessing the impact on cumulative life chances and life expectancy of reducing or eliminating death from particular causes. The competing risks approach in fact originated in historical debate over the impact of smallpox vaccination on population composition, and now a major focus is on assessing the impact on life chances of reducing mortality from the major degenerative diseases.

If the total death rate from all causes at age or time t is $h(t)$ then the probability that someone alive at the start of the ith age interval (x_i, x_{i+1}) will survive to the end of the interval is

Table 9.11 Small area deaths and population.

Age band	Three year deaths	Population at mid-period
0	2	80
1–4	1	326
5–9	0	409
10–14	0	340
15–19	1	311
20–24	1	408
25–29	1	596
30–34	2	475
35–39	1	333
40–44	0	269
45–49	2	249
50–54	3	172
55–59	2	180
60–64	4	136
65–69	14	123
70–74	16	175
75–79	24	114
80–84	27	75
85+	31	82
Total	132	4853

Table 9.12 Small area life table, fixed effects Poisson (number surviving from theoretical $l_0 = 100,000$).

	Mean	SD	2.5%	Median	97.5%
l_1	99170	581	97700	99290	99890
l_2	98760	700.3	97070	98890	99730
l_3	98760	700.5	97070	98890	99730
l_4	98760	701.5	97060	98890	99730
l_5	98240	873.4	96190	98380	99510
l_6	97840	952.7	95640	97980	99280
l_7	97570	989.2	95330	97710	99100
l_8	96880	1092	94460	97010	98680
l_9	96400	1187	93820	96520	98370
l_{10}	96400	1188	93820	96520	98370
l_{11}	95130	1481	91850	95270	97570
l_{12}	92400	2159	87620	92640	95960
l_{13}	90710	2434	85330	90920	94800
l_{14}	86370	3140	79640	86630	91800
l_{15}	71490	4461	62370	71670	79770
l_{16}	61410	4502	52370	61500	70050
l_{17}	43150	4448	34600	43140	51990
l_{18}	23380	3825	16390	23210	31380

Table 9.13 Estimated life expectancies by model (ages 0, 1, 5, 10, 15, 20, ...,85).

	Fixed effects Poisson		Random effects Poisson		Fixed effects binomial		Random effects binomial	
	Mean	SD	Mean	SD	Mean	SD	Mean	SD
E[1]	76.0	1.3	75.4	1.4	76.0	1.3	75.6	1.3
E[2]	75.7	1.3	75.0	1.3	75.6	1.2	75.2	1.3
E[3]	71.9	1.2	71.4	1.3	71.9	1.2	71.5	1.2
E[4]	66.9	1.2	66.5	1.3	66.9	1.2	66.7	1.2
E[5]	61.9	1.2	61.5	1.3	61.9	1.2	61.8	1.2
E[6]	57.3	1.2	56.9	1.2	57.2	1.1	57.2	1.1
E[7]	52.5	1.2	52.2	1.2	52.5	1.1	52.5	1.1
E[8]	47.6	1.2	47.4	1.2	47.6	1.1	47.7	1.1
E[9]	42.9	1.2	42.7	1.2	42.9	1.1	43.0	1.1
E[10]	38.1	1.1	38.0	1.2	38.1	1.1	38.3	1.1
E[11]	33.1	1.1	33.1	1.2	33.1	1.1	33.4	1.1
E[12]	28.6	1.1	28.5	1.1	28.5	1.0	28.8	1.0
E[13]	24.3	1.1	24.3	1.1	24.3	1.0	24.6	1.0
E[14]	19.7	1.0	19.8	1.1	19.7	1.0	20.0	1.0
E[15]	15.6	1.0	15.7	1.0	15.6	0.9	15.8	0.9
E[16]	13.3	0.9	13.4	0.9	13.3	0.8	13.5	0.8
E[17]	10.1	0.9	10.2	1.0	10.1	0.8	10.3	0.8
E[18]	8.3	1.1	8.5	1.1	8.2	0.9	8.4	0.9
E[19]	8.2	1.5	8.4	1.5	8.1	1.2	8.3	1.2

$$p_i = \int_{x_i}^{x_i+1} h(u)du$$

and the probability of dying in that interval is $q_i = 1 - p_i$. As discussed above, this total hazard in the presence of C simultaneous and independent risks of death is

$$h(t) = h_1(t) + \ldots h_C(t).$$

Suppose we make the simplifying assumption that in the ith age interval (x_i, x_{i+1}) the ratio

$$r_{ij} = h_j(t)/h(t) \qquad (9.10)$$

is independent of t within each interval but may differ according to the interval i, and specific risk j concerned. Then let Q_{ij} denote the crude probability of death from cause j in the presence of all other risks. Since by (9.10) the ratio of the risk-specific hazard to the total hazard is assumed constant within a given age interval, this constant also necessarily equals the ratio of the parallel probabilities of dying, namely Q_{ij} and q_i (Chiang, 1968). So

$$r_{ij} = h_j(t)/h(t) = Q_{ij}/q_i$$

and it follows that

$$Q_{i1} + Q_{i2} + \ldots Q_{iC} = q_i.$$

The net probability of death in (x_i, x_{i+1}) is defined for the (hypothetical) situation when cause j is the only cause of death present. It may be written

$$q_{ij} = 1 - \exp[- \int_{x_i}^{x_{i+1}} h_j(u)du]$$

and using the above assumption (9.10) this becomes

$$\begin{aligned} q_{ij} &= 1 - \exp\{-h_j(t)/h(t) \int_{x_i}^{x_{i+1}} h(u)du\} \\ &= 1 - p_i^{r_{ij}}. \end{aligned} \qquad (9.11)$$

The net probability of death when cause j is *eliminated* is by the same principle obtained by integrating over the hazard $h(t) - h_j(t)$, namely

$$\begin{aligned} q_{i.j} &= 1 - \exp\left[- \int_{xi}^{x_{i+1}} \{h(u) - h_j(u)\}du\right] \\ &= 1 - p_i^{(qi - Q_{ij})/q_i}. \end{aligned} \qquad (9.12)$$

We are frequently concerned with estimating $q_{i.j}$ from current mortality data over a defined period, with observed data consisting of mid-period population estimates by age P_i and total registered deaths D_{ij} by age and cause occurring in that period.

The estimate of $q_{i.j}$ is nevertheless based on the life table likelihood principles. As noted above, the likelihood for l_i survivors at the start of the ith interval can be defined in terms of the d_i deaths occurring during that interval and the $l_{i+1} = l_i - d_i$ survivors who remain. The respective binomial probabilities for the event totals are q_i and p_i. In the presence of C possible causes of death the likelihood is similarly multinomial with events $\{d_{i1}, d_{i2}, \ldots, d_{iC}, l_{i+1}\}$ and respective probabilities $\{Q_{i1}, Q_{i2}, \ldots Q_{iC}, p_i\}$. In this case the maximum likelihood estimate of Q_{ij} would be d_{ij}/l_i.

This likelihood is also the basis for deriving the cause- specific net probabilities of death from current mortality data via a Bayesian method. Thus the events $\{D_{i1}, D_{i2}, \ldots D_{iC}, P_i - D_i\}$ are multinomial with probabilities which are respectively the age-specific death rates M_{ij} for causes $j = 1, , \ldots, C$ and survival rates by age $S_i = 1 - M_i$. Alternatively we can use separate binomials for each cause and calculate the survival rate as the remainder.

Given estimates of M_i or M_{ij} at each MCMC iteration we can then use the usual approximation to convert central to life table death rates q_i and Q_{ij}. Thus for an interval of length b_i, let a_i denote the average number of years (assumed to be) lived by someone dying in the interval. For an abridged five-year age group life table, the age bands 0–1, 1–4, 5–9, 10–14, etc. are often used. In this case the successive b_i are 1, 4, 5, 5, ... and the successive a_i are typically taken as 0.1, 2, 2.5, 2.5,... with the value of a_1 reflecting the concentration in infant mortality shortly after birth. Then, letting $c_i = a_i/b_i$

$$\hat{q}_i = b_i M_i/[1 + (1 - c_i)b_i M_i]$$

and

$$\hat{Q}_{ij} = b_i M_{ij}/[1 + (1 - c_i)b_i M_i].$$

The net probabilities $q_{i.j}$ after cause j is eliminated are then estimated from (9.12).

To calculate life expectancies if cause j is eliminated we may derive cumulative survival probabilities, $p_{\alpha\beta.j}$ for the proportion of people who will survive from x_α to x_β if cause j is eliminated:

$$p_{\alpha\beta.j} = \prod_{i=\alpha}^{\beta-1}(1 - q_{i.j}).$$

The life expectancy at x_α when risk j is eliminated is then

$$E_{\alpha j} = a_\alpha + \sum_{\beta \geq \alpha}[c_\beta p_{\alpha\beta.j}]$$

where $c_\beta = b_{\beta-1} - a_{\beta-1} + a_\beta$.

Example 9.21 Male cardiovascular and renal mortality, USA 1960 We illustrate this cause-elimination approach using the data presented by Chiang (1968) on male deaths in 1960 in the USA from all causes combined, and from cardiovascular and renal (CVR) diseases in particular (Table 9.14).

Table 9.14

	White males			White females		
		Deaths			Deaths	
Age	Population 1960	All	Cardiovascular renal diseases	Population 1960	All	Cardiovascular renal dseases
0–1	1794784	48063	228	1719014	34416	132
1–5	7063044	7409	153	6788716	5791	108
5–10	8191158	4408	177	7876239	3019	121
10–15	7488562	3847	208	7188766	2214	196
15–20	5893946	7308	355	5772421	2901	273
20–25	4657470	7755	481	4822377	2916	414
25–30	4725480	7182	768	4839982	3459	602
30–35	5216424	9039	1808	5379640	5215	955
35–40	5461528	13803	4444	5708902	8396	1727
40–45	5094821	21336	9125	5298273	12622	3252
45–50	4850486	34247	16796	4988493	18268	5405
50–55	4314976	50716	26812	4462148	24696	8726
55–60	3774623	66540	36907	3986830	32340	13913
60–65	3100045	85890	49649	3419584	46710	24216
65–70	2637044	108726	65609	3031276	65834	38539
70–75	1972947	119269	75371	2339916	85018	55304
75–80	1214577	109193	73057	1545378	96133	67579
80–85	591251	83885	58713	829736	91771	68073
85–90	235566	49502	36133	367538	65009	49817
90–95	56704	18253	13604	100350	29407	22826
95+	12333	4219	3136	24331	8173	6356
DK	0	267	106	0	170	99
Total	78347769	860857	473640	80489910	644478	368633

The two competing causes of death are then all deaths apart from those due to CVR $(j = 1)$, and CVR deaths $(j = 2)$ themselves. The interest is in the extension of life which would result from eliminating CVR diseases as risks of death.

We may adopt a Dirichlet prior for the multinomial data $\{M_{i1}, M_{i2}, S_i\}$ in relation to population at risk P_i; alternatively we can use separate beta priors on M_{i1}, M_{i2} etc. and obtain S_i as the remainder. There may well be substantive grounds for an informative prior (e.g. using 1959 death patterns). Estimated life table probabilities with CVR present q_i and with CVR eliminated $q_{i.2}$ are given in Table 9.15.

We find that the male expectancy at birth would be raised by over ten years if the CVR risk were eliminated (from 67.3 to 79) – compare E.m[] and Enet.m[] in Program 9.21. The effect if CVR were eliminated is more pronounced at older ages; the expected length of life at age 70 $(i = 16)$ would be doubled. It is notable that all estimated quantities have high precision due to the large population and death totals being used.

Table 9.15 Probabilities of dying

	CVR present			CVR eliminated	
	Mean	SD		Mean	SD
$q1$	0.02615	0.00012	$q1.2$	0.02603	0.00011
$q2$	0.00419	0.00005	$q2.2$	0.00410	0.00005
$q3$	0.00269	0.00004	$q3.2$	0.00258	0.00004
$q4$	0.00257	0.00004	$q4.2$	0.00243	0.00004
$q5$	0.00618	0.00007	$q5.2$	0.00588	0.00007
$q6$	0.00829	0.00009	$q6.2$	0.00778	0.00009
$q7$	0.00757	0.00009	$q7.2$	0.00677	0.00008
$q8$	0.00863	0.00009	$q8.2$	0.00691	0.00008
$q9$	0.01256	0.00010	$q9.2$	0.00853	0.00009
$q10$	0.02072	0.00014	$q10.2$	0.01191	0.00010
$q11$	0.03468	0.00019	$q11.2$	0.01782	0.00013
$q12$	0.05709	0.00025	$q12.2$	0.02733	0.00018
$q13$	0.08442	0.00031	$q13.2$	0.03852	0.00022
$q14$	0.12960	0.00040	$q14.2$	0.05686	0.00029
$q15$	0.18690	0.00051	$q15.2$	0.07877	0.00036
$q16$	0.26260	0.00062	$q16.2$	0.10600	0.00047
$q17$	0.36700	0.00087	$q17.2$	0.14040	0.00065
$q18$	0.52370	0.00124	$q18.2$	0.19950	0.00114
$q19$	0.68880	0.00179	$q19.2$	0.27050	0.00199
$q20$	0.89190	0.00292	$q20.2$	0.43270	0.00561

Table 9.16 Life expectancies (ages 0,1,5,10,15,20,...85).

	Total			CVD eliminated		
	Mean	2.5%	97.5%	Mean	2.5%	97.5%
E_1	67.3	67.3	67.3	79.0	78.7	79.3
E_2	68.1	68.1	68.1	80.1	79.8	80.4
E_3	64.4	64.3	64.4	76.4	76.1	76.7
E_4	59.5	59.5	59.6	71.6	71.3	71.9
E_5	54.7	54.7	54.7	66.8	66.5	67.1
E_6	50.0	50.0	50.0	62.1	61.8	62.4
E_7	45.4	45.4	45.4	57.6	57.3	57.9
E_8	40.7	40.7	40.7	53.0	52.7	53.3
E_9	36.1	36.0	36.1	48.3	48.0	48.6
E_{10}	31.5	31.5	31.5	43.7	43.4	44.0
E_{11}	27.1	27.1	27.1	39.2	38.9	39.5
E_{12}	23.0	22.9	23.0	34.9	34.6	35.2
E_{13}	19.2	19.2	19.2	30.8	30.5	31.1
E_{14}	15.7	15.7	15.8	26.9	26.6	27.3
E_{15}	12.7	12.7	12.7	23.4	23.0	23.8
E_{16}	10.0	10.0	10.0	20.2	19.8	20.6
E_{17}	7.7	7.7	7.7	17.3	16.8	17.7
E_{18}	5.7	5.7	5.7	14.7	14.2	15.2
E_{19}	4.2	4.2	4.2	12.7	12.1	13.3
E_{20}	3.1	3.0	3.1	11.3	10.5	12.2
E_{21}	2.9	2.9	3.0	11.5	10.2	12.9

CHAPTER 10

Bayesian Estimation and Model Assessment

10.1 INTRODUCTION

The typical process of Bayesian estimation, whether by analytical methods, numerical integration, or modern simulation and sampling techniques, involves using the likelihood to move from the prior to posterior knowledge. Often likelihoods are complex and intractable using conventional estimation methods, and are difficult to compute or maximise over many parameters. Sometimes maxima are at the edge of the parameter space, as in finite mixture models or change point analysis in time series. MCMC methods provide an approach to simulating the posterior distributions in complex multiparameter problems without resorting to a search for the maximum likelihood solution – though the posterior mean around which repeated sampling concentrates generally approximates the maximum likelihood estimate in large data sets. There is a wide provenance for MCMC methods, drawing on computational mathematics, statistical developments such as EM, data augmentation and the bootstrap, and scientific applications in physics and image reconstruction.

Suppose we have a set of p parameters $\underset{\sim}{\theta} = (\theta_1, \ldots, \theta_p)$ and our interest is in the marginal distribution of one parameter θ_1. In a Bayesian analysis, a prior density $\pi(\underset{\sim}{\theta})$ is placed on $\underset{\sim}{\theta}$, reflecting prior knowledge about the parameters (e.g. the effect of a risk factor in increasing chances of an illness) or modelling assumptions (e.g. that the errors in a health outcome map are spatially correlated). The joint posterior density $P(\underset{\sim}{\theta}\,|\,y)$ is then proportional to the product of the likelihood $L(y\,|\,\underset{\sim}{\theta})$ and the prior information or modelling assumptions. This applies even in the case where the prior contains essentially no information; for example, when uniform but proper prior densities are placed on the parameters $\underset{\sim}{\theta}$, or when some more specific form of reference prior is used. These are intended to be neutral priors that make essentially no impact on the analysis.

We may be interested in certain parameter values especially; in a multilevel analysis of educational results we might be interested in the correlation between a school's average exam performance and its average attainment gain for pupils of a given ability on entry. In a medical trial we might be interested in a treatment effect after controlling for comorbidities and random influences on the outcome. So some of the estimated parameters are of secondary concern and essentially 'nuisance variables'. Thus we require the marginal posterior density for θ_1 which is obtained from integrating out the joint posterior density $P(\theta_1, \ldots, \theta_p \mid y)$ with respect to all the parameters except the one which is the focus of interest:

$$P(\theta_1 \mid y) \propto \int_{\theta_2} \int_{\theta_3} \cdots \int_{\theta_p} P(\underset{\sim}{\theta} \mid y) d\theta_2 d\theta_3 \ldots d\theta_p.$$

Alternatively, we might be interested in a function of the model parameters and possibly also the data $G(\theta, y)$. An example might involve a school's rank in a league table of performance scores after adjusting (a) for average intake ability and (b) for differences in reliability between scores by pooling strength through a Bayesian procedure. We might then be interested in a 'league table' comparison of actual and adjusted ranks. Thus our posterior expectation of this function is

$$E[G(\theta, y)] = \int_{\theta_1} \int_{\theta_2} \int_{\theta_3} \cdots \int_{\theta_p} G(\theta, y) P(\theta \mid y) d\theta_1 d\theta_2 d\theta_3 \ldots d\theta_p.$$

Numerical methods to obtain these quantities include Gauss–Hermite quadrature (Naylor and Smith, 1982) or the Laplace transform. The latter is applicable when the joint posterior is approximately multivariate normal and is based on a Taylor series expansion of the log of the posterior density around the posterior mode (Tierney and Kadane, 1986).

An alternative strategy is stochastic sampling based on Markov chain kernels. A Markov chain describes an idealised pattern of movement or transitions through a set of states, with a wide range of real-world applications; typical examples include residential moves, and changes of occupational and marriage state. In a Markov chain process the move from the state i occupied at time t to the state j at time $t+1$ depends only on state i. The change of state between time $(t, t+1)$ thus depends only on the current state and is independent of any previous history of moves. Eventually the probabilities $p[i,j]$ in the transition matrix between states reach a stationary or equilubrium matrix which is independent of the initial state (the state at time zero).

Translated into moves by parameter vectors between successive iterations of a Monte Carlo Markov chain sampling scheme, this means that the change in parameter value between iterations t and $t+1, p[\theta^{(t)}, \theta^{(t+1)}]$ depends only on the parameter value at time t, $\theta^{(t)}$. After a large number of iterations, a stationary distribution of the parameters is reached which is independent of the initial parameter values (at iteration 0), $\underset{\sim}{\theta}^{(0)}$. Hence convergence of MCMC methods is equivalently known as stationarity.

Unlike optimisation methods aiming to find parameter values yielding a unique maximum or minimum of a function, Markov chain Monte Carlo methods are stochastic and converge to probability distributions. The fundamental sampling algorithm of the BUGS package is the Gibbs sampling method for parameter sampling and estimation – in the sense of estimating the distribution of parameters. Suppose the parameter set $\underset{\sim}{\theta}$ has joint posterior density $P(\underset{\sim}{\theta} | y)$, often also called the target density in MCMC applications, then Gibbs sampling works by sampling from the conditional posterior density of each parameter given all the others and the data. So given the sampled value of the parameter vector $\underset{\sim}{\theta}^{(t)}$ at iteration t, the value of $\underset{\sim}{\theta}$ at iteration $t+1$ is obtained by drawing successively from the following distributions

$$\begin{aligned}
\theta_1^{(t+1)} &\sim f_1(\theta_1 | \theta_2^{(t)}, \theta_3^{(t)}, \ldots, \theta_p^{(t)}, y) \\
\theta_2^{(t+1)} &\sim f_2(\theta_2 | \theta_1^{(t+1)}, \theta_3^{(t)}, \ldots, \theta_p^{(t)}, y) \\
\theta_3^{(t+1)} &\sim f_3(\theta_3 | \theta_1^{(t+1)}, \theta_2^{(t+1)}, \ldots, \theta_p^{(t)}, y) \\
&\ldots \\
\theta_p^{(t+1)} &\sim f_p(\theta_p | \theta_1^{(t+1)}, \theta_2^{(t+1)}, \ldots, \theta_{p-1}^{(t+1)}, y).
\end{aligned}$$

Since parameters may be sampled in one–dimensional space, restrictions on them may be imposed, most simply lower and upper bounds (L_j, U_j) on the range for θ_j. Thus a more general scheme than the above can be specified of the form

$$\theta_j^{(t+1)} \sim f_j(\theta_j | \theta_1^{(t+1)}, \theta_2^{(t+1)}, \ldots, \theta_{j-1}^{(t+1)}, \theta_{j+1}^{(t)}, \ldots, \theta_p^{(t)}, L_j, U_j, y).$$

Such repeated sampling generates a dependent (autocorrelated) sequence of draws, which subject to certain regularity conditions, eventually forgets the starting value $\theta^{(0)}$ and converges to stationary sampling from the joint posterior density. Assuming in a particular application that convergence does eventually occur to a proper stationary distribution, it is therefore important to measure the rate of convergence.

In practice, the number of iterations taken for convergence to the stationary distribution depends on several factors:

- the complexity of the problem (problems with few parameters converge faster than those with many);
- the sample size (convergence is faster for smaller data sets);
- the way the parameters are expressed (convergence may be quicker for transformed versions of parameters);
- whether the prior and likelihood are conjugate;
- the sampling scheme adopted, including whether parameters are sampled individually or in groups (or blocks);
- the identifiability of the model from the data at hand, and the sampling characteristics of the data likelihood (e.g. single vs multiple modes);
- the closeness of the starting value $\theta^{(0)}$ to that of the stationary distribution.

In simulation via MCMC methods, one creates random draws of the parameter set θ from the posterior density $P(\underset{\sim}{\theta} | y)$ but they are not independent. Suppose we run a single long series $t = 1, \ldots, T$ of simulations, choosing T sufficiently large that $\underset{\sim}{\theta}^{(T)}$ is essentially a draw from $P(\underset{\sim}{\theta} | y)$. This is known as a burn-in period during which any dependence on the starting value is lost (typical values for T might be 500 or 1000). If we continue the sampling for an additional S iterations our aim is to obtain a true sample from the observed posterior density, since under stationary sampling the iterations

$$\underset{\sim}{\theta}^{(T+1)}, \underset{\sim}{\theta}^{(T+2)}, \ldots, \underset{\sim}{\theta}^{(T+S)}$$

are marginally distributed as $P(\underset{\sim}{\theta} \mid Y)$. Despite this, successive samples are still likely to be positively correlated, with samples close together more alike than those widely separated. This reduces the effective sample size, so that if draws of a certain parameter θ_k are highly correlated a larger sample is needed. If there is a lag 1 autocorrelation r_1 among the S sample draws of parameter θ_k then the effective sample size is $S(1 - r_1^2)$. Parameter estimates obtained from autocorrelated samples are still consistent, so that draws $z^{(t)}$ of a function $g(\theta^{(t)})$ of the parameters remains a consistent estimate of $E[g(\theta)]$ in the target density. However the variance of the sample average is not S^{-1} times the variance of a single iterate $z^{(t)}$, and involves covariances among the entire set of iterates $z^{(T+1)}, z^{(T+2)}, \ldots, z^{(T+S)}$. Hence the estimate

$$\bar{z} = S^{-1} \sum_{T+1}^{T+S} z^{(t)}$$

becomes less precise than one based on an independent sample.

One way to lessen the impact of dependence is via thinning or subsampling; posterior estimates are based on selecting only S' iterations separated by k intervening iterations. The gap k is chosen sufficiently large to make the retained values approximately independent. The resulting estimates will be more efficient but the extra computation is not necessarily justified: the average of all $kS' = S$ iterates is more precise than S' iterates based on retaining only every kth sample (Geyer, 1992).

An alternative strategy both for reducing the impact of dependence and ensuring convergence is to run Q parallel chains from diverse starting values (Gelman and Rubin, 1992). If stationarity is judged to have occurred by iteration T, on the basis of convergence diagnostics of variation within and between chains, then a further set of S iterations are run and the posterior summaries based on combining over chains from the second half of this set. In BUGS, the *GR diag* command provides plots of within- and between-chain variation in the traces and their ratio, using the methods of Brooks and Gelman (1998).

Assessment of convergence is only one among several issues in implementing Bayesian models; so also is the choice among MCMC methods (Besag *et al.*, 1995). The default option in BUGS for well-behaved problems (with log-concave densities) is Gibbs sampling. However, other Metropolis type methods are invoked for nonstandard models. The original Metropolis sampling scheme samples parameter values from a symmetric proposal density (such as the uniform or normal) and compares the posterior likelihood of the newly sampled parameter `p(θ.new|y)` with that of the current parameter estimate, `p(θ.current|y)`; the new value is automatically accepted as the updated parameter estimate if `p(θ.new|y)` exceeds `p(θ.current|y)`, and is otherwise accepted with probability `r=p(θ.new|y)/p(θ.current|y)`. Hastings (1970) extends the Metropolis scheme to allow for nonsymmetric proposal densities.

The rate at which new values are accepted in Metropolis–Hastings sampling is known as the acceptance rate. Gibbs sampling is in fact a special case of the Metropolis–Hastings algorithm, such that the acceptance rate is 100% (Gelman *et al.*, 1995, p. 328). In BUGS the slice-sampling algorithm is used for non log-concave densities on a parameter with prior defined on a restricted range (Neal, 1997). For an unrestricted prior, the Metropolis algorithm is invoked and is based on a symmetric normal proposal distribution, with standard deviation tuned to achieve an acceptance rate of between 20% and 40%.

Our focus here is on implementation of a range of modelling techniques via the Bayesian paradigm, and we move on to consider model evaluation against the assumed background of estimation involving stochastic sampling from posterior densities. Markov chain Monte Carlo methods are currently in a continuously evolving state, and there are many excellent reviews of alternative methods and of convergence issues – as well as continually updated web sites. Among recommended reviews are Brooks (1998), Brooks and Roberts (1998), and Sahu and Roberts (1999).

10.2 MODEL COMPARISON AND FIT

In the classical or sampling theory approach for testing alternative hypotheses M_0 and M_1 a test statistic is chosen and the hypothesis accepted or rejected according to whether the test statistic falls in a prespecified critical region. Often a hypothesis takes the form of a sharp or point hypothesis, for example, that a particular parameter or set of parameters have value zero. The Bayesian approach is likely to be cast more in terms of comparing models for the data rather than testing hypotheses; if a specific decision needs to be taken to reject or accept one or other hypothesis then a loss function may need to be established to express the costs of making the wrong choice.

The established method for comparing competing models in a Bayesian framework involves comparing posterior odds after estimating models separately with the current data y. Posterior odds assessment involves considering the marginal likelihoods of different models and the resulting Bayes factors. Our focus in previous chapters on parameter estimation, by sampling from full conditional densities or posterior likelihoods, involved the simplification

$$P(\theta|y) \propto P(y|\theta)\pi(\theta)$$

where $P(y|\theta)$ is the likelihood of the data given θ, and $\pi(\theta)$ specifies the prior probability densities of the parameters.

In model choice we shift back to consider the full Bayes formula,

$$P(\theta|y) = P(y|\theta)\pi(\theta)/P(y)$$

where $P(y)$ is the marginal likelihood. We can see that consideration of the marginal likelihood may imply different conclusions than just considering the value of θ that maximises the likelihood $L(\theta | y) \equiv P(y | \theta)$. Thus, by re-expressing the Bayes formula, the marginal likelihood can be written

$$P(y) = P(y | \theta)\pi(\theta)/P(\theta | y)$$

or following a log transform as

$$\log[P(y)] = \log[P(y | \theta)] + \log[\pi(\theta)] - \log[P(\theta | y)].$$

The terms $\log[\pi(\theta)] - \log[P(\theta | y)]$ act as a penalty to favour parsimonious models, whereas a more complex model virtually always leads to a higher log-likelihood $\log[P(y | \theta)]$.

Suppose we consider choosing between two models, M_0 and M_1. We can either fit them separately and consider their relative fit in terms of summary statistics, or attempt simultaneously to choose between models M_k as well as estimating their parameters $\theta_k | M_k$; for equal prior model probabilities, the best model is then the one chosen most frequently (i.e. with highest posterior probability of being selected). One can use such a simultaneous method to attach weights to each model $w_k = \Pr(\text{True model} = k | y)$, and form an overall average of the parameter θ by combining estimates from separate models according to these weights.

Whatever approach is adopted, whether separate or simultaneous model fitting, the posterior probabilities of each model being correct involve the marginal likelihoods $P(y | M_0)$ and $P(y | M_1)$. The marginal likelihood is thus the probability of the data y given a model $M_k (k = 0, 1)$, and is obtained by averaging over the priors assigned to the parameters in that model. The comparison of models is then based on the ratio of marginal likelihoods, or Bayes factor, of model M_1 as against model M_0. This is

$$B_{10} = P(y | M_1)/P(y | M_0).$$

This resembles a likelihood ratio except that the densities $P(y | M_k)$ are obtained by integrating over parameters rather than maximising, with

$$P(y | M_k) = \int P(y | M_k, \theta_k)\pi(\theta_k | M_k) d\theta_k, \quad k = 0, 1. \tag{10.1}$$

In the integral (10.1), θ_k is the set of parameters estimated under model M_k, and has prior density $\pi(\theta_k | M_k)$ under this model. $P(y | M_k, \theta_k)$ is the sampling distribution of y given model k and the values of the parameters included in it. There is no necessary constraint in these model comparisons that models 0 and 1 are nested with respect to one another – an assumption often necessarily made in classical tests of goodness of model fit.

Note, however, that marginal likelihood approaches are not the only possible Bayesian model fit technique. Some fit measures are based on posterior predictive checks which adopt some features of classical p-tests, or on posterior densities of the likelihoods $P(y | M_k, \theta_k)$ themselves. These options are discussed below.

The Bayes factor expresses the support given by the data for one or other of the models, in a similar way to the conventional likelihood ratio. Taking twice the log of the Bayes factor gives the same scale as the conventional deviance and likelihood ratio statistics. Approximate values for interpreting B_{10} and $2\log_e B_{10}$ are as in Table 10.1.

For large data sets differences in the log marginal likelihoods are the natural measure of model comparison, as probabilities themselves become numerically intractable.

The posterior probability of a model can be obtained from the prior probability and the marginal likelihood via the formula

$$P(M_k \mid y) = P(M_k)P(y \mid M_k)/P(y) \tag{10.2}$$

where $P(M_k)$ is the prior weight on model M_k. It follows that

$$\begin{aligned} P(M_1 \mid y)/P(M_0 \mid y) &= P(y \mid M_1)P(M_1)/[P(y \mid M_0)P(M_0)] \\ &= [P(y \mid M_1)/P(y \mid M_0)][P(M_1)/P(M_0)], \end{aligned}$$

namely that the posterior odds on model M_1 being correct equal the Bayes factor times the prior odds on M_1.

In the case of two alternative models the denominator $P(y)$ in (10.2) is the total likelihood of the data over both models

$$P(y) = P(M_1)P(y \mid M_1) + P(M_0)P(y \mid M_0).$$

If the two models are assigned equal prior probabilities, so that $P(M_0) = P(M_1)$, the posterior probability that model 1 is correct is

$$P(M_1 \mid y) = P(y \mid M_1)/[P(y \mid M_1) + P(y \mid M_0)].$$

This reasoning extends naturally to a situation where $K > 2$ models are being compared. Unless the posterior probability of one model alone is overwhelming, we may average over parameter or function values obtained from different models. This is most likely to happen in small data sets where we are considering regression models with a common model structure, having the same links, same random effects structure and same type of regression (e.g. both linear), and the issue concerns inclusion or exclusion of certain covariates. In larger samples, differences in marginal likelihood (or more usually, log marginal likelihoods) tend to be such as to favour one model choice conclusively.

Table 10.1 Guidelines for Bayes factors.

B_{10}	$2\log_e B_{10}$	Interpretation
Under 1	Negative	Supports M_0
1 to 3	0 to 2	Weak support for M_1
3–20	2–6	Support for M_1
20–150	6–10	Strong evidence favouring M_1
Over 150	Over 10	Very strong support for M_1

Given equal prior probabilities on the models, the weights in this average would be

$$w_k = P(M_k \mid y) = P(y \mid M_k)/[(P(y \mid M_1) + P(y \mid M_2) + \ldots + P(y \mid M_K)].$$

The posterior mean for parameter T would thus be an average over models

$$E(T \mid y) = \sum w_k E(T_k \mid y, M_k) = \sum_k w_k t_k$$

where $t_k = E(T_k \mid y, M_k)$ is the posterior mean under model M_k. The posterior variance is obtainable as

$$\text{Var}(T \mid y) = \sum_k [\text{var}(T_k \mid y, M_k) + t_k^2] - \{E(T \mid y)\}^2.$$

In the case where there is model uncertainty, these results show that selecting a single model will overstate the precision of parameters and other functions derived from assuming that model is the only correct one (i.e. with weight $w_k = 1$).

10.3 SEPARATE MODEL ESTIMATION

10.3.1 Overall fit measures

Criteria for fit based on marginal likelihood approximations

Suppose we compare a set of possible models for y by successive and separate estimation of each. In the following the notation $\{\theta, M\}$ should then be taken as implicitly referring to separate parameter sets and models $\{\theta_k, M_k\}$. As noted above, the most fundamental Bayes model assessment procedures rest on estimates of the marginal likelihood. As for other functions $G(\theta, y)$ of model parameters and data, we can apply numerical quadrature or the Laplace approximation to obtain an estimate of the marginal likelihood. Raftery (1996) and Raftery and Richardson (1995) review approximations to the Bayes factor itself based on the Laplace approximation. These are intended to exploit standard output from packages such as GLIM or SAS, in terms of the maximum likelihood estimate, the deviance and the inverse of the observed information matrix. Thus writing

$$P(y \mid M) = \int P(y \mid \theta, M)\pi(\theta \mid M)d\theta \tag{10.3}$$

we expand the log of the integrand in (10.3), namely $g(\theta) = \log[P(y \mid \theta, M)\pi(\theta \mid M)]$ by Taylor series about $\tilde{\theta}$, the posterior mode of the parameter vector which maximizes $g(\theta)$. Because $g'(\tilde{\theta}) = 0$ by definition, this expansion gives

$$g(\theta) \approx g(\tilde{\theta}) + 1/2(\theta - \tilde{\theta})g''(\tilde{\theta})(\theta - \tilde{\theta}).$$

Substituting in (10.3) and remembering $g(\tilde{\theta})$ is constant, this gives

$$P(y|M) \approx \exp(g(\widetilde{\theta})) \int \exp[1/2(\theta - \widetilde{\theta})g''(\widetilde{\theta})(\theta - \widetilde{\theta})]d\theta. \quad (10.4)$$

The integrand in (10.4) is proportional to a multivariate normal with precision

$$A = [-g''(\widetilde{\theta})].$$

This leads to the marginal likelihood approximation[1]

$$P(y|M) \approx \exp(g(\widetilde{\theta}))(6.2832)^{p/2}|A|^{-0.5}$$

or equivalently

$$\log P(y|M) \approx \log P(y|\widetilde{\theta}, M) + \log\pi(\widetilde{\theta}|M) + p/2\log(6.2832) - 1/2\log|A|.$$

In data sets with n large, $\widetilde{\theta} = \hat{\theta}$ where $\hat{\theta}$ is the MLE. Also $A \approx nI$ where I is the expected information matrix for a single observation, which means $|A| = n^p|I|$. Suppose the prior $\pi(\theta|M)$ is multivariate normal with mean $\hat{\theta}$ and precision I (i.e. the prior is equivalent to a single extra observation). With the aid of this assumption

$$\begin{aligned}\log P(y|M) &\approx \log P(y|\hat{\theta}, M) + \log\pi(\hat{\theta}|M) + p/2\log(6.2832) - 1/2\log|A| \\ &= \log P(y|\hat{\theta}, M) + [1/2\log|I| - p/2\log(6.2832)] + \\ &\qquad p/2\log(6.2832) - p/2\log n - 1/2\log|I| \\ &= \log P(y|\hat{\theta}, M) - p/2\log n.\end{aligned}$$

This quantity is known as the Bayes information criterion (BIC), with

$$\text{BIC} = \log P(y|\hat{\theta}, M) - p/2\log n. \quad (10.5)$$

This criterion penalises models which improve fit but at the expense of more parameters, and so serves as a measure to assess model parsimony (Raftery, 1995). The appropriate definition of the sample size n is discussed by Raftery (1995). For example, in an $I \times J$ contingency table of counts m_{ij} the sample size would not be IJ but the sum $\sum\sum m_{ij}$. For further discussion on the utility of the BIC approximation and the appropriate definition of n a recent review is provided by Weakliem (1999).

For large samples, the Laplace method can also be used to approximate a Bayes factor on model 1 vs model 2. Thus if

$$B_{12} = P(y|M_1)/P(y|M_2)$$

then

$$\begin{aligned}\log B_{12} &= \log P(y|M_1) - \log P(y|M_2) \\ &\approx \log P(y|\hat{\theta}_1, M_1) - \log P(y|\hat{\theta}_2, M_2) - \log n(p_1 - p_2)/2.\end{aligned}$$

So

$$2\log B_{12} \approx G^2 - v\log n \quad (10.6)$$

[1] We set $2\pi = 6.2832$ in order to reserve the notation $\pi(\theta)$ for the prior.

where G^2 is the likelihood ratio comparing the models for $v = p_1 - p_2$ degrees of freedom. When the comparison model M_s is the saturated model then the test for M_k against M_s involves the GLM deviance for model k:

$$\text{BIC}_k = \text{Deviance}(M_k) - v_k \log n. \tag{10.7}$$

It can be seen that (10.5) assumes knowledge of a maximum likelihood solution. This may be approximated (see Raftery, 1996; Gelman *et al.*, 1996) by that θ giving the maximum $L^{(t)}_{\max}$ of the log-likelihood values $L^{(t)} = \log P(y \mid \theta^{(t)}, M)$ observed in a substantial MCMC sampling run. An alternative criterion for use in ad hoc measures of the BIC type is the average $\bar{L}$ of the sampled log-likelihoods, leading to the measure (Carlin and Louis, 1997, Chapter 6):

$$\text{BIC}' = \bar{L} - \frac{p}{2} \log n. \tag{10.8}$$

Approximations such as (10.5) have improved validity for large sample sizes, and are most straightforward in models containing only fixed effects, such as linear or general linear regression models where the only parameters are regression coefficients, and possibly residual variances. A problem with both the Laplace and BIC approximations is that in complex hierarchical models involving latent class probabilities or random effects, the true dimensionality (number of parameters p) of the model is not known, and also that the number of parameters may increase with sample size n.

Again an approximation may be invoked, based on an estimate of the effective number of parameters. Spiegelhalter et al (1999) propose that deviances $D^{(t)}$ be monitored in the same way as log-likelihoods, these deviances being appropriate to the form of sampling density (Gaussian, Poisson, etc.). Suppose also that the components ϕ of θ defining the means of y_i are distinguished from the remainder, giving $\theta = (\phi, \phi_C)$. Then the average deviance $\bar{D}$ from the sample run to estimate θ may be compared with the deviance $D(\bar{\phi})$, based on calculating the deviance at the posterior mean of ϕ, $\bar{\phi}$. In BUGS this calculation requires a secondary 'run', with a single iteration, and using a suitably reduced program to calculate the deviance after summarising an initial model fit to find $\bar{\phi}$. Thus the difference

$$p.eff \approx \bar{D} - D(\bar{\phi})$$

provides an estimate of the effective number of parameters which may be used in BIC type fit measure.

Estimates of the marginal likelihood

Another approach forms an estimate of the marginal likelihood (e.g. its mean value and variance) from the output of iterative Monte Carlo sampling schemes. In practice we often focus on the log marginal likelihood. Omitting the model index for simplicity, our goal is to estimate

$$P(y) = \int P(y \mid \theta) \pi(\theta) d\theta.$$

The most direct method to estimate $P(y)$ would then seem to be to sample a series of values of length T from the prior $\pi(\theta)$ itself, namely $\theta^{(1)}, \theta^{(2)}, \ldots, \theta^{(T)}$. This series would be of much greater length than the typical MCMC run since the prior is not being updated. Then the estimate of the marginal likelihood is the average of the likelihoods $L^{(t)} = P(y|\theta^{(t)})$ at each of these sampled points,

$$\hat{P}_1(y) = T^{-1} \sum L^{(t)}.$$

This is a feasible approach for priors which are relatively close to the region of highest likelihood. If the prior and likelihood are possibly discordant, one option is to use importance sampling to improve sampling of θ so that it is concentrated in areas of high likelihood. This means sampling $\theta^{(t)}$, $t = 1, \ldots, T$, from a density $g(\theta)$ known as the importance function. The marginal likelihood is then

$$P(y) = \int P(y|\theta)\ \pi(\theta)\ d\theta = \int L(\theta)\ g(\theta)\ [\pi(\theta)/g(\theta)] d\theta.$$

With samples $t = 1, \ldots, T$ of θ now taken from g, and with $w_t = \pi(\theta^{(t)})/g(\theta^{(t)})$, this implies the estimator

$$\hat{P}(y) = \left[\sum_t w_t L(\theta^{(t)})\right] \Big/ \left[\sum_t w_t\right].$$

We can write this estimator generically as

$$\hat{P} = \|L\pi/g\|_g / \|\pi/g\|_g.$$

An obvious choice for g as the importance density is the posterior density $P(\theta|y) \propto L(\theta)\pi(\theta)$, or generically $g = L\pi$. The estimator is then seen to be

$$\hat{P}_2(y) = 1\Big/\left[T^{-1} \sum \{1/L(\theta^{(t)})\}\right],$$

namely the harmonic mean of the likelihood values from a standard MCMC run to produce posterior estimates of θ itself. This estimator may be unstable if by chance a few low-likelihood values are present in the sampling output; the impact of such aberrant cases can be monitored by batching the MCMC output (e.g. in bands of 5000 iterations) to assess stability in the harmonic mean. We might then average over batches, or perhaps form some robust estimate of the mean.

Other options are discussed by Raftery (1996) and Newton and Raftery (1994). They include one which uses an importance sampling function based on combined samples from prior $\pi(\theta)$ and posterior $P(\theta|y)$. Thus define

$$g(\theta) = \delta\pi(\theta) + (1 - \delta)P(\theta|y) \tag{10.9}$$

with $0 < \delta < 1$, and typically δ small for numeric stability. They also propose a synthetic estimator, based on (10.9), which actually avoids sampling from the prior density. Thus suppose T values of the likelihood are available from an MCMC output. Then sampling from the prior can be avoided by imagining

$\delta T/(1-\delta)$ further values of θ are notionally sampled from the prior with likelihood values exactly equalling their expectation $P(y)$. The resulting estimator is obtained via a linear iterative scheme (relevant FORTRAN code is contained in the Appendix to this chapter).

More computationally involved is to initially estimate $P(\theta|y)$ from a pilot run (e.g. using a univariate or multivariate normal or Student t approximation, some type of histogram smoothing, or more complex mixtures of normal approximations in the presence of skewness of multimodality). We would then re-estimate the model to draw θ from

$$H(\theta) = \delta\pi(\theta) + (1-\delta)P(\theta|y)$$

with separate sampling both from the posterior $P()$ of θ and from its prior $\pi()$. We then evaluate the probabilities of the drawn values in terms of the same prior densities $\pi(\theta)$ used in the pilot run, and in terms of the parameters in the estimated posterior densities $\hat{P}(\theta|y)$.

For example, suppose a logit regression model for $i = 1, \ldots, n$ binary outcomes y_i just contained regression parameters $b_j, j = 1, \ldots, p$ and that the priors on these parameters were $\mathrm{N}(M_j, T_j^{-1})$. Further assume the (estimated) posteriors $\hat{P}(b_j|y)$ of the parameters, as obtained from the previous sampling run, were taken as approximately univariate normal and so can be summarised in the form $\mathrm{N}(B_j, R_j^{-1})$. Then developing an importance sample in a repeat model estimation run (using the same priors as in the first run) could be coded in BUGS as follows:

```
for (j in 1:p) { # sample from prior without updating
        b.pi[j] ~ dnorm(M[j],T[j])
        # sample parameters updated by Bernoulli likelihood
        b[j] ~ dnorm(M[j],T[j])
# sample parameters from mixture of prior and posterior in new run
        b.m[j] <- del*b.pi[j] + (1-del)*b[j]}
# estimated posterior density of regn coeffs (from previous run) has
  mean B[],
# prec'n R[].
# Derive probs of prior sample and of mixture of prior and posterior
        log.pi[j] <- 0.5*log(T[j]/6.28) -0.5*T[j]*pow(b.pi[j]-
          M[j],2)
        log.P[j] <- 0.5*log(R[j]/6.28)-0.5*R[j]*pow(b[j]-
          B[j],2)
for (i in 1:n) { nu.m[i] <- b.m[1] + b.m[2]*x[i,1]...+b.m[p]*x[i,p]
        LL.m[i] <- y[i]*log(phi(nu.m[i])) + (1-y[i])*
          log(1-phi(nu.m[i]))
# sampling for new model
        y[i] ~ dbern(nu[i])
        nu[i] <- b[1] + b[2]*x[i,1]... +b[p]*x[i,p]}
T.pi <- sum(log.pi[])
T.P <- sum(log.P[])
```

```
# weights for importance sample of D.m
w <- exp(T.pi)/(del*exp(T.pi) +(1-del)*exp(T.P))
# total log-likelihood of b.m[]
ll.m <- sum(LL.m[])
```

The total log-likelihoods and the weights (`ll.m` and `w` in the above code) would be sampled and the weighted average of the exponentiated log-likelihoods calculated, for example in a spreadsheet. This gives an estimator

$$\hat{P}_3(y) = \left[\sum_t w_t L(\theta^{(t)})\right] \Big/ \left[\sum_t w_t\right].$$

10.3.2 Predictive fit measures

Cross-validation approaches

An indirect approach to obtaining the marginal likelihood involves predictive cross-validation. Cross-validation methods are well established in frequentist statistics and involve predictions of a subset y_r of y when only the complement of y_r, denoted $y_{(r)}$ is used to update the prior. Thus if we left out observation 1, $y_{(1)}$ would consist of observations $2, \ldots, n$. Suppose an $n \times p$ covariate matrix x is also partitioned into x_r and $x_{(r)}$. Then even if π, and hence also $P(y)$, is improper, the conditional predictive density

$$P(y_r \mid y_{(r)}) = P(y)/P(y_{(r)}) = \int P(y_r \mid \theta, y_{(r)}, x)\pi(\theta \mid y_{(r)})d\theta$$

is proper because the posterior based on $y_{(r)}$, namely $\pi(\theta \mid y_{(r)})$ is proper. Then $P(y)$ is defined equivalently by the set $P(y_r \mid y_{(r)})$, and Geisser and Eddy (1979) suggest the product

$$\hat{P}_4(y) = \prod_{r=1}^{n} P(y_r \mid y_{(r)}) \tag{10.10}$$

of cross-validation densities or conditional predictive ordinates (CPO), obtained by leaving one case $r = 1, \ldots, n$ out at a time, as an estimate for the marginal likelihood. The ratio of $\hat{P}_4$ for two models M_0 and M_1 is then a surrogate for the Bayes factor B_{01}.

Cross-validation methods have a broader role in model checking than simple overall model fit, namely in terms of identifying influential cases, outliers and other model discrepancies. Gelfand *et al.* propose several checking functions involving comparison of the actual observations with predictions $\hat{y}_r$ from $P(y_r \mid y_{(r)})$. The simplest is the prediction error

$$g_{1r} = y_r - \hat{y}_r$$

with expectation

$$d_{1r} = y_r - \mathrm{E}(y_r \mid y_{(r)}).$$

If $\sigma_r^2 = \mathrm{var}(y_r \mid y_{(r)})$ then a standardised checking function is

$$e_{1r} = d_{1r}/\sigma_r$$

and $D_1 = \sum(e_{1r})^2$ can be used as an index of overall model fit. Under approximate normality, 95% of the e_{1r} should be within -2 to $+2$, and systematic patterns (e.g. as revealed by plots against covariates) indicate model inadequacy.

Another check $g_{2r} = I(\hat{y}_r \leq y_r)$ is simply whether the prediction exceeds or is less than the actual observation y_r, so involving the step() function in BUGS. The expectation is $d_{2r} = \Pr(\hat{y}_r \leq y_r \,|\, y_{(r)})$ and in an adequate model these would be uniformly distributed with average around 0.5. A large number of d_{2r} close to 0 or 1 shows actual observations in the tails of the predictive density. An overall index of fit is then $D_2 = \sum(d_{2r} - 0.5)^2$.

A third possible check involves assessing whether the prediction is contained in a small interval $(y_r - \varepsilon, y_r + \varepsilon)$ around the true value. The function

$$g_{3r} = I(y_r - \varepsilon \leq \hat{y}_r \leq y_r + \varepsilon)/2\varepsilon$$

then has an expectation

$$d_{3r} = P(y_r \,|\, y_{(r)})$$

(i.e. the CPO) when ε tends to zero. The product D_3 of the d_{3r}, namely

$$\hat{P}_5(y) = \prod_{r=1}^{n} d_{3r}$$

is an estimate of the marginal likelihood, as in (10.10) above, and so leads to a pseudo-Bayes factor (Gelfand *et al.*, 1992; Gamerman, 1998). As for the total marginal likelihood we can approximate the statistics d_{3r} without needing to actually exclude case r and carry out n separate estimations (Gelfand and Dey, 1994). Thus if we monitor the inverse likelihood of each case for T iterations after a burn-in period, then an estimate of the CPO is obtained as

$$\text{CPO}_i^{-1} = T^{-1} \sum_{t=1}^{T} [P(y_i \,|\, \theta^{(t)}, B_i)]^{-1}.$$

Summary predictive fit and posterior predictive checks

Predictive cross-validation is a robust procedure in models where omission of cases r one by one does not lead to complex estimation issues but may be difficult to implement in samples with many cases, or in models involving random effects or latent mixtures. Carlin and Louis (1996) propose predictive measures of a different sort, based on sampling 'new' observations y_{new} or replicates of the observed data y. These predictions use the estimated means for case r at iteration t to produce predictions of y from the appropriate sampling density. For example if the y_r, $r = 1, \ldots, n$, were Normal, and $m_r = \beta x_r$ were the means for case r in a regression model, we would predict new values at each iteration as follows

$$y_{r.\text{new}}^{(t)} \mid \beta, s^2 \sim \text{N}(m_r^{(t)}, s^{2(t)}).$$

More generally we predict a vector of new observations $y_{\text{new}}^{(t)}$ given the current sampled value of a parameter set $\theta^{(t)}$, with the goal of deriving the predictive density

$$P(y_{\text{new}} \mid y) = \int P(y_{\text{new}} \mid \theta) P(\theta \mid y) d\theta.$$

We then apply standard criteria such as the checking functions above, or overall fit measures (sum of squares, deviances, etc.) comparing the actual and replicated observations. Thus for Poisson data with observed counts y we could sample $y_{r.\text{new}}$ from the Poisson means of the model and then derive the deviance measure

$$D_{\text{new}} = D(y_{\text{new}}, y) = 2 \sum_r \{y_r \log(y_r / y_{r.\text{new}} \mid y)\}.$$

We could first find the average over an MCMC run of the $D_{\text{new}}^{(t)}$ to provide a measure D_1, known as the expected predictive deviance (EPD). We may also derive a deviance D_2 calculated at the average value of the $y_{r.\text{new}}$. Then Carlin and Louis show how the difference $D_1 - D_2$ may be interpreted as a predictive corrected fit measure, approximately equal (for Poisson data) to

$$E\left\{ \sum_r [y_{r.\text{new}} - \text{E}(y_{r.\text{new}} \mid y)]^2 / \text{E}(y_{r.\text{new}} \mid y) \right\}.$$

Another posterior predictive approach for model assessment is proposed by Gelman *et al.* (1996), developing work by Rubin and Stern (1994). Denote the actual data as $y_{\text{obs}} = y$. For a realised discrepancy measure $D(y_{\text{obs}};\theta)$, such as the deviance or chi-square, a reference distribution P_{R} is derived from the joint distribution of y_{new} and θ:

$$P_{\text{R}}(y_{\text{new}}, \theta) = P(y_{\text{new}} \mid \theta) P(\theta \mid y_{\text{obs}}).$$

The realised value of the discrepancy $D(y_{\text{obs}};\theta)$ may then be located within its reference distribution, for example by a tail probability

$$p_b(y_{\text{obs}}) = P_{\text{R}}[D(y_{\text{new}};\theta) > D(y_{\text{obs}};\theta) \mid y_{\text{obs}}].$$

In practice this involves calculating $D(y_{\text{new}}^{(t)}, \theta^{(t)})$ and $D(y_{\text{obs}}, \theta^{(t)})$ in an MCMC run of length T and then calculating the proportion of samples for which $D(y_{\text{new}}^{(t)}, \theta^{(t)})$ exceeds $D(y_{\text{obs}}, \theta^{(t)})$. A recent application of the posterior predictive p-value method in a Bayes structural equation model is presented by Scheines *et al.* (1999).

10.3.3 Posterior likelihood comparison methods

A slightly controversial approach to model assessment, though arguably within the fully Bayes arena, involves direct consideration of posterior distributions of the data likelihoods $L(\theta_k \mid y) = P(y \mid \theta_k, M_k)$ and of log-likelihood ratios

$\text{LR} = L(\theta_0\,|\,y)/L(\theta_1\,|\,y)$. Thus Dempster (1997) proposed an 'inferential pairs' (γ, k) rule, such that we compare likelihoods against a threshold k, with k small, e.g. $k = 0.1$ or $k = 0.05$, and accept M_1 if the likelihood ratio

$$\text{LR} = P(y\,|\,\theta_0, M_0)/P(y\,|\,\theta_1, M_1)$$

is less than k with a high probability γ. For example we might set γ at 0.8 or 0.6. We could set up the comparison indicator within an MCMC run, and assess γ as the empirical rate for which the indicator is true.

Aitkin (1997) proposes a development on this where we vary k over a set of possible values such as 0.1, 0.2, 0.3, 0.4, 0.5 and 1 and assess changes in the posterior probability π_k that $\text{LR} < k$. The test that $\text{LR} < 1$ is equivalent to a version of the standard P test, and is the least conservative criterion, possibly leading to overstatement of the evidence against M_0. If, however, we wish to obtain stronger evidence against M_0 we take a small value such as $k = 0.1$. He cites the case of a mean of $\bar{y} = 0.4$ obtained from a sample of $n = 25$ cases from a normal population with known variance 1. The null model M_0 specifies is that the normal mean μ is zero, $\mu_0 = 0$. The P test involves assessing the probability

$$\text{LR}(\mu_0)/\text{LR}(\mu) < 1$$

and leads to a probability of 0.046 on M_0 being true, and a high (i.e. possibly overstated) probability of the alternative M_1 that $\mu \neq 0$. By comparison the more stringent test

$$\text{LR}(\mu_0)/\text{LR}(\mu) < 0.2$$

leads to a probability on M_0 being true of 0.327.

A further extension of this approach leads to the posterior Bayes factor approach of Aitkin (1991), which argues that LR comparisons and Bayes factors based on them, are less subject to distortion by prior assumptions than the conventional Bayes Factor. For example, the latter is not defined for improper priors whereas LR comparisons via the (γ, k) rule are. A worked example below illustrating these issues involves data on counts of brain damage in children following administration of an allegedly defective vaccine.

10.4 SIMULTANEOUS MODEL ASSESSMENT

Carlin and Chib (1995) propose a simultaneous model selection procedure sampling over the joint space defined by model indicators and the parameters of each model. Thus consider several possible models $M_1, \ldots, M_K$, with the best model index M as an additional parameter to derive as well as the posterior densities of the parameters $\theta_1, \ldots, \theta_K$ of the model themselves. Suppose the dimensions of the models, in terms of stochastic parameters defining them, are $(d_1, \ldots, d_K)$. Then we can envisage the simultaneous model selection problem as involving estimation of a superparameter $\theta.S = \{\theta_1, \ldots, \theta_K\}$ and the index $M = j$ indicating a specific model chosen with $j\varepsilon\{1, \ldots, K\}$ at any iteration within the MCMC run. For simplicity it is assumed that the parameters in different models are non-overlapping, that is they have no common component. There may be circumstances where certain parameters (e.g. regression intercepts) may be treated as common across different models.

The joint density of the data, the superpopulation parameter $\theta.S$, and the model index is

$$P(y, \theta.S, j) = P(y \mid \theta.S, j)\ p(\theta.S \mid j)\ P(j) \tag{10.11}$$

where $P(j)$ is the prior probability that model j is the true one. Typically we might take

$$P(1) = P(2) = \ldots P(K) = 1/K.$$

Given non-overlapping parameters,

$$P(y \mid \theta.S, j) = P(y \mid \theta_j, j).$$

Also we assume that the parameters of different models are conditionally independent given that one of them is selected, so that the second component in the joint density expansion is

$$P(\theta.S \mid j) = \prod_{i=1}^{K} P(\theta_i \mid j). \tag{10.12}$$

The prior $P(\theta_j \mid j)$ within the product (10.12) is the usual one (the 'true' prior) specifying prior assumptions on the parameters of model j when it is selected. The prior $P(\theta_i \mid j)$ for $i \neq j$ is known as a pseudo-prior and specifies the prior assumptions made about the parameters of model i given that another model (j) is selected. This prior is somewhat counterintuitive but needed if the chain is to switch between models. The full conditional for parameter θ_j is then proportional to

$$P(y \mid \theta_j, j)\ P(\theta_j \mid j)$$

when model j is chosen, but

$$P(\theta_j \mid i)$$

when model i is chosen. Usually we would take common pseudo-priors $P(\theta_j \mid i)$ across all $i \neq j$.

Carlin and Chib (1995) recommend using separate model estimates from pilot runs to provide appropriate parameters for the pseudo-priors. Thus an application of this procedure below to the onion bulb growth data of Gelfand *et al.* (1992) involves choosing between a Gompertz and logistic growth model for the onion bulb evolution through time. We carry out separate pilot runs with a nonlinear regression to estimate Gompertz growth curve parameters θ^{G} and another to estimate logistic parameters θ^{L}, with estimated precisions T^{G} and T^{L}. We then use a precise (i.e. informative) prior based on these pilot estimates as the pseudo-prior and a (considerably) less precise prior centred on these estimates as the true prior. Thus the true prior for the Gompertz (when the Gompertz model is selected) might be

$$\theta \sim \mathrm{N}(\theta^{\mathrm{G}}, C/T^{\mathrm{G}})$$

with $C = 1000$ (say) and the pseudo prior for the Gompertz parameters when the logistic model is selected would be

$$\theta \sim \mathrm{N}(\theta^{\mathrm{G}}, 1/T^{\mathrm{G}}).$$

An alternative approach to simultaneous model assessment is based on the reversible jump (RJ) MCMC algorithm of Green (1995). This involves defining dummy parameters u and v which ensure matching dimensions to enable 'reversible jumps' between models $\{j, k\}$ with p_j and p_k parameters. Suppose the current model is j with parameters θ_j. Then the RJ algorithm proposes a new model k with probability $B_{j,k}$ and generates u and v from a proposal densities $q(u|\theta_j, j, k)$ and $q(v|\varphi_k, k, j)$. If φ_k is the parameter set in model k, then set $(\varphi_k, v) = g_{jk}(\theta_k, u)$ where g_{jk} is a function chosen to ensure $p_j + \dim(u) = p_k + \dim(v)$. The move from model j to model k is accepted with a probability calculated by comparing the posterior densities (given the data) of the move from j to k as against staying in j.

Example 10.1 Separate model assessment via cross-validation: Onion bulb growth Gelfand *et al.* (GDC, 1992) present data on the evolution through time of the dry weight of onion bulbs. These data are also considered by Gamerman (1997). For times $X_t = 1, 2, \ldots, 15$, the growth data are

$$Y_t = (16.08, 33.83, 65.8, 97.2, 191.55, 326.2, 386.87, 520.53, 590.03, \\ 651.92, 724.93, 699.56, 689.86, 637.56, 717.41).$$

Alternative models under consideration for these data by GDC are the Gompertz and logistic with respective forms

$$Y_t = \alpha_1 \exp(-\alpha_2\{\alpha_3^{Xt}\}) + u_{1t}$$

with u_t unstructured errors, $u_{1t} \sim \mathrm{N}(0, 1/\tau_1)$, and

$$Y_t = \beta_1(1 + \beta_2\beta_3^{Xt})^{-1} + u_t$$

with $u_{2t} \sim \mathrm{N}(0, 1/\tau_2)$.

We first apply the predictive discrepancy measures of GDC themselves. For the third measure we take the normal density of y_r with mean $\hat{y}_r$. We obtain posterior means of g_{1r}, g_{2r} and g_{3r} for cases $r = 1, \ldots, n$ from 10 000 iterations with 500 burn-in (see *Program 10.1 Onion Bulb Growth*). In a subsequent calculation, we combine these over cases into the summary measures outlined above. This gives the results in Table 10.2.

For the first two statistics lower values show better fit, while for the third higher values show better fit. The ratio of D_3 for model 2 to model 1 forms a pseudo-Bayes factor which at approximately 30 is strongly in favour of the logistic. Detailed results for these statistics show the largest discrepancies for cases 11 and 14. For the Gompertz there are also problems in poor fit to the first few observations.

Table 10.2 Onion bulb growth, summary fit measures.

Statistic	Gompertz	Logistic
D_1	244	61.7
D_2	1.91	1.97
D_3	1.5E-34	4.5E-33

Example 10.2 Joint space simultaneous model assessment: Onion bulb growth We next consider the supermodel approach to selection between the above two models for the onion bulb growth data. This entails initial 'pilot' runs (separate model estimations) to develop appropriate pseudo-priors. Following GDC, we adopt parameter transforms

$$A_1 = \alpha_1, A_2 = \log(\alpha_2), A_3 = \text{logit}(\alpha_3),$$
$$B_1 = \beta_1, B_2 = \log(\beta_2), B_3 = \text{logit}(\beta_3).$$

Then running the Gompertz model, with noninformative priors on the parameters $(A_j, j = 1, 2, 3)$ we obtain mean estimates (with standard deviations)

$$A_1 = 722\ (21.9), A_2 = 2.57\ (0.29), A_3 = 0.54\ (0.14) \text{ and } \tau_1 = 0.00088\ (0.00036).$$

Similarly running the logistic model, also with vague priors on $(B_j, j = 1, 2, 3)$ we obtain estimates

$$B_1 = 702(14.8), B_2 = 4.5\ (0.37), B_3 = -0.008\ (0.29) \text{ and } \tau_2 = 0.0014\ (0.00052).$$

Note that the third logistic regression parameter might initially seem nonsignificant but in the original scale, β_3 is estimated as 0.5 with standard deviation 0.03.

We use the pilot runs to select initial means and precisions for the pseudo-priors of the supermodel approach. Thus if the Gompertz is model 1 and the logistic is model 2, the pseudo-prior of the Gompertz parameter A_2 under the logistic model 2 is taken as having mean 2.57 and precision $F^2/(0.29^2)$ with $F = 1$. The true prior of A_2 under the Gompertz model 1 is also taken as having mean 2.57 but precision $F^2G^2/(0.29^2)$, where $G < 1$. This corresponds to a low-precision prior such as we would adopt in usual modelling, with typical values being $G = 0.01$ or $G = 0.005$. Given the pilot estimates of the precisions τ_1 and τ_2, their pseudo-priors are set at $G(6, 6800)$ and $G(7, 5000)$. The resulting model (see *Program 10.2 Onion Bulb SuperModel Selection*) is initialised with default settings (for the true priors) of $\tau_1 = \tau_2 = 1$ and (model node) $j = 1$.

Taking $(F, G) = (1, 0.05)$, then $(F, G) = (1, 0.01)$ and finally $(F, G) = (1, 0.005)$ gives posterior probabilities, over single runs of 20 000 iterations, favouring the logistic model. Thus $P(j = 2)$ is estimated as 0.958, 0.956 and 0.958 respectively with the trace plots on the model indicator j showing a regular movement between models. However, lower values of G, such as in the pair $(F, G) = (1, 0.001)$ show less mixing between chains. A single run of 100 000 iterations with this value of (F, G) shows the chain trapped in selecting initially model 1 until about iteration 35 000, when the chain does start to mix between models though selecting model 2 much more frequently. The second half of the run of 100 000 then gives a probability on the logistic model of 0.957. The prior model odds were equal, so converting this to a Bayes factor gives $0.957/0.043 = 22.2$, broadly similar to the result of the cross-validation approach.

To obtain estimates of the Gompertz ($j = 1$) and logistic ($j = 2$) growth parameters from this approach we would average over the conditional posterior values for θ_j, obtained from selecting only those iterations when the model index took the appropriate value. Given the scarcity of iterations when $j = 1$ is selected, this may require a long run to generate a precise estimate.

Example 10.3 Separate model assessment via marginal likelihood fit measures: Simulated Gaussian mixture Raftery (1996) compares various model selection approaches to Normal density latent mixture problems with (a) simulated data and (b) a real example involving stellar populations. Here we follow his analysis under option (a) in simulating $n = 100$ points y_r from a Normal mixture with two latent groups, with respective means $\mu_j (j = 1, 2)$ of 0 and 6, respective variances V_j of 1 and 4, and equal prior masses ρ_j of 0.5. Raftery compared a Laplace approximation to the marginal likelihood with the harmonic mean estimator $\hat{P}_2$ above, and with the BIC approximation (10.5). He was especially concerned with identifying the true solution of $C = 2$ groups against a $C = 3$ group solution, which was by chance also suggested for this small sample despite its mode of generation. However, he found no clear superior estimator of the marginal likelihood and hence no unambiguous estimate of the Bayes factor.

Obtaining the harmonic mean estimate from an MCMC run involves monitoring the log-likelihood L, and extracting values from a long run of values $L^{(t)}$ (e.g. 5 or 10 thousand iterations). Then a subsequent secondary analysis in a spreadsheet, say, is needed to find the likelihoods at each iteration, the inverse likelihood, the average of the inverse likelihoods, and then the reciprocal of this quantity. Then we would take the log of the latter as the estimate of log $\hat{P}_2$. Direct monitoring of the relevant quantities in BUGS is not possible because of numeric underflow, and relevant calculations may even become problematic in a spreadsheet for large sample sizes and highly negative log-likelihoods. For larger samples it may be possible to calculate a Bayes factor even if the harmonic mean estimate of the marginal likelihood is not obtainable directly from the sampled log-likelihoods for numeric reasons. Adding a suitable constant in the exponentiation from log-likelihoods (L_0, L_1) to likelihoods ($f_0 = \exp L_0, f_1 = \exp L_1$) will not affect conclusions from comparing resulting harmonic means. For example, if log-likelihoods $L^{(t)}$ were in the range -1250 for models M_0 and M_1 we might add 1250 to them before exponentiating.

Raftery (1996) assumed the number of parameters in the two- and three-group mixtures as five and eight; these are the different group means and variances, and the free group probabilities (one in the $C = 2$ case, and two in the $C = 3$ case). One might in fact question whether the effective number of parameters is larger than this in latent class models; for example, in moving from a two-group to three-group model it may be argued that rather than three extra parameters, there are something between three and $n+3$ extra parameters – because there are now n extra latent class probabilities.

Here we calculate two likelihoods, one based on the actual observations, the other on the 'complete' data, including the group indicators $z_r^{(t)}$ with values $z_r \in \{1, 2\}$ assigned to each case at each iteration. The observed data likelihood for case r at iteration t is

$$f_r^{(t)} = \rho_1^{(t)} \mathrm{N}(y_r \mid \mu_1^{(t)}, V_1^{(t)}) + \rho_2^{(t)} \mathrm{N}(y_r \mid \mu_2^{(t)}, V_2^{(t)})$$

while the complete likelihood is

$$h_r^{(t)} = \mathrm{N}(y_r \mid \mu_{zr}^{(t)}, V_{zr}^{(t)}).$$

We could apply the BIC criterion with either likelihood. We also calculate the EPD measure of Carlin and Louis (1997) with the discrepancy between $y_{r.\mathrm{new}}$ and y_r being the total sum of squares. Because the BIC criteria have more validity for large samples we also generate a sample of $n = 500$ points from the same mixture.

Estimates of these fit measures are based on 15 000 iterations after a 1000 iteration burn-in. The harmonic mean estimate of the marginal likelihood and the BIC based on the observed data likelihood both suggest little advantage in the three-group over the two-group solution for the smaller sample. The Bayes factor favouring the two-group solution is about 3.3. As an illustration of fluctuations in the harmonic mean estimator, its logged value for the $C = 2$ case and for iterations 1000–6000 is -245.06. For iterations 6 to11 thousand it is -245.82 and for iterations 11–16000 it is -245.66. For the larger sample ($n = 500$) the Bayes Factor is around 2 in favour of the $C = 2$ groups solution. The complete data BIC and EPD measures favour the three-group solution but are based on the assumption that the case membership of group indicators are known. A compromise between the complete and observed likelihood fit measures is indicated and might involve estimating the effective number of parameters (see Spiegelhalter *et al.*, 1999). Table 10.3 shows the fit measure estimates over iterations 1000–16000 in both samples.

Example 10.4 Vaccine Risk via posterior likelihood Ratio comparisons Aitkin (1997) analyses counts of brain damage in children following administration of a trial vaccine. Four cases had been observed when the standard risk assumption implied a Poisson mean of $\mu_0 = 0.9687$. We can consider alternative interval or point hypotheses relating to the parameters underlying the observed data. Consider the alternative interval hypotheses on the underlying Poisson mean μ, M_0:$\mu \leq 0.9687$ and M_1:$\mu > 0.9687$.

One Bayesian approach specifies equal prior probabilities on M_0 and M_1, whether point or interval in form, and considers the Bayes factor on these

Table 10.3 Gaussian mixture model fits, posterior means of fit measures, and maxima of likelihoods, $L(\theta)$ and $L(\theta, z)$.

	No of groups		
$n = 100$	1	2	3
Max $L(\theta)$	−261.8	−240	−239.1
Max $L(\theta, z)$	−261.8	−174.7	−159.1
Margnl LKD	−263.8	−245.5	−246.7
BIC (θ)	−266.4	−251.5	−257.5
BIC (θ, z)	−266.4	−186.2	−177.5
EPD (θ, z)	1113	487.6	416.0
Parameters	2	5	8
$n = 500$			
Max $L(\theta)$	−1312	−1205	−1204.0
Max $L(\theta, z)$	−1312	−873.8	−808.1
Margnl LKD	−1314.6	−1210.3	−1211.0
BIC (θ)	−1316.6	−1216.5	−1222.4
BIC (θ, z)	−1316.6	−885.3	−826.5
EPD (θ, z)	5588	2171	1812.0
Parameters	2	5	8

hypotheses following adoption of diffuse priors on μ. For example, taking $\mu \sim G(0.0001, 0.0001)$ is approximately equivalent to assuming

$$P(\mu) \propto 1/\mu$$

whereas taking $\mu \sim G(1, 0.0001)$ is approximately equivalent to a uniform prior taking

$$P(\mu) \propto 1.$$

A hundred thousand iterations with the uniform prior (see *Program 10.4 Vaccination Risk*) gives a probability R that $\mu > 0.9687$ of 0.9967. The Bayes factor, $(1 - R)/R$, on M_0 is then $0.0033/0.9967 \approx 0.0033$. However taking $\mu \sim G(0.0001, 0.0001)$ gave a probability R on M_1 of 0.983, with Bayes factor 0.017. Such contrasts show potential sensitivity of the Bayes factor to prior assumptions.

This sensitivity applies more so if M_0 is a point hypothesis, $M_0{:}\mu = \mu_0$, and the alternative model M_1 is taken as an unspecified mean rate $\mu \neq \mu_0$. The Bayes factor is sensitive to the values adopted for a_1, a_2 in the gamma prior for μ, $\mu \sim G(a_1, a_2)$, and is undefined if the prior is improper, with one or both of a_1 or a_2 zero. As an illustration of the alternative likelihood ratio perspective, we can consider the posterior probabilities that $\text{LR} < 1$ and $\text{LR} < 0.1$ for the point null hypothesis M_0 and its alternative. The likelihood function is proportional to

$$\exp(-\mu)\mu^x$$

where $x = 4$, and the likelihood ratio for M_0 as against M_1 is then

$$\text{LR} = [\exp(-\mu_0)\mu_0^x]/[\exp(-\mu)\mu^x].$$

If LR is compared with the threshold of 1 (in line with conventional testing), the posterior probabilities favouring M_1 are overwhelmingly high. We use both thresholds with the alternative diffuse priors $\mu \sim$ G $(0.0001, 0.0001)$ (prior A) and $\mu \sim \text{G}(1, 0.0001)$ (prior B).

With runs of 100 000 iterations, the posterior probability that LR < 0.1 using prior A is 0.976 and that LR < 0.1 is 0.58 (see the indicators `PP[]` in Program 10.4). Under prior B the respective posterior probabilities are virtually the same, at 0.975 and 0.578. Note that the more stringent test is less convincing in showing excess risk. Aitkin (1997) demonstrates the relative insensitivity of these posterior probabilities with values $\{a_1, a_2\}$ anywhere in the ranges $0 < a_1 < 1$ and $0 < a_2 < 0.5$.

Example 10.5 Pediatric pain exposure: Checking Box–Cox regression using CPO measures Weiss (1994) considers data on response times by children to a pain exposure (hand immersion in cold water), and the impact on response times in seconds of the child's coping mechanism for pain (binary), and a treatment variable with three levels. Response times y_i are considered in relation to a six-level factor combining coping type and treatment, and to baseline response time B_i obtained prior to the treatment being delivered. The coping types are attenders (A), corresponding to children who pay attention to the pain, and distracters (D), for children who tend to think of other things during the exposure. The treatments were a control (i.e. no treatment, N), counselling to attend (A) and counselling to distract (D). The six coping–treatment groups, denoted G_i for child i, are here arranged as AA, AD, AN, DA, DD and DN.

Weiss considers a Box–Cox transform for both y_i and B_i, with the same power λ being applied to both. So we have a model (for $i = 1, \ldots, 61$ children)

$$Y_i^\lambda = \alpha + \beta[G_i] + \gamma B_i^\lambda$$

where $\beta_1 = 0$ and $\beta_2, \ldots, \beta_6$ are fixed effects. As in Chapter 4, we apply the case identifier procedure in BUGS, with the likelihood probability $P(y_i \mid \theta)$ for case i, where $\theta = (\alpha, \beta, \gamma, \lambda)$, determined by the Box–Cox likelihood. Thus a possible coding is

```
For (i in 1:61) {case[i] <-1
case[i] ~ dbern(f[i])
mu[i] <-a+ b[gr[i]] + c *pow(B[i], lam)
d[i] <- yt[i]-mu[i]
d2[i] <- d[i]*d[i]
f[i] <- sqrt(tau)*abs(lam)*pow(y[i],lam-1)*exp(- d2[i]*tau*0.5)/
  sqrt(6.28) }
```

Note that this likelihood assumes a transformation y^λ rather than $(y^\lambda - 1)/\lambda$ (see Aitkin *et al.*, 1989).

Weiss is interested in cross-validatory procedures to investigate outliers and considers the conditional predictive ordinate diagnostic, based on excluding case i from the likelihood

$$P(y_i \,|\, y_{[i]}) = P(y)/P(y_{[i]}) = \int P(y_i \,|\, \theta, y_{[i]})\pi(\theta \,|\, y_{[i]})d\theta.$$

Cases with low CPO are poorly fit by the model. The CPO can be approximated by a harmonic mean estimate (analogous to one estimate of the marginal likelihood), based on T samples from an MCMC run using all cases (i.e. without needing to actually exclude cases):

$$\text{CPO}_i^{-1} = T^{-1}\sum_{t=1}^{T}[P(y_i \,|\, \theta^{(t)}, B_i)]^{-1}.$$

Weiss also considers 18 predictive densities of response times for new cases defined by each of the six possible coping–treatment combination and by three 'new' baseline times of 6, 24 and 120 seconds.

We find, as Weiss did, that cases with ID numbers 15 and 58 have very low CPOs; though the CPO for case 30 is slightly higher than obtained by Weiss (see *Program 10.5 Pediatric Pain*). We also obtain a higher value of λ than Weiss, namely around 0.36 (SD = 0.06), whereas Weiss cites a value of 0.14 (SD = 0.11). See Table 10.4.

The predictive densities are obtained for y^λ rather than y with transformation of the posterior means of $z = y^\lambda$ then giving estimated means $z^{1/\lambda}$ in the original scale. The latter show that only a distracter coping style enhanced by a distracter treatment (as in the DD group) consistently increases response times over the baseline. See Table 10.5.

Table 10.4 CPOs for pediatric pain model.

Child ID	Median of CPO^{-1}	100*CPO
61	74.0	1.352
33	188.6	0.530
53	45.0	2.225
35	39.1	2.557
7	52.6	1.902
13	32.5	3.078
54	60.1	1.665
3	51.2	1.952
31	57.2	1.749
10	46.2	2.165
34	87.3	1.146
28	57.8	1.731
60	33.2	3.015
57	38.6	2.590
41	32.6	3.072
47	39.2	2.548

Table 10.4 *cont.*

Child ID	Median of CPO^{-1}	100*CPO
21	60.4	1.655
26	36.4	2.750
24	37.4	2.676
59	44.6	2.242
19	121.5	0.823
2	24.7	4.045
23	21.1	4.751
46	84.9	1.179
45	42.2	2.368
16	47.5	2.103
25	31.6	3.166
6	29.4	3.399
55	31.6	3.161
37	35.7	2.803
27	51.9	1.926
49	55.2	1.813
43	25.3	3.957
56	113.7	0.880
4	77.7	1.287
50	30.5	3.277
44	32.0	3.128
17	40.5	2.469
36	64.0	1.564
11	37.2	2.685
15	60780.0	0.002
58	4427.0	0.023
9	118.6	0.843
1	163.9	0.610
22	153.6	0.651
38	88.1	1.135
52	216.8	0.461
48	89.9	1.112
8	88.6	1.129
29	76.5	1.308
42	71.2	1.404
30	604.9	0.165
20	300.1	0.333
14	83.8	1.193
12	76.8	1.301
32	21.1	4.748
51	29.8	3.361
5	40.4	2.475
18	27.9	3.589
40	30.6	3.272
39	38.9	2.574

Table 10.5 Predicted follow up response times.

Coping/Treatment combination (baseline response times)	Y (follow-up response times)	Y^{λ}	SD (Y^{λ})
AA (6)	13.7	2.6	1.1
AA (24)	28.8	3.4	1.3
AA (120)	88.7	5.1	1.9
AD (6)	12.1	2.5	1.1
AD (24)	26.3	3.3	1.2
AD (120)	83.3	5.0	2.0
AN (6)	10.8	2.4	1.1
AN (24)	24.0	3.2	1.2
AN (120)	78.0	4.9	1.9
DA (6)	9.6	2.3	1.1
DA (24)	21.5	3.1	1.2
DA (120)	72.9	4.8	1.8
DD (6)	34.8	3.7	1.4
DD (24)	60.2	4.5	1.7
DD (120)	146.2	6.2	2.4
DN (6)	7.2	2.1	1.0
DN (24)	17.9	2.9	1.1
DN (120)	65.0	4.6	1.8

Example 10.6 Binary outcome model assessment via marginal likelihood, AIC and BIC Chib (1995) considers alternative models for the presence or not of prostatic nodal involvement in a sample of 53 cancer patients. The prediction is in terms of two continuous variates (x_1 = age, x_2 = level of serum acid phosphate) and three binary indicators. These were positive X-ray exam or not (x_3), large tumor or not (x_4) and seriousness of the pathology grading for the tumour, more vs less serious (x_5). Drawing on the work of Collett (1991) he shows the improvement in fit obtained over a constant only model (model 1) by a model involving a constant and x_3 only (model 2), and a more complex model involving a constant, log x_2, x_3 and x_4 (model 3). Finally the most complex model considered adds x_5 to model 3 to give model 4. Chib proposes an approximation method for calculating the marginal likelihood based on the identity

$$P(y) = L(\theta\,|\,y)P(\theta)/P(\theta\,|\,y)$$

where all integrating constants are retained in the likelihood $L(\theta\,|\,y)$, prior $P(\theta)$ and posterior $P(\theta\,|\,y)$. He finds a Bayes factor B_{12} of approximately 25 in favour of model 2 as against model 1, and a smaller advantage ($B_{34} \approx 5.4$) for model 3 vs model 4.

Here we consider evidence produced by the harmonic mean estimate of the marginal likelihood and by the CPO approach via estimates for the quantities

$$P(y_i\,|\,y_{[i]}).$$

We also consider the AIC and BIC measures of (penalised) fit, remembering that for the sample size involved the BIC is best viewed as just a fit measure rather than an approximation to the marginal likelihood. We take long runs of T iterations to obtain approximate estimates of the maximised likelihoods, under each model, namely

$$L_m = \max_{t\varepsilon(1,T)} L^{(t)}.$$

In the modelling situation here, we know the number of parameters p straightforwardly, and so can obtain AIC $= L_m - p$ and BIC $= L_m - 0.5p \log_e(53)$. See Table 10.6.

The analysis via these measures replicates that of Chib (1995) in the high Bayes factor on model 2 as against model 1. However, there seems to be little to choose between models 3 and 4.

A broadly similar conclusion is reached when both posterior and prior are combined on the basis of actual samples from both (see equation (10.9)), as outlined in the BUGS coding above. Again using $\delta = 0.01$ gives a log marginal likelihood on model 4 of -32.0 and on model 3 of -33.6, so slightly favouring the more complex model.

Example 10.7 Infant temperament study As an example of the predictive assessment approach of Gelman *et al.* (GMS, 1996) and Rubin and Stern (RS, 1994), we compare latent class models for a three-way contingency table m_{ijk} with ordered categories on each dimension. For $N = 93$ infants in a study of temperament development there are measures of motor activity at age 4 months ($A = 1, 2, 3, 4$) with higher categories denoting greater activity; of fret/cry activity at 4 months ($C = 1, 2, 3$), and of fear level at 14 months ($F = 1, 2, 3$). Both the papers cited posit latent class models in which A, C and F are imperfect measures of an underlying latent variable X. Within subpopulations defined by X, therefore the observed variables are independent of each other. Let $\pi_{i|x}, \pi_{j|x}$ and $\pi_{k|x}$ define the conditional distributions of

Table 10.6 Nodal involvement, model fits. Estimates of marginal likelihood (ML) and penalised fit measures.

Model	Terms	L_m	log (pseudo ML)[a]	Harmonic mean estimate of ML	Importance sample from mixture[b]	AIC	BIC
1	C	−35.1	−36.1	−36.5	−36.3	−36.1	−37.1
2	$C + x3$	−30.6	−32.7	−32.7	−32.61	−32.6	−34.6
3	$C + \log(x2) + x3 + x4$	−25.8	−30.66	−30.55	−30.37	−29.8	−33.8
4	$C + \log(x2) + x3 + x4 + x5$	−24.8	−31.08	−30.63	−30.34	−29.8	−34.8

[a] Pseudo marginal likelihood estimate based on estimates of individual CPO_I

[b] Using linear iteration scheme as in Newton and Raftery (1994), $\delta = 0.01$

A, C and F for the xth latent class of X, with $x = 1, \ldots, q$. The joint distribution of observed data $\{m_{ijk}\}$ can then be written

$$\pi_{ijk} = \sum_{x=1}^{q} \pi_x \pi_{i|x} \pi_{j|x} \pi_{k|x}.$$

Both GMS and RS present maximum likelihood estimates of 2 to 4 latent class models.

Here we apply fully Bayes estimation to the choice between (1) an independence model for variables A, F and C but without a latent variable present, equivalent to a conventional independent factors log-linear model; and (2) independence between A, F and C within a two-class latent model. Note that for this relatively sparse data, informative priors may affect the posterior estimates of the parameters $\pi_x, \pi_{i|x}, \pi_{j|x}$ and $\pi_{k|x}$. RS present such priors based on subject knowledge. They then augment the observed data by indicators of class membership defined for child c within each cell level (i.e. for $c = 1, \ldots, m_{ijk}$). These indicators Z_{ijkcx} define the joint distribution

$$P(\pi)\left[\prod_{ijk}\prod_{c=1}^{m_{ijk}}\prod_{x=1}^{q} (\pi_x\, \pi_{i|x}\, \pi_{j|x}\, \pi_{k|x})^{Z_{ijkcx}}\right]$$

where $P(\pi)$ is the prior on

$$\pi = \{\pi_x, \pi_{i|x}, \pi_{j|x}, \pi_{k|x}\}.$$

This approach is especially relevant if we have covariates specific to each child. In the absence of these an alternative is to assume latent class indicators g_{ijkx} at the contingency table cell level, with

$$g_{ijkx} \sim \text{ categorical } (\Phi_{ijk}[1:q])$$

and assume a flat Dirichlet prior on each vector Φ_{ijk}. Then

$$m_{ijk} \sim \text{ Poi } (\mu_{ijk})$$

with the selected g_{ijkx} denoted x in the parameterisation

$$\log(\mu_{ijk}) = \gamma_x + \alpha_{ix} + \beta_{jx} + \delta_{kx}$$

(see *Program 10.7 Infant Temperament*). We obtain the estimates given in Table 10.7 for the latent group probabilities π_x and the conditional probabilities of A, C and F given $x = 1, 2$ from the log-linear parameters $\theta = \{\alpha, \beta, \gamma, \delta\}$.

Sampling replicate data $m_{ijk.\text{new}}$ at each iteration we evaluate the GMS tail probability for the two-class latent model using deviances as discrepancy measures, $D(m_{\text{new}};\theta)$ and $D(m, \theta)$. We find a p value for the observed discrepancy of 79 % based on 4500 draws after a 500 iteration burn-in. GMS (1996, p. 754) report a level of 74 % based on a run of 500 draws and conclude that the discrepancy measures are consistent with the two-class model, in the sense of

Table 10.7 Two-group LCA of infant temperament data.

Marginal probabilities of latent classes	Mean	SD
$P(1)$	0.236	0.126
$P(2)$	0.764	0.126
Conditional probs of groups $C = 1, \ldots, 3$ $F = 1, \ldots, 3$ and $M = 1, \ldots, 4$ given latent class 1 or 2		
$P.C(1,1)$	0.436	0.204
$P.C(1,2)$	0.624	0.236
$P.C(2,1)$	0.203	0.118
$P.C(2,2)$	0.148	0.163
$P.C(3,1)$	0.362	0.168
$P.C(3,2)$	0.228	0.172
$P.F(1,1)$	0.245	0.167
$P.F(1,2)$	0.351	0.225
$P.F(2,1)$	0.324	0.184
$P.F(2,2)$	0.263	0.200
$P.F(3,1)$	0.431	0.213
$P.F(3,2)$	0.386	0.297
$P.M(1,1)$	0.143	0.098
$P.M(1,2)$	0.179	0.099
$P.M(2,1)$	0.378	0.224
$P.M(2,2)$	0.434	0.233
$P.M(3,1)$	0.308	0.174
$P.M(3,2)$	0.262	0.238
$P.M(4,1)$	0.171	0.117
$P.M(4,2)$	0.125	0.164
Tail probability (predictive check)		
$p.b$	0.7867	

providing no evidence of lack of fit. By contrast, for a one-class mixture model the p value for the realised discrepancy is about 6.1 %, indicating poor fit.

APPENDIX

FORTRAN code for Estimated Marginal Likelihood, using Importance Sample combining prior and posterior

```
      real*4 L(10000),p(10000),logML,n,P(20)
c sample proportion from prior
      del=0.001
c no of MCMC samples
      m=10000
c input data are samples of log-likelihoods, which are centred before
c exponentiation
```

```
      sL=0.
      do10 i=1,m
      read(5,*)L(i)
10    sL=sL+L(I)
      sL=SL/float(m)
10    p(i)=exp(L(i)-SL)
c number of imaginary further samples from prior
      n = del*float(m)/(1-del)
c initial estimate of Marg LKD
      P(1) = 1
C revised estimates of Marg LKD
      do 2 j=2,10
      ts=0
      bs=0
      do1 i=1,m
      ts=ts+p(i)/(del*P(j-1)+(1-del)*p(i))
1     bs=bs+1.0/(del*P(j-1)+(1-del)*p(i))
c revised estimate at iteration j
      P(j) = (n+ts)/(n/P(j-1)+bs)
C revised log ML
      logML= log(P(j))-SL
2     continue
```

References

Aalen, O (1976) Nonparametric inference in connection with multiple decrement models, *Scandinavian Journal of Statistics, Theory Appl.*, **3**, 15–27.

Abrams, K, Ashby, D and Errington, D (1996) A Bayesian approach to Weibull survival models, *Lifetime Data Analysis*, **2**, 159–174.

Adcock, C (1987) A Bayesian approach to calculating sample sizes for multinomial sampling, *The Statistician*, **36**, 155–159.

Adcock, C (1988) A Bayesian approach to calculating sample sizes, *The Statistician*, **37**, 433–439.

Ahrens, J and Dieter, U (1974) Computer methods for sampling from gamma, beta, Poisson and binomial distributions, *Computing*, **12**, 223–246.

Aitchison, J and Dunsmore, IR (1975) *Statistical prediction analysis*. Cambridge: Cambridge University Press.

Aitkin, M (1979) A simultaneous test procedure for contingency table models, *Applied Statistics*, **28**, 233–242.

Aitkin, M (1991) Posterior Bayes factors, *J. Royal Statistical Society Series B*, 53 (1) 111–142.

Aitkin, M (1997) The calibration of P-values, posterior Bayes factors and the AIC from the posterior distribution of the likelihood, *Statistics and Computing*, **7**, 253–261.

Aitkin, M, Anderson, D, Francis, B and Hinde, J (1989) *Statistical modelling in GLIM*. Oxford University Press.

Akaike, H (1981) Likelihood of model and information criteria, *Journal of Econometrics*, **16**, 3–14.

Albert J and Chib S (1996) Computation in Bayesian econometrics: An introduction to Markov chain Monte Carlo, *Advances in Econometrics*, **11A**, 3–24.

Albert, J (1992) Bayesian estimation of normal ogive item response curves using Gibbs sampling, *J. Educational Statistics*, **17**, 251–269.

Albert, J (1996) *Bayesian computation using Minitab*. Wadsworth.

Albert, J (1996) Bayesian selection of log-linear models. *Canadian Journal of Statistics*, **24** (3), 327–347.

Albert, J and Chib, S (1993) Bayesian analysis of binary and polychotomous response data. *J. American Statistical Association*, **88** (422), 669–679.

Albert, J and Chib, S (1997) Bayesian tests and model diagnostics in conditionally independent hierarchical models, *J. American Statistical Association*, **92**, 916–925.

Albert, J and Pepple, P (1989) A Bayesian approach to some overdispersion models, *Canadian Journal of Statistics*, **17**, (3), 333–444.

Allaire, S, LaValley, M, Evans, S, O'Connor, G, Kelly-Hayes, M, Meenan, R, Levy, D and Felson, D (1999) Evidence for decline in disability and improved health among persons aged 55 to 70 years: The Framingham Heart Study, *American Journal of Public Health*, **89** (11), 1678–1683.

Allison, P (1984) *Event history analysis: Regression for longitudinal event data*. Quantitative Applications in the Social Sciences 46. Sage.

Ameen, JRM and Harrison, PJ (1985) Normal discount Bayesian models. Bayesian statistics 2, Proceedings 2nd International Meeting, Valencia, Spain 1983. Oxford University Press. 271–298.

Analytical Methods Committee (1989) Robust statistics – how not to reject outliers, *Analyst*, **114**, 1699–1702.

Andersen, P, Borgan, O, Gill, R and Keiding, N (1993) *Statistical models based on counting processes*. Springer Series in Statistics. New York: Springer-Verlag.

Andrews, DF and Herzberg, A (1985) *Data. A collection of problems from many fields for the student and research worker*. Springer Series in Statistics. New York: Springer-Verlag.

Anselin, L and Griffith, D (1988) Do spatial effects really matter in regression analysis, *Papers of the Regional Science Association*, **65**, 11–34.

Ansley, C and Kohn, R (1987) Efficient generalized cross-validation for state space models, *Biometrika*, **74**, 139-148.

Antoniak, C (1974) Mixtures of Dirichlet processes with applications to Bayesian nonparametric problems, *Annals of Statistics*, **2**, 1152–1174.

Arora, S and Brown, M (1977) An approach to spatial correlation: An improvement over current practice, *International Regional Science Review*, **2**, 67–78.

Ashby, D, Hutton, J and McGee, M (1993) Simple Bayesian analysis for case-control studies, *The Statistician*, **42**, 385–397.

Ashford, J (1959) An approach to the analysis of data for semi-quantal responses in biological assay, *Biometrics*, **15**, 573–581.

Ashford, J and Sowden, R (1970) Multivariate probit analysis, *Biometrics*, **26**, 535–546.

Bailey, T and Gattrell, A (1995) *Interactive spatial data analysis*, Harlow: Longman.

Baker, S, Rosenberger, W and DerSimonian, R (1992) Closed form estimates for missing counts in two-way contingency tables, *Statistics in Medicine*, **11**, 643–657.

Bandeen-Roche, K, Miglioretti, D, Zeger, S and Rathouz, P (1997) Latent variable regression for multiple discrete outcomes, *J. American Statistical Association*, **92**, (440), 1375–1386.

Barndorff-Nielsen, O and Schou, G (1973) On the parameterization of autoregressive models by partial autocorrelation, *J. Multivariate Analysis*, **3**, 408–419.

Barnett, G, Kohn, R and Sheather, S (1996) Bayesian estimation of an autoregressive model using Markov chain Monte Carlo, *J. Econometrics*, **74**, (2), 237–254.

Bartholomew, D (1980) Factor analysis for categorical data, *J. Royal Statistical Society, Series B*, **42**, 293–321.

Bartholomew, D (1987) *Latent variable models and factor analysis*. Griffin's Statistical Monographs and Courses, No. 40. London: Charles Griffin; New York: Oxford University Press.

Bartholomew, D (1994) Bayes' theorem in latent variable modelling, pp. 41–50 in PR Freeman *et al.* (ed.), *Aspects of uncertainty: A tribute to D. V. Lindley*. Chichester: Wiley.

Bartholomew, D (1996) *The statistical approach to social measurement*. Academic Press.

Bartholomew, D and Knott, M (1999) *Latent variable models and factor analysis*. Kendall's Library of statistics, 7. Chapman & Hall.

Baxter, M (1985) Quasi-likelihood estimation and diagnostics in spatial interaction models, *Environment and Planning*, **17A**, 1627–1635.

Benjamin, B and Pollard, J (1980) *The analysis of mortality and other actuarial statistics*, 2nd ed. Heinemann.

Bennett, S (1983) Log-logistic regression models for survival data, *Applied Statistics*, **32**, 165–171.

Berger, J (1994) An overview of robust Bayesian analysis. *Test* **3**, (1), 5–124.

Berger, J and Deely, J (1988) A Bayesian approach to ranking and selection of related means with alternatives to analysis-of-variance methodology, *J. American Statistical Association*, **83** (402), 364–373.

Berger, J and Yang, R (1994) Noninformative priors for Bayesian testing for the AR(1) model, *Econometric Theory*, **10**, 461–482.

Berger, J (1985) *Statistical decision theory and Bayesian analysis*, 2nd ed. Springer Series in Statistics. New York: Springer-Verlag.

Bernardinelli, L, Pascutto, C, Best, N and Gilks, W (1997) Disease mapping with errors in covariates, *Statistics in Medicine*, **16**, 741–752.

Berndt, E (1990) *The practice of econometrics: Classic and contemporary*. Reading, MA: Addison-Wesley.

Berry, D (1993) A case for Bayesianism in clinical trials, *Statistics in Medicine*, **12**, 1377–1393.

Berry, D (1995) Decision Analysis and Bayesian methods in Clinical trials, Chapter 7 in *Recent advances in clinical trial design and analysis*, ed P Thall. Boston: Kluwer.

Berry, D (1996) *Statistics: A Bayesian perspective*. Duxbury Press.

Besag, J and Higdon, D (1999) Bayesian analysis of agricultural field experiments, *J. Royal Statistical Society, Series B*, **61**, 4.

Besag, J, Green, P, Higdon, D, *et al.* (1995) Bayesian computation and stochastic systems, *Statistical Science*, 10 (1), 3–41.

Besag, J, York, J and Mollié, A (1991). Bayesian image restoration, with two applications in spatial statistics, *Annals of Institute of Statistical Mathematics*, **43**, 1–59.

Best, N (1999) Bayesian ecological modelling, Chapter 14 in *Disease mapping and risk assessment for public health*, eds. A Lawson, A Biggeri, D Boehning, E Lesaffre, J Viel and R Bertollini. Chichester: Wiley.

Best, N, Spiegelhalter, D, Thomas, A and Brayne, C (1996). Bayesian analysis of realistically complex models, *J. Royal Statistical Society, Series A*, **159**, 323–342.

Birkes, D and Dodge, Y (1993) *Alternative methods of regression*. Wiley Series in Probability and Mathematical Statistics. Applied Probability and Statistics. New York: Wiley.

Birkin, M, Clarke, G, Clarke, M and Wilson, A (1994) Applications of performance indices in urban modelling: subsystems frameworks, Chapter 8 in *Modelling the city: Performance, policy, and planning*, eds C Bertuglia and A Wilson. Routledge.

Birnbaum, Z and Saunders, S (1958) A statistical model for life-length of materials, *J. American Statistical Association*, **53**, 151–160.

Bishop, Y, Fienberg, S and Holland, P (1975) *Discrete multivariate analysis: Theory and practice*. London: MIT Press.

Bland, J and Altman, D (1995) Calculating correlation-coefficients with repeated observations, 2. Correlation between subjects, *British Medical Journal* **310** (6980), 633.

Bland, M (1995) *An introduction to medical statistics*. Oxford University Press.

Bloomfield, P (1976) *Fourier analysis of time series: An introduction*. Wiley Series in Probability and Mathematical Statistics. New York: Wiley.

Blyth, C (1986) Approximate binomial confidence limits. *J. American Statistical Association*, **81**, 843–855.

Bock, R and Gibbons, R (1996) High-dimensional multivariate probit analysis, *Biometrics*, **52**, (4), 1183–1194.

Bockenholt, U (1999) Mixed INAR(1) Poisson regression models, *J. Econometrics*, **89**, 317–338.

Bohning, D (2000) *Computer-assisted analysis of mixtures and applications: Meta-analysis, disease mapping, and other applications*. Monographs on Statistics and Applied Probability. Chapman & Hall.

Bohning, D, Dietz, E, Schlattmann, P, Mendonca, L and Kirchner, U (1999) The zero-inflated Poisson model and the decayed, missing and filled teeth index in dental epidemiology, *J. Royal Statistical Society*, **162A**, 195–209.

Boice, J and Monson, R (1977) breast cancer in women after repeated fluoroscopic examinations of the chest, *J. National Cancer Institute*, **59**, 823–832.

Bowmaker, J, Spiegelhalter, D and Mollon, D (1985) Two types of trichromatic squirrel monkey may share a pigment in the red-green region, *Vision Research*, **25**, 1937–1946.

Box, G and Jenkins, G (1970) *Time series analysis: Forecasting and control*, Holden-Day.

Box, G and Tiao, G (1973) *Bayesian inference in statistical analysis*. Addison-Wesley Series in Behavioral Science: Quantitative Methods. Reading, MA: Addison-Wesley.

Box, G and Tiao, G (1992) *Bayesian inference in statistical analysis*. Wiley Classics Library Edition. Wiley.

Bradford-Hill, A (1965) The environment and disease: Association or causation, *Proceedings of the Royal Society of Medicine*, **58**, 295–300.

Breckling, J, Chambers, R, Dorfman, A, Tam, S and Welsh, A (1994) Maximum likelihood inference from sample survey data, *International Statistical Review*, **62**, 349–363.

Breslow, N (1984) Extra-Poisson variation in log-linear models, *Applied Statistics*, **33**, 38–44.

Breslow, N and Day, N (1980) *Statistical methods in cancer research, Vol.1: The analysis of case control studies*. Lyon: International Agency for Research on Cancer.

Breslow, NE and Storer, BE (1985) General relative risk functions for case-control studies, *American Journal of Epidemiology*, **122** (1), 149–162.

Brillinger, D and Preisler, H (1983) Maximum likelihood estimation in a latent variable problem, *Studies in Econometrics, Time Series, and Multivariate Statistics*, commemorating TW Anderson's 65th birthday, pp, 31–65.

Broemeling, L and Cook, P (1993) Bayesian estimation of the mean of an AR process, *J. Applied Statistics*, **20** (1), 25-39.

Broffitt, J (1988) Maximum likelihood alternatives to actuarial estimators of mortality rates, *Transactions of Society of Actuaries*, **36**, 77–122.

Brooks, S (1998) Markov chain Monte Carlo method and its application, *The Statistician*, **47**, 69–100.

Brooks, S and Gelman, A (1998) General methods for monitoring convergence of iterative simulations, *J. Computational and Graphical Statistics*, 7, 434–455.

Brooks, S and Roberts, GO (1998) Assessing convergence of Markov chain Monte Carlo algorithms, *Statistics and Computing*, 8, 319–335.

Brown, P, Le, N and Zidek, J (1994) Inference for a covariance matrix. Chapter 7 in *Aspects of uncertainty: A tribute to D. V. Lindley*, ed. PR Freeman *et al.* Chichester: Wiley. 77–92.

Brown, P, Vannucci, M and Fearn, T (1998) Multivariate Bayesian variable selection and prediction, *J. Royal Statistical Society, Series B*, **60**, 3.

Bryk, A and Raudenbush, S (1992) *Advanced qualitative techniques in the social sciences no 1: Hierarchical linear models: Applications and data analysis methods*. Sage Publications.

Bryman, A and Cramer, D (1994) *Quantitative data analysis for social scientists*. Routledge.

Burkhardt, U, Mertens, T and Eggers, H (1987) Comparison of two commercially available anti-HIV ELISAs, *J. Medical Virology*, **23**, 217–224.

Cameron, MA and Turner, T (1987) Fitting models to spectra using regression packages, *J. Royal Statistical Society, Series* C, **36**, 47–57.

Campbell, M (1996) Spectral analysis of clinical signals: An interface between medical statisticians and medical engineers, *Statistical Methods in Medical Research*, **5**, 51–66.

Carlin, B (1992) A simple Monte Carlo approach to Bayesian graduation, *Transactions of Society of Actuaries*, **44**, 55–76.

Carlin, B and Chib, S (1995) Bayesian model choice via Markov chain Monte Carlo methods, *J. Royal Statistical Society, Series B*, **57** (3), 473–484.

Carlin, B and Louis, T (1996) *Bayes and empirical Bayes methods for data analysis*. London: Chapman and Hall.

Carlin, B, Gelfand, A and Smith, A (1992) Hierarchical Bayesian analysis of change-point problems, *J. Royal Statistical Society, Series C*, **41**, (2), 389–405.
Carlin, B, Polson, N and Stoffer, D (1992) A Monte Carlo approach to nonnormal and nonlinear state-space modeling, *J. American Statistical Association*, **87**, 493–500.
Carter, CK and Kohn, R (1994) On Gibbs sampling for state space models, *Biometrika*, **81** (3), 541–553.
Casella, G (1992) Explaining the Gibbs sampler, *The American Statistician*, **46**; (3) 167–174.
Castle, DJ, Sham, PC, Wessely, S and Murray, RM (1994) The subtyping of schizophrenia in men and women: A latent class analysis. *Psychology of Medicine*, **24** (1), 41–51.
Caussinus, H (1965) Contribution a l'analyse statistique des tableaux de correlation, *Annals of Faculty of Science University of Toulouse, IV, Series 29, 77–183.*
Chatfield, C and Collins, A (1980) *Introduction to multivariate analysis*. London: Chapman and Hall.
Cheng, C and van Ness, J (1998) *Statistical regression with measurement error*. Wiley.
Chetty, V (1968) Bayesian analysis of Haavelmo's models, *Econometrica*, **36**, 582–602.
Chiang, CL (1968) *Introduction to stochastic processes in biostatistics*. New York: Wiley.
Chib, S (1993) Bayes regression with autoregressive errors. A Gibbs sampling approach, *J. Econometrics*, **58**, (3), 275–294.
Chib, S (1995) Marginal likelihood from the Gibbs output. *J. American Statistical Association*, **90**, (432), 1313–1321.
Chib, S (1996) Calculating posterior distributions and modal estimates in Markov mixture models, *J. Econometrics*, **75**, 79–97.
Chib, S and Greenberg, E (1994) Bayes inference in regression models with ARMA(p,q) errors, *J. Econometrics*, **64**, (1–2), 183–206.
Chib, S and Greenberg, E (1998) Analysis of multivariate probit models, *Biometrika*, **85** (2), 347–361.
Child, D (1970) *The essentials of factor analysis*. London: Holt, Rinehart and Winston.
Christensen, R (1997) *Log-linear models and logistic regression*, 2nd ed. Springer Texts in Statistics. New York: Springer.
Chuang, C and Agresti, A (1986) A new model for ordinal pain data from a pharmaceutical study, *Statistics in Medicine*, **5**, 15–20.
Clayton, D (1991) A Monte Carlo method for Bayesian inference in frailty models, *Biometrics*, **47**, 467–485.
Clayton, D and Kaldor, J (1987) Empirical Bayes estimates of age-standardised relative risks for use in disease mapping, *Biometrics*, **43**, 671–681.
Cliff, A and Ord, J (1981) *Spatial processes: Models and applications*. London: Pion.
Cliff, AD and Ord, JK (1975) Model building and the analysis of spatial pattern in human geography. Discussion, *J. Royal Statistical Society, Series B*, **37**, 297–348.
Cochran, W and Cox, G (1957) *Experimental designs*, 2nd ed. New York: Wiley.
Collett, D (1991) *Modelling binary data*. Chapman and Hall.
Collett, D (1993) *Modelling survival data in medical research*. Chapman and Hall.
Collings, B, Margolin, B and Oehlert, G (1981) Analyses for binomial data, with application to the fluctuation test for mutagenicity, *Biometrics*, **37**, 775–794.
Congdon, P (1997) Multilevel and clustering analysis of health outcomes in small areas, *European Journal of Population*, **13** (4), 305–338.
Cowles, M and Carlin, B (1996) Markov chain Monte Carlo convergence diagnostics: A comparative review, *J. American Statistical Association*, **91**, 883–904.
Cox, D (1972) Regression models and life-tables, *J. Royal Statistical Society, Series B*, **34**, 187–220.
Cox, D and Oakes, D (1984) *Analysis of survival data*. Chapman and Hall.
Cox, D and Snell, E (1989) *Analysis of binary data*, 2nd ed. New York: Chapman and Hall.

Cox, DR, Hinkley, DV and Barndorff-Nielsen, OE (1996) *Time series models in econometrics, finance and other fields*. Monographs on Statistics and Applied Probability, 65. London: Chapman and Hall.
Cressie, N (1993) *Statistics for spatial data*, Wiley.
Cressie, N and Read, T (1989) Spatial data analysis of regional counts, *Biometrical Journal*, **6**, 699–719.
Crowder, M (1978) Beta binomial anova for proportions, *Applied Statistics*, **27**, 34–37.
Crowder, M (1978) On concurrent regression lines, *Applied Statistics*, **27**, 310–318.
Crowley, J and Hu, M (1977) Covaraince analysis of heart transplant survival data, *J. American Statistical Association*, **72**, 27–36.
Cryer, J (1986) *Time series analysis*. Boston: Duxbury Press (PWS Publishers).
Daniels, M (1999) A prior for the variance in hierarchical models, *Canadian Journal of Statistics*, **27**, 567–578.
Davies, R (1983) Destination dependence: A re-evaluation of the competing risk approach, *Environment and Planning*, **15A**, 1057–1065.
Deely, J and Smith, A (1998) Quantitative refinements for comparisons of institutional performance, *J. Royal Statistical Society, Series A*, **161**, 5–12.
Dellaportas, P and Smith, A (1993) Bayesian inference for generalized linear and proportional hazards models via Gibbs sampling, *J. Royal Statistical Society, Series C*, **42** (3), 443–459.
Dempster, A (1997) The direct use of the likelihood ratio for significance testing, *Statistics and Computing*, **7**, 247–252.
Dempster, AP, Laird, NM and Rubin, DB (1977) Maximum likelihood from incomplete data via the EM algorithm, *J. Royal Statistical Society, Series B*, **39**, 1–38.
Diebolt, J and Robert, C (1994) Estimation of finite mixture distributions through Bayesian sampling, *J. Royal Statistical Society, Series B*, **56**, (2), 363–375.
Diekmann, A and Mitter, P (1983) The sickle-hypothesis – a time-dependent Poisson model with applications to deviant-behavior and occupational-mobility, *J. Mathematical Sociology*, **9** (2), 85–101, 198.
Diez-Roux, A (1998) Bringing context back into epidemiology: Variables and fallacies in multilevel analysis, *American Journal of Public Health*, **88**: (2), 216–222.
Diggle, P (1990) *Time series: A biostatistical introduction*. Oxford Science Publications.
Diggle, P (1997) Spatial and longitudinal data analysis: Two histories with a common future, pp. 387–402 in *Modelling longitudinal and spatially correlated data*, eds T Gregoire *et al.* Lecture Notes in Statistics 122, New York: Springer.
Diggle, PJ, Tawn, JA and Moyeed, RA (1998) Model-based geostatistics (with discussion). *J. Royal Statistical Society, Series C, Applied Statistics*, **47**, (3), 299–350.
Dobson, A (1990) *An introduction to generalised linear models*. Chapman and Hall, London.
Doll, R (1971) The age distribution of cancer: Implications for models of carcinogenesis, *J. Royal Statistical Society, series A*, **134**, 133–166.
Doll, R and Hill, A (1964) Mortality in relation to smoking: Ten years observations of British doctors, *British Medical Journal*, i, 1399–1410, 1460–1467.
Draper, D (1995) Inference and hierarchical modeling in the social sciences, J. Educational and Behavioral Statistics, 20 (2), 115–147.
Draper, N and Smith, H (1980) *Applied regression analysis*. Wiley Series in Probability and Statistics. Wiley.
Dreze, J and Richard, J (1983) Bayesian analysis of simultaneous equation systems, Chapter 9 in *Handbook of econometrics*, Vol I, eds Z. Griliches and M. Initriligator, North-Holland.
Ducrocq, V and Casella, G (1996) A Bayesian analysis of mixed survival models, *Genetics and Selective Evolution*, **28**, 505–529.
DuMouchel, W (1990) Bayesian meta-analysis, in *Statistical methodology in the pharmaceutical sciences*, ed. D Berry. New York: Dekker.

Dumouchel, W and Duncan, G (1983) Using sample survey weights in multiple regression analyses of stratified samples, *J. American Statistical Association*, **78**, 535–543.

Duncan, O (1961) A socioeconomic index for all occupations, in A. Reiss, *Occupations and social status*. New York: Free Press of Glencoe.

Dunsmore, I and Boys, R (1988) Global vs local screening, pp. 593–599 in *Bayesian statistics*, Vol 3, eds J. Bernardo *et al.* Oxford University Press.

Efron, B and Morris, C (1975) Data analysis using Stein's estimator and its generalizations, *J. American Statistical Association*. **70**, 311–331.

Efron, B and Tibshirani, R (1993) *An introduction to the bootstrap*. Monographs on Statistics and Applied Probability, 57. New York: Chapman & Hall.

Erkanli, A and Gopalan, R (1996) Bayesian nonparametric regression using Gibbs sampling, Chapter 22 in *Bayesian analysis in statistics and econometrics*, eds D Berry *et al.* Wiley.

Escobar, MD and West, M (1998) Computing nonparametric hierarchical models, pp. 1–22 in Dey, Dipak *et al.* (eds), *Practical nonparametric and semiparametric Bayesian statistics*. Lecture Notes in Statistics, 133. New York: Springer.

Evans, A (1976) Causation and disease: The Henle-Koch postulates revisited, *Yale Journal of Biology and Medicine*, **49**, 175–195.

Evans, M, Gilula, Z and Guttman, I (1989) Latent class analysis of two-way contingency tables by Bayesian methods, *Biometrika*, **76** (3), 557–563.

Evans, M, Gilula, Z and Guttman, I (1993) Computational issues in the Bayesian analysis of categorical data: Loglinear and Goodman's RC model. *Statistica Sinica*, **3** (2), 391–406.

Everitt, B (1984) *An introduction to latent variable models*, London: Chapman and Hall.

Everitt, B (1994) *Statistical methods for medical investigation*, 2nd ed., Arnold.

Everitt, B and Dunn, G (1983) *Advanced methods of data exploration and modelling*. Heinemann.

Fahrmeir, L (1992) Posterior mode estimation by extended Kalman filtering for multivariate dynamic generalized linear models, *J. American Statistical Association* **87**, (418), 501–509.

Fahrmeir, L and Knorr-Held, L (1997) Dynamic discrete-time duration models: Estimation via Markov chain Monte Carlo, *Sociological Methodology*, **27**, 417–452.

Fahrmeir, L and Lang, S (1999) *Bayesian inference for generalized additive regression based on dynamic models*, Institut fur Statistik, Universitat Munchen.

Fahrmeir, L and Tutz, G (1994) *Multivariate statistical modelling based on generalized linear models*. Springer Series in Statistics. New York: Springer-Verlag.

Fay, R (1986) Causal model for patterns of nonresponse, *J. American Statistical Association*, **81**, 354–365.

Fayers, P, Ashby, D and Parmar, M (1997) Tutorial in biostatistics: Bayesian data monitoring in clinical trials, *Statistics in Medicine*, **16**, 1413–1430.

Feigl, P and Zelen, M (1965) Estimation of exponential survival probabilities with concomitant information, *Biometrics*, **21** (4), 826–838.

Fernandez, C and Steel, M (1998) On Bayesian modeling of fat tails and skewness, *J. American Statistical Association*, **93** (441), 359–371.

Finney, D (1973) *Statistical methods in bioassay*. New York: Hafner.

Fitzmaurice, G (1998) Regression models for discrete longitudinal data, in *Statistical analysis of medical data*, eds B Everitt and G Dunn. Arnold. 175-197.

Fitzmaurice, G and Lipsitz, S (1995) A model for binary time series data with serial odds ratio patterns, *J. Royal Statistical Society Series C*, **44** 51–56.

Fleming, T and Harrington, D (1991) *Counting processes and survival analysis*. Wiley Series in Probability and Mathematical Statistics: Applied Probability and Statistics Section. New York: Wiley.

Fletcher, R, Fletcher, S and Wagner, E (1982) *Clinical epidemiology*. Baltimore: Williams and Wilkins.

Forster, J and Smith, P (1998) Model-based inference for categorical survey data subject to non-ignorable non-response, *J. Royal Statistical Society, Series B, Statistical Methodology*, **60** (1), 57–70.

Frome, E (1983) The analysis of rates using Poisson regression models, *Biometrics*, **39**, 665–674.

Fuller, W (1976) *Introduction to statistical time series*. New York: Wiley.

Gamerman, D (1997) *Markov chain Monte Carlo: Stochastic simulation of Bayesian inference*. Chapman & Hall.

Gamerman, D (1997) Sampling from the posterior distribution in generalized linear mixed models, *Statistics and Computing*, **7**, 57–68.

Gamerman, D (1998) *Markov chain Monte Carlo*. London: Chapman and Hall.

Gamerman, D and Migon, H (1991) Forecasting the number of AIDS cases in Brazil, *The Statistician*, **40**, 427–442.

Gamerman, D and Smith, A (1996) A Bayesian analysis of longitudinal data studies, in *Bayesian statistics 5*, ed. J Bernardo. Oxford University Press. 587-598.

Gastwirth, J, Johnson, W and Reneau, D (1991) Bayesian analysis of screening data: Application to AIDS in blood donors, *The Canadian Journal of Statistics*, **19**, 135–150.

Gehan, E (1965) A generalized Wilcoxon test for comparing arbitrarily singly-censored samples, *Biometrika*, **52**, 203–223.

Geisser, S (1993) *Predictive inference. An introduction*. Monographs on Statistics and Applied Probability, 55. London: Chapman and Hall.

Geisser, S and Eddy, W (1979). A predictive approach to model selection. *J. American Statistical Association*, **74**, 153–160.

Gelfand, A and Dey, D. (1994) Bayesian model choice: Asymptotics and exact calculations, *J. Royal Statistical Society Series B*, **56**, (3), 501–514.

Gelfand, A, Dey, D and Chang, H (1992) Model determination using predictive distributions with implementations via sampling-based methods, in *Bayesian statistics 4* (eds J. Bernardo *et al.*). Oxford University Press, pp. 147–168.

Gelman, A and Rubin, DB (1992) Inference from iterative simulation using multiple sequences, *Statistical Science*, **7**, 457–511.

Gelman, A, Bois, F and Jiang, J (1996) Physiological pharmacokinetic analysis using population modelling and informative prior distributions, *J. American Statistical Association*, **91**, 1400–1412.

Gelman, A, Carlin, J, Stern, H and Rubin, D (1995) *Bayesian data analysis*. Chapman and Hall.

Gelman, A, Carlin, JB, Stern, HS and Rubin, DB (1998) *Bayesian data analysis*, 2nd edn, Chapman and Hall Texts in Statistical Science Series. London: Chapman & Hall.

Gelman, A, Meng, X and Stern, H (1996) Posterior predictive assessment of model fitness via realized discrepancies (with discussion), *Statistical Sinica*, **6**, (4), 733–807.

George, E and McCulloch, R (1993) On obtaining invariant prior distributions, *J. of Statistical Planning and Inference*, **37** (2), 169–179.

George, E and McCulloch, R (1993) Variable selection via Gibbs sampling, *J. American Statistical Association*, **88** (423), 881–889.

George, E, Makov, U and Smith, A (1993) Conjugate likelihood distributions, *Scandinavian Journal of Statistics*, **20** (2), 147–156.

George, EI, Makov, UE and Smith, A (1994) Fully Bayesian hierarchical analysis for exponential families via Monte Carlo computation, in *Aspects of uncertainty: A tribute to D. V. Lindley*, ed. PR Freeman *et al.*, Chichester: Wiley. 181–199.

Geyer, C (1992) Practical Markov chain Monte Carlo, *Statistical Science* **7** (4), 473–511.

Ghosh, M and Rao, J (1994) Small area estimation: An appraisal, *Statistical Science*, **9**, 56–93.

Gilks, W, Clayton, D, Spiegelhalter, D, Best, N, McNeil, A, Sharples, L and Kirby, A (1993) Modelling complexity: Applications of Gibbs sampling in medicine, *J. Royal Statistical Society Series B*, **55**, No.1, 39–52.

Glass, D (ed.) (1954) *Social mobility in Britain.* London: Routledge.
Glasziou, P and Mackerras, D (1993) Vitamin-A supplementation in infectious-diseases – a meta-analysis, *British Medical Journal*, 306 (6874), 366–370.
Goldberger, A (1972) Maximum-likelihood estimation of regressions containing unobservable independent variables, *International Economics Review*, **13**, 1–15.
Goldstein, H and Spiegelhalter, DJ (1996) League tables and their limitations – statistical issues in comparisons of institutional performance (with discussion). *J. Royal Statistical Society, Series A*, **159**, 385–409.
Goldstein, H, Rasbash, J, Plewis, I, Draper, D, Browne, W, Yang, M, Woodhouse, G and Healy, M (1998) *A users guide to MlwiN*, Multilevel Models Project, Institute of Education, University of London.
Goodman, L (1974) Exploratory latent structure analysis using both identifiable and unidentifiable models. *Biometrika*, **61**, 215–231.
Goodman, L (1979) On quasi-independence in triangular contingency tables, *Biometrics*, **35**, 651–655.
Goodman, L (1981) Criteria for determining whether certain categories in a cross-classification table should be combined, with special reference to occupational categories in an occupational-mobility table, *American Journal of Sociology*, **87** (3), 612–650.
Gordon, K and Smith, A (1990) Modeling and monitoring biomedical time-series, *J. American Statistical Association*, **85** (410), 328–337.
Granger, C (1981) Some properties of time series data and their use in econometric model specification, *J. Econometrics*, **16** (1), 121-130.
Granger, C and Newbold, P (1976) Forecasting transformed series, *J. Royal Statistical Society, Series B*, **38**, 189–203.
Granger, C and Swanson, N (1997) An introduction to stochastic unit-root processes, *J. Econometrics*, **80** (1), 35-62.
Granger, CWJ and Newbold, P (1974) Spurious regressions in econometrics, *J. Econometrics*, **2**, 111-120.
Grant, A (1989) Reporting controlled trials, *British Journal of Obstetrics and Gynaecology*, **96**, 397–400.
Green, P (1995) Reversible jump Markov chain Monte Carlo computation and Bayesian model determination, *Biometrika*, **82** (4), 711–732.
Greenland, S (1998) Causation, pp. 569–572 in *Encyclopaedia of biostatistics*, Vol 1, eds P Armitage and T. Colton, Wiley.
Griffiths, W, Hill, R and Judge, G (1993) *Learning and practising econometrics.* Wiley.
Grigsby, J and Bailey, D (1993) Alternative definitions of functional life expectancy, pp. 121–130 in *Calculation of health expectancies: Harmonization, censensus achieved and future perspectives*, eds J Robine, C Mathers, M Bone, and I Romieu. Colloque INSERM, Vol 226. Paris: INSERM (National Institute of Health and Medical Research).
Grizzle, J and Williams, O (1972) Log linear models and tests of independence for contingency tables, *Biometrics*, **28**, 137–156.
Gunst, R and Hartfield, M (1997) Robust semivariogram estimation in the presence of influential spatial data values, in *Modelling longitudinal and spatially correlated data*, eds T Gregoire *et al.* Lecture Notes in Statistics, **122**. New York: Springer.
Haavelmo, T (1947) Methods of measuring the marginal propensity to consume, *J. American Statistical Association*, **42**, 105–122.
Hachen, D (1988) The competing risks model, *Sociological Methods and Research*, **17**, 21–54.
Hand, D and Crowder, M (1996) *Practical longitudinal data analysis.* Chapman and Hall Texts in Statistical Science Series. London: Chapman and Hall.
Harvey, A (1991) *Forecasting, structural time series models and the Kalman filter*, Cambridge University Press.

Harvey, A and Todd, P (1983) Forecasting economic time series with structural and Box–Jenkins models, *J. Business and Economic Statistics*, **1**, 299–315.

Harvey, AC and Durbin, J (1986) The effects of seat belt legislation on British road casualties: A case study in structural time series modelling (with discussion), *J. Royal Statistical Society, Series A*, **149**, 187–227.

Hastie, TJ and Tibshirani, RJ (1990) *Generalized additive models*. Monographs on Statistics and Applied Probability, 43. London: Chapman and Hall.

Hastings, W (1970) Monte Carlo sampling methods using Markov chains and their applications, *Biometrika* **57**, 97–109.

Hayward, M, Hardy, M and Grady, W (1989) Labor force withdrawal patterns among older men in the United States, *Social Science Quarterly*, **70**, 2 June, 425–448.

Heath, S (1997) Markov chain Monte Carlo methods for radiation hybrid mapping, *J. Computational Biology*, **4** (4), 505–515.

Heckman, J and Singer, B (1984) A method for minimizing the impact of distributional assumptions in econometric-models for duration data, *Econometrica*, **52** (2), 271–320.

Hedges, L (1997) Bayesian meta-analysis, pp. 251–275 in *Statistical analysis of medical data*, eds B Everitt and G. Dunn. London: Arnold.

Hedges, L and Olkin, I (1985). *Statistical methods for meta-analysis*. Academic Press: New York.

Helfenstein, U (1991) The use of transfer-function models, intervention analysis and related time-series methods in epidemiology, *International Journal of Epidemiology*, **20**, (3), 808–815.

Henrici, P (1974) *Applied and computational analysis*, Vol 1. Wiley.

Herbst, AL, Ulfelder, H and Poskanzer, D (1971) Adenocarcinoma of the vagina. Association of maternal stilbestrol therapy with tumor appearance in young women, *New England Journal of Medicine*, 15;284(15), 878–881.

Hirano, K (1998) A semiparametric model for labor earnings dynamics, in *Practical nonparametric and semiparametric Bayesian statistics*, ed. Dey, Dipak *et al.*, Lecture Notes in Statistics 133. New York: Springer. 355–369.

Hoek, H, Lucas, A and van Dijk, H (1995) Classical and Bayesian aspects of robust unit root inference, *J. Econometrics* **69** (1) 27–59.

Holt, D, Smith, T and Tomberlin, T (1979) A model-based approach to estimation for small subgroups of a population, *J. American Statistical Association*, **74**, 405–410.

Hsieh, C and Pugh M (1993) Poverty, income inequality and violent crime: A meta-analysis of recent aggregate data studies, *Criminal Justice Review*, 18, 182–202.

Huerta, G and West, M (1999) Priors and component structures in autoregressive time series models, *J. Royal Statistical Society, Series B, Statistical Methodology*, **61** (4), 881–899.

Hui, S and Berger, J (1983) Empirical Bayes estimation of rates in longitudinal studies. J. American Statistical Association, **78**, 753–760.

Humphreys, K and Carr-Hill, R (1991) Area variations in health outcomes: Artefact or ecology. *International Journal of Epidemiology*, **20**, 251–8.

Ibrahim, J and Kleinman, K (1998) Semiparametric Bayesian methods for random effects models, pp. 89–114 in Dey, Dipak *et al.* (eds), *Practical nonparametric and semiparametric Bayesian statistics*. Lecture Notes in Statistics, 133. New York: Springer.

Jeffreys, H (1961) *Theory of probability*, 3rd ed. International Series of Monographs on Physics. Oxford: Clarendon Press.

Johnson, V and Albert, J (1999) *Ordinal data modelling*. New York: Springer-Verlag.

Johnston, J and DiNardo, J (1997) *Econometric methods* (4th edn). McGraw-Hill.

Jones, C and Marriott, J (1999) A Bayesian analysis of stochastic unit root models, in *Bayesian statistics 6*, eds. JM Bernardo, JO Berger, AP Dawid and AFM Smith. Oxford University Press. 785-794.

Jones, M (1987) Randomly choosing parameters from the stationarity and invertibility region of autoregressive-moving average models, *J. Royal Statistical Society, Series C*, **36**, 134–138.

Joreskog, K and Sorbom, D (1981) Analysis of linear structural relationships by maximum likelihood and least squares methods, Research Report 81–8, University of Uppsala, Sweden.

Joreskog, KG (1973) A general model for estimating a linear structural equation system, in *Structural equation models in the social sciences*, eds AS Goldberger and OD Duncan. New York: Seminar Press.

Joseph, L, Gyorkos, T and Coupal, L (1995) Bayesian estimation of disease prevalence and the parameters of diagnostic tests in the absence of a gold standard, *American Journal of Epidemiology*, **141**, 263–272.

Judge, G, Hill, R, Griffiths, W, Luetkepohl, H and Lee, T (1988) *Introduction to the theory and practice of econometrics*, 2nd ed. New York: Wiley.

Kadane, J (1993) Subjective Bayesian analysis for surveys with missing data, *The Statistician*, **42**, 415–426.

Kafadar and Andrews (1996) in *Statistics in public health*, eds D Stroup and S Teutsch. Oxford University Press.

Kahn, H and Sempos, C (1989) *Statistical methods in epidemiology*. Oxford University Press.

Kalbfleisch, J (1978) Non-parametric Bayesian analysis of survival time data, *J. Royal Statistical Society, Series B*, **40**, 214–221.

Kaldor, JM, Day, NE, Band, P, Choi, NW, Clarke, EA, Coleman, MP, Hakama, M, Koch, M, Langmark, F and Neal, F (1987) Second malignancies following testicular cancer, ovarian cancer and Hodgkin's disease: An international collaborative study among cancer registries, *International Journal of Cancer*, **39** (5), 571–585.

Kandel, D and Yamaguchi, K (1987) Job mobility and drug use: An event history analysis, *American Journal of Sociology*, 92 n(4) 836–878.

Kaplan, EL and Meier, P (1958) Nonparametric estimation from incomplete observations, *J. American Statistical Association*, **53**, 457–481.

Katz, D, Baptista, J, Azen, S and Pike, M (1978) Obtaining confidence intervals for the risk ratio in cohort studies, *Biometrics*, **34**, 469–474.

King, A, Beazley, R, Warren, W and Hankins, C (1988) *Canada Youth and Aids Study*. Ottawa: Federal Centre for AIDS.

Kitagawa, G (1987) Non-Gaussian state-space modelling of nonstationary time series, *J. American Statistical Association*, **82**, 1032–1063.

Kitagawa, G and Gersch, W (1996) *Smoothness priors analysis of time series*. Lecture Notes in Statistics, **116**. New York: Springer.

Klein, L and Goldberger, A (1955) *An econometric model of the United States 1929–52*. Amsterdam: North-Holland.

Kluegl, J, Singleton, R and Starnes, C (1977) Subjective class identifications: A multiple indicator approach, Amererican in Sociological Review **42**, 599–611.

Knorr-Held, L (1997) *Hierarchical modelling of discrete longitudinal data – applications of Markov chain Monte Carlo*. München: Herbert Utz Verlag.

Knox, G (1964) Epidemiology of childhood leukaemia in Northumberland and Durham, *British Journal of Preventive Social Medicine*, **18**, 17–24.

Krzanowski, W (1988) *Principles of multivariate analysis. A user's perspective*. Oxford Statistical Science Series 3. Oxford: Clarendon Press.

Kuo, L and Smith, A (1992) Bayesian computations in survival models via the Gibbs sampler, in *Survival analysis: State of the art*, eds J Klein and P Goel. Kluwer. 11–24.

Lai, DJ and Hardy, RJ (1999) Potential gains in life expectancy or years of potential life lost: Impact of competing risks of death, *International Journal of Epidemiology* **28**: (5), 894–898.

Lai, P (1979) *Transfer function modelling: Relationships between time series variables*, Concepts and Techniques in Modern Geography 22, Institute of British Geographers.

Laird, N (1979) Empirical Bayes methods for two-way contingency tables, *Biometrika*, 65, 581–590.

Laird, N (1982) Empirical Bayes estimates using the nonparametric maximum likelihood estimate for the prior, *J. Statistical Computation and Simulation*, **15**, 211–220.

Laird, N and Ware, J (1982) Random-effects models for longitudinal data, *Biometrics*, **38**, 963–974.

Lancaster, T and Intrator, O (1998) Panel data with survival: Hospitalization of HIV-positive patients, *J. American Statistical Association*, **93** (441), 46–53.

Langford, I, Leyland, A, Rasbash, J and Goldstein, H (1999) Multilevel modelling of the geographical distributions of diseases, *J. Royal Statistical Society Series C, Applied Statistics*, **48**, (2), 253–268.

Lavine, M and West, M (1992) A Bayesian method for classification and discrimination, *Canadian Journal of Statistics* **20** (4), 451–461.

Lee, P (1997) *Bayesian statistics: An introduction*, 2nd ed. Arnold.

Lee, TC Judge, GG and Zellner, A (1968) Maximum likelihood and Bayesian estimation of transition probabilities, *J. American Statistical Association*, **63**, 1162–1179.

Lenk, P (1999) Bayesian inference for semiparametric regression using a Fourier representation, *J. Royal Statistical Society*, **61B**, 863–879.

Leonard, T (1972) Bayesian methods for binomial data, *Biometrika*, 59, 581–589.

Leonard, T (1973) A Bayesian method for histograms, Biometrika, 60, 297–230.

Leonard, T (1975) Bayesian estimation methods for two-way contingency tables, *J. Royal Statistical Society, Series B*, **37**, 23–37.

Leonard, T (1996) On exchangeable sampling distributions for uncontrolled data, *Statistics and Probability Lett*. **26**, (1), 1–6.

Leonard, T and Hsu, J (1994) The Bayesian analysis of categorical data – a selective review, in *Aspects of Uncertainty: a Tribute to DV Lindley*, P Freeman and A Smith (eds). Wiley.

Leonard, T and Hsu, J (1999) *Bayesian methods: An analysis for statisticians and interdisciplinary researchers*, Cambridge University Press.

Leonard, T, Hsu, J, Tsui, K and Murray, J (1994) Bayesian and likelihood inference from equally weighted mixtures, *Annals of the Institute of Statistical Mathematics*, **46** (2), 203–220.

Leroux, B and Puterman, M (1992) Maximum penalized likelihood estimation for independent and Markov dependent mixture models, *Biometrics*, **48**, 545–558.

Lewis, S and Raftery, A (1999) Bayesian analysis of event history models with unobserved heterogeneity via Markov chain Monte Carlo: Application to the explanation of fertility decline, *Sociological Methods and Research*, **28**, 1 Aug, 35–60.

Li, K (1999) Bayesian analysis of duration models: An application to Chapter 11 bankruptcy, *Economics Letters*, **63**, 305–312.

Liang, K and Zeger, S (1986) Longitudinal data-analysis using generalized linear-models, *Biometrika*, **73**: (1), 13–22.

Lindley, D (1979) Analysis of life tables with grouping and withdrawals, *Biometrics*, **35**, 605–612.

Lindsey, J (1992) *The analysis of stochastic processes using GLIM*. Lecture Notes in Statistics, 72. Berlin: Springer-Verlag.

Lindsey, JK (1989) *The analysis of categorical data using GLIM*. Lecture Notes in Statistics, 56. Berlin: Springer-Verlag.

Liska, A (1990) The significance of aggregate dependent variables and contextual independent variables for linking macro and micro theories, *Social Psychology Quarterly*, **53**, 292–301.

Little, R and Schluchter, M (1985) Maximum likelihood estimation for mixed continuous and categorical data with missing values, *Biometrika*, **72**, 497–512.

Little, RJA and Rubin, D (1987) *Statistical analysis with missing data*. Wiley Series in Probability and Mathematical Statistics: Applied Probability and Statistics. New York: Wiley.
Little, T and Gelman, A (1998) Modeling differential nonresponse in sample surveys, *Sankhya*, **60**, 101–126.
Liu, C (1996) Bayesian robust multivariate linear regression with incomplete data. *J. American Statistical Association*, (435), 1219–1227.
Lord, F (1952) The relation of the reliability of multiple-choice tests to the distribution of item difficulties, *Psychometrika*, **17**, 181–194.
Lovison, G (1994) Log-linear modelling of data from matched case-control studies, *J. Applied Statistics*, **21**, 125–141.
MacDonald, I and Zucchini, W (1997) *Hidden Markov and other models for discrete-valued time series*. London: Chapman and Hall.
MacGibbon, B and Tomberlin, T (1989) Small area estimates of proportions via empirical Bayes techniques, *Survey Methodology*, **15**, 237–252.
Machin, D and Gardner, M (1988) Calculating confidence intervals for survival time analyses, *British Medical Journal*, 296 (6633), 1369–1371.
Maddala, G (1981) *Econometrics*. McGraw-Hill.
Maddala, G (1983) *Limited-dependent and qualitative variables in econometrics*. Econometric Society Monographs in Quantitative Economics, 3. Cambridge University Press.
Malec, D, Sedransk, J and Moriarity, C (1997) Small area inference for binary variables in the National Health Interview Survey, *J. American Statistical Association* **92**: (439), 815–826.
Mantel, N and Haenszel, W (1959) Statistical aspects of the analysis of data from retrospective studies, *J. National Cancer Institute*, **22**, 719–748.
Mardia, K., Kent, J. and Bibby, J. (1979) *Multivariate analysis*.
Marriott, J (1988) Reparameterisation for Bayesian inference in ARMA time series. *Bayesian statistics* 3, Proceedings 3rd Valencia International Meeting, Altea, Spain 1987, pp. 701–704.
Marriott, J and Smith, A (1992) Reparameterisation aspects of numerical Bayesian methodology for ARMA models, *J. Time Series Analysis*, **13**, 327–343.
Marriott, J, Ravishanker, N, Gelfand, A and Pai, J (1996) Bayesian analysis of ARMA processes: Complete sampling-based inference under exact likelihoods, in *Bayesian analysis in statistics and econometrics*, eds. D Berry, K Chaloner and J Geweke. New York: Wiley. 243–256.
Matheron, G (1963) Principles of geostatistics, *Economic Geology*, **58**, 1246–1266.
McCullagh, P (1980) Regression models for ordinal data, *J. Royal Statistical Society, Series B*, **42**, 109-142.
McCullagh, P and Nelder, J (1989) *Generalized linear models*, 2nd ed. Monographs on Statistics and Applied Probability, 37. London: Chapman and Hall.
McCulloch, R and Tsay, R (1994) Bayesian-inference of trends-stationarity and difference-stationarity, *Econometric Theory*, **10**, (3–4), 596–608.
McHugh, R (1956) Efficient estimation and local identification in latent class analysis, *Psychometrika*, **21**, 331–347.
McKenzie, E (1988) Some ARMA models for dependent sequences of Poisson counts, *Advanced Applied Probability* **20** (4), 822–835.
McLachlan, G and Basford, K (1988) *Mixture models: Inference and applications to clustering*. Statistics: Textbooks and Monographs 84. New York: Marcel Dekker.
McLachlan, GJ and Krishnan, T (1997) *The EM algorithm and extensions*. New York: Wiley.
McNeil, D, Trussell, T and Turner, J (1977) Spline interpolation of demographic data, *Demography*, **14** (2), 245–252.
Merrill, W and Fox, K (1971) *Introduction to economic statistics*, New York: Wiley.

Migon, H and Gamerman, D (1999) *Statistical inference: An integrated approach*. Arnold.

Miller, R (1984) Evaluation of transformations in forecasting age specific birth-rates, *Insurance Mathematics & Economics*, **3** (4), 263–270.

Molenberghs, G, Goetghebeur, E, Lipsitz, S and Kenward, M (1999) Nonrandom missingness in categorical data: Strengths and limitations, *The American Statistician*, **53** (2), 110–118.

Mollié, A (1996) Bayesian mapping of disease, Chapter 20 in *Markov chain Monte Carlo in practice*, eds W Gilks, S Richardson, and D Spieglehalter. London: Chapman and Hall.

Morimune, K (1979) Comparisons of normal and logistic models in the bivariate dichotomous analysis, *Econometrica*, **47**, 957–975.

Moroney, M (1965) *Facts from figures, 1965*, 3rd edn. Penguin.

Morris, CN and Christiansen, CL (1995) *Hierarchical models for ranking and for identifying extremes, with applications*. With discussion and rejoinder, p. 17 and p.189. Bayesian Statistics 5. Oxford University Press.

Nandram, B and Sedransk, J (1993) Bayesian predictive inference for a finite population proportion: Two stage cluster sampling, *J. Royal Statistical Society*(B), 55 (2), 399–408.

Naylor, J and Smith, A (1982) Applications of a method for the efficient computation of posterior distributions. *J. Royal Statistical Society Series C*, **31**, 214–22.

Naylor, J and Smith, A (1988) Econometric illustrations of novel numerical-integration strategies for Bayesian inference, *J. Econometrics*, **38** (1–2), 103–112.

Neal, RM. (1997) Markov chain Monte Carlo methods based on 'slicing' the density function, Technical Report No. 9722, Department of Statistics, University of Toronto.

Nelson, J (1984) Modeling individual and aggregate victimization rates, *Social Science Research* **13** (4), 352–372.

Newbold, P (1974) The exact likelihood function for a mixed autoregressive-moving average process, *Biometrika*, **61**, 423–426.

Newell, C (1988) *Methods and models in demography*. Wiley.

Newton, M and Raftery, A (1994) Approximate Bayesian inference with the weighted likelihood bootstrap. *J. Royal Statistical Society Series B*, **56**, (1), 3–48.

Nordberg, L (1989) Generalized linear modelling of sample survey data, *J. Official Statistics*, **5**, 223–239.

Nordentoft, M, Breum, L, Munck, L, Nordestgaard, A, Hunding, A and Bjaeldager, P (1993) High mortality by natural and unnatural causes – a 10-year follow-up study of patients admitted to a poisoning treatment center after suicide attempts, *British Medical Journal*, **306**, 1637–1641.

Northridge, M (1995) Public-health methods – attributable risk as a link between causality and public-health action, *American Journal of Public Health*, **85** (9), 1202–1204.

Nusbacher, J, Chiavetta, J, *et al.* (1986) Evaluation of a confidential method of excluding blood donors exposed to human immunodeficiency virus, *Transfusion*, **26**, 539–541.

Oliver, M *et al.* (1992) A geostatistical approach to the analysis of patterns in rare diseases, *J. Public Health Medicine*, **14**, 280–289.

Ord, K (1975) Estimation methods for models of spatial interaction. *J. American Statistical Association*, **70**, 120–126.

Osherson, D, Smith, E, Shafir, E, Gualtierotti, A and Biolsi, K (1995) A source of Bayesian priors, *Cognitive Science*, **19** (3), 377–405.

Pai, J, Ravishanker, N and Gelfand, A (1994) Bayesian-analysis of concurrent time-series with application to regional IBM revenue data, *J. Forecasting*, **13** (5), 463–479.

Park, T and Brown, M (1994) Model for categorical data with nonignorable response, *J. American Statistical Association*, **89**, 44–52.

Paul, S and Banerjee, T (1998) Analysis of two-way layout of count data involving multiple counts in each cell, *J. American Statistical Association*, **99**, 1419–1429.
Percy, D (1992) Priors for seemingly unrelated regressions, *J. Royal Statistical Society*, **54B**, 243–252.
Peto, R, Pike, MC, Armitage, P, Breslow, NE, Cox, DR, Howard, SV, Mantel, N, McPherson, K, Peto, J and Smith, PG (1977) Design and analysis of randomised clinical trials requiring prolonged observation of each patient. II. Analysis and examples, *British Journal of Cancer*, **35** (1), 1–39.
Phillips, M (1993) Contingency tables with missing data, *The Statistician*, **42**, 9–18.
Pitt, M and Shephard, N (1999) Analytic convergence rates and parameterization issues, *J. Time Series Analysis*, **20** (1), 63–86.
Pocock, S (1982) Interim analyses for randomised clinical trails: The group sequential trials, *Biometrics*, **38**, 153–162.
Pocock, S, Hughes, M and Lee, R (1987) Statistical problems in the reporting of clinical trials, *New England Journal of Medicine*, **317**, 426–432.
Pole, A, West, M and Harrison, J (1994) *Applied Bayesian forecasting and time series analysis*. Chapman and Hall Texts in Statistical Science Series. London: Chapman & Hall.
Pollock, S (1993) Smoothing with cubic splines, Departmental Reference Paper 291, Economics Department, Queen Mary and Westfield College, University of London.
Potthoff, R and Roy, S (1964) A generalized multivariate analysis of variance model useful especially for growth curve problems, *Biometrika*, **51**, 313–326.
Prentice, R (1976) Use of logistic model in retrospective studies, *Biometrics*, 32, 599–606.
Press, S J and Shigemasu, K (1989). Bayesian inference in factor analysis, Chapter 15 in *Contributions to Probability and Statistics, Essays in honor of Ingram Olkin*, eds LJ Glesser New Yok: Springer verlag.
Quantin, C, Moreau, T, Asselain, B, Maccario, J and Lellouch, J (1996) A regression survival model for testing the proportional hazards hypothesis, *Biometrics* **52**, (3), 874–885.
Raab, G and Donnelly, C (1999) Information on sexual behaviour when some data are missing, *Applied Statistics*, **48**, 117–133.
Raftery, A (1995) Bayesian model selection in social research, in *Sociological methodology*, ed. P Marsden Oxford: Blackwell.
Raftery, A (1996) Approximate Bayes factors and accounting for model uncertainty in generalised linear models, *Biometrika* 83, (2), 251–266.
Raftery, A (1996) Hypothesis testing and model selection, pp. 163–187 in *Markov chain Monte Carlo in practice*, eds WR Gilks *et al.*, London: Chapman & Hall.
Raftery, A and Richardson, S (1995) Model selection for generalized linear models via GLIB, with application to epidemiology, in *Bayesian biostatistics*, eds DA Berry and DK Strangl. New York: Dekker.
Raftery, A, Madigan, D and Hoeting, J (1993) Model selection and accounting for model uncertainty in linear regression models. Technical Report no. 262, Department of Statistics, University of Washington.
Rasch, D (1960) Probleme der Varianzanalyse bei ungleicher Klassenbesetzung, *Biometrische Zeitschrift*, **2**, 194–203.
Ravishanker, N and Ray, BK (1997) Bayesian analysis of vector ARMA models using Gibbs sampling, *J. Forecasting*, **16**, (3), 177–194.
Richardson S, Stucker, I and Hemon, D (1987) Comparison of relative risks obtained in ecological and individual studies – some methodological considerations, *International Journal of Epidemiology*, **16** (1), 111–120.
Richardson, S (1992) Statistical methods for geographical correlation studies, in *Geographical and environment epidemiology: Methods for small area studies*, eds. P Elliott, J Cuzick, D English and R Stern. Oxford: Oxford University Press.

Richardson, S (1996) Measurement error, Chapter 22 in *Markov chain Monte Carlo methods in practice*, eds WR Gilks, S Richardson and DJ Spiegelhalter. Chapman and Hall.
Richardson, S and Green, P (1997) On Bayesian analysis of mixtures with an unknown number of components (with discussion), *J. Royal Statistical Society, Series B*, **59**, (4), 731–792.
Rindskopf, D and Rindskopf, W (1986) The value of latent class analysis in medical diagnosis, *Statistics in Medicine*, **5**, 21–27.
Ritov, Y and Gilula, Z (1991) The order-restricted RC model for ordered contingency tables: Estimation and testing for fit, *Annals of Statistics*, **19**, (4), 2090–2101.
Robert, C (1997) On Bayesian analysis of mixtures with an unknown number of components – Discussion, *J. Royal Statistical Society, Series B*, **59** (4) 758–792.
Roeder, K (1990) Density estimation with confidence sets exemplified by superclusters and voids in the galaxies, *J. American Statistical Association*, **85**, 617–624.
Rothenberg, T (1974) Bayesian analysis of simultaneous equations models, Chapter 9.3 in *Studies in Bayesian Econometrics and Statistics*, eds S Fienberg and A Zellner. North-Holland.
Rothman, K (1986) *Modern epidemiology*. Boston: Little Brown.
Rothman, KJ, Fyler, DC, Goldblatt, A and Kreidberg, M (1979) Exogenous hormones and other drug exposures of children with congenital heart disease, *American Journal of Epidemiology*, **109** (4), 433–439.
Royston, P and Thompson, S (1992) Model based screening by risk with applications to Downs syndrome, *Statistics in Medicine*, **11**, 257–268.
Rubin, D (1976) Inference and missing data. With comments, *Biometrika*, **63**, 581–592.
Rubin, D (1987) *Multiple imputation for nonresponse in surveys*. New York: Wiley.
Rubin, D, Stern, H and Vehovar, V (1995) Handling 'don't know' survey responses: The case of the Slovenian plebiscite, **90**, 822–828.
Rubin, DB and Stern, HS (1994) Testing in latent class models using a posterior predictive check distribution, in *Latent variables analysis: Applications for developmental research*, eds. A von Eye and CC Clogg. Thousand Oaks, CA: Sage. 420–438.
Sahu, S and Roberts, G (1999). On convergence of the EM algorithm and the Gibbs sampler, *Statistics Computing*, **9**, 55–64.
Salk, L (1973) The role of the heartbeat in the relations between mother and infant, *Scientific American*, **228** (5), 24–29.
Schafer, J (1997) *Analysis of incomplete multivariate data*. London: Chapman and Hall.
Scheines, R, Hoijtink, H and Boomsma, A (1999) Bayesian estimation and testing of structural equation models, *Psychometrika* **64** (1), 37–52.
Schwartz, S (1994) The fallacy of the ecological fallacy: The potential misuse of a concept and the consequences, *American Journal of Public Health*, **84**(s), 819–824.
Schwarz, G (1978) Estimating the dimension of a model, *Annals of Statistics*, **6**, 461–464.
Seeber (1984) in *Stochastic modelling of social processes*, eds A Diekmann and P Mitter. Orlando: Academic Press.
Seltzer, MH, Wong, WH and Bryk, AS (1996) Bayesian analysis in applications of hierarchical models: Issues and methods, *J. of Educational and Behavioural Statistics*, **21** (2), 131–167.
Sethuraman, J (1994) A constructive definition of Dirichlet priors, *Statistica Sinica*, **4**, 639–650.
Shiue, W and Bain, L (1982) Experiment size and power comparisons for two-sample Poisson tests, *J. Royal Statistical Society, Series C*, **31**, 130–134.
Silcocks, P (1994) Estimating confidence limits on a standardized mortality ratio when the expected number is not errorfree, *J. Epidemiological Communications* **H 48** (3), 313–317.
Silliman, N (1997) Hierarchical selection models with applications in meta-analysis. *J. American Statistical Association* **92**, (439), 926–936.

Sivia, D (1996) *Data analysis: A Bayesian tutorial.* Oxford University Press.

Smith, A and Gelfand, A (1992) Bayesian statistics without tears: A sampling–resampling perspective, *The American Statistician*, **46** (2), 84–88.

Smith, M and Kohn, R (1996) Nonparametric regression using Bayesian variable selection, *J. Econometrics*, **75**, (2), 317–334.

Smith, T (1983) On the validity of inferences from nonrandom samples, *J. Royal Statistical Society, Series A*, **146**, 394–403.

Smith, WC, Shewry, MC, Tunstall-Pedoe, H, Crombie, IK and Tavendale, R (1989) Concomitants of excess coronary deaths: Major risk findings from 10,359 men and women in the Scottish Heart Health Study, *Scottish Medical Journal*, **34**, 550–555.

Smith, WC, Tunstall-Pedoe, H, Crombie, IK and Tavendale, R (1990) Cardiovascular disease in Edinburgh and North Glasgow – a tale of two cities, *J. Clinical Epidemiology*, **43**, 637–643.

Snedecor, G and Cochran, W (1989) *Statistical methods*, 8th ed. Ames, IA: Iowa State University Press.

Spiegelhalter, D and Freedman, L (1986) A predictive approach to selecting the size of a clinical trial, based on subjective clinical opinion, *Statistics in Medicine*, **5**, 1–13.

Spiegelhalter, D and Marshall, E (1998) Comparing institutional performance using Markov chain Monte Carlo methods, in *Recent advances in the statistical analysis of medical data* (eds B Everitt and G Dunn), 229–250. London: Edward Arnold.

Spiegelhalter, D, Best, N and Carlin, B (1999) Bayesian deviance, the effective number of parameters, and the comparison of arbitrarily complex models, manuscript, Medical Research Council Biostatistics Unit, Cambridge.

Spiegelhalter, DJ, Freedman, LS and Parmar, MKB (1994) Bayesian approaches to randomised trials, *Journal of the Royal Statistical Society, Series A*, **157**, 357–387.

Spitzer, W (1998) Thromboembolism and the pill: The saga must end. *Human Reproduction*, **13** (5), 1117–1118.

Stasny, E (1991) Hierarchical models for the probabilities of a survey classification and non-response – an example from the National crime survey, *J. American Statistical Association*, **86**, (414), 296–303.

Stern, H, Arcus, D, Kagan, J, Rubin, D and Snidman, N (1995) Using mixture-models in temperament research, *International Journal of Behavioral Development*, **18** (3), 407–423.

Stroud, T (1987) Bayes and Empirical Bayes approaches to small area estimation, in R Platek, J Rao, C Sarndal and J Singh (eds), *Small area statistics*. New York: Wiley.

Stroud, T (1991) Hierarchical Bayes predictive means and variances with application to sample survey inference. *Communications in Statistics, Theory Methods*, **20** (1), 13–36.

Stroud, T (1994) Bayesian analysis of binary survey data, *Canadian Journal of Statistics*, **22** (1), 33–45.

Sun, D, Tsutakawa, R and Speckman, P (1999) Posterior distribution of hierarchical models using CAR(1) distributions, *Biometrika*, **86** (2), 341–350.

Susarla, V and Ryzin, J (1976) Nonparametric Bayesian estimation of survival curves from incomplete observations, *J. American Statistical Association*, **71**, 897–902.

Tanner, M (1996) *Tools for statistical inference. Methods for the exploration of posterior distributions and likelihood functions*, 3rd ed. Springer Series in Statistics. New York: Springer-Verlag.

Tarone, R (1981) On summary estimations of relative risk, *J. Chronic Diseases*, **34**, 463–468.

Theil, H (1971) *Principles of econometrics*. Amsterdam: North-Holland.

Thode, H (1997) Power and sample size requirements for tests of differences between two Poisson rates, *The Statistician*, **46** (2) 227–230.

Thompson, S (1993) Controversies in meta-analysis: The case of trials of serum cholesterol reduction, *Statistics in Medicine*, **2**, 173–192.

Thum (1997) Hierarchical linear models for multivariate outcomes, *J. of Educational and Behavioural Research*, **22**, 77–108.

Tierney, L and Kadane, J (1986) Accurate approximations for posterior moments and marginal densities, *J. American Statistical Association*, **81**, 82–86.

Tobler, W (1970) Review of J. Forrester, 'Urban Dynamics', *Geographical Analysis*, **2** (2), 198-199.

Tuma, N *et al.* (1979) Dynamic analysis of event histories, *American Journal of Sociology*, 84 n(4), 820–854.

Turnbull, B (1974) Nonparametric estimation of a survivorship function with doubly censored data, *J. American Statistical Association*, **69**, 169–173.

Turner, D and West, M (1993) Bayesian analysis of mixtures applied to post-synaptic potential fluctuations, *J. Neouroscience Methods*, **47**, 1–21.

Upton, G (1981) Log-linear models, screening, and regional industrial surveys, *Regional Studies*, **15**, 33–45.

Upton, G (1991) The exploratory analysis of survey data using loglinear models, *The Statistician*, **40**, 169–182.

Upton, G and Fingleton, B (1985) *Spatial data analysis by example. Volume 1: Point pattern and quantitative data.* Wiley Series in Probability and Mathematical Statistics. Applied Probability and Statistics. Chichester: Wiley.

Venables, W and Ripley, B (1994) *Modern applied statistics with S-PLUS*. New York: Springer-Verlag.

Verdinelli, I and Wasserman, L (1991) Bayesian analysis of outlier problems using the Gibbs sampler, *Statistics and Computing*, **1**, 105–117.

Vuchinich, S *et al.* (1991) Families and hazard rates that change over time: Some methodological issues in analysing transitions, *Journal of Marriage and the Family*, 53 n(4), 898–912.

Wagstaff, A and Vandoorslaer, E (1994) Measuring inequalities in health in the presence of multiple-category morbidity indicators, *Health Economics* **3** (4) 281–291.

Wahba, G (1978) Improper priors, spline smoothing and the problem of guarding against model errors in regression. *J. Royal Statistical Society Series* B, **40**, 364–372.

Wald NJ and Kennard, A (1998). Routine ultrasound scanning for congenital abnormalitites, *Annals of the New York Academy of Science* **847**, 173–180.

Walker, S and Mallick, B (1997) Hierarchical generalized linear models and frailty models with Bayesian nonparametric mixing, *J. Royal Statistical Society, Series B*, **59**, (4), 845–860.

Walker, S, Damien, P, Laud, P and Smith, A (1999) Bayesian nonparametric inference, *J. Royal Statistical Society, Series B*, **61**, 485–528.

Weakliem, D (1999) A critique of the Bayesian information criterion for model selection, *Sociological Methods and Research*, **27**, 3 Feb, 359–397.

Wecker, W and Ansley, C (1983) The signal extraction approach to nonlinear regression and spline smoothing, *J. American Statistical Association*, **78**, 81–89.

Wedel, M, Desarbo, W and Bult, J (1993) A latent class Poisson regression-model for heterogeneous count data, *J. Applied Econometrics*, **8** (4), 397–411.

Weiss, R (1994) Pediatric pain, predictive inference, and sensitivity analysis, *Evaluation Review*, 18 (6), 651–776.

Weiss, R (1997) Bayesian sample size calculations for hypothesis testing, *The Statistician*, **46** (2) 185–191.

Welch, BL (1938) The significance of the difference between two means when the population variances are unequal, *Biometrika*, **29**, 350–362.

West, M (1984) Outlier models and prior distributions in Bayesian linear regression. *J. Royal Statistical Society, Series B*, **46**, 431–439.

West, M (1992) Mixture models, Monte Carlo, Bayesian updating and dynamic models, *J. Statistical Planning and Inference*, **62**, 325–333.

West, M and Harrison, J (1989) *Bayesian forecasting and dynamic models*, Springer-Verlag.

West, M and Harrison, J (1997) *Bayesian forecasting and dynamic models*, 2nd ed. Springer-Verlag.

West, M, Harrison, J and Migon, H (1985) Dynamic generalised linear models and Bayesian forecasting, *J. American Statistical Association*, **80**, 73–83.

Westfall, P, Johnson, W and Utts, J (1997) A Bayesian perspective on the Bonferroni adjustment, *Biometrika* **84** (2) 419–427.

Wheaton, B, Muthen, B, Alwin, D and Summers, G (1977) Assessing reliability and stability in panel models, in *Sociological methodology*, ed. D Heise. Jossey Bass.

White, E and Hewings, G (1982) Space–time employment modelling – some results using seemingly unrelated regression-estimators, *J. Regional Science*, **22**, (3), 283–302.

Wilcox, R (1983) Measuring mental abilities with latent state models, *American Journal of Mathematical and Management Sciences*, **3**, 313–345.

Wilcox, R (1996) *Statistics for the social sciences*, Academic Press.

Wong, A, Meeker, J and Selwyn, M (1985) Screening on correlated variables: A Bayesian approach, *Technometrics* **27**, 423–431.

Wood, S and Kohn, R (1998) A Bayesian approach to robust binary nonparametric regression. *J. American Statistical Association*, **93**, (441), 203–213.

Woodward, M (1999) *Epidemiology: Study design and data analysis*. Boca Raton: Chapman and Hall/CRC.

Yashin, AI, Manton, KG and Vaupel, JW (1985) Mortality and aging in a heterogeneous population: A stochastic process model with observed and unobserved variables. *Theoretical Population Biology*, **27** (2), 154–175.

Yule, W, Berger, M, Butler, S, Newham, V and Tizard, J (1969) the WPPSI: an empirical evaluation with a British sample, *British Journal of Educational Psychology* **39**, 1–13.

Zeger, S and Karim, M (1991) Generalized linear-models with random effects – a Gibbs sampling approach, *J. American Statistical Association* **86** (413), 79–86.

Zelen, M and Parker, R (1986) Case-control studies and Bayesian inference, *Statistics in Medicine*, **5**, 261–269.

Zellner, A (1996) *An introduction to Bayesian inference in econometrics*. Wiley.

Zellner, A (1998) The finite sample properties of simultaneous equations' estimates and estimators: Bayesian and non-Bayesian approaches, *J. Econometrics*, **83**, 185–212.

Zellner, A and Rossi, P (1984) Bayesian analysis of dichotomous quantal response models, *J. Econometrics*, **25**, 365–393.

Zellner, A and Tiao, G (1964) Bayesian analysis of the regression model with autocorrelated errors, *J. American Statistical Association* **59**, 763–778.

Zhang, J, Savitz, DA, Schwingl, PJ and Cai, W (1992) A case-control study of paternal smoking and birth defects, *International Journal of Epidemiology*, **21**, 273–278.

Index

WILEY SERIES IN PROBABILITY AND STATISTICS
ESTABLISHED BY WALTER A. SHEWHART AND SAMUEL S. WILKS

Probability and Statistics Section

*ANDERSON · The Statistical Analysis of Time Series
ARNOLD, BALAKRISHNAN, and NAGARAJA · A First Course in Order Statistics
ARNOLD, BALAKRISHNAN, and NAGARAJA · Records
BACCELLI, COHEN, OLSDER, and QUADRAT · Synchronization and Linearity: An Algebra for Discrete Event Systems
BARNETT · Comparative Statistical Inference, *Third Edition*
BASILEVSKY · Statistical Factor Analysis and Related Methods: Theory and Applications
BERNARDO and SMITH · Bayesian Theory
BILLINGSLEY · Convergence of Probability Measures, *Second Edition*
BOROVKOV · Asymptotic Methods in Queuing Theory
BOROVKOV · Ergodicity and Stability of Stochastic Processes
BRANDT, FRANKEN, and LISEK · Stationary Stochastic Models
CAINES · Linear Stochastic Systems
CAIROLI and DALANG · Sequential Stochastic Optimization
CONSTANTINE · Combinatorial Theory and Statistical Design
COOK · Regression Graphics
COVER and THOMAS · Elements of Information Theory
CSÖRGŐ and HORVÁTH · Weighted Approximations in Probability Statistics
CSÖRGŐ and HORVÁTH · Limit Theorems in Change Point Analysis
*DANIEL · Fitting Equations to Data: Computer Analysis of Multifactor Data, *Second Edition*
DETTE and STUDDEN · The Theory of Canonical Moments with Applications in Statistics, Probability, and Analysis
DEY and MUKERJEE · Fractional Factional Plans
*DOOB · Stochastic Processes
DRYDEN and MARDIA · Statistical Shape Analysis
DUPUIS and ELLIS · A Weak Convergence Approach to the Theory of Large Deviations
ETHIER and KURTZ · Markov Processes: Characterization and Convergence
FELLER · An Introduction to Probability Theory and Its Applications, Volume 1, *Third Edition*, Revised; Volume II, *Second Edition*
FULLER · Introduction to Statistical Time Series, *Second Edition*

*Now available in a lower priced paperback edition in the Wiley Classics Library.

FULLER · Measurement Error Models
GHOSH, MUKHOPADHYAY, and SEN · Sequential Estimation
GIFI · Nonlinear Multivariate Analysis
GUTTORP · Statistical Inference for Branching Processes
HALL · Introduction to the Theory of Coverage Processes
HAMPEL · Robust Statistics: The Approach Based on Influence Functions
HANNAN and DEISTLER · The Statistical Theory of Linear Systems
HUBER · Robust Statistics
HUŠKOVÁ, BERAN, and DUPAČ · Collected Works of Jaroslav Hájek—With Commentary
IMAN and CONOVER · A Modern Approach to Statistics
JUREK and MASON · Operator-Limit Distributions in Probability Theory
KASS and VOS · The Geometrical Foundations of Asymptotic Inference
KAUFMAN and ROUSSEEUW · Finding Groups in Data: An Introduction to Cluster Analysis
KELLY · Probability, Statistics, and Optimization
KENDALL, BARDEN, CARNE, and LE · Shape and Shape Theory
LINDVALL · Lectures on the Coupling Method
McFADDEN · Management of Data in Clinical Trials
MANTON, WOODBURY, and TOLLEY · Statistical Applications Using Fuzzy Sets
MARDIA and JUPP · Directional Statistics
MUIRHEAD · Aspects of Multivariate Statistical Theory
OLIVER and SMITH · Influence Diagrams, Belief Nets, and Decision Analysis
*PARZEN · Modern Probability Theory and Its Applications
PEÑA, TIAO, and TSAY (editors) · A Course in Time Series Analysis
PRESS · Bayesian Statistics: Principles, Models, and Applications
PUKELSHEIM · Optimal Experimental Design
RAO · Asymptotic Theory of Statistical Inference
RAO · Linear Statistical Inference and Its Applications, *Second Edition*
RAO and SHANBHAG · Choquet-Deny Type Functional Equations with Applications to Stochastic Models
ROBERTSON, WRIGHT, and DYKSTRA · Order Restricted Statistical Inference
ROGERS and WILLIAMS · Diffusions, Markov Processes, and Martingales, Volume I: Foundations, *Second Edition*, Volume II: Itô Calculus
RUBINSTEIN and SHAPIRO · Discrete Event Systems: Sensitivity Analysis and Stochastic Optimization by the Score Function Method
RUZSA and SZEKELEY · Algebraic Probability Theory
SALTELLI, CHAN and SCOTT (editors) · Sensitivity Analysis
SCHEFFÉ · The Analysis of Variance
SEBER · Linear Regression Analysis
SEBER · Multivariate Observations
SEBER and WILD · Nonlinear Regression
SERFLING · Approximation Theorems of Mathematical Statistics
SHORACK and WELLNER · Empirical Processes with Applications to Statistics
SMALL and McLEISH · Hilbert Space Methods in Probability and Statistical Inference
STAPLETON · Linear Statistical Models
STAUDTE and SHEATHER · Robust Estimation and Testing
STOYANOV · Counterexamples in Probability
TANAKA · Time Series Analysis: Nonstationary and Noninvertible Distribution Theory
THOMPSON and SEBER · Adaptive Sampling

*Now available in a lower priced paperback edition in the Wiley Classics Library.

Probability and Statistics Section (Continued)
WELSH · Aspects of Statistical Inference
WHITTAKER · Graphical Models in Applied Multivariate Statistics
YANG · The Construction Theory of Denumerable Markov Processes

Applied Probability and Statistics Section

ABRAHAM and LEDOLTER · Statistical Methods for Forecasting
AGRESTI · Analysis of Ordinal Categorical Data
AGRESTI · Categorical Data Analysis
ANDERSON, AUQUIER, HAUCK, OAKES, VANDAELE, and WEISBERG · Statistical Methods for Comparative Studies
ARMITAGE and DAVID (editors) · Advances in Biometry
*ARTHANARI and DODGE · Mathematical Programming in Statistics
ASMUSSEN · Applied Probability and Queues
*BAILEY · The Elements of Stochastic Processes with Applications to the Natural Sciences
BARNETT and LEWIS · Outliers in Statistical Data, *Third Edition*
BARTHOLOMEW, FORBES, and McLEAN · Statistical Techniques for Manpower Planning, *Second Edition*
BATES and WATTS · Nonlinear Regression Analysis and Its Applications
BECHHOFER, SANTNER, and GOLDSMAN · Design and Analysis of Experiments for Statistical Selection, Screening, and Multiple Comparisons
BELSLEY · Conditioning Diagnostics: Collinearity and Weak Data in Regression
BELSLEY, KUH, and WELSCH · Regression Diagnostics: Identifying Influential Data and Sources of Collinearity
BHAT · Elements of Applied Stochastic Processes, *Second Edition*
BHATTACHARYA and WAYMIRE · Stochastic Processes with Applications
BIRKES and DODGE · Alternative Methods of Regression
BLISCHKE and PRABHAKAR MURTHY · Reliability: Modeling, Prediction, and Optimization
BLOOMFIELD · Fourier Analysis of Time Series: An Introduction, *Second Edition*
BOLLEN · Structural Equations with Latent Variables
BOULEAU · Numerical Methods for Stochastic Processes
BOX · Bayesian Inference in Statistical Analysis
BOX and DRAPER · Empirical Model-Building and Response Surfaces
*BOX and DRAPER · Evolutionary Operation: A Statistical Method for Process Improvement
BUCKLEW · Large Deviation Techniques in Decision, Simulation, and Estimation
BUNKE and BUNKE · Nonlinear Regression, Functional Relations, and Robust Methods: Statistical Methods of Model Building
CHATTERJEE and HADI · Sensitivity Analysis in Linear Regression
CHERNICK · Bootstrap Methods: A Practitioner's Guide
CHILÈS and DELFINER · Geostatistics: Modeling Spatial Uncertainty
CHOW and LIU · Design and Analysis of Clinical Trials: Concepts and Methodologies
CLARKE and DISNEY · Probability and Random Processes: A First Course with Applications, *Second Edition*
*COCHRAN and COX · Experimental Designs, *Second Edition*
CONGDON · Bayesian Statistical Modelling
CONOVER · Practical Nonparametric Statistics, *Second Edition*

---

*Now available in a lower priced paperback edition in the Wiley Classics Library.

Applied Probability and Statistics (Continued)

CORNELL · Experiments with Mixtures, Designs, Models, and the Analysis of Mixture Data, *Second Edition*

*COX · Planning of Experiments

CRESSIE · Statistics for Spatial Data, *Revised Edition*

DANIEL · Applications of Statistics to Industrial Experimentation

DANIEL · Biostatistics: A Foundation for Analysis in the Health Sciences, *Sixth Edition*

DAVID · Order Statistics, *Second Edition*

*DEGROOT, FIENBERG, and KADANE · Statistics and the Law

DODGE · Alternative Methods of Regression

DOWDY and WEARDEN · Statistics for Research, *Second Edition*

DUNN and CLARK · Applied Statistics: Analysis of Variance and Regression, *Second Edition*

*ELANDT-JOHNSON and JOHNSON · Survival Models and Data Analysis

*FLEISS · The Design and Analysis of Clinical Experiments

FLEISS · Statistical Methods for Rates and Proportions, *Second Edition*

FLEMING and HARRINGTON · Counting Processes and Survival Analysis

GALLANT · Nonlinear Statistical Models

GLASSERMAN and YAO · Monotone Structure in Discrete-Event Systems

GNANADESIKAN · Methods for Statistical Data Analysis of Multivariate Observations. *Second Edition*

GOLDSTEIN and LEWIS · Assessment: Problems, Development, and Statistical Issues

GREENWOOD and NIKULIN · A Guide to Chi-squared Testing

*HAHN · Statistical Models in Engineering

HAHN and MEEKER · Statistical Intervals: A Guide for Practitioners

HAND · Construction and Assessment of Classification Rules

HAND · Discrimination and Classification

HEDAYAT and SINHA · Design and Inference in Finite Population Sampling

HEIBERGER · Computation for the Analysis of Designed Experiments

HINKELMAN and KEMPTHORNE · Design and Analysis of Experiments, Volume 1: Introduction to Experimental Design

HOAGLIN, MOSTELLER, and TUKEY · Exploratory Approach to Analysis of Variance

HOAGLIN, MOSTELLER, and TUKEY · Exploring Data Tables, Trends and Shapes

HOAGLIN, MOSTELLER, and TUKEY · Understanding Robust and Exploratory Data Analysis

HOCHBERG and TAMHANE · Multiple Comparison Procedures

HOCKING · Methods and Applications of Linear Models: Regression and the Analysis of Variables

HOGG and KLUGMAN · Loss Distributions

HOSMER and LEMESHOW · Applied Logistic Regression

HØYLAND and RAUSAND · System Reliability Theory: Models and Statistical Methods

HUBERTY · Applied Discriminant Analysis

HUNT and KENNEDY · Financial Derivatives in Theory and Practice

JACKSON · A User's Guide to Principal Components

JOHN · Statistical Methods in Engineering and Quality Assurance

JOHNSON · Multivariate Statistical Simulation

JOHNSON & KOTZ · Distributions in Statistics

JOHNSON, KOTZ, and BALAKRISHNAN · Continuous Univariate Distributions, Volume 1, *Second Edition*

JOHNSON, KOTZ, and BALAKRISHNAN · Continuous Univariate Distributions, Volume 2, *Second Edition*

---

*Now available in a lower priced paperback edition in the Wiley Classics Library.

Applied Probability and Statistics (Continued)

JOHNSON, KOTZ, and BALAKRISHNAN · Discrete Multivariate Distributions
JOHNSON, KOTZ, and KEMP · Univariate Discrete Distributions, *Second Edition*
JUREČKOVÁ and SEN · Robust Statistical Procedures: Asymptotics and Interrelations
KADANE · Bayesian Methods and Ethics in a Clinical Trial Design
KADANE and SCHUM · A Probabilistic Analysis of the Sacco and Vanzetti Evidence
KALBFLEISCH and PRENTICE · The Statistical Analysis of Failure Time Data
KELLY · Reversibility and Stochastic Networks
KHURI, MATHEW, and SINHA · Statistical Tests for Mixed Linear Models
KLUGMAN, PANJER, and WILLMOT · Loss Models: From Data to Decisions
KLUGMAN, PANJER, and WILLMOT · Solutions Manual to Accompany Loss Models: From Data to Decisions
KOTZ, BALAKRISHNAN and JOHNSON · Continuous Multivariate Distributions, Volume 1, *Second Edition*
KOVALENKO, KUZNETZOV, and PEGG · Mathematical Theory of Reliability of Time-Dependent Systems with Practical Applications
LACHIN · Biostatistical Methods: The Assessment of Relative Risks
LAD · Operational Subjective Statistical Methods: A Mathematical, Philosophical, and Historical Introduction
LANGE, RYAN, BILLARD, BRILLINGER, CONQUEST, and GREENHOUSE · Case Studies in Biometry
LAWLESS · Statistical Models and Methods for Lifetime Data
LAWSON · Statistical Methods for Spatial Epidemiology
LEE · Statistical Methods for Survival Data Analysis, *Second Edition*
LEPAGE and BILLARD · Exploring the Limits of Bootstrap
LEYLAND and GOLDSTEIN (editors) · Multilevel Modelling of Health Statistics
LINHART and ZUCCHINI · Model Selection
LITTLE and RUBIN · Statistical Analysis with Missing Data
LLOYD · The Statistical Analysis of Categorical Data
MAGNUS and NEUDECKER · Matrix Differential Calculus with Applications in Statistics and Econometrics, *Revised Edition*
MALLER and ZHOU · Survival Analysis with Long Term Survivors
MANN, SCHAFER, and SINGPURWALLA · Methods for Statistical Analysis of Reliability and Life Data
McLACHLAN · Discriminant Analysis and Statistical Pattern Recognition
McLACHLAN and KRISHNAN · The EM Algorithm and Extensions
McLACHLAN and PEEL · Finite Mixture Models
McNEIL · Epidemiological Research Methods
MEEKER and ESCOBAR · Statistical Methods for Reliability Data
*MILLER · Survival Analysis, *Second Edition*
MONTGOMERY and PECK · Introduction to Linear Regression Analysis, *Second Edition*
MYERS and MONTGOMERY · Response Surface Methodology: Process and Product in Optimization Using Designed Experiments
NELSON · Accelerated Testing, Statistical Models, Test Plans, and Data Analyses
NELSON · Applied Life Data Analysis
OCHI · Applied Probability and Stochastic Processes in Engineering and Physical Sciences
OKABE, BOOTS, CHUI, and SUGIHARA · Spatial Tesselations: Concepts and Applications of Voronoi diagrams, *Second Edition*
PANKRATZ · Forecasting with Dynamic Regression Models
PANKRATZ · Forecasting with Univariate Box–Jenkins Models: Concepts and Cases

---

*Now available in a lower priced paperback edition in the Wiley Classics Library.

Applied Probability and Statistics (Continued)

PIANTADOSI · Clinical Trials: A Methodologic Perspective
PORT · Theoretical Probability for Applications
PUTERMAN · Markov Decision Processes: Discrete Stochastic Dynamic Programming
RACHEV · Probability Metrics and the Stability of Stochastic Models
RÉNYI · A Diary on Information Theory
RIGDON and BASU · Statistical Methods for the Reliability of Repairable Systems
RIPLEY · Spatial Statistics
RIPLEY · Stochastic Simulation
ROLSKI, SCHMIDLI, SCHMIDT, and TEUGELS · Stochastic Processes for Insurance and Finance
ROUSSEEUW and LEROY · Robust Regression and Outlier Detection
RUBIN · Multiple Imputation for Nonresponse in Surveys
RUBINSTEIN · Simulation and the Monte Carlo Method
RUBINSTEIN and MELAMED · Modern Simulation and Modeling
RYAN · Statistical Methods for Quality Improvement, *Second Edition*
SCHIMEK (editor) · Smoothing and Regression: Approaches, Computation and Applications
SCHUSS · Theory and Applications of Stochastic Differential Equations
SCOTT · Multivariate Density Estimation: Theory, Practice, and Visualization
*SEARLE · Linear Models
SEARLE · Linear Models for Unbalanced Data
SEARLE, CASELLA, and McCULLOCH · Variance Components
SENNOT · Stochastic Dynamic Programming and the Control of Queuing Systems
STOYAN, KENDALL, and MECKE · Stochastic Geometry and Its Applications, *Second Edition*
STOYAN and STOYAN · Fractals, Random Shapes, and Point Fields: Methods of Geometrical Statistics
SUTTON, ABRAMS, JONES, SHELDON and SONG · Methods for Meta-analysis in Medical Research
THOMPSON · Empirical Model Building
THOMPSON · Sampling
THOMPSON · Simulation: A Modeler's Approach
TIJMS · Stochastic Modeling and Analysis: A Computational Approach
TIJMS · Stochastic Models: An Algorithmic Approach
TITTERINGTON, SMITH, and MARKOV · Statistical Analysis of Finite Mixture Distributions
UPTON and FINGLETON · Spatial Data Analysis by Example, Volume 1: Point Pattern and Quantitative Data
UPTON and FINGLETON · Spatial Data Analysis by Example, Volume II: Categorical and Directional Data
VAN RIJCKEVORSEL and DE LEEUW · Component and Correspondence Analysis
VIDAKOVIC · Statistical Modeling by Wavelets
WEISBERG · Applied Linear Regression, *Second Edition*
WESTFALL and YOUNG · Resampling-Based Multiple Testing: Examples and Methods for p-Value Adjustment
WHITTLE · Systems in Stochastic Equilibrium
WINKER · Optimization Heuristics in Econometrics: Applications of Threshold Accepting
WOODING · Planning Pharmaceutical Clinical Trials: Basic Statistical Principles
WOOLSON · Statistical Methods for the Analysis of Biomedical Data
*ZELLNER · An Introduction to Bayesian Inference in Econometrics

*Now available in a lower priced paperback edition in the Wiley Classics Library.

Texts and References Section

AGRESTI · An Introduction to Categorical Data Analysis
ANDĚL · Mathematics of Chance
ANDERSON · An Introduction to Multivariate Statistical Analysis, *Second Edition*
ANDERSON and LOYNES · The Teaching of Practical Statistics
ARMITAGE and COLTON (editors) · Encyclopedia of Biostatistics. 6 Volume set
BALDING, BISHOP and CANNINGS (editors) · Handbook of Statistical Genetics
BARTOSZYNSKI and NIEWIADOMSKA-BUGAJ · Probability and Statistical Inference
BENDAT and PIERSOL · Random Data: Analysis and Measurement Procedures, *Third Edition*
BERRY, CHALONER, and GEWEKE · Bayesian Analysis in Statistics and Econometrics: Essays in Honor of Arnold Zellner
BHATTACHARYA and JOHNSON · Statistical Concepts and Methods
BILLINGSLEY · Probability and Measure, *Second Edition*
BOX · R. A. Fisher, the Life of a Scientist
BOX, HUNTER, and HUNTER · Statistics for Experimenters: An Introduction to Design, Data Analysis, and Model Building
BOX and LUCEÑO · Statistical Control by Monitoring and Feedback Adjustment
BROWN and HOLLANDER · Statistics: A Biomedical Introduction
CHATTERJEE and PRICE · Regression Analysis by Example, *Third Edition*
COOK and WEISBERG · An Introduction to Regression Graphics
COOK and WEISBERG · Applied Regression Including Computing and Graphics
COX · A Handbook of Introductory Statistical Methods
DILLON and GOLDSTEIN · Multivariate Analysis: Methods and Applications
*DODGE and ROMIG · Sampling Inspection Tables, *Second Edition*
DRAPER and SMITH · Applied Regression Analysis, *Third Edition*
DUDEWICZ and MISHRA · Modern Mathematical Statistics
DUNN and CLARK · Basic Statistics: A Primer for the Biomedical Sciences, *Third Edition*
EVANS, HASTINGS and PEACOCK · Statistical Distributions, *Third Edition*
FISHER and VAN BELLE · Biostatistics: A Methodology for the Health Sciences
FREEMAN and SMITH · Aspects of Uncertainty: A Tribute to D. V. Lindley
GROSS and HARRIS · Fundamentals of Queueing Theory, *Third Edition*
HALD · A History of Probability and Statistics and their Applications Before 1750
HALD · A History of Mathematical Statistics from 1750 to 1930
HELLER · MACSYMA for Statisticians
HOEL · Introduction to Mathematical Statistics, *Fifth Edition*
HOLLANDER and WOLFE · Nonparametric Statistical Methods, *Second Edition*
HOSMER and LEMESHOW · Applied Logistic Recession, *Second Edition*
HOSMER and LEMESHOW · Applied Survival Analysis: Regression Modeling of Time to Event Data
JOHNSON and BALAKRISHNAN · Advances in the Theory and Practice of Statistics: A Volume in Honor of Samuel Kotz
JOHNSON and KOTZ (editors) · Leading Personalities in Statistical Sciences: From the Seventeenth Century to the Present
JUDGE, GRIFFITHS, HILL, LÜTKEPOHL, and LEE · The Theory and Practice of Econometrics, *Second Edition*
KHURI · Advanced Calculus with Applications in Statistics
KOTZ and JOHNSON (editors) · Encyclopedia of Statistical Sciences. Volumes 1 to 9 with Index
KOTZ and JOHNSON (editors) · Encyclopedia of Statistical Sciences: Supplement Volume

*Now available in a lower priced paperback edition in the Wiley Classics Library.

Texts and References Section (Continued)

KOTZ, REED, and BANKS (editors) · Encyclopedia of Statistical Sciences: Update Volume 1
KOTZ, REED, and BANKS (editors) · Encyclopedia of Statistical Sciences: Update Volume 2
LAMPERTI · Probability: A Survey of the Mathematical Theory, *Second Edition*
LARSON · Introduction to Probability Theory and Statistical Inference, *Third Edition*
LE · Applied Categorical Data Analysis
LE · Applied Survival Analysis
MALLOWS · Design, Data, and Analysis by Some Friends of Cuthbert Daniel
MARDIA · The Art of Statistical Science: A Tribute to G. S. Watson
MASON, GUNST, and HESS · Statistical Design and Analysis of Experiments with Applications to Engineering and Science
MURRAY · X-STAT 2.0 Statistical Experimentation, Design Data Analysis, and Nonlinear Optimization
PURI, VILAPLANA, and WERTZ · New Perspectives in Theoretical and Applied Statistics
RENCHER · Methods of Multivariate Analysis
RENCHER · Linear Models in Statistics
RENCHER · Multivariate Statistical Inference with Applications
ROBINSON · Practical Strategies for Experimenting
ROSS · Introduction to Probability and Statistics for Engineers and Scientists
ROHATGI and SALEH · An Introduction to Probability and Statistics, *Second Edition*
RYAN · Modern Regression Methods
SCHOTT · Matrix Analysis for Statistics
SEARLE · Matrix Algebra Useful for Statistics
STYAN · The Collected Papers of T. W. Anderson: 1943–1985
TIAO, BISGAARD, HILL, PEÑA, and STIGLER (editors) · Box on Quality and Discovery: with Design, Control, and Robustness
TIERNEY · LISP-STAT: An Object-Oriented Environment for Statistical Computing and Dynamic Graphics
WONNACOTT and WONNACOTT · Econometrics, *Second Edition*
WU and HAMADA · Experiments: Planning, Analysis, and Parameter Design Optimization

WILEY SERIES IN PROBABILITY AND STATISTICS

ESTABLISHED BY WALTER A. SHEWHART AND SAMUEL S. WILKS

Editors
Robert M. Groves, Graham Kalton, J. N. K. Rao, Norbert Schwarz, Christopher Skinner

Survey Methodology Section

BIEMER, GROVES, LYBERG, MATHIOWETZ, and SUDMAN · Measurement Errors in Surveys
COCHRAN · Sampling Techniques, *Third Edition*

---

*Now available in a lower priced paperback edition in the Wiley Classics Library.

Survey Methodology Section (Continued)

COUPER, BAKER, BETHLEHEM, CLARK, MARTIN, NICHOLLS and O'REILLY (editors) · Computer Assisted Survey Information Collection

COX, BINDER, CHINNAPPA, CHRISTIANSON, COLLEDGE, and KOTT (editors) · Business Survey Methods

*DEMING · Sample Design in Business Research

DILLMAN · Mail and Telephone Surveys: The Total Design Method, *Second Edition*

GROVES · Survey Errors and Survey Costs

GROVES and COUPER · Nonresponse in Household Interview Surveys

GROVES, BIEMER, LYBERG, MASSEY, NICHOLLS, and WAKSBERG · Telephone Survey Methodology

*HANSEN, HURWITZ, and MADOW · Sample Survey Methods and Theory, Volume I: Methods and Applications

*HANSEN, HURWITZ, and MADOW · Sample Survey Methods and Theory, Volume II: Theory

KISH · Statistical Design for Research

*KISH · Survey Sampling

KORN and GRAUBARD · Analysis of Health Surveys

LESSLER and KALSBEEK · Nonsampling Error in Surveys

LEVY and LEMESHOW · Sampling of Populations: Methods and Applications

LYBERG, BIEMER, COLLINS, de LEEUW, DIPPO, SCHWARZ, TREWIN (editors) · Survey Measurement and Process Quality

SIRKEN, HERRMANN, SCHECHTER, SCHWARZ, TANUR and TOURNANGEAU (editors) · Cognition and Survey Research

VALLIANT, DORFMAN, and ROYALL · Finite Population Sampling and Inference: A Prediction Approach

---

*Now available in a lower priced paperback edition in the Wiley Classics Library.